Topics in Applied Physics Volume 52

Topics in Applied Physics Founded by Helmut K. V. Lotsch

Sputtering by Particle Bombardment II

Sputtering of Alloys and Compounds, Electron and Neutron Sputtering, Surface Topography

Edited by R. Behrisch

With Contributions by
R. Behrisch G. Betz G. Carter B. Navinšek
J. Roth B. M. U. Scherzer P. D. Townsend
G. K. Wehner J. L. Whitton

With 155 Figures

Springer-Verlag Berlin Heidelberg GmbH 1983

Dr. *Rainer Behrisch*

Max-Planck-Institut für Plasmaphysik, EURATOM Association
D-8046 Garching bei München, Fed. Rep. of Germany

ISBN 978-3-662-31169-1 ISBN 978-3-540-38659-9 (eBook)
DOI 10.1007/978-3-540-38659-9

Library of Congress Cataloging in Publication Data. Main entry under title: Sputtering by particle bombardment. (Topics in applied physics; v. 47,). Includes bibliographical references and indexes. Contents: 1. Physical sputtering of single-element solids – v. 2. Sputtering of alloys and compounds, electron and neutron sputtering, surface topography. 1. Sputtering (Physics) 2. Solids – Effect of radiation on. 3. Surfaces (Physics) – Effect of radiation on. I. Behrisch, Rainer. II. Andersen, Hans Henrik. III. Title: Particle bombardment. IV. Series. QC176.8.S72S68 81-4313

Originally published by Springer-Verlag Berlin Heidelberg New York Tokyo in 1983
Softcover reprint of the hardcover 1st edition 1983

2153/3130-543210

Preface

Sputtering, or the erosion of the surface of a solid due to the impact of energetic particles, is of considerable interest from the standpoint of understanding the fundamental processes involved as well as for the many applications of sputtering in science and technology. In three volumes of this *Topics in Applied Physics* series, an attempt has been made to collect and review the present knowledge of sputtering phenomena. The first book, "Sputtering by Particle Bombardment I" (TAP 47), gives an introduction to collisional sputtering theory and a summary of the measured sputtering yields obtained for polycrystalline single-element solids and for some single crystals during bombardment, primarily with noble gas and hydrogen ions. This second volume emphasizes the sputtering of multicomponent systems, chemical and surface effects, and sputtering by particles other than ions.

In most applications and practical situations, multicomponent solids are sputtered. The major effects observed for such cases are preferential sputtering and compositional changes in the surface layers. These processes are treated in some detail in Chap. 2 of the present volume by G. Betz and G.K. Wehner which includes a table summarizing the systems investigated to date.

If the incident ions interact chemically with the target atoms forming stable compounds or volatile molecules, the sputtering yields may be decreased or increased and generally show a strong dependence on temperature. The present knowledge of these phenomena is treated by J. Roth in Chap. 3.

Sputtering by electrons and photons is of importance in insulating materials and has been investigated primarily in ionic crystals. The major effects occur via excitation and ionization processes in the solids and are reviewed by P. Townsend in Chap. 4. Such processes also occur during ion bombardment, but have yet to be investigated in detail.

Neutron sputtering effects are usually small and thus difficult to measure. This field is of interest in reactor technology and is reviewed in Chap. 5 by R. Behrisch.

Ion implantation and nonuniform surface removal generally lead to a large modification of the surface structure and topography. During bombardment with medium to high-Z atoms, the surface topography is generally dominated by erosion, as treated by G. Carter, B. Navinšek, and J. L. Whitton in Chap. 6. For light ions and high energies, gas implantation and accumulation can lead to blistering, flaking, and finally spongy structures, as shown by B. M. U. Scherzer in Chap. 7.

In the third volume, "Sputtering by Particle Bombardment III" the present knowledge of the angular, energy, mass and charge state distributions of sputtered particles will be presented and the large variety of applications for the sputtering process will be outlined.

Writing such a book is not possible as a monograph, but only in collaboration with leading scientists in the different areas of sputtering research. Each contribution represents the personal view of the authors, but an attempt has been made to integrate the different styles and emphases of the contributions in a coherent way also by including many internal cross references as well as references to Vol. I (TAP 47), by applying a consistent symbol utilization, and by providing a cumulative subject and author index.

A major difficulty in such a collaboration is appropriate timing and the many delays which often cannot be avoided. I would particularly like to thank Gottfried Wehner for persevering with his contribution which was originally agreed to in the Hofbräuhaus in München.

I appreciate the kind support and patience of all the authors who contributed to this book. I gratefully acknowledge, also, the help of several colleagues through their critical comments, and Springer-Verlag for their pleasent collaboration. This book is intended to be the basis and stimulation for basic research and applied work in this exciting field of physics.

Garching, July 1983 *Rainer Behrisch*

Contents

Contributors

Behrisch, Rainer
Max-Planck-Institut für Plasmaphysik, EURATOM Association,
D-8046 Garching bei München, Fed. Rep. of Germany

Betz, Gerhard
Institut für Allgemeine Physik, Technische Universität Wien,
A-1040 Wien, Austria

Carter, George
Department of Electronic and Electrical Engineering, University of Salford,
Salford M54 WT, Great Britain

Navinšek, Boris
J. Stefan Institute, University of Ljubljana,
YU-61001 Ljubljana, Yugoslavia

Roth, Joachim
Max-Planck-Institut für Plasmaphysik, EURATOM Association,
D-8046 Garching bei München, Fed. Rep. of Germany

Scherzer, Bernhard M. U.
Max-Planck-Institut für Plasmaphysik, EURATOM Association,
D-8046 Garching bei München, Fed. Rep. of Germany

Townsend, Peter D.
School of Mathematical and Physical Sciences, University of Sussex,
Brighton, Sussex, BN1 9QH, Great Britain

Wehner, Gottfried K.
Department of Electrical Engineering, University of Minnesota,
Minneapolis, MN 55455, USA

Whitton, James L.
Physical Laboratory II, H. C. Ørstedt Institute, Universitetsparken 5,
DK-2100 Copenhagen Ø, Denmark

1. Introduction and Overview

Rainer Behrisch

Sputtering phenomena of single-element metals can be described by elastic-collision cascades initiated by the incident particles in the surface layers. For multicomponent solids and nonmetals and/or bombardment with ions which react chemically with the atoms of a solid sputtering is influenced by several additional processes. Multicomponent solids show preferential sputtering, i.e. the partial sputtering yields of the different species are not proportional to their atomic concentrations on the surface. During bombardment with chemically reacting ions a compound layer may build up with different surface binding energies and thus different sputtering yields. For nonmetals also the energy transferred to electronic excitation and ionisation contributes to sputtering. Ion implantation and erosion changes the surface topography. Grooves, ridges, and pyramids and/or blistering, exfoliation and a spongy surface may develop.

1.1 Background

If matter in two extreme states as a hot plasma or a beam of energetic particles and a solid are brought together, several processes are initiated in the area of interaction. A fraction of the particles are backscattered from the surface layers, the others are slowed down in the solid and may be trapped or may diffuse to the surface or into the bulk. Electrons and photons can be emitted and atoms from the solid may be released at the surface. This last process is named *sputtering*. Finally, the implantation of particles and the erosion at the surface by sputtering lead to a modification of the composition and structure of surface layers and of the surface topography. Generally all these processes are interrelated and several have to be included for the understanding of sputtering phenomena.

Sputtering was first discovered in electric gas discharges more than 125 years ago. Cathode material was observed to be deposited on the surrounding glass walls [1.1–3], thus the name cathode sputtering can still be found in the literature. It took about 50 years for the physical processes causing sputtering to be recognised [1.4, 5]. Within the last 25 years a large amount of sputtering yield data have been measured applying improved experimental conditions, as mass analysed ion beams, high vacuum and well characterized materials [Ref. 1.6, Chaps. 4 and 5] and a quantitative description of the collisional processes leading to sputtering was developed ([Ref. 1.6, Chaps. 2 and 3] and [1.7]). This was also stimulated by a broad application of sputtering in surface and thin film technology [1.8].

Generally, several additional processes may contribute to the erosion of a surface during particle bombardment. At very high current densities and/or insufficient cooling of the target, the solid may be macroscopically heated and surface atoms may evaporate. Implanted gas atoms may accumulate in the surface layers to form bubbles and may finally cause the formation of blisters or flakes breaking away surface layers (Chap. 7). These two processes are, however, not included in sputtering. With sputtering we shall define the particle-induced surface erosion in the limits of low fluences and low current densities [Ref. 1.6, Chap. 2].

In the first volume of the three-book series *Sputtering by Particle Bombardment* [1.6] the theoretical basis for the understanding of sputtering phenomena in amorphous, polycrystalline and single crystalline solids is presented together with an overview about the experimental information on sputtering yields from single-element solids. These results apply primarily to bombardment of metals with noble gas ions.

This second volume deals with sputtering yields and compositional changes of alloys (Chap. 2) and the modifications of the sputtering yield during bombardment with ions which react chemically with the atoms of the solid (Chap. 3). Sputtering phenomena during electron bombardment (Chap. 4) and neutron bombardment (Chap. 5) are treated each in a chapter and finally today's knowledge about changes of the surface layer and the surface topography due to implantation and sputtering are reviewed (Chap. 6 and 7).

1.2 Sputtering Yields

The total erosion in sputtering is measured by the *sputtering yield*, Y, defined as the average number of atoms removed from the surface of a solid per incident particle. The incident particles may be ions, neutral atoms, neutrons, electrons or photons. For bombardment with monoatomic molecular ions each incident atom is counted. In sputtering experiments with molecules of different elements we shall define the yield per incident molecule. In counting the removed atoms only those of the solid are included while reflected or reemitted atoms are not taken into account. This is different for selfsputtering, i.e. for bombardment with the same ions as the solid, where this distinction cannot be made in an experiment. A selfsputtering yield "one" means that on the average one atom is removed or reflected per incident ion.

In sputtering of multicomponent materials we shall define a *partial sputtering yield*, Y_i, at the surface given by the average number of atoms i removed per incident ion, with $Y=\Sigma Y_i$. For comparison with the sputtering yields of the corresponding single-element solid, we shall further define a *component sputtering yield* Y_i^c by $Y_i^c = Y_i/c_i^s$, c_i^s being the relative atomic concentration of element i on the surface (Chap. 2).

These definitions of the sputtering yields are meaningful only if the number of atoms removed is proportional to the number of incident particles. This has indeed been observed when the total ion fluence was low enough so that the surface layers did not change during the bombardment. Deviations are found for large incident particles such as molecular ions at energies >20 keV, where spike effects [Ref. 1.6, Chaps. 2 and 4] can cause a dependence of the yield on the number of atoms per molecule.

The particles removed from a solid surface by sputtering are emitted with a broad distribution in energy, E_1, in different excitation and charge states, q, at all exit angles Ω. This is described by the *differential sputtering yields* $\partial Y/\partial E_1$, $\partial^2 Y/\partial \Omega^2$ and Y_q with $Y = \Sigma Y_q$. In many experiments only those atoms emitted into a given angular interval are measured, sometimes only those with a given energy and charge state, i.e. $\partial^3 Y_q/\partial E_1 \partial \Omega^2$ are determined. This quantity is mostly not directly proportional to the sputtering yield, because the energy, angular-, excitation-, and charge state distributions of the emitted particles also depend on the different parameters entering the yields.

1.3 Sputtering Investigations for Single-Element Solids

Sputtering has been investigated in most detail for the bombardment of metals with noble-gas ions and hydrogen ions [1.6–25]. Sputtering yields may lie between zero and about 10^4 atoms per ion, but are typically 1–5 except for the lightest ions, where they are 10^{-2} to 10^{-1}. The yields depend on ion energy, mass and angle of incidence, the mass of the target atoms, the crystallinity and the crystal orientation of the solid and on the surface binding energies of the target, but they are usually nearly independent of the temperature [1.26–28]. Below a threshold energy which is about 20–40 eV for normally incident ions, almost no sputtering takes place. Above this threshold the yields increase with incident energy and reach a broad maximum in the energy region of 5–50 keV. The decrease of the sputtering yield at higher energies is related to the larger penetration of the ions into the solid and the lower energy deposition in the surface layers. At the same energy, larger-mass ions usually give larger sputtering yields than lighter ions [Ref. 1.6, Chaps. 2–5].

For ion bombardment at oblique angle of incidence the sputtering yield increases monotonically with increasing angle of incidence up to a maximum near 60–80°. This holds for amorphous solids and for polycrystalline materials with many randomly oriented crystallites in the bombarded area. The location of the maximum depends on the bombarding particle energy and mass and the surface topography.

For single crystals the sputtering yields depend also on the direction of particle incidence relative to the crystal orientation. For bombardment close to the major three close packed crystal axes the yields are about a factor of 2–5 lower than for other directions of incidence. These minima are superimposed on the steady increase in the sputtering yield with increasing angle of incidence. Furthermore for the most close-packed directions the maximum in the energy dependence of the sputtering yield is shifted towards lower energies [Ref. 1.6, Chaps. 3 and 5].

1.4 Theoretical Models for Sputtering

The sputtering of metals can be described by elastic collisions in the surface layers of a solid. The processes are the same as those leading to radiation damage in the bulk of a metal and they take place far from thermal equilibrium. If an incident ion with sufficient energy collides with an atom of the solid, it can create a primary knockon atom. For the collision cascades started by this primary knockon atom several regimes must be considered [Ref. 1.6, Chaps. 2 and 4].

— The *single knockon regime* applies for ion bombardment at low energies, i.e. close to the sputtering threshold and for light ions, where mostly only small energies are transferred to target atoms. In this regime only a few collisions can occur, and also primary knockon atoms can contribute significantly to sputtering.

— The *linear cascade regime* applies for bombardment with ions of medium to large atomic number in the keV energy region. Here larger collision cascades can develop, however, moving target atoms collide only with target atoms at rest.

— The *spike regime* is reached for bombardment of large atomic mass targets with large mass ions or molecular ions at energies around 20 to 80 keV. The cascades leading to sputtering are very dense and a major part of the atoms within the cascade volume are released from their lattice site and set in motion. At the larger energies a splitting of the cascades into dense subcascades has been observed [1.29, 30].

In crystalline material the probability for collisions of the incoming ions to create primary knockon atoms as well as the development of collision cascades are, in addition, influenced by the crystal structure due to the channeling, the blocking or shadowing and the focusing effects [Ref. 1.6, Chaps. 3 and 5].

For a qualitative description of the sputtering process, the collision cascades are generally treated as a sequence of separate binary collisions. Here the parameters involved in the collisions between atoms, i.e. the potentials or the differential cross-sections, the energy loss to electrons, the structure and crystal orientation of the solid and the surface potentials have to be known. Generally a compromise between the best known values of these quantities and those analytically treatable has to be made ([Ref. 1.6, Chaps. 2 and 3] and [1.7]).

In an amorphous solid the collision cascades have been described by a Boltzmann transport equation [1.7, 31–37]. First-order asymptotic solutions have been obtained by *Sigmund* for the linear cascade regime [1.7]. These analytical formulae show the dependence of the yields on the different parameters and provide good reference data. Polycrystalline solids with randomly oriented crystallites can be approximated to first order by an amorphous solid. Thus the formulae derived for amorphous materials mostly give values in reasonable agreement with the sputtering yields measured for polycrystalline targets ([Ref. 1.6, Chaps. 2 and 4] and [1.7]).

The cascades leading to sputtering have also been followed by computer simulation programs both for amorphous and crystalline solids ([Ref. 1.6, Chap. 3] and [1.38–43]). These calculations are mostly time consuming, because the yields have to be calculated for each set of parameters separately. The results which have been obtained with reasonable input parameters agree mostly well with measured values; in addition, useful details about the cascades can be obtained for comparison with the assumptions made in analytical formulae [1.38–43].

1.5 Sputtering of Multicomponent Targets

During particle bombardment of multicomponent solids and/or with other than noble gas ions, sputtering is determined by several additional processes (Chaps. 2 and 3).

Experimentally, the major effect observed in most cases is *preferential sputtering* which means that different mass atoms of the solid are not sputtered proportionally to their atomic concentrations at the surface. As a consequence, the surface concentration of a virgin target is changed during ion bombardment until a surface concentration is reached from which the atoms are sputtered with partial yields proportional to the relative atomic concentrations in the bulk.

Several processes can contribute to preferential sputtering, (Chap. 2) and [1.44, 45]. In elastic collision cascades the energy is generally not equally distributed among different mass atoms in the solid. Further, lower-mass atoms have larger ranges in the solid than heavier atoms, so that lighter atoms are depleted in the solid at the depth of maximum energy deposition and are preferentially sputtered at the surface.

Besides these differences in the collision cascades, preferential sputtering is also influenced by different binding energies of different atoms at the surface, by segregation of one element at the surface and by diffusion in the implanted layers of the solid. Multicomponent materials have only recently been investigated in more detail, but there are still no decisive experiments to distinguish between the different processes contributing to preferential sputtering, (Chap. 2) and [1.44].

The situation becomes even more complex and is less investigated for the case where the incident ions interact chemically with the atoms of the solid

(Chap. 3). Besides the changes in the surface composition due to the implanted atoms, a compound surface layer can be formed, i.e. a layer with different composition having different surface binding energies than the original material. This compound layer may then also show preferential sputtering. If the reaction products are gaseous molecules, like CH_4, or H_2O, they may thermally desorb from the surface. Such chemical effects mostly enhance the sputtering yields, but in some cases also reduce the yields (Chap. 3).

1.6 Sputtering of Nonmetals

In sputtering experiments on nonmetals the measured yields are generally different than expected from collisional theory. While the yields for oxides may be lower or slightly larger than expected (Chap. 2), the yields may be considerably higher for ionic crystals [1.46–57] for condensed gases [1.58–75] or materials like sulfur [1.76]. Here besides the energy transferred to target atoms by elastic collisions, also the energy transferred to electrons producing *electronic excitation and ionisation* can contribute to atomic displacements. Depending on the material and its temperature, long-lived excitation states may diffuse and may have sufficiently long lifetimes to allow their energy to be transferred to atomic motion leading to sputtering at the surface [1.57, 72, 76, 77].

Electronic excitation and ionisation can be produced in insulators also by low energy electron and photon bombarment. In this case sputtering occurs only due to excitation and ionisation because the energy transfer to nuclei by elastic collisions is too small to cause collisional sputtering. Thus sputtering by excitation and ionisation can be investigated separately (Chap. 4).

The details of sputtering by electronic processes, especially the energy transfer from an excitation or ionisation state to lattice atoms which finally causes sputtering are still not yet fully understood, (Chap. 4) and [1.77, 78]. But as for sputtering by elastic collision cascades here also three regimes may be considered depending on the incident particle energy and the excitation density in the solid [Ref. 1.6, Chap. 2]. At low energies in the threshold regime, ionizing or dissociating events are isolated in space. Such events may be generated by low energy ions or electrons or *uv* photons ($\lesssim 100$ eV). The linear ionization cascade regime presents the case where high-energy electrons are produced by high energy incident particles ($\gtrsim 100$ eV). Finally, heavy ions at intermediate velocity ($v \gtrsim e^2\hbar$) may create dense ionization spikes.

The topic ion sputtering of insulators is of special interest in astrophysics [1.79–81] and also in nuclear waste isolation. It has, however, not yet been investigated in sufficient detail experimentally and the importance of the different underlying physical processes are still under discussion [1.76–78].

1.7 Neutron Sputtering

Experimentally sputtering yields during neutron bombardment are only very little investigated as the available neutron fluences are still relatively low. The few yields measured are of the order of 10^{-5} atoms/neutron in forward direction and agree reasonably with the values predicted by collision theory. The release of radioactive primary knockon atoms has been investigated in more detail, because detection methods are much more sensitive (Chap. 5).

Neutron sputtering had been of some interest in fusion research due to the possible introduction of wall atoms as impurities in a fusion plasma [1.82, 83]. The emission of radioactive primary knockon atoms will contribute to introduce radioactivity into the cooling channel in a high-energy neutron flux environment.

1.8 Surface Topography

Generall the removal of atoms from a surface by sputtering does not occur uniformly over the bombarded area. Furthermore, implanted ions can modify the surface structure. Thus, during sputtering a surface topography develops which is mostly different from that of the original state (Chaps. 6 and 7).

On a flat single-crystal surface, small traces of impurities or defects from the surface treatment and/or introduced by the particle bombardment may cause locally different erosion rates. Surface features started by such processes can grow and be modified during prolonged particle bombardment. Generally, micrometer-size pyramids, ridges, grooves and holes develop, depending on the starting conditions, the orientation of the crystal relative to the incident beam, the crystal temperature, and the kind of ions and their energy (Chap. 6).

For a polycrystalline surface the different sputtering yields of differently oriented crystallites cause the crystallites to become visible. At the grain boundaries additional grooves or smooth transitions develop, depending on the orientation of the crystallites and the angle of incidence of the ion beam, (Chap. 6) and [1.84].

The development of these surface features due to erosion is still not fully understood. The major effects are differences in the sputtering yield for differently oriented surfaces and crystallites. However, also redeposition of sputtered atoms and sputtering by ions reflected from the other areas of the surface may contribute (Chap. 6).

For bombardment with low atomic mass ions such as H or He and with other gaseous ions at energies above about 50 keV a major part of the incident ions will be implanted and come to rest in the solid. Depending on the target material and the temperature a large atomic concentration may build up. This may be up to 5 implanted hydrogen atoms or 0.3 implanted helium atoms for one target atom. These high gas concentrations mostly segregate into small bubbles and a high stress parallel to the surface builds up. At a critical fluence the implant concentration gets too high and surface layers with a thickness corresponding to the implanted layer may break off, i.e. blistering, flaking or exfoliation is observed on the surface (Chap. 7). During further ion bombardment other generations of blisters may occur, but finally the blistered layers are gradually sputtered away and generally a sponge-like surface develops. The transition to such a spongy structure without blistering can also be obtained by ion bombardment with a distribution in energies, or by simultaneous surface erosion, so that the maximum concentration of implanted atoms develops within a few atomic layers of the surface (Chap. 7).

1.9 Conclusion

The topics treated in this volume deal mostly with sputtering processes which are under active investigation and the theoretical description has mostly not yet advanced to a clear and closed picture. However, an attempt is made to summarise the status of the experimental results and the different theoretical interpretations offered.

References

1.1 W.R.Grove: Philos. Mag. **5**, 203 (1853)
1.2 J.P.Gassiot: Philos. Trans. R. Soc. London **148**, 1 (1858)
1.3 J.Plücker: Ann. Physik (Leipzig) **103**, 88 and 90 (1858)
1.4 V.Kohlschütter: Verh. Dtsch. Phys. Ges. **4**, 228 and 237 (1902)
1.5 J.Stark, G.Wendt: Ann. Physik (Leipzig) **38**, 921, 941 (1912)
1.6 R.Behrisch (ed.): *Sputtering by Particle Bombardment* I, Topics Appl. Phys., Vol. 47 (Springer, Berlin, Heidelberg, New York 1981)
1.7 P.Sigmund: Phys. Rev. **184**, 383 (1969)
1.8 R.Behrisch (ed.): *Sputtering by Particle Bombardment* III, Topics Appl. Phys. (Springer, Berlin, Heidelberg, New York) to be published
1.9 A.Güntherschulze: J. Vac. Sci. Technol. **3**, 360 (1953)
1.10 G.K.Wehner: Ad. Electron. Electron Phys. **7**, 239 (1955)
1.11 R.Behrisch: Ergebn. Exakten Naturwissenschaften **35**, 295 (1964)
1.12 M.Kaminsky: *Atomic and Ionic Impact Phenomena on Metal Surfaces*, Struktur und Eigenschaften der Materie in Einzeldarstellungen (Springer, Berlin, Heidelberg, New York 1965)
1.13 G.Carter, J.S.Colligon: *Ion Bombardment of Solids* (Heinemann, London 1965)
1.14 R.S.Nelson: *The Observation of Atomic Collisions in Crystalline Solids* (North-Holland, Amsterdam 1968)
1.15 N.V.Pleshivtsev: *Cathode Sputtering* (Atomisdat, Moscow 1968)
1.16 M.W.Thompson: *Defects and Radiation Damage in Metals* (Cambridge University Press, Cambridge 1969)
1.17 H.Oechsner: Appl. Phys. **8**, 185 (1975)

1.17a G.K.Wehner: In *Methods in Surface Analysis*, Vol. 1, ed. by A.W.Czanderna (Elsevier, Amsterdam 1975)
1.18 P.D.Townsend, J.C.Kelly, N.E.Hartley: *Ion Implantation, Sputtering, and Their Applications* (Academic Press, London 1976)
1.19 Physics of Ionized Gases, Proc. Yugoslav Symp. (SPIG): In 7, ed. by V.Vujnovic (Zagreb 1974), H.H.Andersen, p. 361, in 8 ed. by B.Navinsek (Lubljana 1976), G.Carter, p. 281, H.Oechsner, p. 461, V.E.Yurasova, p. 493, in 9 ed. by R.K.Janev (Beograd 1978), M.W.Thompson, p. 289, J.L.Whitton, p. 335, in 10 ed. by M.Matić (Dubrovnik 1980), J.Fine, p. 379, H.H.Andersen, p. 421
1.20 N.H.Tolk, J.C.Tully, W.Heiland, C.W.White (eds.): *Inelastic Ion Surface Collisions* (Academic Press, New York 1972)
1.21 R.Kelly (ed.): *Workshop on Inelastic Ion Surface Collisions* (North-Holland, Amsterdam 1979)
1.22 E.Taglauer, W.Heiland (eds.): *Inelastic Particle-Surface Collisions*, Springer Ser. Chem. Phys., Vol. 17 (Springer, Berlin, Heidelberg, New York 1981)
1.23 *Secondary Ion Mass Spectrometry* SIMS II, ed. by A.Benninghoven, C.A.Evans, Jr., R.A.Powell, R.Shimizu, H.A.Storms, Springer Ser. Chem. Phys., Vol. 9 (Berlin, Heidelberg, New York 1979)
Secondary Ion Mass Spectrometry SIMS III, ed. by A.Benninghoven, J.Giber, J.Làszló, M.Riedel, H.W.Werner, Springer Ser. Chem. Phys., Vol. 19 (Springer, Berlin, Heidelberg, New York 1982)
1.24 P.Varga, G.Betz, F.P.Viehböck (eds.): *Proc. Symp. on Sputtering* (SOS) (Techn. Univ. Wien, 1980)
1.25 *Atomic Collisions in Solids*, Proc. Ninth Intern. Conf. Lyon (1981), ed. by J.Remieux, J.-C.Poiczat, M.J.Gaillard, Nucl. Instrum. Methods **194** (1982), Y.Yamamura, p. 515, M.Szymonski, p. 523, P.Sigmund, A.Oliva, G.Falcone, p. 541, U.Littmark, S.Fedder p. 607
1.26 J.Bohdansky, J.Roth, A.P.Martinelli, Fourth Internat. Conf. Solid Surfaces, Paper 344 (Cannes 1980)
1.27 K.Besoke, S.Berger, W.O.Hofer, U.Littmark: Radiat. Eff. **66**, 35 (1982)
1.28 W.O.Hofer, K.Besoke, B.Stritzker: Appl. Phys. A**30**, 83 (1983)
1.29 K.L.Merkle, L.R.Singer, R.K.Hart: J. Appl. Phys. **34**, 2800 (1963)
1.30 K.L.Merkle: In *Radiation Damage in Metals*, ed. by N.L.Petersen, S.D.Harkness (American Society for Metals, Metals Park, Ohio 1975) p. 58
1.31 G.Leibfried: J. Appl. Phys. **30**, 13800 (1959)
1.32 J.Lindhard, V.Nielsen, M.Scharff, P.V.Thomsen: K. Dan. Vidensk., Selsk. Mat. Fys. Medd. **33**, No. 10 (1963)
1.33 P.H.Dederichs: Phys. Status Solidi **10**, 303 (1965)
1.34 M.T.Robinson: Philos. Mag. **12**, 145, 741 (1965)
1.35 J.B.Sanders: Thesis, Univ. Leiden (1968)
1.36 M.W.Thompson: Philos. Mag. **18**, 377 (1968)
1.37 P.Sigmund: Rev. Roum. Phys. **17**, 823, 969, 1079 (1972)
1.38 D.P.Jackson, *Proc. Symp. on Sputtering* (SOS), ed. by P.Varga, G.Betz, F.P.Viehböck (Techn. Univ., Wien 1980) p. 1
1.39 J.P.Biersack, L.Haggmark: Nucl. Instrum. Methods **174**, 257 (1980)
1.40 R.Behrisch, G.Maderlechner, B.M.U.Scherzer, M.T.Robinson: Appl. Phys. **18**, 391 (1979)
1.41 M.Hou, M.T.Robinson: Appl. Phys. **17**, 295 (1978); **18**, 381 (1979)
1.42 D.E.Harrison, Jr.: Radiat. Eff. **70**, 1 (1983)
1.43 M.T.Robinson: J. Appl. Phys. **54**, 2650 (1983)
1.44 H.H.Andersen: In [Ref. 1.19, Vol. 10]
1.45 P.Sigmund, A.Oliva, G.Falcone: In [1.25]
1.46 G.M.Batanov: Sov. Phys. – Solid State **3**, 471 (1961); **4**, 1306 (1963)
1.47 B.Navinsek: J. Appl. Phys. **36**, 1678 (1965)
1.48 N.Itoh: Nucl. Instrum. Methods **132**, 201 (1976)
1.49 J.P.Biersack, E.Santner: Nucl. Instrum. Methods **132**, 229 (1976)
1.50 H.Overeijnder, R.R.Tol, A.E.de Vries: Surf. Sci. **90**, 265 (1976)
1.51 H.Overeijnder, M.Szymonski, A.Haring, A.E.de Vries: Phys. Status Solidi B **81**, K 11 (1977)

1.52 H.Overeijnder, M.Szymonski, A.Haring, A.E. de Vries: Radiat Eff. **36**, 63 (1978)
1.53 M.Szymonski, H.Overeinjner, A.E. de Vries: Radiat Eff. **36**, 189 (1978)
1.54 H.Overeinjnder, A.Haring, A.E. de Vries: Radiat Eff. **37**, 205 (1978)
1.55 M.Szymonski, A.E. de Vries: *Proc. Symp. on Sputtering* (SOS), ed. by P.Varga, G.Betz, F.P.Viehböck (Techn. Univ., Wien 1980) p. 761
1.56 M.Szymonski, A.E. de Vries: Radiat Eff. **52**, 9 (1980)
1.57 J.P.Biersack, E.Santner: Nucl. Instrum. Methods **198**, 29 (1982)
1.58 S.K.Erents, G.M.McCracken: J. Appl. Phys. **44**, 3139 (1973)
1.59 G.M.McCracken: Vacuum **24**, 463 (1974)
1.60 N.Hilleret, R.Calder: Proc. 7th Intern. Vac. Congr. & 3rd Intern. Conf. Solid Surfaces, ed. by R.Dobrozemsky, F.Rüdenauer, F.P.Viehböck, A.Breth (private publishers, Vienna 1977) p. 227
1.61 W.K.Chu, M.Braun, J.A.Davies, N.Matsunami, D.A.Thompson: Nucl. Instrum. Methods **149**, 115 (1978)
1.62 W.L.Brown, L.J.Lanzerotti, J.M.Poate, W.M.Augustyniak: Phys. Rev. Lett. **40**, 1027 (1978)
1.63 W.L.Brown, W.M.Augustyniak, E.Brody, B.Cooper, L.J.Lanzerotti, A.Ramirez, R.Evatt, R.E.Johson: Nucl. Instrum. Methods **170**, 321–325 (1980)
1.64 W.L.Brown, W.M.Augustyniak, L.J.Lanzerotti, R.E.Johnson, R.Evatt: Phys. Rev. **45**, 1632–1635 (1980)
1.65 R.W.Ollerhead, J.Bøttiger, J.A.Davies, J.L'Ecuyer, H.K.Haugen, M.Matsunami: Radiat. Eff. **49**, 203–212 (1980)
1.66 M.Szymanski, U.Paschke, R.Pedrys, A.Haring, R.A.Haring, H.E.Roosendall, F.W.Saris, A.E. de Vries: In [Ref. 1.24, p. 312]
1.67 P.Børgesen, J.Schou, H.Sørensen: In [Ref. 1.24, p. 822]
1.68 J.E.Griffith, R.A.Weller, L.E.Seiberling, T.A.Tombrello, Radiat. Eff. **51**, 223–232 (1980)
1.69 L.E.Seiberling, J.E.Griffith, T.A.Tombrello: Radiat. Eff. **52**, 201–210 (1980)
1.70 V.Pirronello, G.Strazzulla, G.Foti, E.Rimini: Nucl. Instrum. Methods **182/183**, 315 (1981)
1.71 F.Besenbacher, J.Bøttiger, O.Graversen, J.L.Hansen, H.Sørensen: Nucl. Instrum. Methods **191**, 221–234 (1981)
1.72 P.Børgesen, J.Schou, H.Sørensen, C.Claussen: Appl. Phys. A**29**, 57 (1982)
1.73 R.Pedrys, R.A.Haring, A.Haring, F.W.Saris, A.E. de Vries: Phys. Lett. A**82**, 371 (1981); Radiat. Eff. **64**, 81 (1982)
1.74 L.E.Seiberling, C.K.Meins, B.H.Cooper, J.E.Griffith, M.H.Mendenhall, T.A.Tombrello: Nucl. Instrum. Methods **198**, 17 (1982)
1.75 C.K.Meins, J.E.Griffith, Y.Quiu, M.H.Mendall, L.E.Seiberling, T.A.Tombrello: BAP 34, Radiat. Eff. (1983)
1.76 D.Fink, J.Biersack: Radiat. Eff. **64**, 89 (1982)
1.77 Proc. 1st Intern. Conf. on Radiat. Effects in Insulators, Arco, Lago di Garda, Italy (1981), ed. by P.Mazzoldi, K.Rössler: Radiat. Eff. **64/65** (1982)
1.78 P.K.Haff: BAP–22, Radiat. Eff. (1983)
1.79 G.K.Wehner, C.KenKnight, D.L.Rosenberg: Planet Space Sci. **11**, 885 (1963)
1.80 T.A.Tombrello: In *Proc. VII Intern. Conf. Atomic Coll.* in Solids (Moscow State Univ. Publ. House 1980) p. 35
1.81 T.A.Tombrello: Radiat. Eff. **65**, 389 (1982)
1.82 R.Behrisch, B.B.Kadomtsev: Plasma Physics and Contr. Nucl. Fusion Research, Proc. 5th Intern. Conf. Tokyo 1974, IAEA Vienna II (1976) p. 229
1.83 M.Kaminsky: Plasma Physics and Contr. Nucl. Fusion Research, Proc. 5th Intern. Conf. Tokyo 1974, IAEA Vienna II (1976) p. 287
1.84 W.Hauffe: Dissertation B, Technical University of Dresden DDR (1978) and Proc. Intern. Conf. Ion. Beam Modifications of Materials, Budapest (1978) p. 1079

2. Sputtering of Multicomponent Materials

Gerhard Betz and Gottfried K. Wehner

With 25 Figures

This chapter deals with sputtering of multicomponent systems, i.e., materials consisting of more than one element such as alloys or compounds. The major effects concern nonstoichiometric removal of surface atoms leading to changes in surface composition depending on ion mass, ion energy, target temperature, and ion fluence. Only for high ion fluences does the composition of the sputtered material become equal to that of the bulk target material. Furthermore, the angular and energy distribution of the different constituents may be different, especially at oblique ion incidence. In addition, a complex surface topography may develop for high ion fluences.

After a short historical introduction and review of applications in Sect. 2.1, the sputtering process for multicomponent materials is characterized and the different types of sputtering yield are defined in Sect. 2.2.

Section 2.3 deals with ion bombardment of single phase alloys and of compounds which contain no high vapor pressure components. Changes in surface composition due to preferential sputtering of one component, the development of an altered layer and changes in the total sputtering yield with composition for such materials are reviewed. Experimental results are explained by differences in the evolution of the collision cascade, by different surface binding energies of the constituents, by recoil implantation and ion beam mixing and by thermal or radiation induced diffusion and segregation.

In Sect. 2.4, compounds with a high vapour pressure component such as oxides are considered. For such compounds not only compositional changes, but also structural changes have been frequently observed under ion bombardment, in which a crystalline phase is converted into an amorphous or different crystalline phase of different composition. For these materials, collision-spike sputtering and sputtering via electronic processes also have to be taken into account.

Section 2.5 deals with multiphase solids. While for single phase systems theoretical approaches evolving partly from single element sputtering considerations are pertinent, this is not the case for ion bombardment of multiphase materials. Initially for a very low bombardment fluence, each phase or grain should obey similar laws as in the case of compounds or single phase materials. For a higher ion fluence, however, the experimental results are complex and difficult to understand due to the development of a pronounced surface topography.

2.1 Historical Review and Need for Data

2.1.1 Historical Remarks

Sputtering phenomena, i.e., the erosion of a solid due to energetic particle bombardment, have been known for more than 125 years [2.1]. The majority of investigations were made on elemental targets, while systematic investigations of the sputtering of multicomponent targets such as of alloys and compounds have become a subject for study only within the last 10–15 years. It has been known for a long time that sputtering yields of single element targets are sensitive to impurities and small additions of another element [2.2, 3], but the main aim of the investigations until the early 1960's was to avoid these effects and to determine sputtering yields of pure elements and single crystals in order to understand the underlying physical principles of this process.

However, in many sputtering experiments, surface contamination by residual gas atoms from the environment cannot be avoided, in particular at low bombarding ion current densities. Furthermore, part of the bombarding ions are implanted and trapped in the surface and near-surface region of the solid, so that after some time, always a two-component material is sputtered. In fact, almost all sputtering investigations, especially in sputtering yield measurements, are performed with very high particle fluences which causes the target to become saturated with the implanted ions. While the rather low obtainable concentration and inertness of rare gas atoms can probably be neglected in most cases (but not always [2.4]), this is definitely not true for sputtering with other than noble gas ions. For example, the extensive yield measurements of *Almén* and *Bruce* [2.5] show conspicious yield minima for bombardment with other than rare gas ions. Due to the implantation effect, the formation of compounds or alloys or the precipitation of the implanted species occurs and one always deals finally, in fact, with a multicomponent system. It was shown that the yield maxima and minima do not appear when they are determined for very low ion fluences where embedment of ions is still negligible [2.6, 7]. Implanted ions mostly cause a decrease of the sputtering yield but the yield may also increase if the ions form a volatile compound with the target atoms. These changes in the sputtering yield due to compound formation in the implanted layer are the topic of Chap. 3 by *J. Roth* and will not be discussed here. We will only deal with sputtering of given multicomponent targets, i.e., materials already consisting of several different atom species such as alloys and compounds. The ion implantation effects which also occur in these systems are not considered. The understanding of sputtering of a given multicomponent target is a precondition to comprehending the modifications which ion implantation may cause in sputtering.

Already in a rather early sputtering investigation of multicomponent materials, a change in the near surface composition as well as in the total sputtering yield was observed as a consequence of ion bombardment. In 1929, *Asada* and *Quasebarth* [2.8] found for Hg^+ ion bombardment of Cu (which contained minute amounts of Au), an enhancement in the Au sputtering yield relative to the bulk composition. They explained this observation with the formation of a Au enriched layer at the target surface which was continously replenished by diffusion from the bulk.

A change in surface composition due to ion bombardment was first clearly observed by *Gillam* [2.9] in 1959. He investigated near-surface composition changes due to bombardment by noble gas ions with energies of up to 5 keV for Cu_3Au and Ag–Pd alloys by transmission electron diffraction. For Cu_3Au he found an altered layer with a thickness of up to 40 Å which was enriched in Au. This was the first measurement of an altered layer thickness and the only one for the next 10 years. For Ag–Pd alloys of low Pd content, no surface composition changes were observed, but for alloys with a higher Pd concentration, again a layer enriched in Ag was formed and this enrichment increased with increasing bulk Pd content of the alloy. The main difficulty in these early measurements was that no method was known for reliable composition analyses of such thin layers.

In 1967, *Patterson* and *Shirn* [2.10] studied sputtering of Ni–Cr alloys and found an enrichment of Cr in the surface. They presented a simple analytical model, based on different sputtering yields of the different components, for explaining the observed composition changes under ion bombardment. With various modifications this model has been used often since then to analyze measured experimental data.

For sputtering of Nickel base alloys, Aluminium alloys and steels as studied by *Wehner* [2.11] under 0.1–0.5 keV Hg^+ bombardment, total sputtering yields similar to the yields of the components were generally found, while *Hanau* [2.2] observed lower yields for an impure Al probe (4% Cu, 0.5% Mg, 0.5% Mn) than for 99.99% pure Al.

No general predictions on the changes in surface composition or total sputtering yields could be derived from these early measurements. However, they showed that during the initial bombardment a composition change in the target surface and of the sputtered atoms occurs, and only at high bombardment fluences does the composition of the sputtered deposit become identical to that of the original alloy [2.10, 12–14] as required by conservation of matter.

Güntherschulze [2.15] was the first to observe cone formation in sputtering of certain (probably not very pure) metals and observed a reduction in the sputtering yield after such rough surface conditions were obtained. In 1969, *Anderson* [2.16], using a spectroscopic emission technique, was the first to demonstrate the large amount of sputtering required for multiphase solids to reach equilibrium. When sputtering a nonmixable Ag–Cu alloy, it took about 1 μm thickness of material removal at 100 eV Ar^+ ion bombardment to reach steady-state, i.e., until the composition of the sputtered material became identical to that of the target.

Much progress in the investigation of sputtering multicomponent materials was made when sensitive surface composition analysis methods became available in the late sixties, e.g., Auger Electron Spectroscopy (AES), Secondary

Ion Mass Spectroscopy (SIMS), Ion Scattering Spectroscopy (ISS), Rutherford Backscattering Spectroscopy (RBS), and Electron Spectroscopy for Chemical Analysis (ESCA), which is identical to X-ray Photoelectron Spectroscopy (XPS) [2.17]. With these methods, the composition of the surface and the selvedge ("near surface region") can be determined and any changes during sputtering of a multicomponent target can be monitored.

Surface composition changes under ion bombardment were observed for most multicomponent materials analyzed. As an example, modifications in the composition of the lunar surface by the solar wind bombardment were already predicted by *Wehner* et al. [2.18] in 1963, long before lunar rocks were available for analysis. Recently the existence of finely distributed metallic iron in lunar samples and their surface enrichment in Fe, Ti, and Si has been attributed to preferential sputtering by the protons and other ions of the solar wind [2.19].

2.1.2 Applications

In nearly all practical situations where sputtering is either employed or attempts are made to minimize it, multicomponent materials are used [Ref. 2.20, Chaps. 4, 6, 7]. Thus the knowledge of overall sputtering yields, changes in surface composition and in the composition of the sputtered atoms for these materials is important.

Multicomponent materials are sputtered for surface cleaning, ion etching (ion milling), depth profiling in combination with surface sensitive analytical techniques, thin film deposition of alloys, or of chemical or intermetallic compounds including insulators and organics, and in the search for materials with low sputtering yields in thermonuclear fusion devices.

A number of surface analysis techniques (AES, ESCA, ISS) would not be nearly as useful as they are if one did not have a method for in situ surface cleaning and microsectioning for performing thin film composition versus depth analyses. Thus one needs to know many details such as the composition of an altered layer (or the dose or time constant for its formation) which arises as a result of the ion bombardment, the sputtering yield of this material, the sputtering situation when one reaches an interface during depth profiling, and the development of topographical etch features. Beyond doubt, sputtering can provide a wealth of "artifacts" which may seriously limit quantitative analysis with these methods.

SIMS, of course, would not exist without sputtering. It differs from the other methods insofar as the sputtered material is used for analysis and not the surface which is left behind. After sputtering a large amount of material, one reaches a steady-state condition in which the sputtered material has the same composition as the bulk. Knowledge of the composition of the altered layer would then be hardly of interest. However, one should know the fluence which is required for reaching steady-state. Before reaching steady-state and in the so-called static SIMS [2.21] where one applies such a small dose that one usually is far from an equilibrium situation, a knowledge of the sputtering yields of various components is vital for quantitative measurements.

A peculiar, not yet well understood phenomenon can arise when even small amounts of certain seed metals, either provided internally or externally, are present at a target during sputtering [2.3, 22]. This situation can lead to a radical change in surface topography, namely, cone formation. Such cone formation can lead to a substantial reduction in the sputtering yield. This is especially of much concern in the deposition of films by co-sputtering from different targets.

In the deposition of multicomponent films, the substrate bias, intentionally or unintentionally, can seriously alter the film composition. Therefore, a special need exists for data on resputtering yields of components during film deposition in a situation which usually involves rather low bombarding ion energies. Ion bombardment induced alterations of the surface composition (bias-sputtering) has been used intentionally in thin film sputter-deposition technology to achieve a certain degree of control over the composition and quality (purity, ohmic resistance, hardness) of sputter-deposited thin films [2.23].

The search for materials with low sputtering yields has been a long-standing goal for many plasma and gas discharge devices. The possibility exists that certain multicomponent materials have lower yields than monoatomic targets. Presently, much effort is devoted to this research in respect to the first wall material in thermonuclear fusion devices [2.24].

2.2 Basic Considerations

In this section a short characterization of the sputtering processes for different kinds of multicomponent materials is presented and relevant definitions are given.

2.2.1 Multicomponent Materials

Multicomponent materials can be classified according to their structure into single-phase and multiphase materials. *Single-phase* materials are characterized by having the same composition of elements uniformly distributed. They may be single or polycrystalline or amorphous. In polycrystalline single-phase materials, different grains can have different orientations and also segregation of one component or of impurities at the grain boundaries can exist. Into this group belong single-phase alloy systems which are fully miscible for all concentrations such as Au–Cu or Ag–Au–Pd; further alloys with compositions where only one phase is present such as Cr–Pd for Pd concentrations above 65 at. %; and finally, compounds and amorphous substances.

We will further distinguish between single-phase solids, which consist only of low vapor pressure elements, and those containing also high vapor pressure elements like oxides or nitrides.

Multiphase materials are characterized by consisting of grains with different composition and possibly different crystal structure. The majority of alloys and also mixtures of different materials, like pressed powders or materials consisting of a single-phase matrix with precipitates of another compound, belong in this group. Some alloys may be a single-phase or a multiphase material depending on the composition. An intermetallic compound such as Al_2Au is a single-phase material, whereas an alloy with a slightly deviating compositon such as $Al_{0.7}Au_{0.3}$ consists of two phases (Al_2Au and Au grains).

The following three sections are divided up according to the different types of materials and their specific responses to ion bombardment. In Sect. 2.3, single-phase alloys and compounds with no high vapor pressure component are considered, whereas Sect. 2.4 deals with compounds having a high vapor pressure component, mostly oxides. Multiphase materials are dealt with in Sect. 2.5. The distinction between single and multiphase materials which has been overlooked sometimes in the past is an important one, from an experimental as well as theoretical point of view.

2.2.2 Characterization of the Sputtering Process for Multicomponent Materials

In the sputtering of multicomponent materials, the following effects have been found:

If a virgin multicomponent surface is subjected to particle bombardment, the different components are generally not sputtered in proportion to their concentration on the surface of the material. This phenomena is called *preferential sputtering*. Several effects can cause this preferential sputtering, e.g. mass differences of the target atoms influencing the spread of the collision cascade, different surface binding energies or chemical bonding differences between the target constituents, recoil implantation and also thermal or radiation induced segregation have been quoted as explanations for the preferential removal of one component. For compounds containing a high vapor pressure element, spike effects also have to be taken into account.

The preferential removal of one component from the surface leads to the formation of the so-called "*altered layer*" which is a near-surface region with a stoichiometry different from the bulk composition. The depth of the altered layer is generally thicker than the depth from which sputtered atoms originate (topmost two to three atomic layers) and mostly comparable to the range of the incident ions. At sufficiently low temperatures, when thermal diffusion is not important, the altered layer stays at a finite thickness and steady-state conditions can be reached, where the amount of material sputtered from each species becomes proportional to the bulk concentration. The thickness of the altered layer and the fluences needed to establish the dynamic equilibrium depend on the mechanisms which extend the altered composition at the surface to greater depth. Such processes can be collisional effects due to the incoming ions like recoil implantation and casacade mixing, or effects like thermal or radiation enhanced diffusion, as well as thermal segregation and radiation induced segregation.

The *sputtering yields* for the target constituents are generally different from those of the pure constituents at the same concentration and the total sputtering yield of the material is, in general, not a simple superposition of the yields of the pure elements. Very little data are available on these sputtering yields and especially their dependence on fluence before steady-state conditions are reached.

The *angular* and *energy distributions* of sputtered constituents are, in general, different from those of the atoms in elementary target sputtering. Their dependence on target constituents and bombardment conditions has hardly been investigated so far.

Finally, the *surface topography* which develops in the sputtering of multicomponent materials and its influence on the sputtering behaviour is even less understood than for elemental single crystals.

For *multiphase materials*, the crystallites with different composition and often different crystal structure show a different response under ion bombardment, called *selective sputtering*. At very low fluences each phase or grain behaves similarly to a single-phase material. For higher fluences a pronounced surface topography develops, like recessed high sputtering yield grains, cones, pyramides and ripples. Only very few experimental results and no theoretical approaches are available, as discussed in Sect. 2.5. In a very simplified picture, the behaviour of multiphase materials under ion bombardment can be understood in terms of the contamination of high sputter yield phases by material from low sputter yield phases, thus effectively reducing the overall yield and also initiating cones and other topography features. A distinction between single and multiphase materials is the fluence needed to reach the equilibrium surface composition. For single-phase systems, equilibrium is reached, in general, after removing a layer of thickness of the order of the penetration depth of the bombarding ions, while for multiphase systems, equilibrium will only be reached after removing a much thicker layer, usually of the order of a few grain diameters.

2.2.3 Definitions of Total, Partial, and Component Sputtering Yield

For multicomponent materials different types of sputtering yields have to be defined.

The *total sputtering yield* Y as for mono-element targets is defined as the average number of sputtered atoms per incident particle. Each removed atom is counted independently of its mass.

A *partial sputtering yield* Y_i of component i of a multicomponent material can be defined as the average number of sputtered atoms of component i per incident ion.

Total and partial sputtering yields are then related by

$$Y = \sum_i Y_i . \tag{2.1}$$

A *component sputtering yield* Y_i^c is defined as the partial sputtering yield Y_i divided by the equilibrium surface concentration c_i^s of component i during sputtering:

$$Y_i^c = Y_i / c_i^s . \tag{2.2}$$

With this definition, the trivial dependence of the partial sputtering yield on concentration (i.e., the partial yield goes to zero if the concentration goes to zero) is separated. The component yield Y_i^c can be compared to that of the pure element in order to observe matrix effects on the yield. A correction for different atomic densities in the alloy and the pure components is also included sometimes [2.25]. Both Y_i and Y_i^c are often denoted as partial sputtering yields in the literature, but one has to clarify if either Y_i^c or Y_i is designated as partial sputtering yield.

Partial sputtering yields and also component sputtering yields generally change if the surface concentration changes during ion bombardment. At sufficiently low temperatures, where bulk diffusion is not important, an equilibrium situation will always develop under particle bombardment where the target looses material in its bulk composition. For such steady-state conditions, a simple relation can be given between component sputtering yields and bulk and surface concentrations. The derivation for a binary target proceeds as follows [2.10, 26].

At steady-state the ratio of the number of sputtered atoms n_A of component A to n_B of component B must be equal to the ratio of the bulk concentrations c_A/c_B:

$$n_A / n_B = c_A / c_B . \tag{2.3}$$

As the ratio of the number of sputtered atoms n_A/n_B is equal to the ratio of the partial sputtering yields Y_A/Y_B per definition of the sputtering yield, one obtains from (2.2, 3) the following relation for steady-state between the surface concentrations c_A^s, c_B^s and the bulk concentrations c_A, c_B (see also [Ref. 2.27, Eq. (2.5.13)])

$$Y_A^c / Y_B^c = (c_A / c_B) / (c_A^s / c_B^s) . \tag{2.4}$$

Equation (2.4) allows the experimental determination of the ratio of the component sputtering yields and their dependence on concentration for a given binary system by measuring the surface as well as the bulk composition, i.e., the changes in surface composition under ion bombardment. Furthermore, these can be compared to the yields of the pure elements.

The dependence of the surface composition on ion fluence, leading for high fluences (steady-state) to (2.4), was originally derived by *Patterson* and *Shirn* [2.10] in 1967. In 1976, *Werner* et al. [2.28a] proposed a model for the time dependence of the surface composition similar to that of [2.10], specifying explicitly that sputtering occurs only from the momentary top layer. Also *Van Oostrom* [2.29a] developed a similar model in 1976, in which he considered the sequential sputtering of discrete monolayers of the target surface and determined the number of layers which must be removed in order to approach steady-state.

One problem associated with the definition of the component sputtering yield and with such measurements is the definition of the surface concentration in (2.2). Strictly speaking the surface composition should refer to the properly weighted average composition of the layer from which sputtering occurs, i.e. the top 2 to 3 atomic layers ([Ref. 2.27, Chap. 2] and [2.28b, c and 29b]):

$$c_i^s = \frac{1}{Y_i} \int_0^\infty c_i(x) y_i(x) dx \tag{2.4a}$$

if $y_i(x)dx$ is the contribution to Y_i from a layer dx at depth x and $c_i(x)$ the depth dependent concentration of species i.

At least for higher bombarding ion energies, this region is much thinner than the altered layer (Sect. 2.3.3). Depending on the probing depth of the surface sensitive method which is used to determine c^s, one can find different values for the "surface composition" and thus different component sputtering yield ratios.

Winters and *Coburn* [2.30] were the first to consider an altered layer much thicker than the escape depth of the sputtered particles, assuming but not calculating diffusion processes. Their model shows that the thickness of the altered layer is not always a good measure of the depth to which one needs to sputter-etch in order to reach steady-state. For example, in a dilute binary alloy in which the sputtering yield of the dilute component is much smaller than the yield of the second component, one may have to sputter-etch to a depth many times the thickness of the altered layer before a sufficient amount of the dilute component has accumulated near the surface to establish steady-state.

In addition, there are terminology discrepancies. Besides preferential sputtering, the terms "differential" or "selective" sputtering are also used by different authors to describe the preferred emission of one target component, thus creating a surface different in composition from the bulk. Furthermore, some authors use the term "preferential" with respect to the bulk composition of the solid [2.31, 32]. With the latter definition, preferential sputtering occurs only during the time until the altered layer is established and steady-state conditions are reached, because afterwards material is removed according to its bulk composition. We will here, however, follow *Andersen's* definition [2.33] of *preferential sputtering* which is with respect to the composition of the material from which the sputtered species originate, i.e., the outermost two to three atomic layers. With this definition, preferential sputtering also takes place during steady-state as the composition of the material sputtered from the surface is generally not equal to the composition of the first few atomic layers of the surface.

Selective sputtering is used to describe the phenomena occuring in ion bombardment of multiphase solids, while *differential sputtering* yields refer to the emission into given energy or angular intervals.

2.3 Single Phase Alloys and Compounds with Only Low Vapor Pressure Components

In this section, sputtering is considered mainly of solid solutions, i.e., either single-phase systems or those composition regions of other alloys where solid solutions exist. In addition, compounds like silicides and carbides as well as semiconductor and intermetallic compounds are included, for which none of the constituents has a high enough vapor pressure to allow thermal evaporation to affect the surface composition and the sputtering rate. For compounds with only low vapor pressure components, precipitation of one component due to rare gas ion bombardment cannot be excluded in principle, however, this has rarely been observed or described so far [2.34a, b]. Precipitation of the implanted species or formation of new phases [2.35, 36] which can occur under ion bombardment of other than rare gas ions, like in ion implantation, will be discussed in Chap. 3.

After a short discussion of experimental methods in Sect. 2.3.1, Sect. 2.3.2 will deal with surface composition changes, while in Sect. 2.3.3, composition changes deeper inside the material, i.e., the altered layer and its formation are discussed.

2.3.1 Experimental Methods

In order to investigate the changes in the surface and near-surface composition of alloys and compounds under ion bombardment, most of the surface sensitive analytical methods like AES, ISS, XPS, and SIMS and also RBS have been employed. Recently, a collector method in combination with RBS was also used [2.37]. Preferential sputtering effects and component sputtering yields can be deduced from comparing the composition of the sputtered flux with the composition of the surface layer from which sputtered particles are emitted. In Fig. 2.1, the information depth of different surface analysis methods is compared with the depth from which sputtered particles are emitted. Except for very low energy bombarding ions, this is generally smaller than the layer thickness for which an altered composition has been observed.

AES, ISS, XPS, and SIMS are semiquantitative methods and a calibration with pure element standards [2.38] can only sometimes correct for matrix

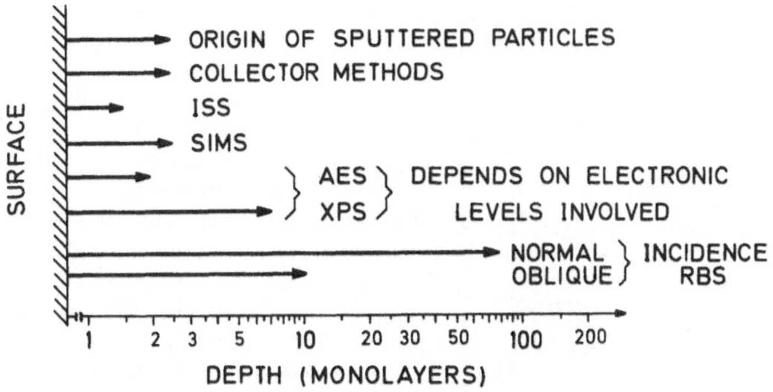

Fig. 2.1. Comparison between the depth from which sputtered particles are emitted and the information depth for different techniques used to measure changes in surface composition

effects which influence the signal. Thus, mostly only relative changes can be measured and it is desirable to start from a clean surface with a composition equal to the bulk composition. For solid solutions one of two different methods has usually been employed by different authors [2.39–42] to create such a surface.

i) *Fracturing the samples* in the UHV-system prior to analysis. If the fracture is transgranular, e.g., through the grains, the surface created will have a composition equal to the bulk. For an intergranular fracture, segregation to the grain boundaries can give rise to a surface different in composition to the bulk.

ii) *Scribing the samples* with a stainless steel or diamond tip will also generally result in a surface with composition equal to the bulk for solid solutions.

Provided the same composition is found for these quite different methods, it can be assumed that the newly created surface has a composition equal to the bulk. So far this has been found for all alloys forming solid solutions. However, different surface compositions were found for alloys of the nonmixable Ag–Cu system [2.39] and also for those Ag–Au–Cu alloys [2.43] which do not form solid solutions.

Comparing the Auger signal ratio of the alloy components for a scribed or fractured surface (prop. to c_A/c_B) with the signal ratio from the sputtered surface (prop. c_A^s/c_A^s), the component yield ratio Y_A^c/Y_B^c can be obtained with (2.4). Figures 2.2a, b give two different representations of the experimental results obtained in this manner for Ag–Pd alloys of different compositions under 2 keV Ar ion bombardment [2.44]. Surface enrichment of Pd can be seen. The component sputtering yield ratio Y_{Ag}^c/Y_{Pd}^c calculated using (2.4) as a function of alloy composition ranges from 1.6 up to 3.4, as shown in both figures, and is compared to experimental [2.41] and calculated [2.45] values of the sputtering yield ratio Y_{Ag}/Y_{Pd} for the pure elements. Y_{Ag}^c/Y_{Pd}^c can be readily seen in Fig. 2.2a as the difference between the sputtered surface and the scribed

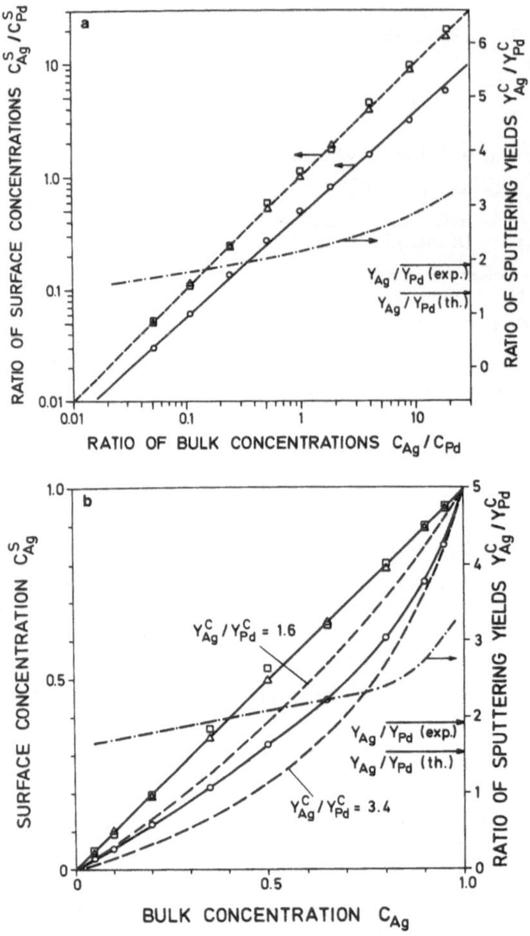

Fig. 2.2. (**a**) Ratio c^s_{Ag}/c^s_{Pd} of measured surface concentrations using AES [Ag(351/355 eV) and Pd(326/330 eV) transitions] versus ratio of bulk concentrations c_{Ag}/c_{Pd} for Ag–Pd alloys: (□) fractured, (△) scribed, (○) after 2 keV Ar ion bombardment. Also shown is the component sputtering yield ratio Y^c_{Ag}/Y^c_{Pd} as obtained with (2.4) from the measured changes in surface composition. Measured [2.41] and calculated [2.45] sputtering yield ratios of the pure constituents are also indicated (Exp. data from [2.44]).

(**b**) Using the experimental data of Fig. 2.2a, the Ag surface concentration c^s_{Ag} is plotted for fractured (□), scribed (△) and sputtered (○) alloys versus the Ag bulk concentration c_{Ag}. The component sputtering yield ratio Y^c_{Ag}/Y^c_{Pd} as a function of alloy composition and experimental and calculated sputtering yield ratios of the pure elements are also presented. The dashed lines are the calculated Ag surface concentrations with (2.4) for the limiting cases of $Y^c_{Ag}/Y^c_{Pd}=1.6$ or 3.4, respectively. The experimental values for c^s_{Ag} after ion bombardment vary between these values

or fractured bulk composition [2.44] and in Fig. 2.2b, the surface concentration of Ag is shown to vary between the values calculated for $Y^c_{Ag}/Y^c_{Pd}=1.6$ and 3.4 using (2.4).

As the mean escape depth of the Auger electrons varies from a few Å for low energy electrons up to some 10 Å for electrons from high energy Auger transitions, the depth of the surface layer whose composition is determined varies. Therefore, depending on which Auger transitions of a given element are employed, slightly different component sputtering yield ratios will be deduced with (2.4) if no corrections are made. On the other hand, these differences in the escape depth of the Auger electrons for different energies have been employed for determining the thickness of the altered layer by using high and low energy Auger transitions [2.46, 47].

If a surface composition equal to the bulk composition cannot be produced at the start of the measurements, AES still allows the measurement of the relative changes in surface composition due to variation in energy or the type of bombarding ions [2.48].

Ion Surface Scattering (ISS) has been employed similarly to AES to determine relative changes in surface composition [2.48, 49a, b]. As ISS is specific to the first monolayer only, no information on the depth of the altered layer can be deduced. While the information depth of AES using high energy Auger spectra is deeper than the surface layer from which sputtered atoms are ejected, it is smaller for ISS.

X-ray Photoelectron Spectroscopy (XPS) can give additional information on the formation of new phases or compounds and has been mainly used for the analysis of compounds and oxides (Sect. 2.4).

With SIMS, only few investigations have been performed [2.50] to study explicitly surface composition changes under ion bombardment. However, the dependence of the secondary ion yield on composition is not well known. This is crucial for the interpretation of the results and therefore only under special conditions can definite conclusions be drawn.

A very successful technique in surface and thin film analysis is RBS because it gives quantitative results up to a depth of a few 1000 Å. However, its depth resolution is generally only 30–300 Å (Fig. 2.1). With this method the thickness and concentration gradients in the altered layer, the surface enrichment and the total and partial sputtering yields have also been determined [2.51, 52]. Figure 2.3 shows a RBS analysis of a PtSi film before and after 80 keV Ar ion bombardment [2.53]. Pt enrichment and also the depth of the enriched layer were determined. Because of the limited depth resolution of RBS, this method is especially useful in studying sputter induced composition changes for higher

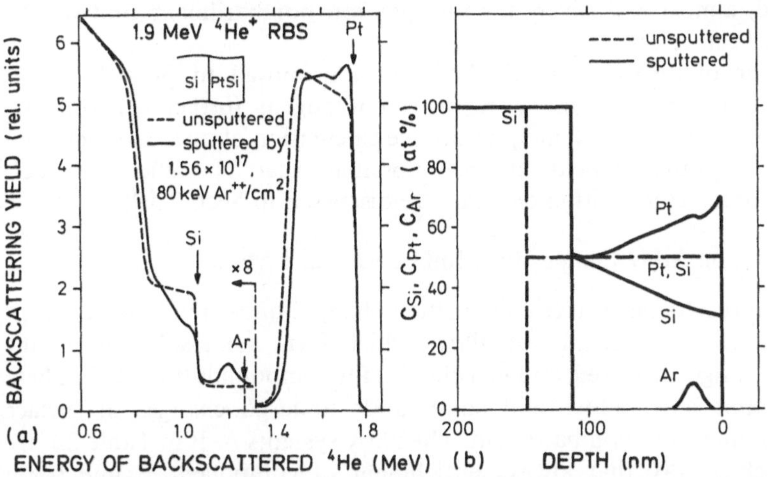

Fig. 2.3. (**a**) Backscattering spectra of an 80 keV Ar^{2+} sputtered and unsputtered PtSi sample. The unsputtered spectrum is from a 1460 Å thick film and the sputtered spectrum from an originally 2200 Å thick film after sputtering away 960 Å. Pt enrichment and Si depletion after ion bombardment as compared to the unsputtered film can be observed [2.53]. (**b**) Schematic representation of the concentration profiles as obtained from the RBS-spectrum

primary ion energies (>10 keV) where the altered layers are thicker. Furthermore, only alloys or compounds with components of quite different masses can be analysed. Surface compositions given by this method are therefore averages for a layer thickness of at least 50–100 Å. Thus, for example, no surface enrichment was found for InP, GaP, and GeSi using RBS [2.51] but this does not exclude the possibility that the concentration in the top few monolayers may have been changed.

Andersen et al. [2.37] recently collected sputtered atoms on carbon foils as a function of ion bombardment-fluence. The deposits were afterwards analyzed by RBS in order to measure the fluence-dependent changes in the composition.

To measure total sputtering yields, the same methods can be employed as for elemental targets. Angular distributions of the different constituents of a multicomponent target have been measured by condensing the sputtered material on a collector at different ejection angles, followed by analyses with AES [2.54] or RBS, or for carbon using the nuclear reaction [$^{12}C(d,p)^{13}C$] [2.24, 55].

2.3.2 Surface Composition Changes Due to Ion Bombardment

In the following we deal with the compositional changes of single phase alloys and compounds in the first 1–3 atomic layers which are measured by surface techniques such as AES, SIMS or ISS. This is also the range from which sputtered atoms originate. Changes in the composition of a layer of thickness of the range of the implanted ions or deeper will be treated in Sect. 2.3.3.

A large amount of data for different alloys and compounds has been accumulated, almost exclusively for rare gas ion bombardment, and will be discussed in Sect. 2.3.2a.

A number of different models have been proposed to predict surface enrichment, but most of these were proved wrong as further experimental results were collected. In fact, no quantitative calculation of surface enrichment or component sputtering yields has been possible so far. The different models predicting surface composition changes are discussed in Sect. 2.3.2b.

a) Dependence on Alloy Composition, Ion Energy, and Mass

Experimental results on surface composition changes under ion bombardment are summarized in Table 2.1. for alloys which form solid solutions in the composition range measured, and in Table 2.2 for compounds (Sect. 2.7). These data show that most alloys and compounds undergo changes in surface composition under ion bombardment. The alloy systems A–B in Table 2.1 are listed in such a way that always enrichment of component B under ion bombardment was observed. Column 2 lists the atomic weight ratio M_A/M_B and the ratio of the heats of sublimation U_A/U_B of the constituents. In Columns 5 and 6, component yield ratios as derived from measured surface composition changes using (2.4) are compared with sputtering yield ratios of the pure constituents. In a few cases, absolute component and partial sputtering yields have also been measured. This subject will be discussed in Sect. 2.3.4.

Systematic studies on alloy composition were performed on a number of binary alloy systems. Using AES and comparing scribed or fractured surfaces with sputtered ones, the alloy systems Au–Cu, Au–Ag, Au–Pd, Ag–Pd, Cr–Fe, Cu–Pd, Cu–Pt, Ni–Fe, and Ni–Pt were studied under 1 or 2 keV Ar ion bombardment [2.39, 40, 44]. The thickness of the surface layer analyzed with the Auger transitions used (see Sect. 2.3.1) in these measurements varies between 3–5 Å for Ag and Pd up to 20 Å for Pt and Au. However, for the latter two elements, lower energy Auger transitions also exist and were employed for analysis, yielding, however, only slightly different results [2.39, 40]. Up to 10 different compositions for each alloy system were measured. In each case surface enrichment of one component (except for Au–Cu with no enrichment of either component) was found, but mostly no pronounced changes of the component sputtering yield ratio with alloy bulk composition were observed.

For ternary systems only few investigations are reported in the literature. The ternary system Ag–Au–Cu has been studied under 2 keV Ar ion bombardment [2.43] using AES and the same elemental Auger transitions as for the corresponding binary systems. This alloy system forms solid solutions except for alloys rich in Ag and Cu. Surface enrichment of Au and Cu was found under ion bombardment. The surface concentration ratio Au/Cu, however, remained unchanged from the bulk concentration ratio. The component sputtering yield ratios Y^c_{Ag}/Y^c_{Cu} and Y^c_{Ag}/Y^c_{Au} were found to be about 2.5 independent of the alloy composition as can be seen in Fig. 2.4. These results agree well with the results for the binary system Ag–Au where Au enrichment was found [2.40] and the system Au–Cu where no change in surface composition [2.40] during sputtering was observed. For alloys of the ternary system Ag–Au–Pd, the surface enrichment found was also in agreement with the results of the corresponding binary systems [2.56].

Changes in surface composition for Ag–Au–Pd alloys under 2 keV Ar^+ ion bombardment are summarized in Fig. 2.5 and indicated by arrows. For this alloy system, the surface always has a larger Pd and a lower Ag content than the bulk due to sputtering, whereas the Au content can be equal, larger or smaller than for the bulk, depending on the Ag/Pd bulk concentration ratio, as seen from Fig. 2.5.

For austenitic stainless steels and Inconel alloys [2.57–60], no or almost no enrichment of any component was found using AES in agreement with results for the corresponding binary systems.

For all these metal alloy systems, surface enrichment is observed of that component, which has the lower elemental sputtering yield [cf. Table 2.1 (Sect. 2.7)]. Thus, surface enrichment can be predicted qualitatively from the sputtering yields of the pure constituents.

A few investigations were also concerned with the changes in surface composition with ion energy and type of bombarding rare gas ions. For a Cu-40 at.% Ni film, an increase in the Ni enrichment with ion energy from

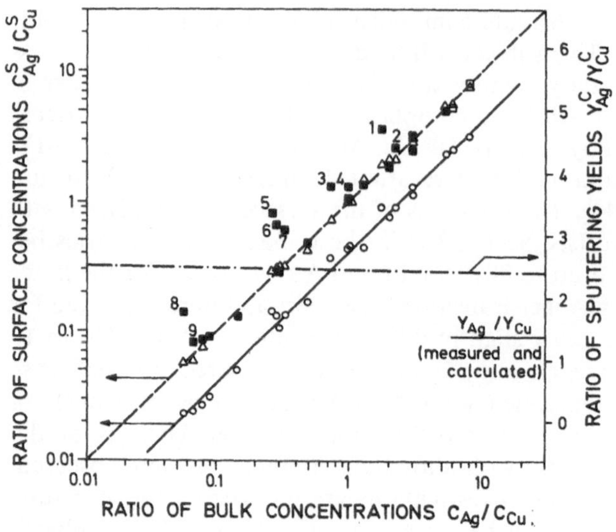

Fig. 2.4. Ratio c^s_{Ag}/c^s_{Cu} of measured surface concentrations with AES (Ag-351/355 eV and Cu-920 eV Auger transitions) versus ratio of bulk concentrations c_{Ag}/c_{Cu} for Ag–Au–Cu alloys: (□) fractured, (△) scribed, (○) after 2 keV Ar ion bombardment. Data for fractured alloys with a composition outside the single-phase region are indicated by numbers. Most of them show pronounced Ag enrichment after fracturing. Also shown is the component sputtering yield ratio Y^c_{Ag}/Y^c_{Cu} for single phase alloys and measured [2.40] and calculated [2.45] sputtering yield ratios of the pure constituents are indicated. Note: alloys are plotted according to their c_{Ag}/c_{Cu} bulk ratio independent of the Au content [2.43]

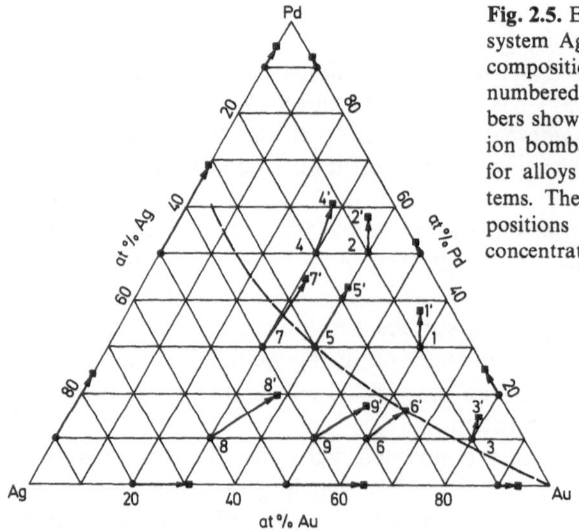

Fig. 2.5. Equilibrium diagram for the ternary system Ag–Au–Pd. Numbered dots indicate composition of alloys analysed. Arrows from numbered dots to squares with primed numbers show the changes in composition under ion bombardment. Unnumbered arrows are for alloys of the corresponding binary systems. The dashed line indicates bulk compositions of alloys with no change in Au concentration due to sputtering [2.56]

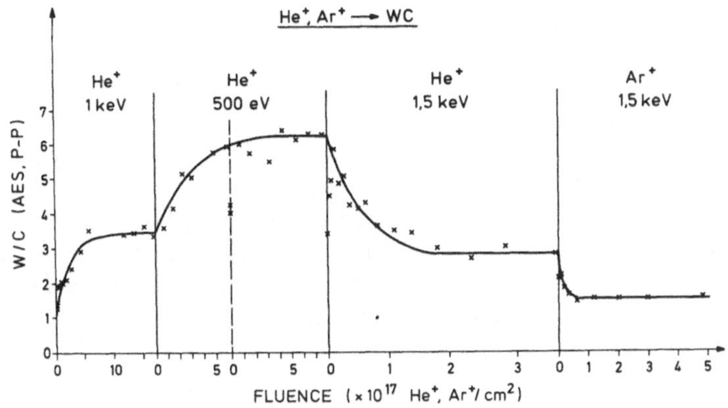

Fig. 2.6. Surface composition ratio W/C (Auger peak-to-peak intensity ratios) for WC for different ion bombardment conditions. W surface enrichment is most pronounced for 500 eV He^+ bombardment and decreases with increasing ion energy and higher mass ions (Ar^+) [2.48]

44 at.% Ni at 500 eV to 62 at.% Ni at 2 keV was observed [2.61]. For ion energies below 500 eV, *Tarng* and *Wehner* [2.62] observed for a Cu–Ni alloy a strong increase in surface enrichment of Ni near threshold energies. This was explained by the lower threshold energy for sputtering of Cu than Ni, thus dramatically increasing the component yield ratio Y^c_{Cu}/Y^c_{Ni} from 1.6 under 200 eV to about 7 under 35 eV Ar^+ ion bombardment.

A systematic study of 9 binary alloy systems (Au–Cu, Au–Ag, Au–Pd, Ag–Pd, Cr–Fe, Cu–Pd, Cu–Pt, Ni–Fe, Ni–Pt) and also 5 compounds ($AlNi_3$, AlFe, $AlCr_2$, AlAu, $AlAu_2$) was performed for 0.5–5 keV Ne, Ar, and Xe bombardment using AES [2.63, 64]. In this energy range, the changes of the surface enrichment were small and often lay within the error of the analysis method (AES). Generally increased enrichment of the lighter component with increasing ion energy could be found for all these alloys and all primary ions used. The changes from one type of rare gas ion to another were generally small and no general trend could be derived. Also, no changes in the near surface composition (about top 100 Å) with ion energy (20–80 keV) and ion type (Ne, Ar, Kr, Xe) were found within the experimental uncertainties using RBS on alloys, metallic compound phases and intermetallic silicides [2.51].

However, pronounced changes in surface composition were observed by *Taglauer* and *Heiland* [2.48] for carbides (metal enrichment in TaC and WC) with the energy of the bombarding He ions (500–1500 eV), and also comparing He and Ar ion bombardment (Fig. 2.6) using AES and ISS for analysis.

b) Models Predicting Surface Composition Changes

The experimental results (compare also Sect. 2.7) show that for many alloys and compounds, especially for all single phase metal alloys, the observed surface enrichment is mostly in agreement with the sputtering yields of the pure elements, i.e., the component with the lower elemental yield becomes enriched under ion bombardment. However, this correlation between observed surface enrichment and elemental sputtering yields is by no means quantitative. Furthermore, there are some compounds, especially those with a large mass difference in the constituents, where enrichment of the heavier mass constituent is observed [2.48, 51, 64] and not enrichment of the low sputtering yield component. Such a correlation with mass differences was also observed for sputtering of chemisorbed nitrogen from W and Mo surfaces [2.65a].

Various attempts have been made to explain these sputter induced changes in surface composition. The effects which can be responsible for the surface composition changes under ion bombardment can be divided into mass effects and surface binding effects.

Mass Effects: If a *collision cascade* is initiated in a solid with atoms of different masses, the momentum and energy will be distributed differently to each constituent resulting in different ejection probabilities for the atoms. This will, in general, cause preferential sputtering of the lighter component and therefore surface enrichment of the heavier component. Three collisional regimes can be distinguished [Ref. 2.27, Chap. 2]:

— the linear cascade regime for medium and heavy mass ions at energies in the keV range;
— the spike regime for heavy ions and energies above some ten keV;
— the single knock-on regime for very low energies near the sputtering threshold or for light ions.

Recoil implantation and *recoil mixing* will cause, especially at higher bombardment fluences, additional compositional changes in the surface as well as deeper inside the target.

Bonding Effects: Differences in the *surface binding energy* of the constituents will result in a preferential emission of the component with the lower binding energy and therefore cause enrichment of the component with the higher surface binding energy.

One component may be enriched on the surface due to *surface segregation* and this component will then be preferentially removed by sputtering. This process cannot produce a changed surface composition under steady-state sputtering conditions and will be discussed further in Sect. 2.3.3c.

None of these effects has been treated in enough detail yet to predict even qualitatively all the experimental data collected. The different effects can give opposite results and generally a combination of mass effects and bonding effects will act. Depending on mass and surface binding energy differences between the components, ion energy and mass and target temperature, either mass or bonding effects can dominate. Generally for alloys and compounds with components of not too different masses, the surface binding energies of the constituents should play the dominant role, while enrichment in the heavier components is found mostly for systems with quite different constituent masses. In the following, the different models will be discussed.

i) Mass Effects: Within the limits of the linear cascade theory, i.e., at intermediate and higher ion energies, enrichment of the heavier components has been predicted by several authors ([Ref. 2.27, Chap. 2] and [2.53, 65b–d, 66]), if the surface binding energies for both components are similar. More energy is retained by the lighter mass atoms and they have a larger range than the heavier mass atoms, but even if both components get the same energy, the lighter mass atoms have a larger range than the heavier mass atoms in the solid ([2.65b] and [Ref. 2.20, Chap. 2]). In this energy region, the following equation was derived for the ratio of the component sputtering yields in a binary alloy due to the cascade [Ref. 2.27, Eq. (2.5.10)]

$$Y_A^c/Y_B^c = (M_B/M_A)^{2m}(U_B/U_A)^{1-2m} \tag{2.5}$$

with $0 \lesssim m \lesssim 0.2$ depending on ion energy and M_i and U_i being the mass and surface binding energy of the atoms of component *i*. In the derivation of this equation, an undisturbed surface composition equal to the bulk composition is assumed, i.e., it is valid only in the limit of zero fluence. After prolonged ion bombardment, secondary effects due to a changed composition near the surface might become important [2.67]. In (2.5), the dependence on mass differences is rather small and surface enrichment can be reversed if the surface binding energy differences dominate.

In a model by *Kelly* [2.68], linear cascade sputtering is assumed up to but excluding the top monolayer. This allows a changed surface composition to be taken into account and also assumes that the bulk composition up to but excluding the top monolayer remains unchanged even after prolonged ion bombardment. The energy transfer to atoms A or B in the top monolayer is then considered following the calculations of *Winters* and *Sigmund* [2.69] for sputtering of adsorbed layers. He arrived at the following expression for the component sputtering yield ratio:

$$Y_A^c/Y_B^c = [(c_A + c_B\gamma)/(c_B + c_A\gamma)]\,(U_B/U_A) \tag{2.6}$$

with $\gamma = 4M_AM_B/(M_A+M_B)^2$ the maximum energy transfer factor. Neglecting further the influence from the surface binding energies enrichment should result in the less abundant component [2.68].

For high energy ($E \gtrsim 50$ keV) incident heavy ions, i.e., for the spike regime, a stronger dependence on the mass ratio than for the linear cascade regime (2.5) was obtained by *Sigmund* [Ref. 2.27, Eq. (2.5.12)]

$$Y_A^c/Y_B^c = (M_B/M_A)^{1/2}\exp[(U_B - U_A)/k_BT] \tag{2.7}$$

with a spike temperature T dependent on beam and target parameters.

Assuming energy equipartition between the components, i.e., spike conditions, enrichment in the heavier constituents was predicted by *Haff* [2.65b] because the lighter mass atoms have a longer effective free path in the solid and therefore can be ejected from further beneath the surface than heavier atoms [2.51, 53]. Describing the movement of the low energy secondaries in the spike in terms of a random walk process, he obtained a $(M_B/M_A)^{1/4}$ dependence of the component sputtering yield ratio.

In the single knock-on regime, i.e., at low energies close to threshold, sputtering is dominated by primary and secondary recoils and the few collisions occuring are generally not sufficient to acquire momentum randomization as in the linear cascade regime. In this region, the projectile-to-target atom mass ratio can play a larger role than in the other regimes.

Firstly, for approximately equal masses of the bombarding ions and one of the target atom constituents, a very efficient energy transfer is possible. Hence, sputtering of these target atoms is more efficient and the surface becomes depleted in this target component [2.33].

Further, for ion bombardment of targets with large differences in the mass, another mechanism comes into play [2.54], i.e., under normal ion incidence, two successive binary collisions are necessary to reverse the momentum of the primary knock-on to give backsputtering. A light target component may have its momentum inverted by colliding with a heavy component atom underneath, but not vice versa. Hence, sputtering of the light component should be favored.

Finally, for bombardment by very light ions (H^+, He^+) but ion energies not necessarily close to threshold, target atoms are predominantly sputtered in a direct collision with an ion which has been backscattered from the interior. As the energy transfer factor γ for a collision between an ion and a target atom is larger for the lighter component of a solid, this leads to preferential sputtering of the lighter target atoms [2.70]. This was demonstrated for He^+ and H^+ ions on TaC and WC [2.48, 71] with strong Ta or W enrichment, as shown already in Fig. 2.6. This model predicts that the depletion of the lighter target component becomes stronger with decreasing energy [2.24, 48]. It explains the extremely strong depletion of carbon for carbides under light ion bombardment for energies close to the sputtering threshold of the pure metals but still above the sputtering threshold for carbon [2.24, 48].

The mass effects discussed up to now lead to a surface depletion of the lighter target component and are caused by a preferential emission of the lighter atoms. In addition, the surface composition is changed by *recoil implantation* [2.72–75] where predominantly the depth distribution of primary knock-on atoms in the target is considered. *Kelly* and co-workers [2.68, 72–74] predicted enrichment in the heavier component due to recoil implantation. Be–Cu [2.76] with a surface consisting only of Cu under ion bombardment is an example where recoil effects have been assumed to be dominant due to the high mass ratio $M_{Cu}/M_{Be}=7$ [2.77]. Linear cascade sputtering alone cannot explain this extreme Cu enrichment.

Sigmund [2.75], however, predicted for recoil implantation an initial enrichment of the heavier atoms on the surface and of the lighter component at a greater depth. At high fluences, the surface layers are sputtered away and the region enriched with the lighter component reaches the surface. If preferential sputtering is neglected the lighter component should now be enriched at the surface.

Recoil implantation is not generally sufficient to explain surface compositional changes at high fluences. It will, however, produce compositional changes beneath the surface layer from which sputtered particles are emitted. The arguments are similar to those for surface composition changes due to surface segregation and will be discussed in more detail in Sect. 2.3.3c.

ii) Bonding Effects: The influence of the surface binding energies on the ratio of the component sputtering yields has already been included in (2.5–7). In general, the surface binding energies of the different components are, however, not independent of the alloy composition. The binding energy of a surface atom in a material depends on its position and on the type of neighboring atoms. If the atom sits alone on top of a surface plane or on an edge, it is more loosely bound than inside a flat plane. For elemental targets, the heat of sublimation was generally taken as a mean value for the surface binding energy [Ref. 2.27, Chap. 3]. But calculations for some fcc and bcc single-crystal metal surfaces have shown that they may be greater than the measured heats of sublimation [2.78].

Surface binding energies for metals and alloys can, however, be determined from the measurement of the energy distribution of the sputtered atoms if in the sputtering regime of linear cascades ([2.79] and [Ref. 2.27, Chap. 2]).

Szymonski et al. [2.79] concluded from energy distribution measurements of sputtered atoms from an AuAg alloy that there was a decrease in the surface binding energy of silver U_{Ag} from 3.1 on pure Ag to 2.1 eV in the alloy and for Au, a decrease from 3.8 to 3.3 eV. Also for Cu–Zn alloys, the energy distributions of the sputtered Cu and Zn atoms indicate changes of the surface binding energies of Cu and Zn atoms up to a factor of 2 with alloy composition [2.80]. *Oechsner* and *Bartella* [2.81] found for a 88 at. % Ni–W alloy, a small increase of U_{Ni} from 4.5 to 4.7 eV and a large decrease of U_W from 8.7 to 5.6 eV, in the alloy. From surface enrichment under ion bombardment and measured total sputtering yields of Cu–Ni alloys, surface binding energies of Cu and Ni as a function of alloy composition were calculated [2.40]. No strong deviations of the surface binding energies from the values of the pure components were found in this case, however, both are lower in the alloy than for the pure constituents.

For a random binary alloy the surface binding energies U_A and U_B have been expressed by *Kelly* [2.68, 77] as functions of the nearest-neighbor bond strengths U_{AA}, U_{BB}, and U_{AB}. Thus, one obtains for U_A:

$$U_A = -c_A^s Z_s U_{AA} - c_B^s Z_s U_{AB}\,, \tag{2.8a}$$

where Z_s is the surface coordination number. Attributing cohesion solely to nearest-neighbor interactions [2.82], U_{AA}, U_{BB}, and U_{AB} can be expressed in a first-order approximation as functions of ΔH_A and ΔH_B, the heats of atomization for each element, resulting in

$$\frac{U_B}{U_A} \cong \frac{(1+c_B^s)\Delta H_B + (1-c_B^s)\Delta H_A}{(1+c_A^s)\Delta H_A + (1-c_A^s)\Delta H_B}. \tag{2.8b}$$

This ratio of the surface binding energies can be introduced in (2.5–7). Using (2.6) *Kelly* obtained the following result for linear cascade sputtering [2.6, 77]:

$$\frac{Y_A^c}{Y_B^c} = \frac{c_A + \gamma c_B}{c_B + \gamma c_A} \frac{(1+c_B^s)\Delta H_B + (1-c_B^s)\Delta H_A}{(1+c_A^s)\Delta H_A + (1-c_A^s)\Delta H_B} \tag{2.9}$$

predicting enrichment of the component with the higher surface binding energy or heat of atomization.

The enrichment of one component predicted by the surface binding energy argument is in agreement with experimental results for many alloys. However, it was found that enrichment after (2.9) is in most cases too small as compared to the experimentally observed changes [2.77].

2.3.3 The Altered Layer Formed by Ion Bombardment

Strictly speaking, *preferential sputtering* means different ejection probabilities of the different sputtered atoms. With this definition, surface composition changes due to recoil implantation or surface segregation as mentioned in the previous section do not belong to preferential sputtering. Furthermore, preferential sputtering effects only cause composition changes in a surface layer with a thickness corresponding to the depth from which sputtered particles are emitted.

Experimental results show, however, that compositional changes occur to much larger depths, typically comparable to the range of the incident ions. This layer of changed composition is called the *altered layer*.

Changes under ion bombardment over such depth ranges must be caused, in addition to preferential sputtering, by transport processes such as

- thermal diffusion,
- radiation enhanced diffusion,
- recoil implantation or cascade mixing,
- thermal surface segregation,
- radiation enhanced segregation.

Thermal diffusion and radiation enhanced diffusion can only counteract compositional changes produced by preferred sputtering of one component at the surface and thus increase the thickness of the zone of altered composition. Recoil implantation and cascade mixing as well as surface segregation and radiation induced solute segregation, however, can produce compositional changes by their own and thus enforce or weaken existing surface compositional changes due to preferred emission of one component, provided they are rapid enough.

a) Experimental Results

The two methods used to determine the altered layer, its thickness and its development with ion fluence are Rutherford Backscattering Spectroscopy (RBS) and Auger Electron Spectroscopy (AES), the latter using the variation in escape depth of the Auger electrons with energy. The experimentally determined altered layer thicknesses as obtained by different authors for alloys are summarized in Table 2.3 (Sect. 2.7).

Using RBS, *Liau* et al. [2.51, 53] studied the development of the altered layer as a function of ion fluence and the dependence of the thickness of this layer on ion energy (10–80 keV) for metal alloys such as AuAg, and for intermetallic silicides such as PtSi. They found that the thickness of the altered layer agrees roughly with the maximum penetration depth of the primary ions (Fig. 2.7). The surface enrichment of one component increases linearly with fluence (Fig. 2.8) until steady-state conditions are reached at approx. 2×10^{17} ions cm^{-2} for 80 keV Ar^{2+} on PtSi. The removal of about one to two altered layer thicknesses was necessary to reach steady-state conditions [2.51, 53]. The results are explained in terms of preferred emission of the lighter

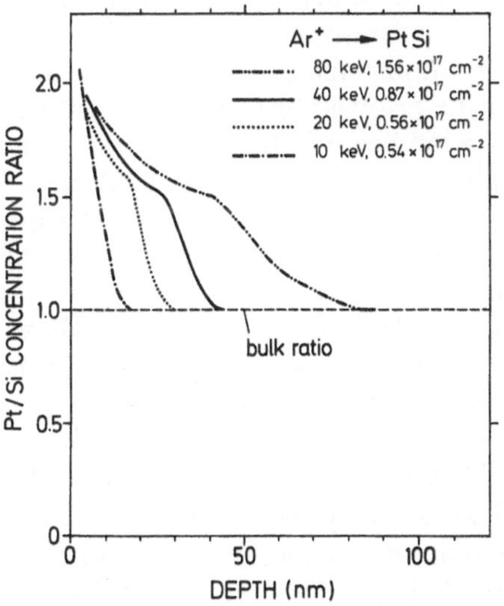

Fig. 2.7. Steady-state Pt/Si profiles of 10, 20, 40, and 80-keV Ar^+ sputtered PtSi samples from RBS measurements. The thickness of the altered layer increases with bombarding ion energy, while the surface composition within the depth resolution limits of RBS is independent of ion energy [2.53]

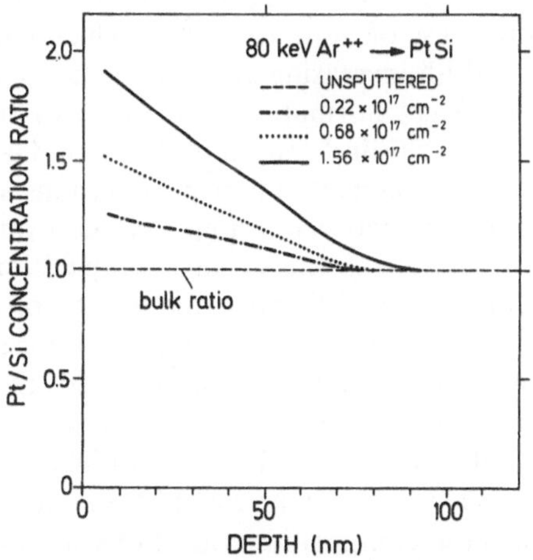

Fig. 2.8. Depth profiles of Pt/Si concentration ratio for 80-keV Ar^{2+} sputtered samples measured with RBS. Pt enrichment increases with fluence, while the altered layer thickness is independent of the fluence [2.53]

component and radiation enhanced diffusion which creates a composition gradient over the range of the primary ions. For Ar sputtering of PtSi, evidence of a repetitive Ar bubble formation process was also found with RBS [2.53, 83].

While RBS measurements are restricted to measurements of thick altered layers (>10 nm) and consequently high bombarding ion energies for sputtering, AES is restricted to measurements of thin alterd layers and therefore generally low bombarding ion energies (Sect. 2.3.1).

Experimentally, the altered layer of Cu–Ni alloys was analyzed by several authors [2.25, 46, 47, 84–86] using AES and employing Auger transitions of different energies and thus different escape depths for the same element (HEAES and LEAES).

Watanabe et al. [2.46] compared low energy Auger spectra (LEAES) and high energy Auger spectra (HEAES) of Cu and Ni in Cu–Ni alloys under ion bombardment. Approximating the altered layer by a step function, they found values between 4 Å for Ni rich and 30 Å for Cu rich alloys under 700 eV Ar^+ bombardment in good agreement with the results for Cu–Ni according to the kinetic model (Sect. 2.3.3c) [2.25]. Assuming a Gaussian distribution function for the altered layer, similar depths and diffusion constants in the range of $10^{-19}-10^{-16}$ cm^2/s were obtained. For Cu–Pt and Ni–Pt alloys, altered layer thicknesses of the same magnitude were also obtained for 2 keV Ar^+ bombardment by the same technique assuming an exponential concentration decrease of the enriched component in the altered layer [2.40].

Contrary to this, *Goto* et al. [2.47, 84, 86] obtained for a 52 at.% Ni–Cu alloy under 0.5–2 keV Ar^+ sputtering, the same composition from LEAES and HEAES data after corrections for peak overlapping, indicating an altered layer thickness of more than 20 Å which cannot be resolved by the Auger method.

With increasing angle of incidence of the ion beam (500 eV Ar^+) relative to the target normal, the thickness of the altered layer was found to decrease [2.85]. This indicates that the altered layer thickness is related in this case to the penetration depth of the primary ions and mainly caused by kinetic processes of collisions between ion and target atoms or radiation enhanced processes.

Measuring the surface composition of Ag–Pd, Cu–Pt, and Ag–Au alloys with AES [2.87] when the energy of the bombarding ions (Ne, Ar or Xe) is changed from 0.5 to 5 keV and vice versa, a transition period was found where the surface composition showed rapid changes until a new equilibrium composition was reached, as shown in Fig. 2.9. These dynamic changes can be attributed to altered layers of different thickness and different ion fluences are needed to reach a new equilibrium for different ion energies. From the analysis of these composition variations it was found that the altered layer thickness agrees well with the calculated mean projected range of the primary ions at 500 eV or 5 keV, respectively, and furthermore, the ion fluence needed to reach equilibrium corresponds to the removal of one altered layer thickness. These results agree with the RBS measurements at much higher energies indicating that, in general, the altered layer thickness is comparable with the range of the primary ions, and steady-state conditions are reached after a fluence necessary

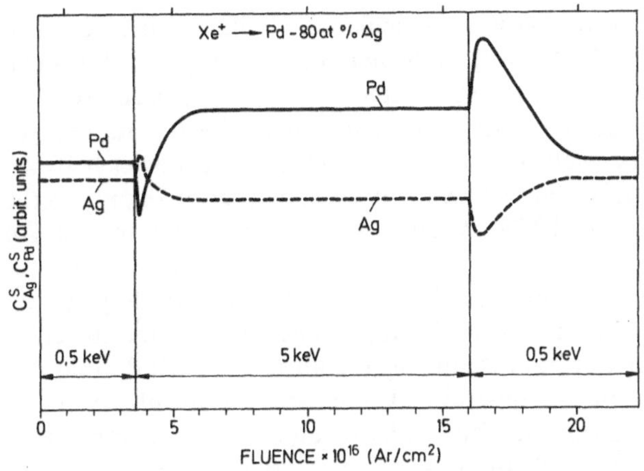

Fig. 2.9. Changes in the Ag and Pd surface concentrations c^s_{Pd} and c^s_{Ag} with fluence and Xe ion energy for an 80 at. % Ag–Pd alloy measured by AES [2.87]

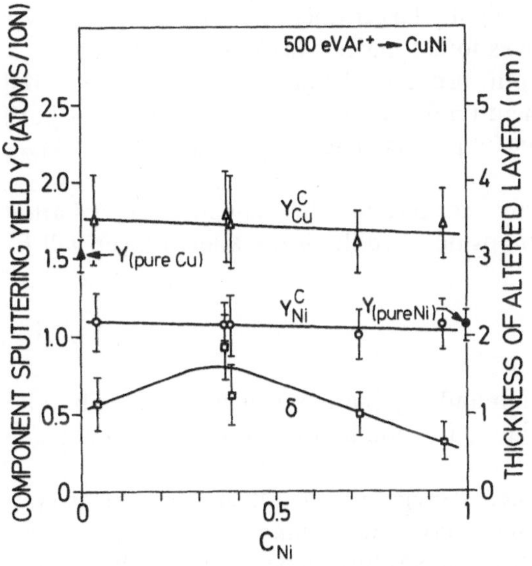

Fig. 2.10. Altered layer thickness δ and component sputtering yields of Cu and Ni in their alloy for Cu–Ni alloys as a function of composition under 500 eV Ar^+ sputtering. Also indicated are measured sputtering yield values for pure Cu and Ni [2.25]

to remove about one altered layer thickness. Using a Monte-Carlo based computer code to simulate alloy sputtering, *Roush* et al. [2.88a, b] were able to simulate dynamic composition changes such as those shown in Fig. 2.9. Such dynamic composition changes under variation of the primary ion energy were observed recently for other alloys and compounds, too [2.88c–e].

The dependence on fluence of the Cu/Ni surface concentration ratio was measured with AES for Cu–Ni alloys under 0.5–2 keV Ar^+ ion bombardment [2.25]. From this and the measured total sputtering yields, *Ho* et al. [2.25] were able to determine with their kinetic model (Sect. 2.3.3c) the surface enrichment, altered layer thickness and also the component sputtering yields for various alloy compositions (Fig. 2.10). An altered layer thickness of about 10 Å for

500 eV increasing to 20–25 Å for 2 keV was calculated. Similar results were also obtained for Ag–Au alloys [2.89], while *Yabumoto* et al. [2.90] found a much larger altered layer thickness using the same kinetic model. Furthermore, they proposed an altered layer profile with no enrichment in the top monolayer followed, however, by a Au enriched zone, to explain the discrepancies between ISS measurements [2.49] (no surface enrichment observed, sensitive only to the top monolayer) and AES results [2.39, 89, 90], with Au enrichment for Ag–Au alloys.

XPS has also been used to study the altered layer for Ag–Pd alloys under 500 eV Ar^+ bombardment [2.91a]. From the appearance of a double structure of the Pd $3d_{5/2}$ peak under ion bombardment, the authors conclude that the altered layer is formed by the segregation of a separate alloy phase at the surface which is different from the bulk phase, and with a composition significantly enriched in Pd. XPS analysis indicated that the altered layer thickness increased from one monolayer for a 5 at. % Pd–Ag alloy to about 20 Å for a 93 at. % Pd–Ag alloy. A slightly larger thickness for the altered layer but otherwise similar behaviour was found under 500 eV N_2^+ sputtering.

Using AES and sputter depth profiling *Varga* and *Taglauer* [2.91b] observed for light ion bombardment (He) and at low ion energies (0.3–4 keV) also good agreement between mean projected ion range and altered layer thickness for carbides.

Additional experimental results concerning the influence of radiation induced segregation onto the composition profile of the altered layer will be dealt with in Sect. 2.3.3c.

b) Temperature Dependence

Some of the effects, such as diffusion and segregation, which can contribute to the formation of an altered layer due to ion bombardment, should show strong temperature dependence.

In order to determine how much segregation or diffusion processes contribute to the thickness of the altered layer and changes in surface composition for different ion fluences, the temperature dependence of the surface composition and altered layer thickness was measured for Cu–Ni alloys by AES. At room temperature, Cu–Ni alloys show strong Ni enrichment under ion bombardment [2.25, 46]. At $-170\,°C$, only negligibly less Ni enrichment at the surface was measured as compared to room temperature under 0.5–2 keV Ar^+ bombardment [2.86]. This was taken as evidence that kinetic processes are dominant in preferential sputtering for these parameters and this system. Increasing the temperature up to 300 °C gave more pronounced Ni enrichment. For still higher temperatures ($\geqq 400\,°C$), however, this trend was reversed and Cu enrichment was observed [2.92, 93]. Probably surface segregation [2.94] and/or diffusion of Cu to the surface at these temperatures was larger under the given experimental conditions than the depletion due to preferential sputtering. This means that an equilibrium state was no longer

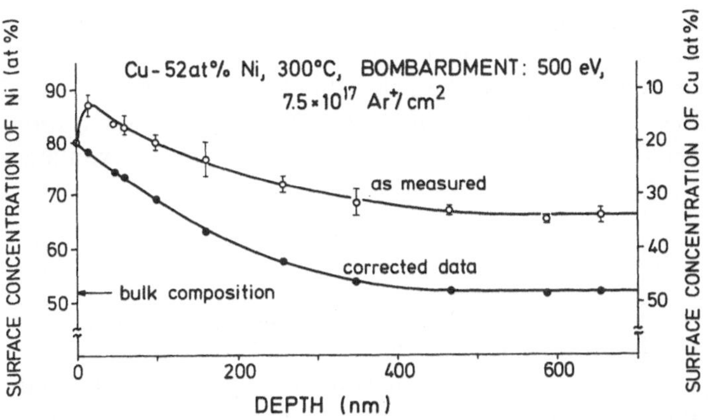

Fig. 2.11. Composition versus depth profile of a 52 at. % Ni–Cu alloy as obtained from sputter-depth profiling with 2 keV Ar ions at room temperature using AES and after correction of the surface composition data for preferential sputtering. The sample was previously bombarded with 7.5×10^{17} Ar^+ cm^{-2} at 500 eV at 300 °C, which has produced an altered layer of about 3000–4000 Å [2.92]

attained in sputtering at high temperatures, i.e., the material sputtered off may no longer have the same composition as the bulk. Such deviations from stoichiometry in sputtering were verified by measuring the composition of the sputter-deposited films which became more Cu enriched with increasing temperature [2.92]. Surface recession competes with radiation enhanced diffusion or segregation of Cu on the surface resulting in an excess of sputtered Cu atoms. Diffusion coefficients as high as 10^{-13} cm^2/s at 400 °C were estimated. Furthermore, this leads to the formation of a Cu depleted layer submicrons deep beneath the surface at temperatures as low as 300 °C. For example, Fig. 2.11 shows a depth profile [2.92] after 7.5×10^{17} Ar^+/cm^{-2} (500 eV) sputtering at 300 °C. The profile was measured by sputtering and AES at room temperature and shows a 3–4000 Å thick Cu depleted layer. For 200 °C, the layer was less than 200 Å and for 400 °C about 3 μm deep [2.92].

For higher current densities and ion energies (5 keV), *Rehn* et al. [2.95] observed for a 60 at. % Cu–Ni alloy always the same Ni enrichment in the top 2–3 monolayers after the equilibrium state was reached, independent of the target temperature between 50 °C and 600 °C. The ion fluences needed to reach equilibrium were about 10^{17} Ar^+ cm^{-2} below 300 °C, but above 300 °C, increasingly higher fluences up to 10^{19} Ar^+ cm^{-2} at 600 °C were necessary. Comparing LEAES and HEAES, the authors observed for temperatures above 300 °C that the subsurface is much more enriched in Ni than the uppermost atomic layers. This was also verified by AES and sputter-depth profiling at room temperature for samples previously ion bombarded, until an equilibrium state was reached at temperatures above 300 °C. Similar to *Shikata* et al. [2.92], Ni enrichment was found, the maximum amount being observed, however, beneath the surface. Altered layer thicknesses up to 3 μm for 600 °C were observed. The results were also explained in terms of radiation enhanced segregation of Cu on the surface thus giving rise to a Ni enriched layer.

An altered layer of about 300 Å, much thicker than the penetration depth of the primary 1 keV Xe ions, was also found in Al–Cu (2–11 at. %) alloys at room temperature by *Chu* et al. [2.96]. The results have been interpreted assuming a very high diffusivity for these alloys at room temperature.

c) Altered Layer Models

A number of models have been proposed to describe the transient stage until the equilibrium altered layer, as well as its thickness and composition profile, are formed. In all models given, component sputtering yields for the alloy constituents are assumed which lead to a composition change at the surface. The models then describe the propagation of the changed surface composition deeper into the material by thermal diffusion and/or radiation enhanced diffusion including erosion of the target due to sputtering [2.97]. Furthermore, thermal surface segregation and radiation induced segregation have been considered. Collisional processes like recoil implantation or cascade mixing have not yet been considered in detail [2.75, 98, 99]. Most of these models agree in one respect, that for steady-state conditions ($t \to \infty$), the final surface composition is only given by the component sputtering yields as given with (2.4) (Sect. 2.2.3), which does not include any diffusion and assumes that sputtering occurs only from the momentary top atom layer [2.10, 28a, 29a]. However, the models give transient stages of different surface compositions which are determined by the interplay between the different component sputtering yields of the alloy constituents, the erosion of the target and different diffusion processes.

An altered layer can be formed under ion bombardment by a combination of preferential sputtering at the surface and a transport mechanism.

Thermal diffusion can give compositional changes over a depth much greater than the range of the primary particles. Such extremely deep altered layers have been observed, for example, at elevated temperatures for Cu–Ni alloys [2.92, 95].

In 1976, *Pickering* [2.100] was the first to incorporate the effects of diffusion into his model and he calculated the thickness and concentration profile of the altered layer for steady-state. The principal assumption is that the component sputtering yields are so different, that for steady-state the surface concentration of the high yield component is zero. Then he used Fick's second law to determine the diffusion fluxes of material toward the surface being sputtered. An effective thickness δ of the depletion zone in the alloy: $\delta = D/u$ (D: diffusion coefficient, u: erosion rate) was obtained, which is ion flux dependent.

A kinetic model was developed by *Ho* et al. [2.25] for the time-dependent changes of the average composition of the altered layer. The composition profile with depth is approximated by a step function. A mass balance equation is set up similar to the *Patterson* and *Shirn* model [2.10] considering composition changes not only in the top monolayer, but also in a layer of constant but altered composition. From the variation of the surface concentration ratio with time as measured with AES and the measured total sputtering yield, this model allows the surface enrichment, altered layer thickness and also the component sputtering yield for an alloy to be determined (Fig. 2.10).

During the last few years, several papers have appeared in the literature which present mathematical approaches for the time dependence of the composition profile of the altered layer [2.101–103]. Papers by *Arita* et al. [2.101, 102] show solutions of the problem after several simplifications for the limiting cases, where the diffusion rate is much lower than the sputter rate and vice versa.

Radiation enhanced diffusion: under ion bombardment, many lattice atoms are displaced in the course of the collision cascade creating vacancies, interstitials and finally larger defects, all of which enhance diffusion. Thus, a layer up to the depth of the primary ions is in a highly mobile state and diffusion coefficients can be much larger than for the undamaged material.

Ho [2.97] later extended the kinetic analysis of preferred sputtering [2.25] by including (radiation-enhanced) diffusion. Starting from an initially homogeneous target, the mass balance is considered within the altered layer and on the sputtered surface, assuming that concentration gradients are the only driving forces for diffusion and sputtering is the only effect that depletes the surface of a component. The solutions for steady-state shows that the composition of the altered layer varies in a simple exponential manner with the distance from the surface and the effective altered layer thickness δ is equal to the diffusion coefficient divided by the velocity of the receding surface, as was already predicted by *Pickering* [2.100]. Estimates of the magnitude of diffusivity from measurements [2.25] yield values of the order of $10^{-16}\,\mathrm{cm^2/s}$ for Cu–Ni alloys, indicating that sputter damage is extremely effective in enhancing the diffusivity within the altered layer.

A different model was adopted by *Webb* et al. [2.104, 105] in which, similar to *Pickering* [2.100] and contrary to *Ho* [2.97], diffusion enhancement of one species only is considered, but this enhancement is promoted only over a fixed depth.

An altered layer can be also formed under ion bombardment by the following processes, which do not necessarily need preferential sputtering.

Collisional processes such as *recoil implantation* and *cascade mixing* can give, in principle, composition changes over the depth of the collision cascade [2.75, 98, 99]. No direct experimental evidence has yet been given of altered layer formation, which could be clearly attributed to collisional processes.

Surface forces: in thermal equilibrium the surface composition of an alloy is generally different from the bulk [2.106, 107]. Thus, *equilibrium segregation* can possibly enhance or reduce preferential sputtering effects.

Comparing surface enrichment due to ion bombardment and due to *surface segregation*, for a multitude of metal alloys just the opposite behaviour is found, i.e., if in an alloy A–B enrichment of A due to surface segregation is observed, then under ion bombardment, depletion of A was found. According to a model by *Kelly* [2.108], an alloy showing surface segregation will lose the segregated component preferentially under ion bombardment, provided replenishment of the surface in the segregating component is sufficiently rapid at the temperatures where the investigation is performed. This is claimed to be the case by postulating that ion bombardment leads to segregation even at room temperature because the bombardment injects point defects.

This process alone, however, cannot produce a changed surface composition under steady-state conditions. Let us assume that the ejection probabilities for the sputtered atom species are equal, i.e., no preferential sputtering occurs. Therefore, after steady-state conditions have been reached, the average composition of the surface layer from which sputtering occurs (1–2 atomic layers) must be equal in composition to the bulk. However, beneath the surface a region depleted in the segregating component will exist. As most surface composition measurements have been performed with methods (such as AES) which probe deeper than the top monolayer, a changed surface composition (i.e., depletion in the segregating component) would have been claimed. Thus, a knowledge of the exact depth profile would be necessary to distinguish surface segregation from preferential sputtering proper.

The same argument is also valid for recoil implantation which cannot produce a changed surface composition by itself in the equilibrium state too.

As surface segregation can be understood in terms of chemical bonding [2.109], the differences in the heat of atomization of the components enter in *Kelly's* model in a similar way to preferential sputtering models which predict surface enrichment on the basis of differences in the surface binding energies. Thus, without information on the exact depth profiles, it cannot be excluded that the good qualitative agreement with experimental results is accidental.

Radiation enhanced segregation: defects are produced under ion bombardment and defect fluxes to sinks (regions with low or zero defect concentration) are considered the basic driving force for solute segregation [2.110, 111]. Thus, radiation-induced solute segregation to defect sinks, which are on the surface and in the bulk (beyond the defect region), can cause a redistribution of the alloy constituents in the near-surface region in addition to preferential sputtering [2.110a, 111a].

The effects of radiation-induced solute segregation and preferential sputtering were combined in a model for dilute binary alloys by *Lam* et al. [2.110a, b, 111a]. The surface composition was calculated as a function of sputter time, ion flux and temperature. The dominance of radiation-induced solute segregation over preferential sputtering effects is shown for certain temperature regions in the transition stage before steady-state conditions are reached. Such calculated surface composition changes with time are shown in Fig. 2.12a for a Ni-1 at. % Cu alloy, and in Fig. 2.12b for a Ni-1 at. % Mo alloy. The surface concentration of the alloy components are predicted to go generally through maxima and minima with time. For steady-state, preferential sputtering is found to be dominant and the final surface composition is predicted to be determined only

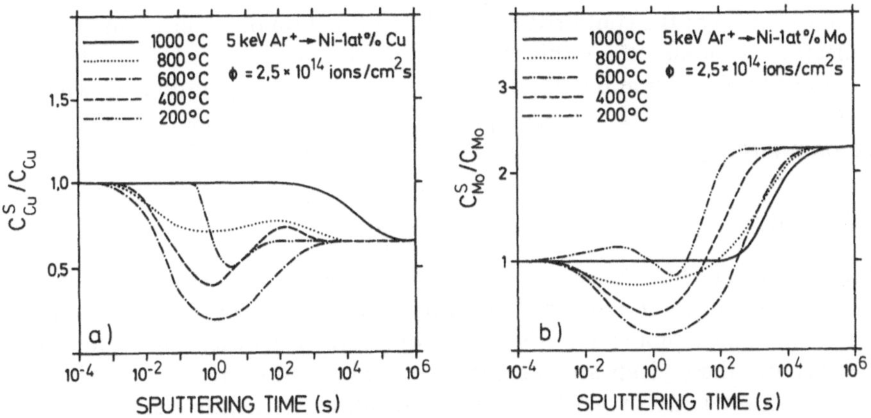

Fig. 2.12. (**a**) Calculated time dependence of the Cu concentration at the surface of a Ni-1 at. % Cu alloy sputtered with 5 keV Ar ions for different temperatures; (**b**) for a Ni-1 at. % Mo alloy. The steady-state surface composition is determined by the assumed component sputtering yield ratio but is independent of temperature. However, before equilibrium is reached the surface composition can go through maxima and minima [2.111a]

by the component sputtering yields as in (2.4) (Sect. 2.2.3). Experimental evidence of radiation-enhanced segregation for Cu–Ni alloys [2.95] and other alloys [2.110c] at elevated temperatures has been discussed in Sect. 2.3.3b.

Generally an exponential depth profile is assumed for enrichment or depletion of a component under preferential sputtering. Contradicting results from AES and ISS measurements have lead to the conclusion, that due to segregation (especially at elevated temperatures) for some alloy systems the altered layer consists of the topmost atomic layer enriched in one component, followed by the much thicker rest of the altered layer depleted in the same species [2.111b–d, 241]. In such a case ISS measurements will indicate enrichment but AES measurements depletion of the same component due to different sampling depths. Consequently the question if preferential sputtering exists has to be decided using (2.4a), i.e. taking into account the contributions from all atomic layers contributing to sputter emission. For example, AES measurements indicate for Cu–Ni alloys always Ni enrichment. *Swartzfager* et al. [2.111d], however, obtained a depth profile which shows for the top monolayer no enrichment of either component followed by a Ni enriched altered layer. In this case the sample was sputtered at 200 °C and after quenching to −90 °C sputter depth profiling was performed using ISS. With the atom-probe field-ion microscope (APFIM) [2.111e] similar results were obtained for Cu–Ni alloys as well as for Ag–Au and Ag–Pd alloys using ISS [2.111d].

It has been pointed out by *Sigmund* et al. [2.65d] that if a strong composition gradient exists within the sputter escape depth, the angular distribution of the component in which the surface is depleted, will be more forward pointed. Thus surface segregation under ion bombardment can be also studied by measuring the composition of the sputtered material as a function of the ejection angle by a collector technique. Using this technique *Andersen* et al. [2.111f–h] observed for a number of alloys and compounds segregation of the component, which is usually found to be the depleted species in AES measurements, especially at high bombarding ion energies (20 keV and higher).

2.3.4 Component and Total Sputtering Yields

While many data have been measured on the sputtering yields for elemental targets especially for bombardment with noble gas or Hg ions, very little is known about the overall sputtering yields of alloys or of other multicomponent materials. But these yields are of great practical interest in sputter deposition of thin films from alloy targets and in composition vs. depth analyses, where surface analysis methods like AES, SIMS or ISS are combined with in situ sputter microsectioning, for establishing a reliable depth scale. Generally, due to preferential sputtering the yields may change with bombarding ion fluence and a steady-state yield value is only obtained for sufficiently high fluences. In addition, small amounts of one metal species in another can reduce the sputtering yield compared to that of the monoelement target and thus can very much change the relationship between sputter time and sputter depth [2.3, 22, 112]. These effects are often connected to surface topography effects, especially for multiphase materials as will be discussed in Sect. 2.5.

In order to measure total sputtering yields of alloys and compounds, the same methods have been employed as for elemental targets. Component sputtering yields have been calculated from the measurement of the total sputtering yield and the changes in surface composition [2.25, 42]. Changes in the partial sputtering yield with ion fluence have been measured by collecting the sputtered material and analysis by RBS [2.37].

In 1969, *Ogar* et al. [2.113] studied sputtering of Cu_3Au single crystals [(100) surface] for 1–14 keV Ar^+ and Hg^+ bombardment. They concluded from an analysis of the sputtered material by a radioactive tracer method, that the total sputtering yield is substantially larger than the yield of either pure element. Furthermore, Au was reported to be sputtered slightly in excess of its stoichiometric abundance. The latter effect was explained by an Au enriched surface layer and continous supply of Au atoms due to diffusion from the bulk. However, in many recent measurements by RBS [2.37, 52, 53], always steady-state conditions have been observed at high fluences, i.e., sputtering was stoichiometric.

For a 60 at. % Ag–Au alloy, *Szymonski* et al. [2.79] also report a higher total sputtering yield of $Y = 17.2$ atoms/ion than for either pure component ($Y_{Ag} = 16.0$, $Y_{Au} = 13.8$) under 6 keV Xe ion bombardment, while *Holloway* [2.42] observed for Au–Cr alloys that the addition of 10% Cr to Au is sufficient to decrease the total sputtering yield Y by 40%.

Recent AES measurements for Au–Cu alloys showed no surface enrichment of either Au or Cu [2.39, 40, 114] and also no changes in the total

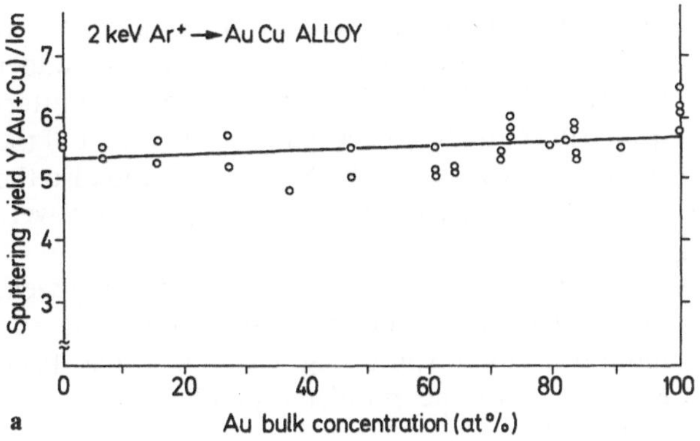

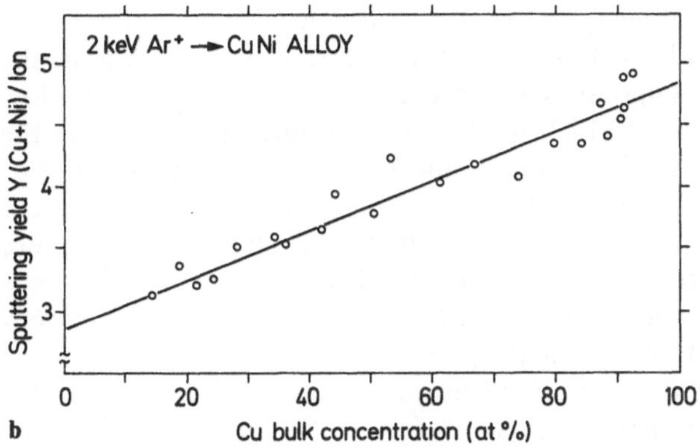

Fig. 2.13. (**a**) Total sputtering yields for Au–Cu alloy films as a function of the Au bulk concentration under 2 keV Ar ion bombardment measured by AES [2.40].

(**b**) Total sputtering yields for Cu–Ni alloy films as a function of their Cu bulk concentration under 2 keV Ar^+ bombardment measured by AES [2.40]

sputtering yield with alloy composition [2.40] which is in agreement with nearly equal sputtering yields for pure Au and Cu under 2 keV Ar ion bombardment (Fig. 2.13a). For Cu–Ni, Au–Cu, and Ag–Au alloys, total sputtering yields for thin film alloys under 2 keV Ar ion bombardment were determined as a function of composition using AES [2.40, 115]. Film preparation was by electron beam evaporation (Au–Cu, Ag–Au) or sputter deposition (Cu–Ni, Ag–Au). A linear increase of the yield from the low yield component to the high yield component was observed for all three systems. Figure 2.13 shows the results for Au–Cu and Cu–Ni. Similar results were obtained by *Ho* et al. [2.25] for Cu–Ni alloys at low ion energies and for Ag–Au at much higher ion energies (200 keV) [2.37].

From these partly contradicting experimental results, no general conclusions can be drawn. However, indications are that for single-phase systems and low bombarding ion energies, the total sputtering yield of the alloy increases monotonically with concentration from the low yield component to the high yield component. Deviations are probably due to the presence of nonmixing trace elements leading to topography effects.

Total yields for ZrC slightly lower than for Zr under Ar ion bombardment [2.116] and identical yields for TiB_2 and ZrB_2 with the metals under 500 eV Cd^+ sputtering were reported [2.117]. For PtSi and NiSi, total sputtering yields under 20 keV Ar^+ sputtering close to and even higher than that of the elemental, high yield component (Pt) were found by RBS [2.52, 53] and also the dependence of the partial and total sputtering yields on Ar ion fluence were studied in the case of PtSi [2.53].

No systematic investigations of the dependence of the total sputtering yield on ion energy have been performed so far, except for low energy, light ion sputtering where a large amount of total sputtering yield data has been accumulated recently in connection with the problems of plasma wall interaction in fusion research [2.24, 55].

Sputtering yields of different carbides in the energy range between 0.1–8 keV under H^+, D^+, and He^+ ion bombardment show that the total yield of these carbides is very similar to the yield of the component having the lower elemental yield. For example, TaC has a sputtering yield very similar to the yield of pure Ta, while the yield of pure C is orders of magnitude higher (for low ion energies), as shown in Fig. 2.14 [2.24, 71]. These results also agree with the high surface enrichment of TaC in Ta as found with AES and ISS [2.48].

In Table 2.4 (Sect. 2.7), total sputtering yields for a number of different compounds are given. In this table, also compounds with a high vapor pressure component are listed and this will be discussed in Sect. 2.4.4.

Concerning component sputtering yields, *Holloway* [2.42] observed strongly concentration dependent component sputtering yields for Au–Cr (up to 20 at. %) alloys under 2 keV Ar ion bombardment. Y^c_{Au} decreases from 7.2 to 5.2 for concentrations varying from 1.1 to 19.5 at. % Cr, while Y^c_{Cr} remains practically constant but is roughly 2.5 times greater than for pure Cr. On the

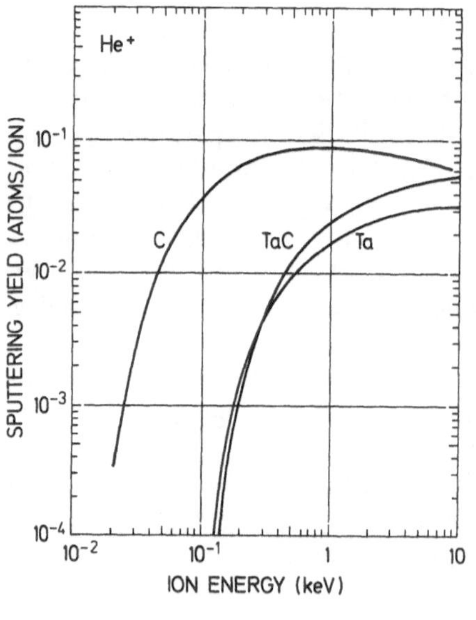

Fig. 2.14. Energy dependence of the sputtering yield of TaC under He ion bombardment.. For comparison the energy dependence of C and Ta under He^+ sputtering is also shown. The sputtering yield of TaC is similar to that of pure Ta, while the sputtering yield for C is orders of magnitude larger [2.71]

other hand, component sputtering yields $Y^c_{Ag} = 3.3$ and $Y^c_{Au} = 1.7$ which are almost independent of the alloy composition for Ag–Au alloys under 1 keV Ar^+ sputtering have been reported by *Ho* et al. [2.89]. Also for Cu–Ni alloys, little variation of the component sputtering yields with composition was found, as seen from Fig. 2.10 [2.25].

Mostly only component sputtering yield ratios are known, which can be deduced with (2.4) from measured surface composition changes. They are listed in Table 2.1 for alloys or can be calculated for compounds with (2.4) from the data given in Table 2.2 (Sect. 2.7).

2.3.5 Angular Distribution of Sputtered Constituents

If an alloy or compound is sputtered the angular distribution of the sputtered atoms can be different for each component. Especially at low bombarding energies near the threshold for sputtering where primary and secondary recoils are dominant, mass differences of the alloy constituents lead to large differences in the collisional energy transfer. At higher ion energies in the range of linear cascade theory, different energy distributions for components of different mass in the solid were also predicted by *Andersen* and *Sigmund* [2.66]. At leaving the surface, i.e., overcoming the surface binding energy, these differences should show up in different angular distributions [2.65d] as well as in the energy distributions. Differences in the angular distribution are of much practical interest, for example, in thin film deposition by sputtering and surface analysis by SIMS.

Angular distributions for sputtering of multicomponent targets have been measured by collecting the emitted atoms on a collector followed by quantitative analysis of the different collected materials. Besides RBS, AES and nuclear reactions, neutron activation was also applied because of its high sensitivity. Not many experimental results are available so far and several important questions are still open, like how does the angular distribution of the components change with bombardment fluence, what is the influence of topography effects on the results obtained so far, variation of the angle of ion incidence or dependence on ion mass and energy.

For low primary ion energies, *Olson* et al. [2.54] have considered atoms sputtered by a "collision cascade" of only 2 collisions in the case of normal ion incidence. This is the minimum number of collisions leading to a sputtered atom because of conservation of momentum. In the case where the primary knock-on atom with mass M_1 leaves the target after a collision with another target atom of mass M_2, they show that if M_1 is smaller than M_2, the ejection probability in the direction of the target normal is higher than for the case of M_1 being larger than M_2. This is because a lighter mass atom can be backscattered from a heavier one but not vice versa. In the case where the knocked-on atom is emitted, ejection in the direction of the target normal is not possible. Therefore a preferred ejection of the lighter mass atoms in the target normal direction occurs.

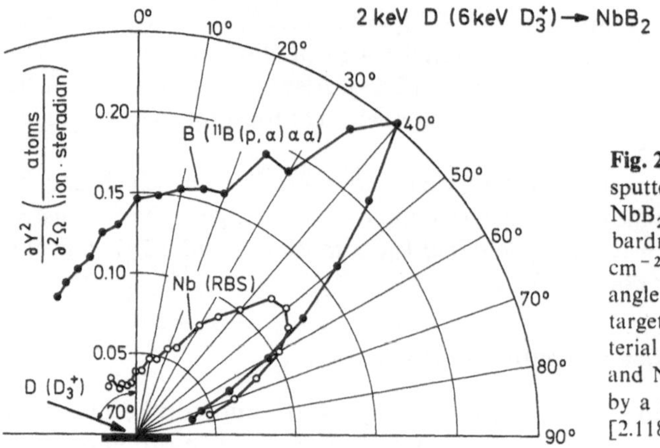

Fig. 2.15. Angular distribution of sputtered Nb and B atoms from a NbB_2 target subjected to a bombardment with 2×10^{19} C_3^+ ions cm^{-2} at 6 keV under an oblique angle of incidence of 70° to the target normal. The sputtered material was collected on Al foils and Nb measured by RBS and B by a nuclear reaction [B(*p*, α)αα] [2.118]

Experimental evidence is given for a 18 at. % Fe–Ni alloy which showed preferred Fe emission, a 55 at. % Ni–Cu alloy which showed preferred Ni emission, and a 50 at. % Ag–Au alloy which showed preferred Ag emission in the direction of the target normal, all under 100 eV Hg^+ ion bombardment [2.54]. The emitted material was collected on a semicircular collector strip and afterwards analyzed by AES.

For an angle of ion incidence of 70° to the target normal and also in the single knock-on regime for sputtering, *Roth* and *Bodhansky* [2.118] observed for 6 keV D_3^+ (equivalent to 2 keV D^+) ion bombardment of NbB_2 preferred forward emission of the heavier component Nb (Fig. 2.15). This result can also be explained by a momentum transfer argument similar to the case of near threshold energies and normal incidence by *Olson* et al. [2.54]. However, for NbB_2 under normal ion incidence, no differences in the emission distributions of the different components were observed. Analysis of the collected sputtered material was performed with RBS for Nb and nuclear reaction for B. Similar results indicating preferred forward emission of the lighter component under an oblique angle of incidence were reported for other compounds [2.118].

For the lighter isotopes of elemental targets, a preferred ejection probability in the direction of the surface normal was also reported for Cu, Mo, W and even for U under 100 eV Hg^+ ion bombardment and normal ion incidence [2.119, 120]. For higher ion energies, where atoms are ejected as a result of a larger number of successive collisions, this effect disappears. For Mo and Cu targets, the preferred normal ejection of the lighter isotope was no longer observable at 200–300 eV Hg ion bombardment [2.54]. In particular, the surface binding energies are equal for isotopes, therefore these results are a clear indication that for energies near the sputtering threshold, collisional energy transfer effects dominate.

For normal ion incidence and higher ion energies, the few experimental results available so far are contradictory. For the Fe–Ni and Ag–Au alloys discussed above, the effect was reversed at higher ion energies (200–1000 eV)

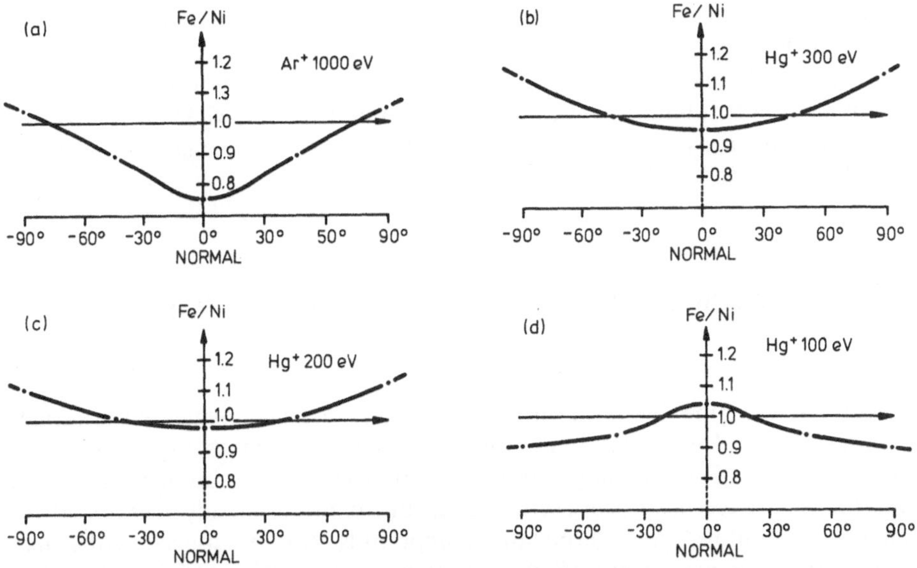

Fig. 2.16a–d. The angular distribution of the ratio of the peak-to-peak amplitudes of the 651 eV Fe to the 848 eV Ni Auger peak, normalized with respect to the target Fe/Ni peak ratio. The targets are sputtered under normal ion incidence: (**a**) 1000 eV Ar^+ ion bombardment; (**b**) 300 eV Hg^+ ion bombardment; (**c**) 200 eV Hg^+ ion bombardment; (**d**) 100 eV Hg^+ ion bombardment [2.54]

leading to preferred normal ejection of the heavier components Ni or Au, respectively, while no change in preferred normal ejection of Ni was observed for the Ni–Cu alloy [2.54]. Figure 2.16 shows the angular distribution of the concentration ratio of the sputtered material for a Fe–Ni alloy for different ion energies [2.54]. For Fe–Ni and Ni–Cu alloys, preferential ejection of Ni with up to 30% enrichment was also found under 1 keV Ar ion bombardment [2.121]. While the strong mass effects related to primary and secondary recoils disappear for higher ion energies, as seen for the isotopes, the experimental results indicate that for alloys, factors other than mass become dominant and determine the angular distribution of the components. However, with the few experimental results available so far, it is too early to make any general conclusions for bombarding energies not very close to the threshold. While in Fe–Ni and Cu–Ni the components have similar mass, this is not the case for Ag–Au and while strong preferential sputtering exists for Ag–Au [2.40, 89] and Cu–Ni [2.61], no preferential sputtering was found for Fe–Ni alloys [2.58, 59, 63, 122]. In addition, the angular distributions have been measured for high ion fluences where topography effects may have influenced the observed results and indeed were observed for oblique angles of ion incidence [2.121].

The angular distribution of the sputtered constituents for compound single crystals has only been studied for a few binary semiconductor crystals [2.123, 124]. The polar faces A(111) and B($\overline{1}\overline{1}\overline{1}$) of InSb were bombarded with 15 keV Ne ions and the emitted material collected and analysed by AES

[2.123]. The observed spot patterns exhibited strong anisotropy in composition. For example, sputtering the A(111) surface, preferred emission was observed for In in the (110) and (113) directions and for Sb in the (111) direction. Studying different binary semiconductor compounds of types $A^{III}B^{V}$ and $A^{II}B^{VI}$ [2.124], anisotropy was only observed for those with a cubic lattice and near-equal mass of the constituents.

2.4 Oxides and Other Compounds with One High Vapor Pressure Component

In order to describe the sputtering process and especially changes in the surface composition and the formation of the altered layer for oxides, halogenides and other compounds containing at least one high vapor pressure element, other mechanisms have been proposed in addition to those already mentioned for single-phase alloys and intermetallic compounds. An enrichment of the high vapor pressure component above a certain concentration can lead to spontaneous desorption [2.48], which means that this component will show less enrichment. As the desorption depends on temperature, collisional spikes can largely contribute to the erosion and compositional changes. Furthermore, especially for oxides, structural and compositional changes have been frequently observed under ion bombardment at room temperature where a crystalline phase is converted into an amorphous or different crystalline phase of different composition.

2.4.1 Experimental Techniques

Besides compositional changes, a major interest for these compounds are additional investigations in respect to ion bombardment induced changes of the structure and the chemical bonds. Experimental techniques most frequently employed are:

The *surface analysis techniques* like AES, ISS, and RBS which are also used for the analysis of alloys and compounds containing no high vapor pressure component. These methods can detect, however, only compositional changes at (ISS, AES) or near (AES, RBS) the surface [2.40, 48, 51] as already discussed in Sect. 2.3.1, and cannot give chemical or structural information. AES has an additional disadvantage for compounds containing a high vapor pressure component because composition changes due to the analyzing electron beam have been observed for oxides, halogenides, carbonates and silicates [2.125–128].

Only the surface sensitive technique XPS allows compositional changes to be detected and, in addition, from chemical shifts, the determination of newly formed compounds or phases is possible [2.129].

Reflection or *transmission electron diffraction* gives structural information and is used to identify an ion-bombardment induced transformation into an amorphous form or a new crystalline phase [2.9, 130–134]. By this method, however, only changes occuring in layers much thicker than 1 nm can be observed as distinct from more surface sensitive methods like XPS.

Dissolution experiments: if the "altered layer" due to its changed chemical composition dissolves much quicker in a given solvent than the bulk, its thickness can be determined in combination with ion implantation of a radioactive tracer element of known implantation profile [2.134, 135].

Sheet conductivity changes can indicate phase changes and/or loss of a compound component such as, for example, if the oxide is reduced to the metal [2.130, 131].

Except for the surface sensitive techniques (ISS, AES, XPS), all the other techniques listed can only give information on the average composition or structure of the altered layer if it is sufficiently thick, i.e., for higher bombarding ion energies. One cannot detect by these methods a depth profile for the compositional and/or chemical changes which might exist at or very close to the surface due to different ejection probabilities of the alloy constituents.

2.4.2 Structural Changes Under Ion Bombardment

Ion bombardment of oxides and other nonmetallic compounds may cause both amorphisation or crystallisation of a near surface layer [2.136], while metals and alloys are normally found to resist amorphisation at room temperature. For Si and alloys and compounds containing Si, a transition temperature exists below which amorphisation occurs [2.137]. Amorphisation, crystallization and stoichiometric changes are generally observed to begin under tens of keV ion bombardment at fluences of about 10^{13}–10^{16} ions cm^{-2}, while below 10^{13} ions cm^{-2}, only the formation of individual point defects and defect clusters occurs [2.136].

Different groups of oxides and nonmetallic compounds have been distinguished and mechanisms for their structural changes have been proposed by *Kelly* and co-workers and will be summarized in this section. These compounds have been divided into three groups according to their structural behaviour under ion bombardment [2.136]:

i) oxides and compounds which remain or become *crystalline* under ion bombardment, e.g., alkalihalides [2.138], NbO [2.139], MgO [2.138, 145], and UO_2 [2.136];

ii) oxides and compounds which remain *amorphous* or for which *amorphisation* of an initial crystalline phase occurs, e.g., for arsenides like GaAs [2.140], GaP [2.141], SiC [2.142, 143], SiO_2 [2.144], and Cr_2O_3 [2.145];

iii) oxides for which at first at an intermediate ion fluence, *amorphisation* and loss of oxygen occurs and finally at high ion fluences, the oxygen deficient

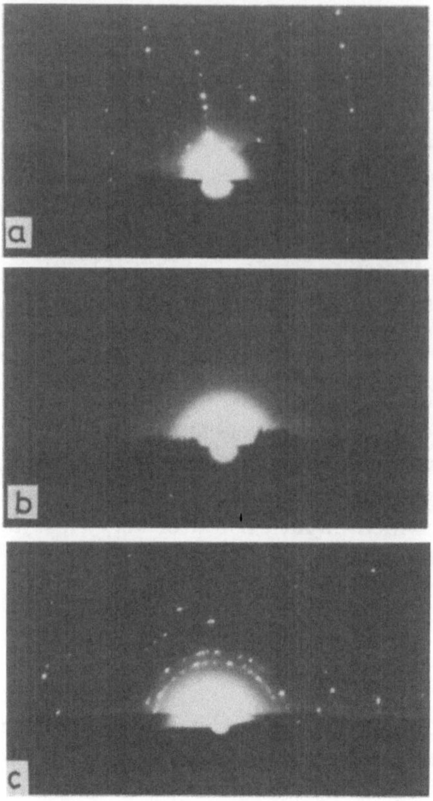

Fig. 2.17a–c. Reflection electron diffraction patterns taken at 80 kV of sintered Nb_2O_5 pellets under the following conditions: (**a**) the single crystal pattern of Nb_2O_5 before ion bombardment; (**b**) amorphous halos after a fluence of 4×10^{15} ions cm^{-2} of 35 keV oxygen ions; (**c**) the polycrystalline pattern of NbO after a fluence of 2×10^{17} ions cm^{-2} of 35 keV oxygen ions [2.133]

zone deepens by diffusion or cascade mixing and a *new phase of a lower oxide* precipitates randomly. This behaviour was observed for some transition metal oxides like MoO_3 [2.130, 132], Nb_2O_5 [2.133, 134], TiO_2 [2.135], V_2O_5 [2.130, 132], and Ta_2O_5 [2.131]. The oxides Ta_2O_5, V_2O_5, and Nb_2O_5 which belong to this group also belong to the most easily amorphized substances known [2.131]. For example, in Nb_2O_5, amorphisation under 35 keV Kr^+ or O_2^+ ion bombardment starts at a fluence of approx. 5×10^{13} ions cm^{-2} and nucleation of NbO after a fluence of approx. 1×10^{16} ions cm^{-2}. After a fluence of 2×10^{17} ions cm^{-2}, a 30 nm thick layer of crystalline NbO has formed by random nucleation (Fig. 2.17) [2.133, 134].

Two different mechanisms for such structural changes have been proposed by *Naguib* and *Kelly* [2.136]:

a) The impacting ion creates a hot, disordered region which is equivalent to a liquid surrounded by a crystal. This region cools rapidly and crystallisation sets in when the temperature falls below the melting point T_m. Depending on the velocity of the crystallisation front and the velocity of cooling down, a temperature ratio criterion was found in good agreement with experimental results:

$$T_c/T_m < 0.3 \text{ (crystallisation)}$$
$$T_c/T_m > 0.3 \text{ (amorphisation)} \quad (2.10)$$

with T_c the crystallisation temperature in a typical macroscopic experiment.

b) Assuming that disorder is more easily retained in a covalent than an ionic structure, an empirically found bond type criterion based on ionicity was proposed [2.136]. The ionicity i, which measures the degree of the ionic nature of a chemical bond, can be defined as [2.146]

$$i = 1 - \exp[-0.25(X_A - X_B)^2] \quad (2.11)$$

with X_i the electronegativity [2.146].

The criterion predicts amorphisation for ion bombardment of compounds with predominantly covalent bonds ($i < 0.47$) and that compounds with predominantly ionic bonds ($i > 0.59$) remain or become crystalline.

It appears to be a fairly general result that oxides and other nonmetallic compounds undergo structural changes under ion bombardment. However, only a few materials such as, for example, Ta_2O_5 [2.131] have been studied so far in detail and also the underlying mechanisms are not well understood.

2.4.3 Compositional Changes Under Ion Bombardment

In addition to those mechanisms already mentioned for alloys, thermal and spike sputtering effects have been discussed for oxides and halides [2.147]. It was found for a multitude of oxides that oxygen is lost preferentially under ion bombardment if the decomposition pressure of the oxide at assumed thermal spike temperatures is sufficiently high, or if appropriate surface binding energies, identified as partial or total heats of atomisation, are sufficiently low. The loss of oxygen very often causes the formation of lower oxide phases on the surface under tens of keV ion bombardment. The formation of such oxide phases was observed both by reflection and transmission electron diffraction [2.131, 139]. At lower bombarding ion energies (< 10 keV), XPS measurements have often revealed the reduction of the oxide to a mixture of the original oxide, lower oxides (if they exist) and metal [2.129, 148]. For example, XPS measurements revealed that Fe_3O_4 is reduced near the surface under 1 keV Ar ion bombardment to FeO and metallic Fe [2.265]. The experimental results of surface enrichment and on phases formed are summarized in Table 2.5 for a large number of oxides and other compounds containing at least one high vapor pressure component (Sect. 2.7). The experimental techniques used have to be taken into account for the interpretation of the listed surface enrichment. Only ISS, AES, and XPS results are representative for the top few monolayers, while all other techniques probe deeper.

Measurements of altered layer thicknesses show a similar relation to the penetration depth of the primary ions as for alloys and in addition, structural changes (Sect. 2.4.2). Applying a combination of different techniques (RBS, Transmission Electron Microscopy, Reflection Electron Diffraction and measurements of dissolution and sheet conductivity changes), the composition and structure of the altered layer for Ta_2O_5 under 35 keV Kr^+ bombardment has been studied [2.131]. It was found that an 11 nm thick altered layer of amorphous Ta_2O_5 was formed after a fluence of 5×10^{13} ions cm^{-2} which became oxygen deficient with increasing fluence. After a fluence of approx. 10^{17} ions cm^{-2}, a 22 nm thick layer of crystalline δ-Ta–O in amorphous Ta_2O_{5-x} was observed. If $\langle x_D \rangle$ is the mean damage depth using *Winterbon's* tables [2.149], the thickness of the amorphous layer lies between $\langle x_D \rangle$ and 2 $\langle x_D \rangle$, depending on fluence, and the thickness of the lower oxide "altered layer" is approximately 3 $\langle x_D \rangle$ for high fluences, i.e., in the equilibrium state. A combination of preferential oxygen sputtering at the surface, diffusion of the point defects and random nucleation of a phase with lower stoichiometry was proposed for the formation mechanism of the altered layer. Similar altered layer thicknesses showing the same relation to the damage depth were reported for Nb_2O_5 [2.133, 134] and V_2O_5 [2.130, 132]. Furthermore, it was observed that the altered layer thickness increases linearly with ion energy [2.131].

An altered layer thickness equal to the mean projected range of the primary ions was observed for low energy (1–10 keV), light ion (H^+, D^+, He^+, Ne^+) bombardment of Ta_2O_5 measured by RBS (Fig. 2.18) [2.150]. Figure 2.18 shows that the maximum oxygen depletion increases with decreasing energy of the bombarding ions and shifts to smaller depths. The increase of the oxygen concentration near the surface may be a real effect, but may also be caused by the vacuum of 10^{-8} mbar, being still not sufficiently low for such measurements due to adsorption of oxygen and CO from the residual gas. Furthermore, it was

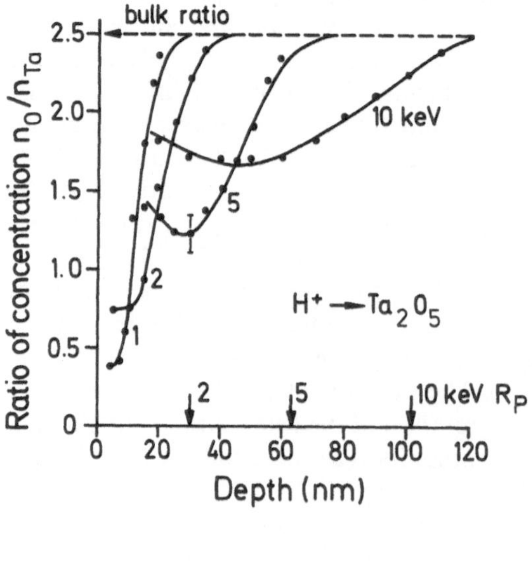

Fig. 2.18. The atomic concentration ratio n_O/n_{Ta} with depth for 1, 2, 5, and 10 keV H ion bombardment of Ta_2O_5 obtained by RBS. The dashed line indicates the bulk ratio. The arrows indicate calculated values of the mean projected range R_p for 2, 5, and 10 keV [2.150]

found that the maximum oxygen depletion is nearly independent of the ion type for a given ion range.

At low bombarding ion energies, extremely small values for the altered layer, much smaller than the mean projected range, were reported for PbO under 400 eV Ar^+ sputtering [2.151a]. Contrary to this *Hofmann* and *Sanz* [2.151b] observed good agreement between mean projected ion range and altered layer thickness ($\approx$2 nm) for 3 keV Ar ion bombardment of different oxides.

The results for the *compositional changes* found for different compound types will now be discussed together with the processes which have been proposed to explain the measured results.

a) Mass Effects

As discussed for alloys in Sect. 2.3.2b, surface enrichment of the heavier component follows from the energy sharing between atoms of different masses in the collision cascade ([2.66] and [Ref. 2.27, Chap. 2]). These effects have often been suggested as an explanation for oxygen deficiency because oxygen is the lighter component in most oxides [2.53, 65]. However, as for solid solutions, counter examples exist which do not show metal enrichment, e.g., MgO [2.152], Cr_2O_3 [2.129, 153, 154] or PbO for Kr^+ and Xe^+ sputtering [2.151]. Also BeO [2.48], which should show oxygen enrichment ($M_O > M_{Be}$), cannot be explained by mass effects. In the case of BeO, an enrichment in O on the surface due to mass effects could probably also lead to spontanous desorption of O [2.48], but certainly this cannot explain the observed weak Be enrichment.

Mass effects seem to be dominant in the single knock-on regime, i.e., for light ions (H^+, He^+) and/or low energy. Expecially for high mass ratios M_{metal}/M_{ion}, similar to carbides, the better mass match between ion and oxygen atoms results in a more efficient collisional energy transfer [2.155]. As an example, surface depletion in oxygen for low energy light ion bombardment of Ta_2O_5 was found to increase with decreasing energy (Fig. 2.18) [2.150, 155, 156] and with decreasing mass of the ions (He, Ar) [2.155].

b) Bonding Effects

For heavier ions and higher energies (>1 keV), bonding effects seem to be dominant obscuring the mass effects, such as, for example, in the case of Ar^+ bombardment of Al_2O_3 or even He^+ sputtering of BeO [2.48]. PbO was found to become most strongly depleted in oxygen under He bombardment and not at all for Kr^+ and Xe^+ sputtering [2.151a].

In the *linear cascade regime* ([2.44] and [Ref. 2.27, Chap. 2]) the sputtering yield is inversely proportional to the surface binding energy U, see (2.5). For oxide sputtering similar to the sputtering of alloys, a problem arises in the definition of the quantity U. An identification of U with the average total ΔH^t and partial ΔH^p heats of atomisation can predict oxygen loss for a given oxide very well [2.135, 136]. As an example, for TiO_2 the total heat of atomization is

$\Delta H^{t} = 6.4$ eV/(gas-atom) and can be obtained from the total atomisation of TiO_2 into the vapor phase:

$$TiO_2(\text{amorph}) = Ti(\text{gas}) + 2\,O(\text{gas}) - 3\,[6.4\,\text{eV}]\,,$$

while the partial heats of atomization ΔH^{p} can be obtained from other possible decomposition processes such as

$$2\,TiO_2(\text{amorph}) = Ti_2O_3(\text{crystalline}) + O(\text{gas}) - [5.1\,\text{eV}]$$

which gives $\Delta H^{p} = 5.1$ eV/(gas-atom). Generally processes with the smallest ΔH are expected to have the highest sputtering yield. Therefore, for TiO_2, oxygen loss and the formation of Ti_2O_3 will occur. Equal sputtering of both components with no surface composition change should be observed if ΔH^{t} is smaller than all possible ΔH^{p}, as is the case for Cr_2O_3, MgO or MnO. With this criterion oxides can be identified in which oxygen is more loosely bound than in other oxides. However, up to now no qualitative estimates of the oxygen depletion exist and the effects are probably too small to account for the experimental observations [2.77].

Thermal evaporation from an *elastic collision spike* has been under discussion for many years. However, the existence of a "spike temperature" and thus the assumption of a thermodynamic equilibrium in the different models ([2.147, 157–161] and [Ref. 2.27, Chap. 2]) is doubtful.

The strong influence of the surface binding energies in *Sigmund's* treatment ([2.161] and [Ref. 2.27, Chap. 2]) of sputtering from elastic collision spikes can be seen from (2.7) (Sect. 2.3.2b). However, the magnitude of the surface binding energies of the metal and the oxygen atoms in a spike are not known for an oxide. Furthermore, it remains doubtful as to whether collisional spike sputtering occurs at the low bombarding ion energies (~ 1 keV) of most experiments. For pure metals, *Sigmund* and *Claussen* [2.161] estimated that substantial spike sputtering should only occur for $Y_{\text{lin}} \gtrsim 10$, with Y_{lin} the calculated sputtering yield in the linear cascade regime.

The importance of collisional spike sputtering for ion bombardment of oxides and halides, even at low bombarding ion energies (~ 1 keV), was discussed by *Kelly* [2.147]. As a rough criterion, *Kelly* estimated for the significance of erosion by spikes,

$$Y_{\text{spike}} \gtrsim 1 \quad \text{if} \quad p_{\text{d}}(T) \gtrsim 10^{2\pm1}\,\text{bar}\,. \tag{2.12}$$

Taking for p_{d} the decomposition pressure of the oxide (which is the partial pressure of oxygen over the oxide) at assumed spike temperatures of 2000–4000 K [2.147, 160], preferential release of oxygen from the surface is predicted correctly for most oxides [2.160]. In Fig. 2.19, oxides with a decomposition pressure greater than 10^2 bar at approx. 3000 K are predicted to lose oxygen preferentially, while those with lower decomposition pressure

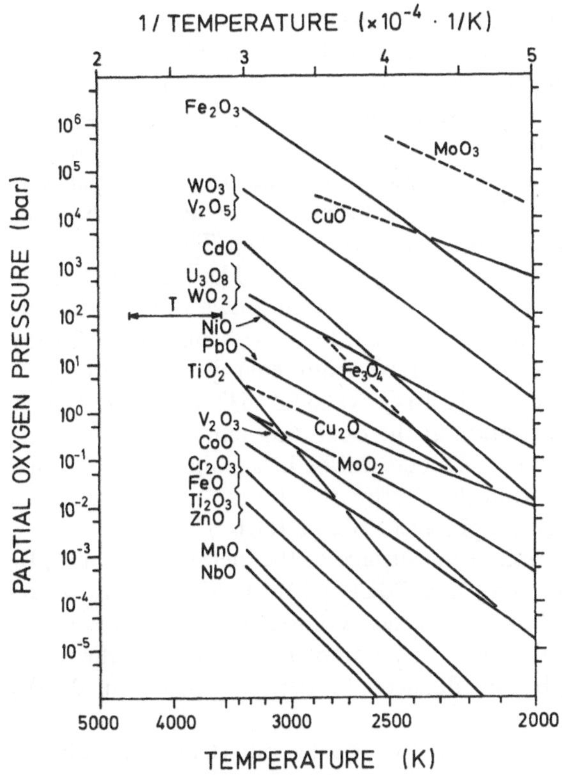

Fig. 2.19. Partial pressure of oxygen over the oxide (decomposition pressure) for oxides. Those oxides lying above PbO show bombardment induced loss of oxygen, whereas most of the others do not. The separation between the two groups occurs at about 10^2 bar for a temperature of about 3000 K, which is assumed to be characteristic for spikes [2.160]

should sputter such that the surface composition remains equal to the bulk composition. Experimental results are still too contradictory to make a clear decision as to whether the proposed criterion for losing or not losing oxygen from oxides is correct. For instance, WO_3 and CuO have been reported to lose oxygen, in agreement with this criterion, but for MoO_2 and FeO, experimental results disagree (Table 2.5). Also no evidence for spike sputtering has been found in the energy distributions of sputtered Mg and Zn atoms [2.80, 162], which should show spike sputtering as a result of their high vapor pressure [2.147].

Whether spike sputtering of oxides exists or not, or taking the heat of atomisation for the surface binding energy is correct, the importance of bonding for oxide sputtering is clearly evident. Oxygen is always preferentially lost if appropriate bonds are sufficiently weak.

c) Sputtering Via Electronic Processes

For halides, additional processes summarized as electronic sputtering may occur [2.147] (see also the following Chap. 4). Essentially, two possible electronic processes which can also occur under electron or photon bombardment are discussed. Direct neutralisation of a surface halogen ion may occur due to interaction with the incident ion at the surface or through an interatomic Auger decay, followed by thermal release of the halogen atom from the surface. The second process is based on the creation of excitations by the incident ion like V_k centers, H centers or bound electron hole pairs, which can diffuse to the surface and result in neutralisation of a halogen ion on the surface, followed by thermal release. Thus, the surface loses halide atoms and becomes enriched in the metal. If the metal is volatile enough to evaporate at the operating temperature, it is called *slow thermal sputtering* [2.147]. The surface composition is given by the yields of the concurring processes. This has been used to explain the velocity distribution of sputtered atoms for NaCl, where a thermal component corresponding to a temperature of 300 K was observed [2.163]. Halides showing this behaviour as inferred from energy distributions of sputtered atoms are RbCl, RbBr, RbI [2.164], NaCl [2.163], NaI [2.164, 165], CdI_2 [2.166]. If the metal vaporisation at the operating temperature is small, the surface becomes metallised and the metal will be lost mainly by collisional sputtering. This was concluded from the energy distribution measurements of the metals for sputtering of AgBr, AgF, and PbI_2 [2.166].

Direct evidence of the reduction of the surface to the metal exists for AgBr [2.167], CuF_2 [2.168], NiF_2 [2.168], and $PdCl_2$ [2.169]. XPS studies of the CuCl and ZnF_2 systems have shown no surface compositional changes under ion bombardment, while for CoF_2 and FeF_2, complete reduction to the metal was observed only after high ion fluences ($>10^{19}$ Ar^+ cm^{-2} at 2 keV) [2.168, 170].

Electron sputtering was also considered for oxides but yields are probably so low that sputtering by other processes dominates [2.147].

2.4.4 Sputtering Yields

Among the compounds with a high vapor pressure component, only for oxides, and to some extent also for alkali halides, have sputtering yields and their dependence on ion mass and energy been studied to any extent. Even for these compounds, fluence dependent measurements have not yet been performed. Yield measurements have been performed with the same techniques as for elemental targets. However, for insulators, as for most of these compounds, charge-up problems have to be circumvented [Ref. 2.27, Chap. 4], for example, by supplying electrons from a heated filament near the target. For oxides, quite often thin films obtained by anodic oxidation were used for the sputtering yield measurements. The sputtering yield data measured for oxides and other compounds by different groups are summarized in Table 4 (Sect. 2.7).

As for alloys, the sputtering yield of an oxide is defined as the total number of emitted *atoms* per incident ion. Sometimes, however, the sputtering yield has also been defined as the number of sputtered molecules per incident ion.

Early measurements indicated a much lower sputtering yield for an *oxide* than for the corresponding pure metal [2.171, 172]. *Schirrwitz* [2.172] observed a sputtering yield coefficient $Y=0.7$ for 5 keV Ar ion bombardment of MgO as compared to $Y=5$ for Mg, and for Al_2O_3 he found $Y=0.1$ which was more than one order of magnitude lower than for pure Al with $Y=2.3$. From such measurements it was generally assumed that oxides have a protective action against sputtering. More recent sputtering yield measurements by *Kelly* and co-workers [2.139, 173, 174] led the authors to the conclusion that the sputtering yields of oxides are in most cases similar to or even higher than those of the corresponding metals. The only exceptions were observed for MgO and Al_2O_3 with sputtering yield ratios of $Y_{MgO}/Y_{Mg}=0.2$ and $Y_{Al_2O_3}/Y_{Al}=0.5$ under 10 keV Kr ion bombardment. Sputtering yields have been measured by *Kelly* et al. [2.139, 173, 174] for a number of oxides (Al_2O_3, MoO_3, Nb_2O_5, SiO_2, SnO_2, Ta_2O_5, TiO_2, V_2O_5, WO_3, ZrO_2) prepared as thin films by anodic oxidation and sputtered with 2–30 keV Kr^+ ions. *Oechsner* et al. [2.175] report similar results for Ta_2O_5 under 100–600 eV Ar ion bombardment.

Sputtering yield coefficients are shown in Fig. 2.20a, b as a function of ion energy for noble gas ion bombardment of SiO_2 and Ta_2O_5. For comparison, measured data for Si and Ta are included. For SiO_2 the sputtering yields are about the same magnitude or slightly lower than for Si, but for Ta_2O_5, generally higher yields than for Ta are observed.

Sputtering yields for low energy (0.05–8 keV), light ion (H, D, He) bombardment of Al_2O_3, BeO, Ta_2O_5, and SiO_2 [2.55, 194] are also higher or similar to those of the pure metals. Only for near threshold energies (below 200 eV) are the oxide sputtering yields, in some cases, lower than those of the metals.

Some oxides (MoO_3, SnO_2, V_2O_5, WO_3) were found to show relatively high sputtering yields and it was proposed that for these oxides, spikes contribute significantly to the total yield [2.139]. The fact that such oxides show a strong dependence of the yield on target temperature, such as for MoO_3, V_2O_5, and WO_3, was also taken as evidence for spike sputtering [2.139].

Figure 2.21a, b show the energy dependence of the sputtering yields obtained for Nb_2O_5 and WO_3 under Kr^+ bombardment [2.173]. They are compared with calculated sputtering yields as obtained from the linear cascade theory by *Sigmund* [2.45] for different values of the surface binding energy U. For Nb_2O_5, reasonable agreement of U, as determined by fitting the theoretical yield curve to the experimental results, and the heat of atomisation are found. This indicates that sputtering of Nb_2O_5 may be explained in terms of linear cascade sputtering. For WO_3, the fit to the experimental data gives a value for U less than half the heat of atomisation which the authors took as evidence for spike sputtering for this oxide. For comparison, Fig. 2.21a, b also show measured and calculated sputtering yields for elemental W and Nb targets. Similar yields for Nb and Nb_2O_5, but higher yields for WO_3 than for W can be observed.

Higher yields than for the pure metals are also reported for nitrides [2.117]. The sputtering yields of these nitrides were also found to be higher than those for the corresponding borides, in spite of the same type of interstitial lattice.

Total sputtering yields of *alkali halides* (single crystal NaCl, KCl, LiF, KBr) under 2–10 keV Ar^+ bombardment have been measured by *Navinšek* [2.197].

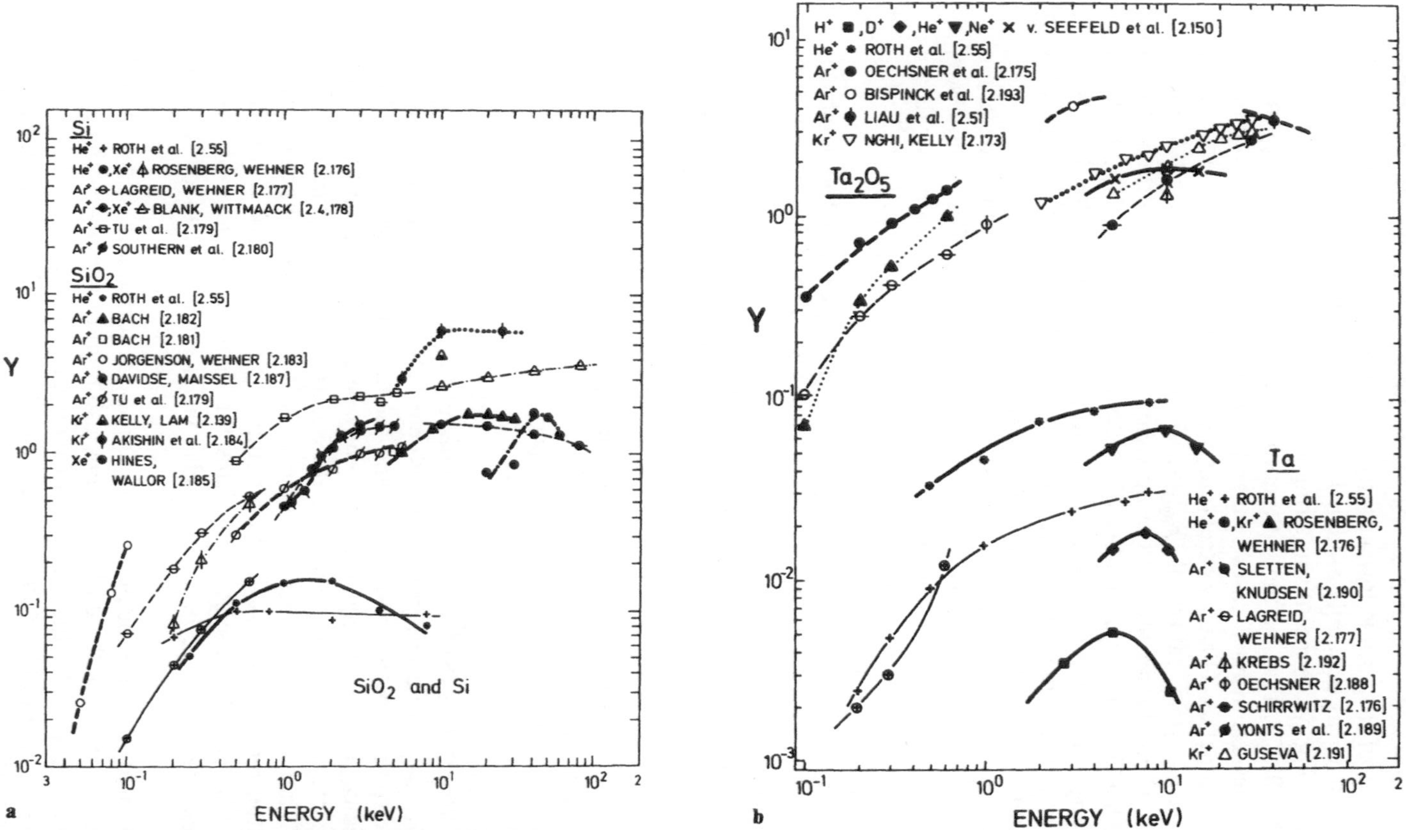

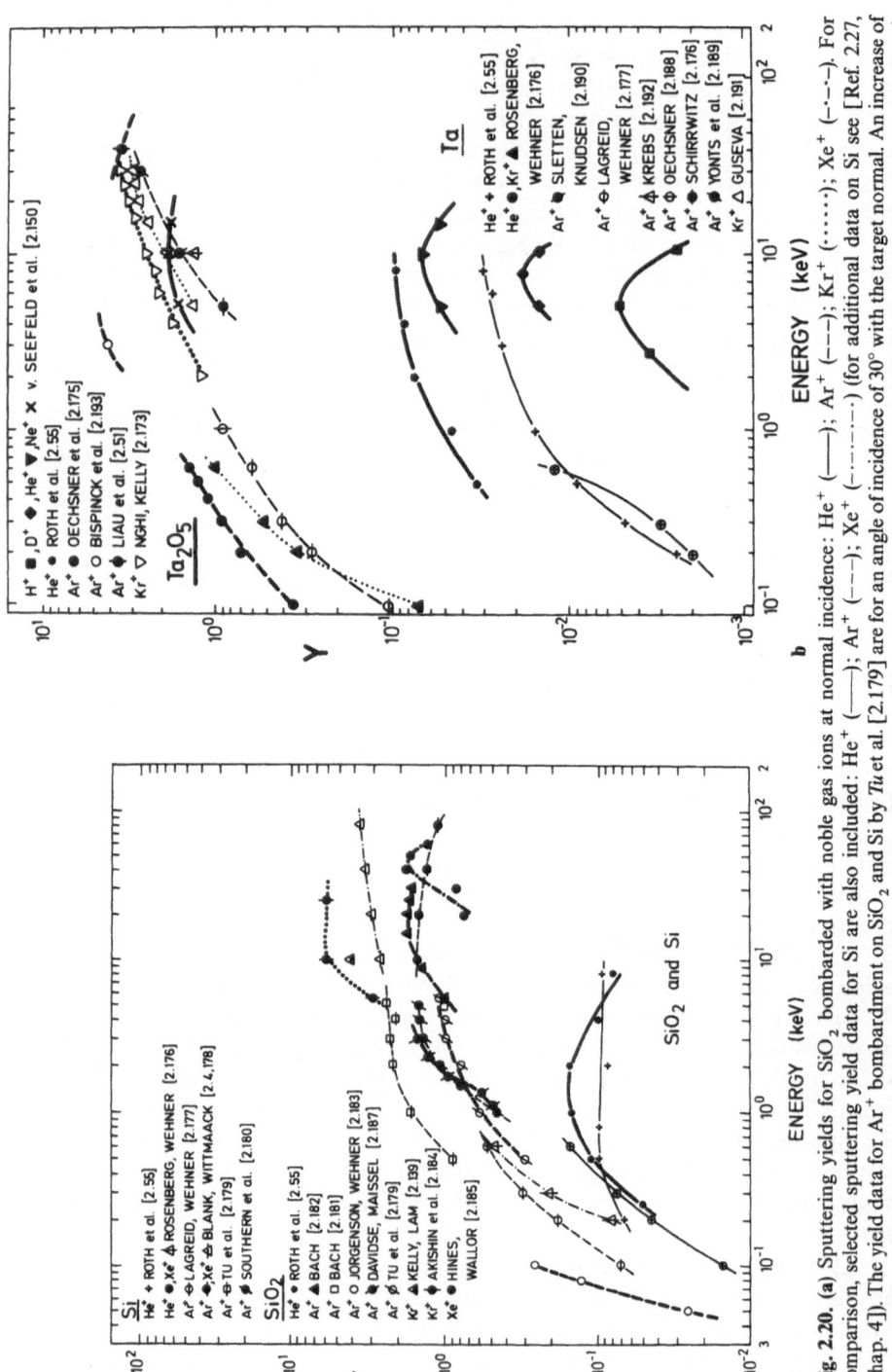

Fig. 2.20. **(a)** Sputtering yields for SiO_2 bombarded with noble gas ions at normal incidence: He^+ (——); Ar^+ (– – –); Kr^+ (· · · · · ·); Xe^+ (–·–·–). For comparison, selected sputtering yield data for Si are also included: He^+ (——); Ar^+ (– – –); Xe^+ (–·–·–·–·) (for additional data on Si see [Ref. 2.27, Chap. 4]). The yield data for Ar^+ bombardment on SiO_2 and Si by *Tu* et al. [2.179] are for an angle of incidence of 30° with the target normal. An increase of the sputtering yield by a factor of 2 as compared to normal incidence was observed in this energy range for Si [2.179] and for SiO_2 under 32 keV Ar^+ bombardment [2.186]. **(b)** Sputtering yields for Ta_2O_5 bombarded with noble gas ions at normal incidence: He^+ (——); Ar^+ (– – –); Kr^+ (· · · · · ·). For comparison, selected sputtering yield data for Ta are also included: He^+ (——); Ar^+ (– – –); Kr^+ (· · · · · ·) (for additional data on Ta see [Ref. 2.27, Chap. 4])

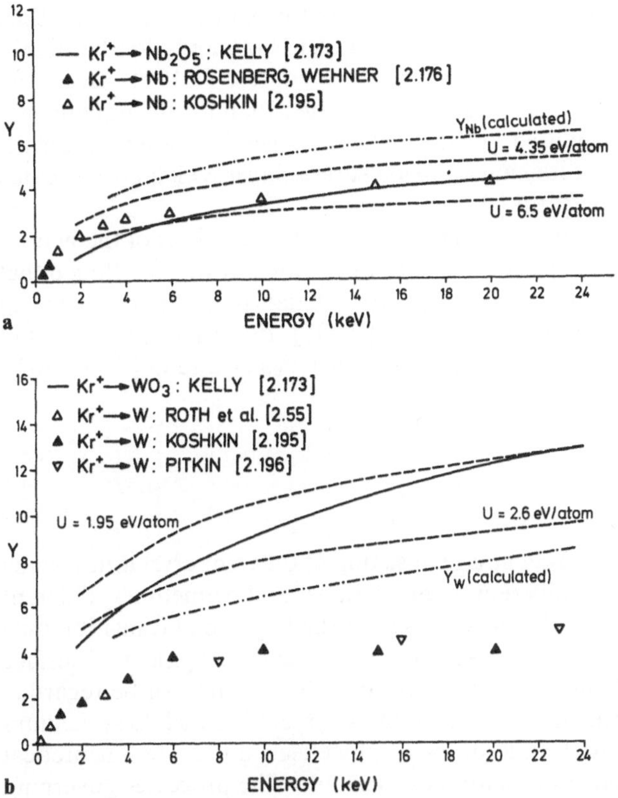

Fig. 2.21a, b. Sputtering yields for Kr^+ bombardment of (**a**) Nb_2O_5 and (**b**) WO_3 thin films as a function of ion energy determined by interference-color change measurements and by the weight loss technique. The dashed lines show calculated values by the linear cascade theory of *Sigmund* [2.45] with the following values for the surface binding energy U:

(**a**) $U = 4.35$ and 6.5 eV/atom, respectively, for Nb_2O_5 which has an average total heat of atomization of 6.8 eV/atom.

(**b**) $U = 1.95$ and 2.6 eV/atom, respectively, for WO_3 which has an average total heat of atomization of 6.3 eV/atom.

For comparison, experimental and calculated ([2.45] and [Ref. 2.27, Chap.4]) sputtering yield data are included for Nb and W, respectively, [2.173]

The sputtering yields of alkali halides were observed to be higher than expected from linear cascasde theory. Energy distribution measurements for NaCl [2.163] indicate the dominance of sputtering via electronic processes (see Sect. 2.4.3). Evaporation of Na at room temperature domimates over collisional sputtering. From measured sputtering yields for alkali halides (MeX, Me = Na, K, Rb, Cs; X = Cl, Br, I), a spike contribution to the total yield was concluded by *Barbashev* et al. [2.198].

The *charged state* of the emitted atoms or molecules (charged or neutral) depends strongly on the nature of the chemical bond. Charged particles represent a much higher part of the sputtered flux for oxides and halides than with metals. For example, in the case of KBr under low energy ion bombardment, only charged particles and no neutrals are emitted [2.199]. Secondary ion yields of oxides can be orders of magnitude larger than for the pure metals. This effect is often routinely used in SIMS where oxygen ions are used for the sputtering of metals, thus producing an oxide surface layer which gives a much higher secondary metal ion yield [2.21].

Molecular products can constitute large (up to 50%) fractions of the neutral sputtered products for oxides [2.200, 201]. In this context it was concluded that the sputtering process is controlled by the partial molar enthalphies of vaporisation of the various sputtered species [2.201]. For Ta_2O_5, Nb_2O_5, and WO_3 the sputtered species are essentially metal Me, MeO, and O [2.175, 202].

2.5 Multiphase Materials

A multiphase material is heterogenous, consisting of different crystallites which represent different phases, different compositions and sometimes different crystal structures. Sputtering of such a solid is marked by a long transition time before steady-state conditions are reached and by the development of surface microtopography. At low fluence, the sputtering phenomena can be regarded as a superposition of all the different single-phase crystallites, while at medium and higher fluences, the surface microtopography effects dominate. Theoretical sputtering models for such systems are not available. The processes governing the evolution of surface topographies are still under discussion even for single-phase, high purity materials (Chap. 6).

If a multiphase material is subjected to ion bombardment, three regimes with increasing fluence will be distinguished in the following.

2.5.1 Ion Bombardment for Low and Medium Fluences

In the limit of *low fluence*, i.e., removal of a few monolayers, for each single-phase grain, preferential sputtering as for solid solutions will exist on a microscopic scale. Enrichment of one component in each of these grains should, in principle, be the same as if the material consisted only of this phase. For example, in a system like Cu–Ag consisting of pure Ag and pure Cu crystallites, initially the Cu grains will sputter like a Cu target and the Ag grains like a Ag target.

For *medium fluences*, i.e., removal of a thickness in the order of a grain diameter, the grains representing a phase with a high sputtering yield will become eroded faster than those with lower yields. Without other processes taking place, the surface will become enriched in low sputtering yield grains. We will call this process, based on selective erosion of different grains, *selective sputtering*. This is distinguished from preferential sputtering, which is based on preferred ejection of the atoms of one of the constituents from a single-phase solid. Also for single-phase materials, differently oriented crystallites are sputtered with different rates, but no changes in composition are obtained for such a selective erosion because all grains have the same composition. Such changes in the surface composition due to selective sputtering were revealed, for example, under 500 eV Ar^+ sputtering of fine grained (particle size 50 Å) MgO/Au cermet films [2.203]. An exponential decrease of the Au concentration on the surface with bombarding fluence was found experimentally. This is in agreement with a much higher sputtering yield of Au than of MgO.

In addition, selective sputtering of a multiphase material will cause at medium ion fluences the development of pronounced microtopographical features like protrusions (cones, pyramides) and etch pits. Such a surface relief is firstly produced by the selective sputtering itself. Further during sputtering of the relief surface, a *seeding* of low yield material from protruding features, such as low sputtering yield phase grains, onto other more recessed, high sputtering yield phase grains can occur. This can lead to cone formation on the high sputtering yield grains in combination with surface migration of these atoms, forming islands. It has been shown that the presence of extremely small amounts of impurities *(internal seeding)* or the continuous deposition of certain foreign atoms at the surface while sputtering *(external seeding)* can result in such strong cone formation [2.3, 22, 204–207, 210d]. External seeding can easily be achieved by subjecting a foil of seed material to ion bombardment in such a way that its sputtered atoms arrive (seed) at the target.

Finally, sputtering alloy targets containing low sputtering yield precipitates also results in the formation of cones on the target surface due to masking of the matrix by the precipitates. For Inconel 718 and Incoloy 800 containing Nb and Ti carbide precipitates (grain size 1.4 μm) in a single phase matrix (grain size 10 μm), it was found that each precipitate grain gives rise to a cone (Fig. 2.22). As a result, the surface density of cones equals the precipitate density [2.208].

Wehner and *Hajicek* [2.3] found for external seeding of Cu targets with Mo atoms that cone formation starts with an arrival rate of one Mo atom per 500 sputtered Cu atoms (Fig. 2.23). Cone formation was also observed for seeding of Ag, Au, and Cu targets with Mo or W atoms. Furthermore, it was recently found that cone formation at least for external seeding, occurs not if seed atoms from a low sputtering yield material are deposited on a higher sputtering yield target but rather if the seed metal has a higher melting point than the target material [2.207]. For sputtering a multiphase material, it is reasonable to assume that if cone formation is found, the low yield material (which should also be the material with the higher melting point) causes internal seeding because due to selective sputtering, this phase will recess more slowly and thus can more easily cause seeding of the more recessed phase grains.

Seeding can also occur due to backdiffusion of sputtered atoms from the gas phase if plasma sputtering with a high gas pressure is used [2.208]. This is

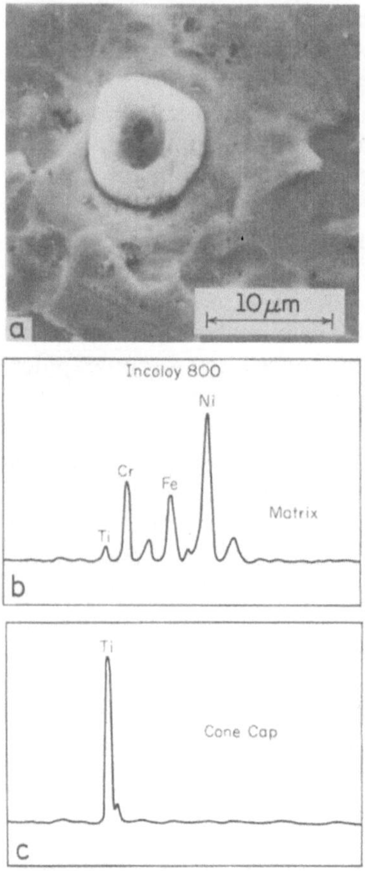

Fig. 2.22a–c

Fig. 2.23a–c

Fig. 2.22. (**a**) A scanning electron micrograph of a cone formed on an Incoloy 800 sample after 3 h of 2 keV Ar^+ sputtering in a glow discharge at 40 mTorr Ar pressure. (**b**) Energy dispersive x-ray spectrum taken from the area outside the cone in (a). (**c**) Energy dispersive x-ray spectrum taken from the cone cap, showing that the cone exists exclusively of TiC [2.208]

Fig. 2.23a–c. Electron scanning micrographs from 3 different areas of a Cu target seeded with Mo atoms during 600 eV Hg^+ sputtering. (**a**) From a zone with an arrival rate of about one Mo atom per 500 sputtered Cu atoms. (**b**) From a zone with a higher arrival rate than the zone in (a). (**c**) From a zone with a still higher arrival rate of about one Mo atom per 20 sputtered Cu atoms [2.3]

typically the case with a high sputtering rate, commercial dc or rf sputtering system.

The importance of surface migration can be concluded from the observation that for external seeding, cone formation can be suppressed by cooling of the sample and thus lowering surface migration [2.207, 209, 210a, b, d]. Scanning Auger and Electron Microprobe measurements of samples sputtered by high ion fluences seem to indicate that such minority species are sometimes concentrated on the cones [2.207–209].

Besides lowering the surface temperature and thus lowering the surface mobility, also sputtering with reactive ions like O or N [2.210c, 211–214] will often suppress the formation of cones. The role of the reactive ions in this context is attributed to the formation of an oxide or nitride layer which impedes the surface migration of atoms. Amorphous oxide or nitride films can furthermore decrease the topographical problems caused by the crystallinity and multi-phase nature of the substrate.

2.5.2 Ion Bombardment for High Fluences

For *high fluence*, the interplay of selective sputtering and the formation of surface topography features will finally lead to steady-state conditions. For selective sputtering being the dominant process, steady-state conditions will be reached after sputter erosion of a layer in the order of a few grain diameters, as has been demonstrated for the case of MgO/Au cermets [2.203]. If topography effects dominate, an equilibrium is reached after development of a quasi steady-state surface topography, i.e., when the rates of cone formation and annihilation are equal.

For two phase Ni-based alloy targets (Ni–Fe–Cr or Ni–Cu matrix, resp., with Nb or Ti carbide precipitates, resp.) under 2 keV Ar^+ sputtering, equilibrium was reached after the removal of 1.0–1.5 μm material thickness, which is in the order of the precipitate (1–1.5 μm) and much smaller then the matrix grain size (10–20 μm) [2.208].

An interesting observation was made for two-phase Nb–NbO alloys under 15 keV Ar^+ sputtering. Both cone and etch pits of similar size were observed and attributed to differently oriented NbO precipitates in Nb crystallites [2.215]. The differences in grain orientation of NbO and Nb result in different relative sputtering yields and thus NbO gives rise to either cone or pit formation.

The development of cones and similar features impedes the use of sputtering for composition vs depth profiling of multiphase materials. On the other hand, sputter etching a surface and simultaneously sputter depositing a lower sputter yield material onto the surface, i.e., external seeding, can be used as a controlled method for texturing a surface [2.209, 216a–c].

2.5.3 Sputtering Yields

A few results on sputtering yield measurements of multiphase materials will conclude this section. Generally, anomalously low sputtering yields have been found in such materials [2.22, 40, 207, 217].

For pure selective sputtering (no topography effects and no surface migration) in a nonmixable two-phase system, the number of atoms n_A and n_B sputtered from the two phases by an ion fluence, it should be $n_A = Y_A c_A^s It$ and $n_B = Y_B c_B^s It$, with Y_A and Y_B being the sputtering yields of the pure elements and c_A^s and c_B^s the fractional surface area of each component. For steady-state

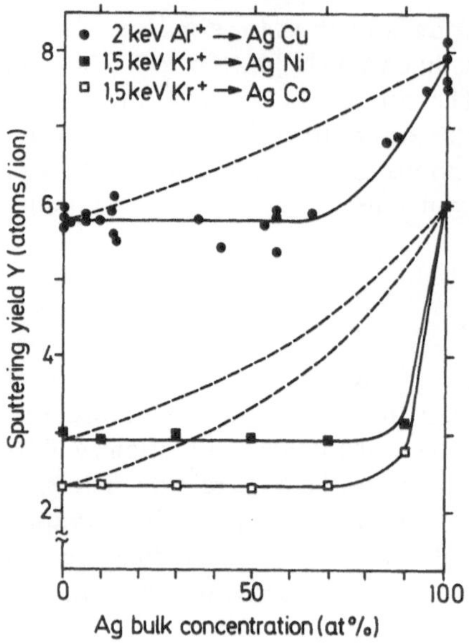

Fig. 2.24. Sputtering yields for 2 keV Ar^+ bombardment of Ag–Cu alloy thin films [2.218]; and for 1.5 keV Kr^+ bombardment of Ag–Ni and Ag–Co pressed powder samples [2.217]. The dashed lines are the dependence of the sputtering yield on composition according to (2.13)

conditions, $n_A/n_B = c_A/c_B = Y_A c_A^s / Y_B c_B^s$ also has to be valid. Therefore, one obtains for the total sputtering yield Y [2.40, 218]:

$$Y = 1/(c_A/Y_A + c_B/Y_B)\,. \tag{2.13}$$

However, sputtering yields measured for such nonmixable systems were found to be mostly much lower than predicted by (2.13).

Experimental results for total sputtering yields for compacted powders (grain size: 50–100 μm) of Ag–Ni and Ag–Co [2.217] and of Ag–Cu [2.40, 218] alloys as a function of composition are shown in Fig. 2.24. In all three cases, the total sputtering yield remains close to the value of the low yield component and increases to this value only for very high concentrations of the high yield component. It was concluded that these low yields are caused by coating the Ag (high sputtering yield) phase with Ni, Co or Cu, respectively, which in combination with an increased surface mobility under ion bombardment leads to cone formation. Evidence was observed of surface diffusion of Ag atoms onto Cu grains in Ag–Cu alloys subjected to ion bombardment, as analysed by AES [2.219]. Furthermore, steady-state conditions for sputtering Ag–Co and Ag–Ni were reached after the removal of a layer much smaller than the average grain diameter [2.217], also indicating the dominance of topography effects and surface migration over pure selective sputtering.

A sputtering yield reduction was also observed for Cu–Mo alloys [2.207]. As little as 5 at. % Mo reduced the total sputtering yield to the value of pure Mo.

Related experiments for sputter removal of a thin Mo layer from Au, Al, and W substrates show that in the case of a W target, Mo removal proceeds

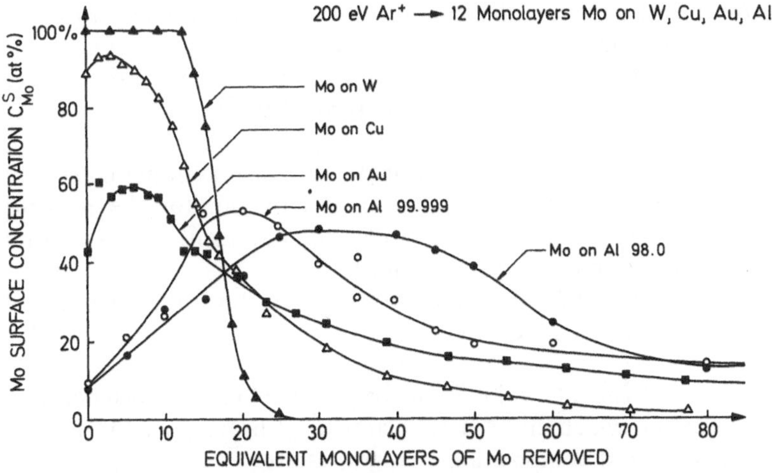

Fig. 2.25. Strength of the Mo Auger signal after deposition of 12 monolayers of Mo on different metal surfaces and their subsequent sputtering with 200 eV Ar^+ ions [2.22]

with the sputtering rate of elemental Mo. However, for the other substrates, the "last" monolayer of Mo is extremely difficult to sputter away [2.22] (Fig. 2.25). The Mo atoms agglomerate by surface migration into multilayered islands and cause cone formation. The sputtering yield of Mo was found to decrease under these conditions by up to a factor of 10^3 as compared to the yield of a pure Mo target under otherwise identical bombardment conditions. Thus the sputtering yield of one metal species from another can be very different from its bulk value, which is of much practical importance in depth-profiling of an interface using surface sensitive techniques.

2.6 Conclusion

In spite of the large amount of experiments performed within the last ten years, the sputtering process for multicomponent materials is by far not fully understood. Even for single-phase metal alloys where preferential sputtering effects, i.e., ion bombardment induced surface composition changes and altered layer thicknesses, have been extensively studied, no single alloy system exists where the exact composition profile of the altered layer and its development with bombardment fluence is known.

It was recently pointed out by *H.H. Andersen* that the ideal experiment has not yet been performed. Such an experiment, performed with a mass analyzed ion beam and under UHV conditions, will consist of:

— measurement of the sputtered flux to establish fluence-dependent partial sputtering yields, for example, by collecting the sputtered material and analysis by RBS;

— simultaneous measurement of the composition of the top monolayer by ISS, of the top few monolayers by AES and of the altered layer depth profile deeper inside the target by RBS to establish the development of the surface and altered layer compositon with bombardment fluence and for steady-state;
— in addition, XPS and Electron Diffraction measurements should be performed, especially for compounds (as for oxides) to monitor the possible formation of new compounds (lower oxides or the metal) and structural changes (amorphisation or recrystallisation);
— altered layer studies at different temperatures to establish the influence of diffusion and segregation processes.

Only from such experiments will it finally be possible to distinguish between the different processes which have been claimed to be responsible for, or contribute to the observed composition changes, as are:

— preferential sputtering in the strict sense, i.e., different ejection probabilities of the different target components;
— recoil implantation and cascade mixing;
— thermal and/or radiation induced diffusion or segregation.

Nevertheless, the experimental results so far indicate that preferential sputtering produces neither always enrichment of the heavier component, nor always enrichment of the element with the lower elemental sputtering yield, as has been claimed consistently in the past. Both differences in the mass and differences in the surface binding energies of the constituents seem to play an important role. Only for low energy, light ion bombardment, i.e., in the single knock-on regime, do mass effects clearly dominate.

Ion bombardment of a compound with no high vapor pressure component, such as carbides and intermetallic compounds, will generally produce an altered layer whose composition at thermal equilibrium will no longer correspond to a single-phase alloy or compound. Thus for such compounds, precipitation of a component cannot be excluded from our present knowledge and indeed has been observed.

In addition, structural changes and formation of new compounds has been frequently observed for compounds containing a high vapor pressure component, e.g., oxides and ionic compounds. Besides semiempirical rules, no theoretical predictions for such processes exist and further investigations are needed.

Considering total sputtering yields for alloys, not enough data are available to decide if the strongly reduced or increased sputtering yield values observed for some alloys are a general result, or rather, if the sputtering yield increases linearly from the yield of the low yield constituent to that of the high sputtering yield element, as observed for some single-phase metal alloys. Also the often-claimed protective action of oxides under ion bombardment seems to be verified only for Al_2O_3 and MgO, while most other oxides have sputtering yields comparable to the corresponding metals.

Finally, the few experimental results on the angular distribution of the different constituents of an alloy or compound do not allow any general conclusions to be drawn, except maybe for the single knock-on regime, where mass effects seem to be dominant due to differently effective energy transfer factors.

For multiphase systems, selective sputtering of the high yield phase grains and for higher bombardment fluences, topography effects dominate. An attempt has been made to give a phenomenological description of the behaviour of such materials under ion bombardment. Especially for multiphase materials, further experiments under defined conditions (UHV, mass analyzed ion beam) are needed to obtain more information and insight into the development of the surface topography features such as cones or etch pits. Total sputtering yields for such materials seem to be strongly influenced by the topography and cannot be predicted at all.

Acknowledgements. The authors gratefully acknowledge many suggestions and remarks by R. Behrisch and P. Braun. One of the authors also wants to express his gratitude to Prof. F. P. Viehböck for his support and continous interest and to R. Kelly for discussion of his work and supplying manuscripts in advance of publication. Support of this work by the Austrian "Fonds zur Förderung der wissenschaftlichen Forschung" is gratefully acknowledged (Project Nr. F 4547).

2.7 Tables

An attempt has been made to compile all available experimental data on ion bombardment induced surface and near surface composition changes for single-phase alloys and compounds, on altered layer thicknesses for single-phase alloys and on total sputtering yields for compounds. The tables contain no data on multiphase materials. The compilation of these tables has been greatly facilitated by the existence of similar tables on surface composition changes by *Coburn* [2.31] and *Kelly* [2.108, 147].

Surface composition changes for single-phase metal alloys are listed in Table 2.1, those for compounds containing no high vapor pressure component in Table 2.2, and those for compounds containing at least one high vapor pressure component in Table 2.5. Table 2.5, in which mainly oxides and ionic compounds are listed, also contains information on the observed formation of new compounds or on structural changes under ion bombardment.

The different experimental techniques used vary in their information depth (see Fig. 2.1). Thus, the listed surface enrichment in the tables is the average enrichment of a near surface region whose thickness is given by the sampling depth of the applied experimental technique.

Altered layer thicknesses for single-phase metal alloys are given in Table 2.3 and *total sputtering yield coefficients* for compounds are listed in Table 2.4.

Table 2.1. Comparison of component sputtering yield ratios Y_A^c/Y_B^c in alloys with measured yield ratios of the pure elements for Ar ion bombardment. Also listed are the observed surface enrichment, the atomic weight ratio M_A/M_B and the surface binding energy ratio U_A/U_B of the alloy constituents. For the surface binding energies the elemental heats of sublimation as tabulated by *Gschneidner* [2.287] were taken. An alloy system where several alloys with different compositions were measured is indicated as A–B, like for Ag–Au. Otherwise the composition of the analyzed alloy is indicated as for $AgAu_{0.19}$

System A–B (composition)	Atomic weight and surface binding energy ratios M_A/M_B	U_A/U_B	Bombarding ion and energy	Surface enrichment of	Component sputtering yield ratio Y_A^c/Y_B^c acc. to (2.4)	Measured sputtering yield ratio of pure elements Y_A/Y_B for Ar^+	Method and Reference (S: scribed, F: fractured)
Ag–Au	0.5	0.78	Ar, 1 keV	Au	1.7–1.8	1.1 (1 keV) [2.17], 1.2 (1 keV) [2.188], 1.3 (600 eV) [2.177], 1.2 (5 keV) [2.172]	AES(S, F) [2.39, 40]
Ag–Au			Ar, 0.5–2 keV	Au	2.0 ± 0.2		AES [2.89]
Ag–Au			Ar, 700 eV	Au	2.3–2.9		AES [2.90]
$AgAu_{0.19}$			Ne, Ar, Kr, Xe, 20–80 keV	Au	1.2 (independent of ion energy and typ)		RBS [2.51]
AgAu			Ar, 15–200 keV	Au	1.6		Collector technique combined with RBS [2.37]
Ag–Au			Ne, 1.5 keV	None	1.0		ISS, [2.49a]
Ag–Au in Ag–Au–Pd			Ar, 2 keV	Au	1.3–1.9		AES (S, F) [2.56]
Ag–Au in Ag–Au–Cu			Ar, 2 keV	Au	2.5		AES (S, F) [2.43]
Ag–Cu in Ag–Au–Cu (single-phase alloys due to Au presence)	1.7	0.84	Ar, 2 keV	Cu	2.5	1.4 (2 keV) [2.40], 1.1 (1 keV) [2.17], 1.3 (1 keV) [2.188], 1.5 (600 eV) [2.177]	AES (S, F) [2.43]
Ag–Pd	1.0	0.76	Ar, 2 keV	Pd	1.6–3.4	1.5 (1 keV) [2.188], 1.9 (2 keV) [2.41], 1.4 (600 eV) [2.177]	AES (S, F) [2.40, 45]
Ag–Pd			Ar, 0.5–2 keV	Pd	2.4–2.5		AES(F) [2.41]
$Ag_{0.26}Pd_{0.74}$			Ar, 1 keV oblique incidence	Pd	1.4		AES (pure standards) [2.38]
$Ag_{0.13}Mn_{0.19}Pd_{0.68}$			Ar, 1 keV oblique incidence	Pd			AES (pure standards) [2.38]
Ag–Pd			Ar, 950 eV	Pd	2.1 (10 at. % Ag)–3.0 (70 at. % Ag)		AES [2.220]
Ag–Pd			Ar, N_2, 500 eV	Pd	1.5–1.6		XPS [2.91]

Ag–Pd in Ag–Au–Pd			Ar, 2 keV	Pd	2.0–2.9		AES (S, F) [2.56]
Al–Cu (2–11 at.% Cu)	0.4	0.95	Ar, Xe, 1–2 keV	Cu	2.0	0.5 (1 keV) [2.188], 0.5 (600 eV) [2.177]	AES and RBS [2.96]
$Al_xGa_{1-x}As$ ($0 \leqq x \leqq 0.95$), solid solution, single crystal			Ar, 1 keV	None			AES [2.221]
Al–Pd[a] (27–77 at.% Al)	0.3	0.86	Ar, 1 keV	Pd	3.0	0.5 (600 eV) [2.177] 0.6 (1 keV) [2.188]	AES [2.223, 2.224]
Al–Si[a] (5–26 at.% Al)	1.0	0.71	Ar, 1 keV	Si	2.6	2.3 (600 eV) [2.177]	AES [2.223, 2.224]
Au–Cr (0–20 at.% Cr)	3.8	0.92	Ne, Ar, 0.5–2 keV	Cr	2.0 (1% Cr)– 1.3 (20% Cr)	4 (2 keV) [2.42], 2.2 (600 eV) [2.177]	AES (S) [2.42]
Au–Ni (6–26 at.% Ni), quenched, solid solutions	3.3	0.85	Ar, 1 keV	Ni	1.2 (6% Ni)– 2.5 (26% Ni)	2 (1 leV) [2.114], 1.8 (600 eV) [2.177]	AES (S) [2.114]
Au–Pd	1.8	0.97	Ar, 2 keV	Pd	1.0–1.4	1.3 (1 keV) [2.188], 1.2 (600 eV) [2.177]	AES (S, F) [2.40, 45]
Au–Pd			Ar, 1.5 keV, oblique incidence	Pd	1.0–3.0		AES [2.225]
Au–Pd			Ne, 2 keV	Au(at. 200° C)			ISS [2.111 d]
Au–Pd in Ag–Au–Pd			Ar, 2 keV	Pd	1.2–1.6		AES (S, F) [2.56]
$Be_{0.13}Cu_{0.87}$	0.1	0.96	Ar, 400 eV	Cu (no Be at surface)		0.35(600 eV) [2.177]	AES [2.76]
Cr–Pd (9–24 at.% Cr)	0.5	1.05	Ar, 1 keV, oblique incidence	None	1.0	0.55(600 eV) [2.177]	AES (pure standards) [2.38]
Cu–Au	0.3	0.93	Ar, 1 keV	None	1.0	1.0 (1 keV) [2.17], 0.9 (1 keV) [2.188], 0.9 (600 eV) [2.177]	AES (S, F) [2.39, 40]
Cu_3Au			Ne, Ar, Kr, Xe, 20–80 keV	Au	1.1 (independent of ion typ and energy)		RBS [2.51]
Cu_3Au			Ar, 25–85 keV	None	1.0		Collector technique combined with RBS [2.37]
Cu–Au (more than 95 at.% Cu)			H_3, 1 keV	Au			SIMS [2.50]

Table 2.1 (continued)

System A–B (composition)	Atomic weight and surface binding energy ratios		Bombarding ion and energy	Surface enrichment of	Component sputtering yield ratio Y_A^c/Y_B^c acc. to (2.4)	Measured sputtering yield ratio of pure elements Y_A/Y_B for Ar^+	Method and Reference (S: scribed, F: fractured)
	M_A/M_B	U_A/U_B					
Cu–Au (43–94 at. Au)			Ar, 1 keV	None	1.0		AES (S) [2.114]
$Cu_{43}Au_{57}$			Ar, 1.5–2 keV	Au	2.0		ISS [2.111 b]
Cu_3Au_{97}			He, 0.15 keV	Au			ISS [2.49b]
Cu–Au in Ag–Au–Cu			Ar, 2 keV	None	1.0		AES (S, F) [2.43]
Cu–Ga (0–15 at.% Ga)	0.9	1.25	Ar, 2 keV	None	1.0		Differential reflectometry [2.226]
Cu–Ni	1.1	0.79	Ar, 0.5–2 keV	Ni (2 keV)	1.7	1.6 (1 keV) [2.17], 1.7 (1 keV) [2.188], 1.5 (600 eV) [2.177], 1.8 (5 keV) [2.172]	AES [2.61]
Cu–Ni			Ar, 0.5–2 keV	Ni	1.7		AES [2.25]
Cu–Ni			Ar, 500 eV	Ni	1.9		AES [2.227]
CuNi			Ar, 500 eV	Ni	1.8		AES [2.26]
$Cu_{0.55}Ni_{0.45}$			Ar, 35–500 eV	Ni	7.0 (35 eV)–1.6 (above 200 eV)		AES [2.62]
CuNi (52 at. %)			Ar, 3 keV	Ni	2.0 analysis depth = 15 Å		HEAES [2.241]
CuNi (52 at. %)			Ar, 3 keV	Ni	3.2 analysis depth = 4 Å		LEAES [2.241]
CuNi (52 at. %)			Ar, 3 keV	Cu	0.6 analysis depth = 1 monolayer		ISS [2.241]
Additional information on Cu–Ni can be found in: [2.46, 47, 84–86, 92, 93, 95, 111d, 111e, 228–240, 294]							
Cu–Pd	0.6	0.90	Ar, 2 keV	Pd	1.5–1.6	1.2 (1 keV) [2.188], 1.0 (600 eV) [2.177]	AES (S, F) [2.40, 45]
$Cu_{0.55}Pd_{0.45}$			Ar, 1 keV, oblique incidence	Pd	1.7		AES (pure standards) [2.38]
Cu–Pd			Ar, 0.5–5 keV	Pd	1.5		AES (pure standards) [2.293]
Cu–Pt	0.3	0.60	Ar, 2 keV Xe, 0.5–5 keV	Pt	1.6–3.0	1.5 (600 eV) [2.177], 1.4 (5 keV) [2.172]	AES (S, F) [2.40]
Fe–Cr (5–25 % Cr)	1.1	1.05	Ne, 2.5 keV	None	1.0	1.0 (600 eV) [2.177] 1.3 (1 keV) [2.242]	ISS [2.243]
Fe–Cr			Ar, Ne, 0.5–2 keV	None	1.0		AES (S) [2.244]

Fe–Cr			Ar, 0.5–5 keV	Minor Cr	0.8–1.0 (2 keV)		AES (S, F) [2.59, 63]
Fe–Cr in Fe–Cr–Ni			Ar, 2 keV	None	1.0		AES (S, F) [2.59]
Fe–Cr in Fe–Cr (0–25 at. %)–Mo (0–4 at. %)			Kr, 2 keV	None	1.0	0.94 (2 keV, Kr) [2.245]	AES (F) [2.245]
Fe–Mo in Fe–Cr (0–25 at. %)–Mo (0–4 at. %)	0.6	0.63	Kr, 2 keV	Mo	2.0	1.6 (2 keV, Kr) [2.245], 1.4 (600 eV) [2.177]	AES (F) [2.245]
$In_{10}Cu_{90}$ (supersaturated, fcc single-phase)	1.8	0.71	Ar, 1–3 keV	Cu	11.0		AES [2.34]
In_xSn_{1-x} ($x=0.02, 0.9$)	1.0	0.80	Ar, 2 keV	Sn			AES [2.286, 289]
Mg–Al	0.9	0.46	Ar, 2 keV	Al		2.0 (5 keV) [2.172]	AES (S, F) [2.222]
Ni–Co (70 %)	1.0	1.01	Ar, 5 keV	Co (minor	1.1	1.1 (600 eV) [2.177]	AES [2.236]
Ni–Cr in Fe–Cr–Ni	1.1	1.08	Ar, 0.5–5 keV	Minor Cr	1.1 (2 keV)	1.2 (600 eV) [2.177]	AES (S, F) [2.59]
Ni_xFe_{1-x} ($x=0.05$–0.95, single-phase, thin films)	1.1	1.03	Xe, 1 keV	None	1.0	1.2 (600 eV) [2.177] 1.2 (1 keV, Xe) [2.122]	AES, XPS [2.122]
Ni–Fe			Ar, 0.5–5 keV	None	1.0		AES (S, F) [2.59, 63]
Ni–Fe in Fe-Cr-Ni			Ar, 0.5–5 keV	Minor Fe	1.1 (2 keV)		AES (S, F) [2.59]
Ni–Fe			Ar, 3 keV	None	1.0		XPS [2.58]
Ni–Pt	0.3	0.76	Ar, 2 keV Xe, 0.5–5 keV	Pt	1.5–1.9	1.0 (600 eV) [2.177] 0.8 (5 keV) [2.172]	AES (S, F) [2.40]
$Ni_{0.88}W_{0.12}$	0.3	0.51	Ar, 2.5 keV	W	2.7	2.5 (600 eV) [2.177], 1.6 (5 keV) [2.172]	AES [2.81]
Pb–In	1.8	0.82	Ar, 0.3 keV	In			AES [2.246]
Pb_xIn ($x=0.002, 0.5, 0.9$)			Ar, 2 keV	In			AES [2.286, 289]
Pb_xSn_{1-x} $x=0.002$ and 0.97	1.7	0.65	Ar, 2 keV	Sn			AES [2.263, 289]

Table 2.1 (continued)

System A–B (composition)	Atomic weight and surface binding energy ratios M_A/M_B	U_A/U_B	Bombarding ion and energy	Surface enrichment of	Component sputtering yield ratio Y_A^c/Y_B^c acc. to (2.4)	Measured sputtering yield ratio of pure elements Y_A/Y_B for Ar^+	Method and Reference (S: scribed, F: fractured)
Pd_xFe_{1-x} ($x=0.06$–0.7), single-phase, thin films	1.9	0.9	Xe, 1 keV	Fe		1.9 (600 eV) [2.177], 2.1 (1 keV, Xe) [2.122]	AES, XPS [2.122, 247]
Pd–Ni	1.8	0.88	Ar, 0.5–2 keV	Ni	1.3–1.5	1.5 (1 keV) [2.188], 1.6 (2 keV) [2.41], 1.6 (600 eV) [2.177]	AES (F) [2.41]
Pd–Pt	0.6	0.67	Ar, 2 keV	Pt		1.5 (600 eV) [2.177]	AES [2.248]
Si–Pd[a] (23–78 at. % Si)	0.3	1.2	Ar, 1 keV	Pd	3.9	0.22 (600 eV) [2.177]	AES [2.223, 224]
Sn–Au (86 at. % Au) (99 at. % Au)	0.6	0.82	Ar, 1.5 keV	Au		0.7 (5 keV) [2.172]	AES [2.249]
U–Nb (6–28 at. % Nb)	2.6	0.71	Ar, 1 keV, oblique incidence	Nb	1.7	1.5 (600 eV) [2.177]	AES (pure standards) [2.38]
Ternary Systems							
Ag–Au–Cu			Ar, 2 keV	Au, Cu			AES (S, F) [2.43]
Ag–Au–Pd			Ar, 2 keV	Au, Pd			AES (S, F) [2.56]
$Ag_{0.13}Mn_{0.19}Pd_{0.68}$			Ar, 1 keV, oblique incidence	Pd			AES (pure standards) [2.38]
$Al_xGa_{1-x}As$ ($0 \leq x \leq 0.95$), solid solution, single crystal			Ar, 1 keV	None			AES [2.221]
Fe–Cr–Ni			Ar, 0.5–5 keV	Minor (Cr + Fe)			AES (S, F) [2.59]
Fe–Cr–Ni (304 SS)			Ar, 3 keV	Minor (Cr + Fe)			XPS [2.58]
Fe–Cr–Ni (77 % Ni, 15.6 Cr; Inconel 600)			Ar, 3 keV	Minor (Cr + Fe)			XPS [2.58]
Fe–Cr–Ni (31 % Ni, 19 % Cr; Inconel 800)			Ar, 3 keV	None			XPS [2.58]
Fe–Cr (0–25 at. %)–Mo(0–4 at. %)			Kr, 2 keV	Mo			AES (F) [2.245]
$Si_{96}Fe_4W_{0.1}$			Ar, Xe, 10–150 keV	Fe, W for energies below 40 keV			RBS [2.283]

[a] Thin film samples prepared by electron beam evaporation, almost completely amorphous.

Table 2.2. Surface enrichment due to ion bombardment for compounds. Also listed are the measured sputtering yield ratios Y_A/Y_B of the pure elements for Ar ion bombardment, the atomic weight ratio M_A/M_B and the surface binding energy ratio U_A/U_B of the compound constituents. For the surface binding energies the elemental heats of sublimation as tabulated by *Gschneidner* [2.287] were taken. Component sputtering yield ratios Y_A^c/Y_B^c can be calculated from the given bulk and surface compositions after (2.4)

System A–B	Atomic weight and surface binding energy ratios		Bombarding ion and energy	Surface enrichment of	Surface composition after ion bombardment	Measured values for Y_A/Y_B of pure elements for Ar^+	Method and reference
	M_A/M_B	U_A/U_B					
AlAu	0.1	0.88	Ne, Ar, Xe, 0.5–5 keV	Au	$Al_{16}Au_{84}$ (2 keV, Ar)	0.5 (600 eV) [2.177], 0.3 (5 keV) [2.172]	AES [2.63, 64]
$AlAu_2$			Ne, Ar, Xe, 0.5–5 keV	Au	Al_9Au_{92} (2 keV, Ar)		AES [2.63, 64]
Al_2Au			Ne, Ar, Kr, Xe, 20–80 keV	Au	$Al_2Au_{1.3}$ [a]		RBS [2.51]
$AlAu_2$			Ne, Ar, Kr, Xe, 20–80 keV	Au	$AlAu_{3.8}$ [a]		RBS [2.51]
$AlCr_2$	0.5	0.82	Ne, Ar, Xe, 0.5–5 keV	Cr	$Al_{14}Cr_{86}$ (2 keV, Ar)	1.0 (600 eV) [2.177]	AES [2.63, 64]
AlFe	0.5	0.78	Ne, Ar, Xe, 0.5–5 keV	Fe	$Al_{45}Fe_{55}$ (2 keV, Ar)	1.0 (600 eV) [2.177], 1.5 (5 keV) [2.172]	AES [2.63, 64]
$AlNi_3$	0.5	0.75	Ne, Ar, Xe, 0.5–5 keV	Ni	$Al_{16}Ni_{84}$ (2 keV, Ar)	0.8 (600 eV) [2.177], 0.8 (5 keV) [2.172]	AES [2.63, 64]
AuSn	1.7	1.22	Ar, 1.5 keV	Au		1.4 (5 keV) [2.172]	AES [2.249]
Cu_3Sn	0.5	1.13	Ar, 2 keV	None	Cu_3Sn	1.0 (5 keV) [2.172]	AES, e^- probe [2.251]
Cu_6Sn_5			Ar, 2 keV	Cu			AES, e^- probe [2.251]
GaAs (100)	0.9	2.24	Ar, 600 eV	Ga	$Ga_{56}As_{44}$		AES [2.29]
GaAs (111)			Ar, 900 eV	Ga (weak)			XPS, AES [2.253]
GaAs			Ar, 0.5–5 keV	Ga (temperature dependent)			AES [2.254]
GaAs (111) and (100) single crystal			Ne, Ar, Kr, Xe 500–2 keV	Ga or As (both weak)			AES [2.125]
GaP	2.2	0.86	Ne, Ar, Kr, Xe, 20–80 keV	None [a]	GaP		RBS [2.51]
GaP			Xe, 80 keV	Ga	$Ga_{1.5}P$		RBS [2.255]
GaP single crystal, (111) Ga and $(\bar{1}\bar{1}\bar{1})$ P face			Ar, 900 eV	Ga			XPS, AES [2.253]
GeSi	2.6	0.83	Ne, Ar, Kr, Xe, 20–80 keV	None [a]	GeSi	2.3 (600 eV) [2.177]	RBS [2.51]
InP	3.7	0.76	Ne, Ar, Kr, Xe, 20–80 keV	None [a]	InP		RBS [2.51]

Table 2.2 (continued)

System A–B	Atomic weight and surface binding energy ratios M_A/M_B	U_A/U_B	Bombarding ion and energy	Surface enrichment of	Surface composition after ion bombardment	Measured values for Y_A/Y_B of pure elements for Ar^+	Method and reference
InSb (100) single crystal	0.9	0.92	Ne, Ar, Kr, Xe, 0.5–2 keV	In or Sb (both weak)[b]			AES [2.125]
NiSi	2.1	0.95	Ne, Ar, Kr, Xe, 20–80 keV	Ni	$Ni_{1.6}Si$[a]	3.3 (20 keV) [2.52]	RBS [2.51]
$NiSi_2$			Ar, Xe, 0.5–2 keV	Strong Ni enrichment decreasing with ion energy			AES [2.88e]
PtSi	7.0	1.25	Ne, Ar, Kr, Xe, 20–80 keV	Pt	$Pt_{2.1}Si$[a]	2.7 (20 keV) [2.52] 2.9 (600 eV) [2.177]	RBS [2.51–53]
PtSi			Ar, Xe, 0.5–2 keV	Strong Pt enrichment decreasing with ion energy			AES [2.88e]
Pt_2Si			Ne, Ar, Kr, Xe, 20–80 keV	Pt	$Pt_{3.3}Si$[a]		RBS [2.51]
PtSn	1.6	1.88	Ar, 500 eV	Pt	$Pt_{70}Sn_{30}$	0.8 (5 keV) [2.172]	AES [2.252]
Pt_3Sn			Ar, 500 eV	Pt	$Pt_{85}Sn_{15}$		AES [2.252]
SiC	2.3	0.63	Ar, 700 eV	None	SiC	3.0 (600 eV) [2.177]	AES [2.256]
SiC			H, 700 eV	C		1.3 (700 eV, H^+) [2.55]	AES [2.256]
TaC	15.0	1.09	He, Ar, 0.3–4 keV	Ta[c]		0.18 (1 keV, He) [2.71],	ISS, AES [2.48]
TaC			H, He, Ar, Xe, 0.3–4 keV	Ta	$Ta_{80}C_{20}$ for 1 keV He^+	0.4 (5 keV) [2.172]	AES [2.91b]
TaC			H, D, He, 0.2–8 keV	Ta			[2.24, 71][d]
TaC			Ar, 0.5–5 keV	Ta			AES [2.88 d]
TiB	4.4	0.83	H, He, Ar, Xe, 0.3–4 keV	None[e]			AES [2.91b]
TiC	4.0	0.66	Ar, 0.5–2 keV	None	TiC		AES [2.17]
TiC			Ar, 0.5–5 keV	None	TiC		AES [2.88 d]
TiC			H, He, Ar, Xe, 0.3–4 keV	None[e]	TiC		AES [2.91b]
WC	15.3	1.17	H, D, He, 0.2–8 keV	W		0.22 (1 keV, He)	[2.24, 71][d]
WC			He, Ar, 0.3–4 keV	W[c]		[2.55], 0.7 (5 keV)	ISS, AES [2.48]
WC			Ne, Ar, 1.5 keV	None	WC	[2.172]	ISS, AES [2.258]
WC			Ar, 1 keV	None	WC		AES [2.259]
WC			Ar, 0.5–5 keV	W			AES [2.88 d]

[a] Independent of ion energy and ion type, surface composition is average composition of **top 100 Å** [2.51] or 40 Å [2.53], respectively.

[b] Depending on ion energy and ion type.

[c] Enrichment for He^+ agrees with yield ratios of pure Ta and C after *Roth* [2.24].

[d] Inferred from yield data similar to pure Ta and C.

[e] For H_2^+ bombardment Ti enrichment due to chemical effects was observed.

Table 2.3. Altered layer thickness for different alloy systems and compounds. If not otherwise indicated, experiments were performed at room temperature and normal ion incidence

System	Bombarding ion and energy	Altered layer thickness	Method and reference
Ag–Au	1 keV, Ar^+	25–30 Å	AES [2.89], kinetic model after *Ho* et al. [2.25]
Ag–Au (48.7 at. %)	0.5–2 keV, Ar^+	15 Å (500 eV)–25 Å (2 keV)	
Ag–Au	700 eV, Ar^+	25 Å (30 at. % Au)–90 Å (70 at. % Au)	AES [2.90], kinetic model after *Ho* et al. [2.25]
Ag–Pd	500 eV, Ar^+ and 500 eV, N_2^+	2 Å (5 at. % Pd)–20 Å (95 at. % Pd)	XPS [2.91a], altered layer forming extra phase
Ag–Pd (20 at. % Pd)	500 eV, Xe^+ 5 keV, Xe^+	5 Å 40 Å	AES [2.87], depth profiling with different ion energies
Al–Cu (2–11 at. %)	1 keV, Xe^+ and Ar^+	200–300 Å	AES and RBS [2.96]
Au–Ni (14–26 at. %)	1 keV, Ar^+	50–75 Å	AES [2.114], kinetic model after *Ho* et al. [2.25]
Cu–Ni (4–94 at. % Cu)	500 eV, Ar^+, oblique incidence	10 Å	AES [2.25], kinetic model after *Ho* et al. [2.25]
Cu–Ni (62 at. %)	0.5–2 keV, Ar^+	10 Å (500 eV)–25 Å (2 keV)	
Cu–Ni (40 at. %)	5 keV, Ar^+ 300–600 °C	1 µm (500 °C) 3 µm (600 °C)	AES [2.95], sputter depth profiling of altered layer at room temperature
Cu–Ni	700 eV, Ar^+	5 Å (Ni rich)–30 Å (Cu rich)	AES (LEAES and HEAES) [2.46]
Cu–Ni	0.5–2 keV, Ar^+ −170 °C up to 30 °C	More than 20 Å	AES (LEAES and HEAES) [2.47, 84, 86]
Cu–Ni (52 at. %)	500 eV, Ar^+, angle of incidence: 0° 60° 0° 60°	15 Å 10 Å 48 Å 10 Å	AES [2.85], kinetic model after *Ho* et al. [2.25] (LEAES and HEAES)

Table 2.3 (continued)

System	Bombarding ion and energy	Altered layer thickness	Method and reference
Cu–Ni	500 eV, Ar^+ at 200 °C at 300 °C at 400 °C	- Less than 200 Å approx. 3500 Å approx. 3 μm	AES [2.92] sputter depth profiling at room temperature
Cu–Pt	2 keV, Ar^+ oblique incidence (70°)	20–30 Å (Cu rich) –5 Å (Pt rich)	AES (LEAES and HEAES) [2.40]
Ni–Pt	2 keV, Ar^+, oblique incidence (70°)	15 Å (Ni rich)– –5 Å (Pt rich)	AES (LEAES and HEAES) [2.40]
PtSi Pt_2Si	20 keV, Ar^+ 40 keV, Ar^+	250 Å 450 Å	RBS [2.51–53] (see also Fig. 2.7) RBS [2.51], Altered layer thickness agrees with penetration depth of bombarding ions; also found for AuAg, Cu_3Au, $AuAl_2$, Au_2Al, Pt_2Si PtSi, and NiSi under 20–80 keV Ne^+, Ar^+, Kr^+, and Xe^+ ion bombardment [2.51]
TaC	0.3–3 keV, He^+	30–70 Å	AES and Ar^+ sputter depth profiling [2.91 b]

Table 2.4. Total sputtering yields of compounds

Compound	Bombarding ion and energy	Sputtering yield Y	Reference and remarks
Al_2O_3	1–3 keV, Ar	0.2 (1 keV)	[2.187], (a)
Al_2O_3	0.1–8 keV, H, D, He	0.17 (1 keV, He)	[2.55], (a)
Al_2O_3	2–10 keV, Kr, O_2	1.2 (5 keV, Kr)	[2.174], (b)
Al_2O_3	5 keV, Ar	0.1	[2.172]
α-Al_2O_3	5 keV, Ar	0.9	[2.181], single crystal (001), (c)
$AuAl_2$	40 keV, Ar	9.1	[2.51], RBS
Au_2Al	40 keV, Ar	12.6	[2.51], RBS
B_4C	0.1–8 keV, H, D, He	0.15 (1 keV, He)	[2.71, 260], (a)
BeO	0.05–8 keV, H, D, He	0.15 (1 keV, He)	[2.55, 194], (a)
$CaCo_3$	5 keV, Ar	1.1	[2.181], single crystal (1011) (c)
CaF_2	5 keV, Ar	0.8	[2.181], single crystal (111), (c)
CsBr	8 keV, Ar	3.8	[2.198], (a)
CsCl	8 keV, Ar	4.3	[2.198], (a)
CsI	8 keV, Ar	3.3	[2.198], (a)
GaSb	0.05–1.5 keV, Ar	2.4 (1 keV)	[2.14], single crystal, different faces, (a)
Glass different types)	5 keV, Ar	About 1	[2.181], (c)
Inconel	0.1–8 keV, H, D, He	0.18 (1 keV, He)	[2.55], (a)
KBr	Ar, 2–10 keV	0.5 (5 keV)	[2.197], (a)
KBr	8 keV, Ar	1.9	[2.198], (a)
KCl	8 keV, Ar	2.3	[2.198], (a)
KCl	2–10 keV, Ar	1.7 (5 keV)	[2.197], (a)
KCl	70–300 keV, H, He, Ar	6.7 (300 keV, Ar)	[2.285], RBS
KI	8 keV, Ar	1.6	[2.198], (a)
LaB_6	100–500 eV, Cd	0.55 (500 eV)	[2.117], (a)
LiF	5 keV, Ar	2.8	[2.181], single crystal (100). (c)
LiF	Ar, 2–10 keV	1.9 (5 keV)	[2.197], (a)
LiF	10–40 keV, Ar, Xe	5 (10 keV, Xe, partial yield for F)	[2.261], RBS
MgO	5 keV, Ar	0.7	[2.172]
MgO	5 keV, Ar	0.82	[2.181], single crystal (100), (c)
MoO_3	10 keV, Kr	9.6	[2.139]
$MoSi_2$	100–500 eV, Cd	0.85 (500 eV)	[2.117], (a)
NaBr	8 keV, Ar	2.2	[2.198], (a)
NaCl	8 keV, Ar	2.6	[2.198], (a)
NaCl	Ar, 2–10 keV	0.8 (5 keV)	[2.197], (a)
NaCl	5 keV, Ar	2.4	[2.181], single crystal (100), (c)
NaI	8 keV, Ar	1.7	[2.198], (a)
Nb_2O_5	100–600 eV, Ar	0.9 (600 eV)	[2.175], SNMS
Nb_2O_5	2–30 keV, Kr	1.0 (2 keV)	[2.173], (a), (b)
NiSi	20 and 40 keV, Ar	4.6, 4.7 (40 keV)	[2.51, 52], RBS
PbS	5 keV, Ar	2.6	[2.181], single crystal (100), (c)
PtSi	0.9 and 20 and 40 keV, Ar	1, 8, 4.7, 4.5	[2.51, 52, 53], RBS
Pt_2Si	40 keV, Ar	4.1	[2.51], RBS
RbBr	8 keV, Ar	2.6	[2.198], (a)
RbCl	8 keV, Ar	3.2	[2.198], (a)
RbI	8 keV, Ar	2.3	[2.198], (a)

Table 2.4 (continued)

Compound	Bombarding ion and energy	Sputtering yield Y	Reference and remarks
SiC	0.1–8 keV, H, D, He	0.13 (1 keV, He)	[2.71, 260], (a)
SiC	5–15 keV, H, D, Ar	0.3 (10 keV, Ar) 0.025 (10 keV, H)	[2.262], (c)
SiO_2	5–30 keV, Ar	1.8 (20 keV)	[2.182], (c)
SiO_2	0.1–8 keV, H, D, He	0.15 (1 keV, He)	[2.55], (a)
SiO_2	5 keV, Ar	1.05	[2.181], single crystal (1011) (c)
SiO_2	1–3 keV, Ar	1.2 (2 keV)	[2.187], (a)
SiO_2	10 keV, Kr	4.2	[2.139]
SnO_2	10 keV, Kr	15.3	[2.139]
SS 304	0.1–8 keV, H, D, He	0.09 (1 keV, He)	[2.55], (a)
TaC	0.4–8 keV, H, D, He	3×10^{-2} (1 keV, He)	[2.71], (a)
Ta_2O_5	3 keV, Ar	4.2 (3 keV)	[2.193]
Ta_2O_5	1–20 keV, H, D, He	0.075 (10 keV, He)	[2.150], RBS, results similar to pure Ta
Ta_2O_5	0.5–8 keV, H, D, He	4.6×10^{-2} (1 keV, He)	[2.55], (a) (see Fig. 2.14)
Ta_2O_5	100–600 eV, Ar	1.4 (600 eV)	[2.175, 202]
Ta_2O_5	2–30 keV, Kr	1.2 (2 keV)	[2.173], (a), (b)
Ta_2O_5	40 keV, Ar	3.6	[2.51], RBS
TiB_2	100–500 eV, Cd	0.45 (500 eV)	[2.117], (a)
TiC	0.1–8 keV, H, D, He	6.8×10^{-2} (1 keV, He)	[2.71, 260], (a)
TiN	100–500 eV, Cd	0.7 (500 eV)	[2.117], (a)
TiO_2	2–10 keV, Kr, O_2	1.5 (5 keV, Kr)	[2.174], (b)
TiO_2	5 keV, Ar	0.9	[2.181], single crystal (001), (c)
VN	100–500 eV, Cd	0.6 (500 eV)	[2.117], (a)
V_2O_5	10 keV, Kr	12.7	[2.139]
WC	0.1–8 keV, H, D, He	3.7×10^{-2} (1 keV, He)	[2.71], (a)
WO_3	2–30 keV, Kr	4.7 (2 keV)	[2.173], (a), (b)
ZnS	5 keV, Ar	2.6	[2.181], single crystal (110). (c)
ZrB_2	100–500 eV, Cd	0.4 (500 eV)	[2.117], (a)
ZrC	5–17 keV, Ar	1.2 (5 keV)	[2.116], (a)
ZrC	0.1–8 keV, H, D, He	4.5×10^{-2} (1 keV, He)	[2.55], (a)
ZrN	100–500 eV, Cd	0.6 (500 eV)	[2.117], (a)
ZrO_2	10 keV, Kr	2.8	[2.139]

[(a) Weight loss method; (b) interference color changes of thin films; (c) measurement of sputtered volume]

Table 2.5. Changes in surface and near surface composition under ion bombardment for oxides and other compounds with at least one high vapor pressure component

System	Enrichment observed of	Bombarding ion and energy	Analysis method	Reference and remarks
AgBr	Ag	6 keV, Xe^+	EDM [a]	[2.166]
AgBr	Ag	4 keV, O_2^+, N_2^+	e^-DIFF [b]	[2.167], reduction to metallic Ag
AgF	Ag	6 keV, Xe^+	EDM [a]	[2.166]
Ag_2O	Ag	400 eV, Ar^+	XPS	[2.129]
Ag_2O_2	Ag	400 eV, Ar^+	XPS	[2.129]
Al_2O_3	–	400 eV, Ar^+	XPS	[2.129]
Al_2O_3	Al	0.3–4 keV, He^+, Ar^+	ISS, AES	[2.48]
Al_2O_3	–	3 keV, Ar^+	XPS	[2.151 b]
Au_2O_3	Au	400 eV, Ar^+, O_2^+	XPS	[2.129], metallic Au formed (also for O_2^+)
BeO	(Be)	0.3–4 keV, He^+, Ar^+	ISS, AES	[2.48] (minor enrichment of Be)
Bi_2O_3	Bi	5 keV, Ar^+	XPS	[2.264]
$CaSiF_6$	–	1 keV, Ar^+	XPS	[2.291]
CdI_2	Cd	6 keV, Xe^+	EDM [a]	[2.166]
CdO	Cd	400 eV, Ar^+	XPS	[2.129]
$Co(CN)_2$	Co	2 keV, Ar^+	XPS	[2.168], complete reduction to metallic Co
CoF_2	(Co)	2 keV, Ar^+	XPS	[2.168], complete reduction to metallic Co after high fluence ($>10^{19}$ ions cm^{-2})
$CoFe_2O_4$	Co, Fe	1 keV, Ar^+	XPS	[2.265], Fe^{III} reduced to Fe^{II} and Fe^0 (minor); reduction of Co^{II} to Co^0
$CoFe_2O_4$	Co, Fe	3 keV, Ar^+	XPS	[2.266], partly red. of Fe^{III} to Fe^{II} (FeO)
CoO	Co	1 keV, Ar^+	XPS	[2.265, 267], partly red. to Co; oxygen bombardment results in Co_3O_4 formation
Co_2O_3	Co	2 keV, Ar^+	XPS	[2.268]
Co_3O_4	Co	1 keV, Ar^+	XPS	[2.267], reduction to CoO
Co_3O_4	Co	1 keV, Ar^+	XPS	[2.265], mainly CoO formed, at higher fluences also Co
$Co(OH)_2$	Co	1 keV, Ar^+	XPS	[2.265], reduction to CoO, at high fluences also partly to metallic Co
CoS	Co	1.2 keV, Ar	XPS	[2.275], reduction to metallic Co and S
$CoSiF_6$	Co	1 keV, Ar^+	XPS	[2.291], enrichment in metallic Co
Cr_2O_3	–	400 eV, Ar^+	XPS	[2.129]
Cr_2O_3	–	3 keV, Ar^+	XPS	[2.153]
CuCN	Cu	2 keV, Ar^+	XPS	[2.168], complete reduction to metallic Cu
CuCl	–	2 keV, Ar^+	XPS	[2.168, 281]
$CuCl_2$	Cu	2 keV, Ar^+	XPS	[2.168, 281], reduction to CuCl
CuF_2	Cu	1.5 keV, Ar^+	XPS	[2.170], complete reduction to metallic Cu
CuF_2	Cu	2 keV, Ar^+	XPS	[2.168, 281], complete reduction to metallic Cu

Table 2.5 (continued)

System	Enrichment observed of	Bombarding ion and energy	Analysis method	Reference and remarks
CuO	Cu	400 eV, Ar^+	XPS	[2.129]
CuO	Cu	1.5 keV, Ar^+	XPS	[2.170], complete reduction to metallic Cu
Cu_2O	Cu	400 eV, Ar^+	XPS	[2.129]
Cu_2O	(O)	0.5–4 keV, Ar^+, Kr^+, Xe^+	AES	[2.288], weak oxygen enrichment
CuS	Cu	1.2 keV, Ar^+	XPS	[2.275], reduction to metallic Cu and S
$CuSiF_6$	Cu	1 keV, Ar^+	XPS	[2.291], reduction to metallic Cu
FeF_2	Fe	1.5–2 keV, Ar^+	XPS	[2.168, 170], complete reduction to metallic Fe after high bombardment fluence ($>10^{19}$ ions cm^{-2})
FeO	–	3 keV, Ar^+	XPS	[2.266]
FeO	Fe	400 eV, Ar^+	XPS	[2.129]
FeO	(Fe)	1 keV, Ar^+	XPS	[2.265, 269], remains mainly FeO; metallic Fe (minor)
FeOOH	Fe	1 keV, Ar^+	XPS	[2.265, 269], mainly FeO formed, metallic Fe(minor)
Fe_2O_3	Fe	1 keV, Ar^+	XPS	[2.265, 269], FeO(mainly), Fe_3O_4 and metallic Fe(minor) formed
Fe_2O_3	Fe	400 eV, Ar^+	XPS	[2.129]
Fe_2O_3	Fe	1.5 keV, Ar^+	XPS	[2.170], complete reduction to metallic Fe
Fe_2O_3	Fe	3 keV, Ar^+	XPS	[2.266], almost complete reduction to FeO
Fe_3O_4	Fe	1 keV, Ar^+	XPS	[2.265, 269], FeO(mainly) and metallic Fe(minor) formed
Fe_3O_4	Fe	3 keV, Ar^+	XPS	[2.153, 266], partly FeO formed
FeS	Fe	1.5 keV, Ar^+	XPS	[2.281], complete reduction to metallic Fe and S for fluences $>10^{19}$ ions cm^{-2}
FeS_2	Fe	1.5 keV, Ar^+	XPS	[2.281], reduction to metallic Fe and S for fluences $>10^{19}$ ions cm^{-2}
FeS_2	O	10 keV, O^+	e^-DIFF[b]	[2.270], formation of Fe_3O_4
$FeSO_4$	Fe	1.5 keV, Ar^+	XPS	[2.281], reduction to FeO and after high fluences partly to metallic Fe
$FeSiF_6$	Fe	1 keV, Ar^+	XPS	[2.291], reduction to FeF_2
Ga_2O_3	–	1 keV, Ar^+	AES	[2.271], single crystal
HfO_2	Hf	3 keV, Ar^+	XPS	[2.151 b], reduction to lower oxide (HfO) and meatllic Hf
HfO_2	–	10–35 keV, Kr^+	e^-DIFF[b]	[2.136], amorphous HfO_2 film becomes crystalline
IrO_2	Ir	400 eV, Ar^+, O^+	XPS	[2.129], metallic Ir formed (also for O_2^+)
$KMnO_4$	K, Mn	2 keV, Ar^+	XPS	[2.268], Mn_2O_3 and K_2O formed
K_2PtCl_6	K, Pt	10 keV, Ar^+	XPS	[2.273], reduction of K_2PtCl_6 to K_2PtCl_4
K_2PtCl_4	–	10 keV, Ar^+	XPS	[2.273]
K_2SiF_6	–	1 keV, Ar^+	XPS	[2.291]
$LaCoO_3$	La, Co	2 keV, Ar^+	XPS	[2.268]
$LaSr(CoO_3)_2$	La, Sr, Co	2 keV, Ar^+	XPS	[2.268]

LiF	Li	100 keV O^-	e^-DIFF[b]	[2.274], fcc Li precipitates
LiF	Li	10–40 keV, Ar^+, Xe^+	RBS	[2.261], Li precipitates
Metalorganics like $K_3(Fe(CN)_6)$, $Co(NH_3)_3(NO_2)_3$, $Na_3(Co(CN_6))$ and so on				
	metal	2 keV, Ar^+	XPS	[2.268], most of them show reduction to lower oxides
MgO	–	1 keV, Ar^+	AES	[2.17, 152]
MnO_2	Mn	2 keV, Ar^+	XPS	[2.268], Mn_2O_3 formed
MoO_2	–	400 eV, Ar^+	XPS	[2.129]
MoO_3	Mo	10–40 keV, Kr^+	e^-DIFF[b]	[2.130, 132] becomes amorphous at intermediate fluences (10^{14} ions cm^{-2}) and recrystallizes to MoO_2 at higher fluences (10^{17} ions cm^{-2})
MoO_3	Mo	400 eV, Ar^+	XPS	[2.129], MoO_2 formed at high fluences, amorph. MoO_x at intermediate fluences
MoO_3	Mo	1–5 keV, Ar^+	XPS	[2.264], formation of MoO_2
MoS_2	Mo	300 eV, Ar^+	AES	[2.276], metallic Mo islands found
NaCl	Na	20 keV, He^+, Ne^+, Ar^+, Kr^+, Xe^+	EDM[a]	[2.163]
NaI	Na	10–30 keV, He^+, Ne^+, Ar^+, Kr^+, Xe^+	EDM[a]	[2.164, 165]
Na_2SiF_6	–	1 keV, Ar^+	XPS	[2.291]
NbO	–	30 keV, Kr^+	e^-DIFF[b]	[2.139, 278], amorphous NbO becomes polycrystalline NbO
NbO	Nb	0.5–2 keV, Ar^+	XPS	[2.292]
NbO_2	Nb	0.5–2 keV, Ar^+	XPS	[2.292]
Nb_2O_5	Nb	0.5–2 keV, Ar^+	XPS	[2.292], reduction to NbO
Nb_2O_5	Nb	3 keV, Ar^+	XPS	[2.151 b], reduction to lower oxides (NbO_2, NbO)
Nb_2O_5	Nb	35 keV, Kr^+, O^+	e^-DIFF[b]	[2.133, 134], amorphisation at intermediate fluences (10^{14}–10^{15} ions cm^{-2}), crystalline NbO phase formed at high fluence (10^{17} ions cm^{-2}) by random nucleation
Nb_2O_5	Nb	1.7–5 keV, Ar^+	XPS	[2.264]
$Ni(CN)_2$	Ni	2 keV, Ar^+	XPS	[2.168], complete reduction to metallic Ni
NiF_2	Ni	2 keV, Ar^+	XPS	[2.168], complete reduction to metallic Ni at high fluences ($>10^{19}$ ions cm^{-2})
NiO	–	3 keV, Ar^+	XPS	[2.153]
NiO	O	400 eV, O^+	XPS	[2.154], Ni_2O_3 formed
NiO	Ni	400 eV, Ar^+	XPS	[2.129, 154], metallic Ni observed
Ni_2O_3	Ni	400 eV, Ar^+	XPS	[2.129, 154], metallic Ni observed
$Ni(OH)_2$	–	400 eV, Ar^+	XPS	[2.129]
$Ni(OH)_2$	Ni	1 keV, Ar^+	XPS	[2.265], partly NiO formed
$Ni(OH)_2$	Ni	3 keV, Ar^+	XPS	[2.153], partial decomposition to NiO
NiS	Ni	1.2 keV, Ar^+	XPS	[2.275], reduction to metallic Ni and S
$NiSiF_6$	Ni	1 keV, Ar^+	XPS	[2.291], enrichment in metallic Ni

Table 2.5 (continued)

System	Enrichment observed of	Bombarding ion and energy	Analysis method	References and remarks
PbI_2	Pb	6 keV, Xe^+	EDM[a]	[2.166]
PbO	Pb	400 eV, He^+, Ne^+, Ar^+	XPS	[2.151a], reduction to metallic Pb with decreasing degree He, Ne, Ar
PbO	–	400 eV, Kr^+, Xe^+	XPS	[2.151a]
PbO	Pb	400 eV, Ar^+	XPS	[2.129]
PbO_2	Pb	400 eV, Ar^+	XPS	[2.129]
$PdCl_2$	Pd	50 keV, Ne^+	RBS	[2.169], almost complete loss of Cl
PdO	Pd	0.2–1 keV, Ar^+	XPS	[2.129], metallic Pd formed
RbBr	Rb	6 keV, Xe^+	EDM[a]	[2.164]
RbCl	Rb	6 keV, Xe^+	EDM[a]	[2.164]
RbI	Rb	6 keV, Xe^+	EDM[a]	[2.164]
RuO_2	Ru	400 eV, Ar^+	XPS	[2.129], metallic Ru formed
RuO_2	Ru	400 eV, O_2^+	XPS	[2.129], Ru_2O_3 formed
Si_3N_4	Si	0.5–2 keV, He^+	AES	[2.282]
Si_3N_4	N	0.5–2 keV, Ar^+	AES	[2.282]
Si_3N_4	Si	2 keV, Ar^+	AES	[2.277]
SiO_2	O	2 keV, Ar^+	RBS	[2.279]
SiO_2	–	400 eV, Ar^+	XPS	[2.129]
SiO_2	(Si)	3 keV, Ar^+	XPS	[2.151 b], loss of oxygen less than 5%
SiO_2	Si	0.5–2 keV, Ar^+	AES	[2.125]
SiO_2	Si	500 eV, Ar^+	XPS	[2.284]
SiO_2	Si	5 keV, Ar^+	XPS	[2.264]
SnO_2	–	400 eV, Ar^+	XPS	[2.129]
SnO_2	–	30 keV, Kr^+	e^-DIFF[b]	[2.139, 278], remains crystalline SnO_2
$SrMnO_3$	Sr, Mn	2 keV, Ar^+	XPS	[2.268]
Ta_2O_5	Ta	5 keV, Ar^+	XPS	[2.264]
Ta_2O_5	Ta	0.5–5 keV, Ar^+	AES, XPS, ISS	[2.156], Ta enrichment increases with decreasing energy
Ta_2O_5	–	0.5–3 keV, He^+	ISS, XPS	[2.280]
Ta_2O_5	Ta	3 keV, Ar^+	XPS	[2.151 b], reduction to lower oxides (TaO_2, TaO) and metallic Ta
Ta_2O_5	Ta	20–80 keV, Ne^+Ar^+, Kr^+, Xe^+	RBS	[2.51], $Ta_{4.5}O_5$ formed independently on ion type and energy
Ta_2O_5	Ta	20–80 keV, Ar^+	e^-DIFF[b], RBS	[2.131], amorphisation starts at fluences of approx. 5×10^{13} ions cm^{-2}. At higher fluences (10^{16} ions cm^{-2}) crystalline δ-Ta–O randomly nucleates in amorphous Ta_2O_{5-x}, with RBS a surface composition of $Ta_2O_{2.2}$ was found

Ta_2O_5	Ta	1–20 keV, H^+, D^+, He^+, Ne^+	RBS, AES	[2.150], (Fig. 18)
Ta_2O_5	–	400 eV, Ar^+	XPS	[2.129]
Ta_2O_5	Ta	0.3–1.8 keV, He^+, Ar^+	ISS, AES	[2.155], enrichment at surface decreases with ion energy, Ta/O ratio larger for He than for Ar
TiO	Ti	0.5–2 keV, Ar^+	AES	[2.128]
TiO_2	Ti	0.5–2 keV, Ar^+	AES	[2.128]
TiO_2	Ti	30 keV, Kr^+	e^-DIFF[b]	[2.135, 139], Ti_2O_3 formed at high fluence ($>10^{16}$ ions cm^{-2}), amorphous TiO_x at intermediate fluences
TiO_2	Ti	2 keV, Ar^+	AES	[2.257]
TiO_2	Ti	5 keV, Ar^+	XPS	[2.264]
TiO_2	Ti	3 keV, Ar^+	XPS	[2.151 b], reduction to lower oxides (Ti_2O_3, TiO)
Ti_2O_3	–	400 eV, Ar^+	XPS	[2.129]
Ti_2O_3	–	30 keV, Kr^+	e^-DIFF[b]	[2.136, 278], amorphous Ti_2O_3 becomes polycrystalline Ti_2O_3
UO_2	–	30 keV, Kr^+	e^-DIFF[b]	[2.136], amorphous UO_2 becomes polycrystalline UO_2
U_3O_8	U	30 keV, Kr^+	e^-DIFF[b]	[2.136, 138], amorphisation at intermediate fluences (10^{14}–10^{15} ions cm^{-2}), at higher ion fluences polycrystalline UO_2 formed
VO	–	30 keV, Kr^+	e^-DIFF[b]	[2.136], remains polycrystalline VO
V_2O_3	–	40 keV, Kr^+	e^-DIFF[b]	[2.130, 136], remains polycrystalline V_2O_3
V_2O_5	V	40 keV, Kr^+	e^-DIFF[b]	[2.130, 132], amorphisation for intermediate fluences (10^{14} ions cm^{-2}), cryst. V_2O_3 formed at high fluences (10^{17} ions cm^{-2})
WO_2	W	400 eV, Ar^+	XPS	[2.129]
WO_3	W	5 keV, Ar^+	XPS	[2.264]
WO_3	W	400 eV, Ar^+	XPS	[2.129], initially reduction to WO_2, for high fluence further reduction to metallic W
WO_3	W	30 keV, Kr^+	e^-DIFF[b]	[2.136, 139], amorphisation at intermediate fluences (10^{14}–10^{15} ions cm^{-2}), $W_{18}O_{49}$ at high fluences (10^{17} ions cm^{-2})
ZnF_2	–	2 keV, Ar^+	XPS	[2.168]
ZnO	–	400 eV, Ar^+	XPS	[2.129]
ZnS	–	1.2 keV, Ar^+	XPS	[2.275]
$ZnSiF_6$	Zn	1 keV, Ar^+	XPS	[2.291], enrichment in metallic Zn
ZrO_2	Zr	5 keV, Ar^+	XPS	[2.264]
ZrO_2	–	2–35 keV, Kr^+	e^-DIFF[b]	[2.272], amorphous ZrO_2 becomes crystalline ZrO_2
ZrO_2	Zr	3 keV, Ar^+	XPS	[2.151 b], reduction to lower oxide (ZrO) and metallic Zr

[a] Inferred from energy distribution measurements of the sputtered atoms (Sect. 2.4.3c).
[b] Transmission or reflection electron diffraction.

References

2.1 W.R.Grove: Philos. Trans. R. Soc. London **1**, 87 (1852)
2.2 R.Hanau: Phys. Rev. **76**, 155 (1949)
2.3 G.K.Wehner, D.J.Hajicek: J. Appl. Phys. **42**, 1145 (1971)
2.4 P.Blank, K.Wittmaack: J. Appl. Phys. **50**, 1519 (1979)
2.5 O.Almen, G.Bruce: Nucl. Instrum. Methods **11**, 279 (1961)
2.6 H.H.Andersen, H.L.Bay: Radiat. Eff. **13**, 67 (1972)
2.7 H.H.Andersen, H.L.Bay: Radiat. Eff. **19**, 139 (1973)
2.8 T.Asada, K.Quasebarth: Z. Phys. Chem. A**143**, 435 (1929)
2.9 E.Gillam: J. Phys. Chem. Sol. **11**, 55 (1959)
2.10 W.L.Patterson, G.A.Shirn: J. Vac. Sci. Technol. **4**, 343 (1967)
2.11 G.K.Wehner: Phys. Rev. **112**, 1120 (1958)
2.12 T.F. Fisher, C.E.Weber: J. Appl. Phys. **23**, 181 (1952)
2.13 D.Haneman: J. Phys. Chem. Sol. **14**, 162 (1960)
2.14 S.P.Wolsky, E.J.Zdanuk, D.Shooter: Surf. Sci. **1**, 110 (1964)
2.15 A.Güntherschulze, W.V.Tollmien: Z. Phys. **119**, 685 (1942)
2.16 G.S.Anderson: J. Appl. Phys. **40**, 2884 (1969)
2.17 G.K.Wehner: "The Aspects of Sputtering in Surface Analysis Methods" in *Methods of Surface Analysis*, ed. by A.W.Czanderna (Elsevier, Amsterdam, Oxford, New York 1975) Chap. 1, p. 5
2.18 G.K.Wehner, C.E.Kenknight, D.Rosenberg: Planet. Space Sci. **11**, 1257 (1963)
2.19 C.T.Pillinger, L.R.Gardiner, A.J.T.Jull: Earth Planet. Sci. Lett. **33**, 289 (1976)
2.20 R.Behrisch (ed.): *Sputtering by Particle Bombardment III*, Topics Appl. Phys. (Springer, Berlin, Heidelberg, New York 1984) to be published
2.21 A.Benninghoven: Surf. Sci. **53**, 596 (1975)
A.Benninghoven, C.A.Evans, Jr., R.A.Powell, R.Shimizu, H.A.Storms (eds.): *Secondary Ion Mass Spectrometry* SIMS II, Springer Ser. Chem. Phys., Vol. 9 (Springer, Berlin, Heidelberg, New York 1979) Part IV
2.22 M.L.Tarng, G.K.Wehner: J. Appl. Phys. **43**, 2268 (1972)
2.23 J.J.Cuomo, P.Chandhari, R.J.Gambino: J. Electron. Mater. **3**, 517 (1974)
2.24 J.Roth: In Proc. Symp. on Sputtering, ed. by P.Varga, G.Betz, F.P.Viehböck, Perchtoldsdorf/Vienna (1980) p. 773
2.25 P.S.Ho, J.E.Lewis, H.S.Wildman, J.K.Howard: Surf. Sci. **57**, 393 (1976)
2.26 R.Shimizu, N.Saeki: Surf. Sci. **62**, 751 (1977)
2.27 R.Behrisch (ed.): *Sputtering by Particle Bombardment* I, Topics Appl. Phys., Vol. 47 (Springer, Berlin, Heidelberg, New York 1981)
2.28a H.W.Werner, N.Warmoltz: Surf. Sci. **57**, 706 (1976)
2.28b G.Falcone, P.Sigmund: Appl. Phys. **25**, 307 (1981)
2.28c H.H.Andersen: In *Advances in Ion Implantation*, ed. by J.M.Poate, J.S.Williams (Academic Press, New York 1982)
2.29a A.van Oostrom: J. Vac. Sci. Technol. **13**, 224 (1976)
2.29b H.H.Andersen: In *The Physics of Ionized Gases SPIG 1980*, ed. by M.Matic (Boris Kidric Inst. of Nucl. Sciences, Beograd 1980)
2.30 H.F.Winters, J.W.Coburn: Appl. Phys. Lett. **28**, 176 (1976)
2.31 J.W.Coburn: Thin Solid Films **64**, 371 (1979)
2.32 J.W.Coburn: J. Vac. Sci. Technol. **13**, 1037 (1976)
2.33 H.H.Andersen: J. Vac. Sci. Technol. **16**, 770 (1979)
2.34a L.Rivaud, I.D.Ward, A.H.Eltoukhy, J.E.Greene: Surf. Sci. **102**, 610 (1981)
2.34b L.Rivaud, A.H.Eltoukhy, J.E.Greene: Radiat. Wff. **61**, 83 (1982)
2.35 S.T.Picraux: In Site Characterization and Aggregation of Implanted Atoms in Materials, ed. by A.Perez, R.Coussement (Plenum Press, New York 1980) pp. 307, 325
2.36 J.A.Borders: In Site Characterization and Aggregation of Implanted Atoms in Materials, ed. by A.Perez, R.Coussement (Plenum Press, New York 1980) p.295

2.37 H.H. Andersen, F. Besenbacher, P. Goddiksen: In Proc. Symp. on Sputtering, ed. by P. Varga, G. Betz, F.P. Viehböck, Perchtoldsdorf/Vienna (1980) p. 446

2.38 L.A. West: J. Vac. Sci. Technol. **13**, 198 (1976)

2.39 W. Färber, G. Betz, P. Braun: Nucl. Instrum. Methods **132**, 351 (1976)

2.40 G. Betz: Surf. Sci. **92**, 283 (1980)

2.41 H.J. Mathieu, D. Landolt: Surf. Sci. **53**, 228 (1975)

2.42 P.H. Holloway: Surf. Sci. **66**, 479 (1977)

2.43 G. Betz, M. Arias, P. Braun: Nucl. Instrum. Methods **170**, 347 (1980)

2.44 G. Betz, J. Marton, P. Braun: Nucl. Instrum. Methods **168**, 541 (1980)

2.45 P. Sigmund: Phys. Rev. **184**, 383 (1969)

2.46 K. Watanabe, M. Hashiba, T. Yamashina: Surf. Sci. **69**, 721 (1977)

2.47 T. Koshikawa, K. Goto, N. Saeki, R. Shimizu: In Proc. 7th Intern. Vacuum Congress and 3rd Intern. Conference on Solid Surfaces, ed. by R. Dobrozemsky, F. Rüdenauer, F.P. Viehböck, A. Breth, Wien (1977) p. 1489

2.48 E. Taglauer, W. Heiland: In Proc. Symp. on Sputtering, ed. by P. Varga, G. Betz, F.P. Viehböck, Perchtoldsdorf/Vienna (1980) p. 423

2.49a G.C. Nelson: J. Vac. Sci. Technol. **13**, 974 (1976)

2.49b G.C. Nelson, R. Bastasz: J. Vac. Sci. Technol. **20**, 498 (1982)

2.50 R. Bastasz, J. Bohdansky: In Proc. Symp. on Sputtering, ed. by P. Varga, G. Betz, F.P. Viehböck, Perchtoldsdorf/Vienna (1980) p. 430

2.51 Z.L. Liau, W.L. Brown, R. Homer, J.M. Poate: Appl. Phys. Lett. **30**, 626 (1977)

2.52 J.M. Poate, W.L. Brown, R. Homer, W.M. Augustyniak, J.W. Mayer, K.N. Tu, W.F. van der Weg: Nucl. Instrum. Methods **132**, 345 (1976)

2.53 Z.L. Liau, J.W. Mayer, W.L. Brown, J.M. Poate: J. Appl. Phys. **49**, 5295 (1978)

2.54 R.R. Olson, M.E. King, G.K. Wehner: J. Appl. Phys. **50**, 3677 (1979)

2.55 J. Roth, J. Bohdansky, W. Ottenberger: Rpt. IPP 9/26 (1979). Max-Planck-Institut für Plasmaphysik, Garching, FRG

2.56 G. Betz, J. Dudonis, P. Braun: Surf. Sci. **104**, L 185 (1981)

2.57 L.E. Davis, A. Joshi: In Surf. Analysis Techniques for Metallurgical Applications, ed. by R.S. Carbonara, J.R. Cuthill, (ASTM Spec. Techn. Publ. 596, Philadelphia, PA 1976) p. 52

2.58 N.S. McIntyre, F.W. Stanchell: J. Vac. Sci. Technol. **16**, 798 (1979)

2.59 M. Opitz, G. Betz, P. Braun: In Proc. 4th Intern. Conf. on Solid Surfaces and 3rd European Conf. on Surface Science, ed. by D.A. Degras, M. Costa, Cannes (1980) p. 1225

2.60 S.D. Dahlgreen, A.G. Graybeal: J. Appl. Phys. **41**, 3181 (1970)

2.61 K. Goto, T. Koshikawa, K. Ishikawa, R. Shimizu: J. Vac. Sci. Technol. **15**, 1695 (1978)

2.62 M.L. Tarng, G.K. Wehner: J. Appl. Phys. **42**, 2449 (1971)

2.63 M. Opitz: Thesis, University of Technology, Vienna (1979)

2.64 M. Opitz, G. Betz, P. Braun: Acta Phys. Acad. Sci. Hung. **49**, 119 (1980)

2.65a H.F. Winters: J. Vac. Sci. Technol. **20**, 493 (1982)

2.65b P.K. Haff: Appl. Phys. Lett. **31**, 259 (1977)

2.65c B.J. Garrison: Surf. Sci. **114**, 23 (1982)

2.65d P. Sigmund, A. Oliva, G. Falcone: Nucl. Instrum. Methods **194**, 541 (1982)

2.66 N. Andersen, P. Sigmund: Mat.-Fys. Medd. Dan Vid. Selsk. **39**, No. 3 (1974)

2.67 P. Sigmund: J. Vac. Sci. Technol. **17**, 396 (1980)

2.68 R. Kelly: Nucl. Instrum. Methods **149**, 553 (1978)

2.69 H.F. Winters, P. Sigmund: J. Appl. Phys. **45**, 4760 (1974)

2.70 R. Behrisch, G. Maderlechner, B.M.U. Scherzer, M.T. Robinson: Appl. Phys. **18**, 391 (1979)

2.71 J. Roth, J. Bohdansky, A.P. Martinelli: Radiat. Eff. **48**, 213 (1980)

2.72 R. Kelly, J.B. Sanders: Nucl. Instrum. Methods **132**, 335 (1976)

2.73 R. Kelly, J.B. Sanders: Surf. Sci. **57**, 143 (1976)

2.74 S. Dzioba, R. Kelly: J. Nucl. Mater. **76/77**, 175 (1978)

2.75 P. Sigmund: J. Appl. Phys. **50**, 7261 (1979)

2.76 T. Koshikawa, R. Shimizu: Phys. Lett. A**44**, 112 (1973)

2.77 R. Kelly: Surf. Sci. **100**, 85 (1980)

2.78 D.P. Jackson: Radiat. Eff. **18**, 185 (1973)

2.79 M. Szymonski, R. S. Bhattacharya, H. Overeijnder, A. E. de Vries: J. Phys. D **11**, 751 (1978)
2.80 M. Szymonski: Appl. Phys. **23**, 89 (1980)
2.81 H. Oechsner, J. Bartella: In Proc. VII. Conf. Atomic Coll. in Solids, Moscow (1977) (Moscow State University Publishing House 1980) Vol.2, p. 327
2.82 R. A. Swalin: *Thermodynamics of Solids*, 2nd Ed. (Wiley, New York 1972) pp. 128, 141
2.83 Z. L. Liau, T. T. Sheng: Appl. Phys. Lett. **32**, 716 (1978)
2.84 K. Goto, T. Koshikawa, K. Ishikawa, R. Shimizu: In Proc. 7th Intern. Vacuum Congress and 3rd Intern. Conf. on Solid Surfaces, ed. by R. Dobrozemsky, F. Rüdenauer, F. P. Viehböck, A. Breth, Wien (1977) p. 1493
2.85 N. Saeki, R. Shimizu: Surf. Sci. **71**, 479 (1978)
2.86 T. Koshikawa, K. Goto, N. Saeki, R. Shimizu, E. Sugata: Surf. Sci. **79**, 461 (1979)
2.87 G. Betz, M. Opitz, P. Braun: Nucl. Instrum. Methods **182/183**, 63 (1981)
2.88a M. L. Roush, T. D. Andreadis, F. Davarya, O. F. Goktepe: Nucl. Instrum. Methods **191**, 135 (1981)
2.88b M. L. Roush, T. D. Andreadis, F. Davarya, O. F. Goktepe: Appl. Surf. Sci. **11/12**, 235 (1982)
2.88c Th. Wirth, V. Atzrodt, H. Lange: Phys. Status Solidi (A)**72**, K 89 (1982)
2.88d H. Störi, G. Betz, P. Braun: In *Symp. Atomic and Surface Physics 1982*, ed. by W. Lindinger, F. Howorka, T. D. Märk, F. Egger (Inst. Atomphysik, Univ. Innsbruck, Austria 1982)
2.88e P. H. Holloway, R. S. Bhattacharya: J. Vac. Sci. Technol. **20**, 444 (1982)
2.89 P. S. Ho, J. E. Lewis, J. K. Howard: J. Vac. Sci. Technol. **14**, 322 (1977)
2.90 M. Yabumoto, K. Watanabe, T. Yamashina: Surf. Sci. **77**, 615 (1978)
2.91a G. J. Slusser, N. Winograd: Surf. Sci. **84**, 211 (1979)
2.91b P. Varga, E. Taglauer: J. Nucl. Mater. **111/112**, 726 (1982)
2.92 M. Shikita, R. Shimizu: Surf. Sci. **97**, L 363 (1980)
2.93 H. Shimizu, M. Ono, K. Nakayama: J. Appl. Phys. **46**, 460 (1975)
2.94 K. Watanabe, M. Hashiba, T. Yamashina: Surf. Sci. **61**, 483 (1976)
2.95 L. E. Rehn, S. Danyluk, H. Wiedersich: Phys. Rev. Lett. **43**, 1764 (1979)
2.96 W. K. Chu, J. K. Howard, R. F. Lever: J. Appl. Phys. **47**, 4500 (1976)
2.97 P. S. Ho: Surf. Sci. **72**, 253 (1978)
2.98 P. Sigmund, A. Gras-Marti: Nucl. Instrum. Methods **168**, 389 (1980)
2.99 U. Littmark, W. O. Hofer: Nucl. Instrum. Methods **168**, 329 (1980)
2.100 H. W. Pickering: J. Vac. Sci. Technol. **13**, 618 (1976)
2.101 M. Arita, G. R. St. Pierre: Trans. Jpn. Inst. Met. **18**, 545 (1977)
2.102 M. Arita, M. Someno: In Proc. 7th Intern. Vacuum Congress and 3rd Intern. Conf. on Solid Surfaces, ed. by R. Dobrozemsky, F. Rüdenauer, F. P. Viehböck, A. Breth, Wien (1977) p. 2511
2.103 R. Collins: Radiat. Eff. **37**, 13 (1978)
2.104 G. Carter, R. Webb, R. Collins, D. A. Thompson: Radiat. Eff. **40**, 119 (1979)
2.105 R. Webb, G. Carter, R. Collins: Radiat. Eff. **39**, 129 (1978)
2.106 H. H. Brongersma, M. J. Sparnaay, T. M. Buck: Surf. Sci. **71**, 657 (1978)
2.107 A. Crucq, L. Degols, G. Lienard, A. Frennet: Surf. Sci. **80**, 78 (1979)
2.108 R. Kelly: In Proc. Symp. on Sputtering, ed. by P. Varga, G. Betz, F. P. Viehböck, Perchtoldsdorf/Vienna (1980) p. 390
2.109 F. L. Williams, D. Nason: Surf. Sci. **45**, 377 (1974)
2,110a N. Q. Lam, G. K. Leaf, H. Wiedersich: J. Nucl. Mater. **88**, 289 (1980)
2.110b N. Q. Lam, H. Wiedersich: Radiat. Eff. Lett. **67**, 107 (1982)
2.110c L. E. Rehn, V. T. Boccio, H. Wiedersich: Surf. Sci. (to be published)
2.111a N. Q. Lam, G. K. Leaf, H. Wiedersich: J. Nucl. Mater. **85/86**, 1085 (1979)
2.111b H. J. Kang, R. Shimizu, T. Okutani: Surf. Sci. **116**, L 173 (1982)
2.111c R. S. Li, T. Koshikawa, K. Goto: Surf. Sci. **121**, L 561 (1982)
2.111d D. G. Swartzfager, S. B. Ziemecki, M. J. Kelley: J. Vac. Sci. Technol. **19**, 185 (1982)
2.111e M. P. Thomas, B. Ralph: J. Vac. Sci. Technol. **21**, 986 (1982)
2.111f H. H. Andersen, J. Chevallier, V. Chernysh: Nucl. Instrum. Methods **191**, 241 (1981)
2.111g H. H. Andersen, B. Stenum, T. Sorensen, H. J. Whitlow: Nucl. Instrum. Methods (to be published)

2.111h H.H.Andersen, V.Chernysh, B.Stenum, T.Sorensen, H.J.Whitlow: Surf. Sci. Lett. **123**, 39 (1982)
2.112 J.Kirschner, H.W.Etzkorn: In Proc. 7th Intern. Vacuum Congress and 3rd Intern. Conf. on Solid Surfaces, ed. by R.Dobrozemsky, F.Rüdenauer, F.P.Viehböck, A.Breth, Wien (1977) p. 2213
2.113 W.T.Ogar, N.T.Olson, H.P.Smith, Jr.: J. Appl. Phys. **40**, 4997 (1969)
2.114 H.G.Tompkins: J. Vac. Sci. Technol. **16**, 778 (1979)
2.115 W.K.Mayer, G.K.Wehner: J. Vac. Sci. Technol. **16**, 808 (1979)
2.116 B.M.Gurmin, T.Martynenko, Yu.A.Ryzhov: Sov. Phys. – Solid State **10**, 324 (1968) [Fiz. Tverd. Tela **10**, 411 (1968)]
2.117 T.P.Martynenko: Sov. Phys. – Solid State **9**, 2887 (1968) [Fiz. Tverd. Tela **9**, 3655 (1967)]
2.118 J.Roth, J.Bohdansky: J. Nucl. Mater. **103**, 339 (1981)
2.119 G.K.Wehner: Appl. Phys. Lett. **30**, 185 (1977)
2.120 G.K.Wehner, R.R.Olson, M.E.King: In Proc. 7th Intern. Vacuum Congress and 3rd Intern. Conf. on Solid Surfaces, ed. by R.Dobrozemsky, F.Rüdenauer, F.P.Viehböck, A.Breth, Wien (1977) p. 1461
2.121 R.R.Olson, G.K.Wehner: J. Vac. Sci. Technol. **14**, 319 (1977)
2.122 Wen-Yaung Lee: J. Vac. Sci. Technol. **16**, 774 (1979)
2.123 S.P.Linnik, M.A.Buleev, V.E.Yurasova, V.I.Zaporozhchenko, V.S.Chernysh: Radiat. Eff. **52**, 191 (1980)
2.124 V.E.Yurasova, V.I.Shulga, I.G.Bunin, B.M.Mamaev, L.N.Nevzorova, A.S.Petrov: Radiat. Eff. **27**, 173 (1976)
2.125 G.E.McGuire: Surf. Sci. **76**, 130 (1978)
2.126 P.Braun, W.Färber, G.Betz, F.P.Viehböck: Vacuum **27**, 103 (1976)
2.127 T.Smith: Surf. Sci. **55**, 601 (1976)
2.128 H.J.Mathieu, J.B.Mathieu, D.E.Mc.Clure, D.Landolt: J. Vac. Sci. Technol. **14**, 1023 (1977)
2.129 K.S.Kim, W.E.Baitinger, J.W.Amy, N.Winograd: J. Electron Spectrosc. Relat. Phenom. **5**, 351 (1974)
2.130 H.M.Naguib, R.Kelly: J. Phys. Chem. Sol. **33**, 1751 (1972)
2.131 D.K.Murti, R.Kelly, Z.L.Liau, J.M.Poate: Surf. Sci. **81**, 571 (1979)
2.132 H.M.Naguib, R.Kelly: Radiat. Eff. **25**, 79 (1975)
2.133 D.K.Murti, R.Kelly: Surf. Sci. **47**, 282 (1975)
2.134 D.K.Murti, R.Kelly: Thin Solid Films **33**, 149 (1976)
2.135 T.E.Parker, R.Kelly: J. Phys. Chem. Sol. **36**, 377 (1975)
2.136 H.M.Naguib, R.Kelly: Radiat. Eff. **25**, 1 (1975)
2.137 E.C.Baranova, V.M.Gusev, Yu.V.Martynenko, C.V.Starinin, I.B.Haibullin: Radiat. Eff. **18**, 21 (1973)
2.138 Hj.Matzke, J.L.Whitton: Can. J. Phys. **44**, 995 (1966)
2.139 R.Kelly, N.Q.Lam: Radiat. Eff. **19**, 39 (1973)
2.140 R.Bicknell, P.L.F.Hemment, E.C.Bell, J.E.Tansey: Phys. Status Solidi A**12**, K 9 (1972)
2.141 H.M.Naguib, W.A.Grant, G.Carter: Radiat. Eff. **18**, 279 (1973)
2.142 Hj.Matzke, M.Königer: Phys. Status Solidi A**1**, 469 (1970)
2.143 R.R.Hart, H.L.Dunlap, O.J.Marsh: Radiat Eff. **9**, 261 (1971)
2.144 Hj. Matzke: Phys. Status Solidi **18**, 285 (1966)
2.145 Hj.Matzke: Can. J. Phys. **46**, 621 (1968)
2.146 S.S.Batsanov: Russ. Chem. Rev. **37**, 332 (1968)
2.147 R.Kelly: Surf. Sci. **90**, 280 (1979)
2.148 T.J.Chuang, C.R.Brundle, K.Wandelt: J. Vac. Sci. Technol. **16**, 797 (1979)
2.149 K.B.Winterborn: *Ion Implantation Range and Energy Deposition Distributions*, Vol. 2 (IFI/Plenum, New York 1975)
2.150 H.v.Seefeld, R.Behrisch, B.M.U.Scherzer, Ph.Staib, H.Schmidl: In Proc. VII. Intern. Conf. Atomic Coll. in Solids, Moscow (1977) (Moscow State University Publishing House 1980) Vol. 2, p. 327
2.151a K.S.Kim, W.E.Baitinger, N.Winograd: Surf. Sci. **55**, 285 (1976)
2.151b S.Hofmann, J.M.Sanz: J. Trace Microprobe Techn. **1**, 213 (1982/83)

2.152 G.K.Wehner: Jpn. J. Appl. Phys., Suppl. 2, **1**, 495 (1974)
2.153 N.S.McIntyre, D.G.Zetaruk: J. Vac. Sci. Technol. **14**, 181 (1977)
2.154 K.S.Kim, N.Winograd: Surf. Sci. **43**, 625 (1974)
2.155 E.Taglauer, W.Heiland: Appl. Phys. Lett. **33**, 950 (1978)
2.156 P.H.Holloway, G.S.Nelson: J. Vac. Sci. Technol. **16**, 793 (1979)
2.157 M.W.Thompson, R.S.Nelson: Philos. Mag. **7**, 2015 (1962)
2.158 R.S.Nelson: Philos. Mag. **11**, 291 (1965)
2.159 R.Kelly: Radiat. Eff. **32**, 91 (1977)
2.160 C.J.Good-Zamin, M.T.Shehata, D.B.Squires, R.Kelly: Radiat. Eff. **35**, 139 (1978)
2.161 P.Sigmund, C.Claussen: In Proc. Symp. on Sputtering, ed. by P.Varga, G.Betz, F.P.Viehböck, Perchtoldsdorf/Vienna (1980) p. 113
2.162 M.Szymonski: Acta Phys. Pol. **A56**, 289 (1979)
2.163 W.Husinsky, R.Bruckmüller: Surf. Sci. **80**, 637 (1979)
2.164 H.Overeijnder, A.Haring, A.E.de Vries: Radiat. Eff. **37**, 205 (1978)
2.165 W.Husinsky, R.Bruckmüller, P.Blum, F.P.Viehböck, D.Hammer, E.Benes: J. Appl. Phys. **48**, 4734 (1977)
2.166 M.Szymonski, H.Overeijnder, A.E.de Vries: Radiat. Eff. **36**, 189 (1978)
2.167 J.-J.Trillat, N.Terao, L.Tertian: Compt. Rend. (Paris) **242**, 1294 (1956)
2.168 L.Yin, T.Tsang, I.Adler: Geochim. Cosmochim. Acta Suppl. **7**, 891 (1976)
2.169 J.L.Whitton, G.Sørensen, J.S.Williams: Nucl. Instrum. Methods **149**, 743 (1978)
2.170 L.I.Yin, S.Ghose, I.Adler: Appl. Spectrosc. **26**, 355 (1972)
2.171 A.Güntherschulze, H.Betz: Z. Phys. **108**, 780 (1938)
2.172 H.Schirrwitz: Beitr. Plasmaphysik **2**, 188 (1962)
2.173 L.Q.Nghi, R.Kelly: Can. J. Phys. **48**, 137 (1970)
2.174 R.Kelly: Can. J. Phys. **46**, 473 (1968)
2.175 H.Oechsner, H.Schoof, E.Stumpe: In Proc 7th Intern. Vacuum Congress and 3rd Intern. Conf. on Solid Surfaces, ed. by R.Dobrozemsky, F.Rüdenauer, F.P.Viehböck, A.Breth, Wien (1977) p. 1497
2.176 D.Rosenberg, G.K.Wehner: J. Appl. Phys. **33**, 1842 (1962)
2.177 N.Laegreid, G.K.Wehner: J. Appl. Phys. **32**, 365 (1961)
2.178 P.Blank: Thesis, München, (1977)
2.179 Yung Yi Tu, T.J.Chuang, H.F.Winters: In Proc. Symp. on Sputtering, ed. by P.Varga, G.Betz, F.P.Viehböck, Perchtoldsdorf/Vienna (1980) p. 337
2.180 A.Southern, W.R.Willis, M.T.Robinson: J. Appl. Phys. **34**, 153 (1963)
2.182 H.Bach: Nucl. Instrum. Methods **84**, 4 (1970)
2.182 H.Bach, I.Kitzmann, H.Schröder: Radiat. Eff. **21**, 31 (1974)
2.183 G.V.Jorgenson, G.K.Wehner: J. Appl. Phys. **36**, 2672 (1965)
2.184 A.I.Akishin, S.S.Vasil'ev, L.N.Isaev: Izv. Akad. Nauk SSSR, Fiz. **26**, 1356 (1962) [Engl. Transl.: Bull. Acad. Sci. USSR, Phys. Ser. **26**, 1379 (1962)]
2.185 R.L.Hines, R.Wallor: J. Appl. Phys. **32**, 202 (1961)
2.186 R.P.Edwin: J. Phys. D **6**, 833 (1973)
2.187 P.D.Davidse, L.I.Maissel: J. Vac. Sci. Technol. **4**, 33 (1967)
2.188 H.Oechsner: Z. Phys. **261**, 37 (1973)
2.189 O.C.Yonts, C.E.Normand, D.E.Harrison: J. Appl. Phys. **31**, 447 (1960)
2.190 G.Sletten, P.Knudsen: Nucl. Instrum. Methods **102**, 459 (1972)
2.191 M.I.Guseva: Radiotekh. Elektron. **7**, 1680 (1962) [Engl. transl.: Rad. Eng. Electron. Phys. **7**, 1563 (1962)]
2.192 K.H.Krebs: In *Atomic and Molecular Data for Fusion* (IAEA, Wien 1977) p. 185
2.193 H.Bispinck, O.Ganschow, L.Wiedmann, A.Benninghoven: Appl. Phys. **18**, 113 (1979)
2.194 J.Roth, J.Bohdansky, R.S.Blewer, W.Ottenberger: J. Nucl. Mater. **85/86**, 1077 (1979)
2.195 V.K.Koshkin: Aeronauti Institut "Sergev Ordzhonikidze" Report 1979
2.196 E.T.Pitkin: Progr. Astronaut. Rocketry **5**, 195 (1961)
2.197 B.Navinšek: J. Appl. Phys. **36**, 1678 (1965)
2.198 S.V.Barbashev, N.A.Poklonova, T.B.Shashkina: All-Russian Conf., 1978, Minsk, p. 38
2.199 A.G.Campbell III, C.B.Cooper: J. Appl. Phys. **43**, 863 (1972)
2.200 J.W.Coburn, E.Taglauer, E.Kay: Proc. 6th Intern. Vac. Congr. Kyoto, Japan, in Jpn. J. Appl. Phys. Suppl. II, **1**, 501 (1974)

2.201 P.A.Finn, D.M.Gruen, D.L.Page: In *Radiation Effects on Solid Surfaces*, ed. by M.Kaminsky, Advances in Chemistry Series, No. 158 (Am. Chem. Soc. 1976) p. 30
2.202 H.Oechsner, H.Schoof, E.Stumpe: Surf. Sci. **76**, 343 (1978)
2.203 V.E.Henrich, J.C.C.Fan: Surf. Sci. **42**, 139 (1974)
2.204 J.L.Vossen: J. Appl. Phys. **47**, 544 (1976)
2.205 I.H.Wilson: Radiat. Eff. **18**, 95 (1973)
2.206 T.Oohashi, S.Yamanaka: Jpn. J. Appl. Phys. **11**, 1581 (1972)
2.207 G.K.Wehner, P.Yurista, S.Bhatia: J. Appl. Phys. (to be published)
2.208 J.E.Greene, B.R.Natarajan, F.Sequeda-Osorio: J. Appl. Phys. **49**, 417 (1978)
2.209 W.R.Hudson: J. Vac. Sci. Technol. **14**, 286 (1977)
2.210a H.R.Kaufman, R.S.Robinson: J. Vac. Sci. Technol. **16**, 175 (1979)
2.210b S.M.Rossnagel, R.S.Robinson: Radiat. Eff. Lett. **58**, 11 (1981)
2.210c K.Tsunoyama, Y.Ohashi, T.Suzuki, K.Tsuruoka: Jpn. J. Appl. Phys. **13**, 1683 (1974)
2.210d R.S.Robinson, S.M.Rossnagel: J. Vac. Sci. Technol. **21**, 790 (1982)
2.211 M.Bernheim, G.Slodzian: Int. J. Mass. Spectrom. Ion Phys. **12**, 93 (1973)
2.212 W.O.Hofer, P.J.Martin: Appl. Phys. **16**, 271 (1978)
2.213 W.O.Hofer, H.Liebl: Appl. Phys. **8**, 359 (1975)
2.214 H.M.Windawi, J.R.Kratzer, C.B.Cooper: Phys. Lett. **59**A, 62 (1976)
2.215 P.F.Tortorelli, C.J.Altstetter: J. Vac. Sci. Technol. **16**, 804 (1979)
2.216a R.S.Berg, G.J.Kominiak: J. Vac. Sci. Technol. **13**, 403 (1976)
2.216b S.M.Rossnagel, R.S.Robinson: J. Vac. Sci. Technol. **20**, 336 (1982)
2.216c H.G.Craighead, R.F.Howard, P.F.Liao, D.M.Tennant, J.E.Sweeney: Appl. Phys. Lett. **40**, 662 (1982)
2.217 S.D.Dahlgren, E.D.McClanahan: J. Appl. Phys. **43**, 1514 (1972)
2.218 G.Betz, W.Färber, P.Braun: In Proc. 7th Intern. Conf. on Atomic Collisions, Moscow (1977) (Moscow State University Publishing House 1980) Vol. 2, p. 50
2.219 G.Betz, P.Braun, W.Färber: J. Appl. Phys. **48**, 1404 (1977)
2.220 F.Garbassi, G.Parravano: Surf. Sci. **71**, 42 (1978)
2.221 J.R.Arthur, J.J.LePore: J. Vac. Sci. Technol. **14**, 979 (1977)
2.222 P.Braun, G.Betz, M.Arias: To be published
2.223 J.E.Lewis, P.S.Ho: J. Vac. Sci. Technol. **16**, 772 (1979)
2.224 P.S.Ho, J.E.Lewis, W.K.Chu: Surf. Sci. **85**, 19 (1979)
2.225 A.Jablonski, S.H.Overbury, G.A.Somorjai: Surf. Sci. **65**, 578 (1977)
2.226 D.B.Dove, C.G.Pantano, J.B.Andrews: J. Appl. Phys. **48**, 2776 (1977)
2.227 H.Shimizu, M.Ono, K.Nakayama: Surf. Sci. **36**, 817 (1973)
2.228 D.T.Quinto, V.S.Sundaram, W.D.Robertson: Surf. Sci. **28**, 504 (1971)
2.229 K.Nakayama, M.Ono, H.Shimizu: J. Vac. Sci. Technol. **9**, 749 (1972)
2.230 M.Ono, Y.Takasu, K.Nakayama, T.Yamashina: Surf. Sci. **26**, 313 (1971)
2.231 H.H.Brongersma, T.M.Buck: Surf. Sci. **53**, 649 (1975)
2.232 C.R.Helms, K.Y.Yu: J. Vac. Sci. Technol. **12**, 276 (1975)
2.233 T.Yamashina, K.Watanabe, Y.Fukuda, M.Hashiba: Surf. Sci. **50**, 591 (1975)
2.234 M.Ono, H.Shimizu, K.Nakayama: Surf. Sci. **52**, 681 (1975)
2.235 T.Narusawa, T.Satake, S.Komiya: J. Vac. Sci. Technol. **13**, 514 (1976)
2.236 H.Goretzki, A.Mühlratzer, J.Nickl: In Proc. 7th Intern. Vacuum Congress and 3rd Intern. Conf. on Solid Surfaces, eds. by R.Dobrozemsky, F. Rüdenauer, F.P. Viehböck, A.Breth, Wien (1977) p. 2387
2.237 N.Saeki, R.Shimizu: Jpn. J. Appl. Phys. **17**, 59 (1978)
2.238 K.Goto, T.Koshikawa, K.Ishikawa, R.Shimizu: Surf. Sci. **75**, L373 (1978)
2.239 H.Shimizu, M.Ono, N.Koyama, Y.Ishida: Jpn. J. Appl. Phys. **19**, L567 (1980)
2.240 H.Shimizu, N.Koyama, Y.Ishida: Jpn. J. Appl. Phys. **19**, L671 (1980)
2.241 T.Okutani, M.Shikita, R.Shimizu: Surf. Sci. **99**, L410 (1980)
2.242 B.D.Sartwell: J. Appl. Phys. **50**, 7887 (1979)
2.243 R.P.Frankenthal, D.L.Malm: J. Electrochem. Soc. **123**, 186 (1976)
2.244 R.P.Frankenthal, D.E.Thompson: J. Vac. Sci. Technol. **16**, 6 (1979)
2.245 H.J.Mathieu, D.Landolt: Appl. Surf. Sci. **3**, 348 (1979)
2.246 S.Berglund, G.A.Somorjai: J. Chem. Phys. **59**, 5537 (1973)
2.247 W.Y.Lee, M.H.Lee, J.M.Eldridge: J. Vac. Sci. Technol. **15**, 1549 (1978)

2.248 F.J.Kuijers, B.M.Tieman, V.Ponec: Surf. Sci. **75**, 657 (1978)
2.249 S.H.Overbury, G.A.Somorjai: J. Chem. Phys. **66**, 3181 (1977)
2.250 M.Szymonski, A.E. de Vries: Radiat. Eff. **54**, 135 (1981)
2.251 Y.Taga, K.Nakajima: Trans. Jpn. Inst. Met. **18**, 535 (1977)
2.252 R.Bouwman, L.H.Toneman, A.A.Holscher: Surf. Sci. **35**, 8 (1973)
2.253 K.Jacobi, W.Ranke: J. Electron Spectrosc. Relat. Phenom. **8**, 225 (1976)
2.254 I.L.Singer, J.S.Murday, L.R.Cooper: J. Vac. Sci. Technol. **15**, 725 (1978)
2.255 Z.L.Liau, J.W.Mayer: J. Vac. Sci. Technol. **15**, 1629 (1978)
2.256 M.Mohri, K.Watanabe, T.Yamashina: J. Nucl. Mat. **75**, 7 (1978)
2.257 S.Thomas: Surf. Sci. **55**, 754 (1976)
2.258 L.L.Tongson, J.V.Biggers, G.O.Dayton, J.M.Bind, B.E.Knox: J. Vac. Sci. Technol. **15**, 1133 (1978)
2.259 P.N.Ross, Jr., P.Stonehart: J. Catal. **39**, 298 (1975)
2.260 J.Bohdansky, H.L.Bay, W.Ottenberger: J. Nucl. Mater. **76/77**, 163 (1978)
2.261 J.Tardy, J.Pivot, J.P.Dupin, A.Cachard: In Proc. 7th Intern. Vacuum Congress and 3rd Intern. Conf. on Solid Surfaces, eds. by R.Dobrozemsky, F.Rüdenauer, F.P.Viehböck, A.Breth, Wien (1977) p. 1481
2.262 M.Mohri, K.Watanabe, T.Yamashina, H.Doi, K.Hayakawa: J. Nucl. Mater. **75**, 309 (1978)
2.263 R.P.Frankenthal, D.J.Siconolfi: Surf. Sci. **104**, 205 (1981)
2.264 R.Holm, S.Storp: Appl. Phys. **12**, 101 (1977)
2.265 T.J.Chuang, C.R.Brundle, K.Wandelt: Thin Solid Films **53**, 19 (1978)
2.266 N.S.McIntyre, D.G.Zetaruk: Anal. Chem. **49**, 1521 (1977)
2.267 T.J.Chuang, C.R.Brundle, D.W.Rice: Surf. Sci. **59**, 413 (1976)
2.268 Y.Umezawa, C.N.Reilly: Anal. Chem. **50**, 1290 (1978)
2.269 C.R.Brundle, T.J.Chuang, K.Wandelt: Surf. Sci. **68**, 459 (1977)
2.270 J.J.Trillat, K.Mihama: C. R. Acad. Sci. (Paris) **248**, 2827 (1959)
2.271 C.C.Chang, B.Schwartz, S.P.Murarka: J. Electrochem. Soc. **124**, 922 (1977)
2.272 H.M.Naguib, R.Kelly: J. Nucl. Mater. **35**, 293 (1970)
2.273 A.Katrib: J. Electron Spectrosc. Relat. Phenom. **18**, 275 (1980)
2.274 D.B.Carrol, H.K.Birnbaum: J. Appl. Phys. **36**, 2658 (1965)
2.275 G.J.Coyle, T.Tsang, I.Adler, L.Yin: J. Electron Spectrosc. Relat. Phenom. **20**, 169 (1980)
2.276 H.C.Feng, J.M.Chen: J. Phys. C **7**, L75 (1974)
2.277 S.Thomas, R.J.Mattox: J. Electrochem. Soc. **124**, 1942 (1977)
2.278 R.Kelly, N.Q.Lam, D.K.Murti, H.M.Naguib, T.E.Parker: In Proc. VI. Yugoslav Symp. and Summer School on the Physics of Ionized Gases (1972), ed. by M.V.Kurepa (Institute of Physics, Beograd 1972) p. 349
2.279 A.Turos, W.F. van der Weg, D.Sigurd, J.W.Mayer: J. Appl. Phys. **45**, 2777 (1974)
2.280 G.C.Nelson: J. Vac. Sci. Technol. **15**, 702 (1978)
2.281 T.Tsang, G.J.Coyle, I.Adler, L.Yin: J. Electron Spectrosc. Relat. Phenom. **16**, 389 (1979)
2.282 R.S.Bhattacharya, P.H.Holloway: Appl. Phys. Lett. **38**, 545 (1981)
2.283 K.Wittmaack, P.Blank: Nucl. Instrum. Methods **170**, 331 (1980)
2.284 E.Görlich, J.Haber, A.Stoch, J.Stoch: J. Solid State Chem. **33**, 121 (1980)
2.285 J.P.Biersack, E.Santner: Nucl. Instrum. Methods **132**, 229 (1976)
2.286 R.P.Frankenthal, D.J.Siconolfi: Surf. Sci. **111**, 317 (1981)
2.287 K.A.Gschneidner, Jr.: Solid. State Phys. **16**, 344 (1964)
2.288 J.Herion, G.Scharl, M.Tapiero: To be published
2.289 R.P.Frankenthal, D.J.Siconolfi: J. Vac. Sci. Technol. **20**, 515 (1982)
2.290 S.Hofmann, J.H.Thomas: J. Vac. Sci. Technol. **B1**, 43 (1983)
2.291 G.J.Coyle, T.Tsang, I.Adler, L.Yin: Surf. Sci. **112**, 197 (181)
2.292 P.C.Karulkar: J. Vac. Sci. Technol. **18**, 169 (1981)
2.293 M.Sundaraman, S.K.Sharma, L.Kumar, R.Krishnan: Nucl. Instrum. Methods **191**, 289 (1981)
2.294 F.Toyokawa, K.Furuya, T.Kikuchi: Surf. Sci. **110**, 329 (1981)

3. Chemical Sputtering

Joachim Roth

With 30 Figures

During bombardment of a solid with energetic ions, projectile atoms generally accumulate in a surface layer. For projectiles other than noble gas ions, the altered chemical composition of the surface layers leads to changes in the sputtering yield.

Most often a reduction of the sputtering yield for the target atoms is observed due to the dissolution of target atoms in the surface layer. Changes of the surface binding energy as well as the development of the collision cascades in the altered layer may lead both to increased or decreased physical sputtering yields. If molecules formed between projectile ions and target atoms have binding energies at the surface low enough to desorb at the temperature of investigation, chemical sputtering occurs and enhanced yields are observed.

After a short review of experimental requirements to determine chemical effects in sputtering, the accumulation of chemically reactive ions in solids is described. Examples for changes in the physical sputtering yield due to the chemical interaction of projectiles and target atoms are given. Chemical sputtering is primarily characterised by strong temperature variations of the sputtering yield and a number of experimental observations indicative for chemical sputtering is given. The atomic processes leading to chemical sputtering consist of several steps such as implantation of reactive ions, molecule formation and molecule desorption. For the case of interaction of hydrogen with carbon, some atomic models for chemical sputtering are discussed.

3.1 Identification of Chemical Effects

Bombardment of a solid surface with other than noble gas ions generally leads to a chemical reaction between the incident ions and the atoms of the solid. Even before sputtering was identified as the removal of surface atoms from a cathode due to the impact of energetic ions from a gas discharge, chemical reactions on the electrodes of a gas discharge tube such as oxidation in an oxygen discharge and reduction in a hydrogen discharge had been described [3.1].

At the time, when the basic mechanisms of physical sputtering were under active discussion, *Kohlschütter* [3.2] proposed in 1912 a process for the erosion of the cathode in a gas discharge where volatile radicals are formed in the inter-

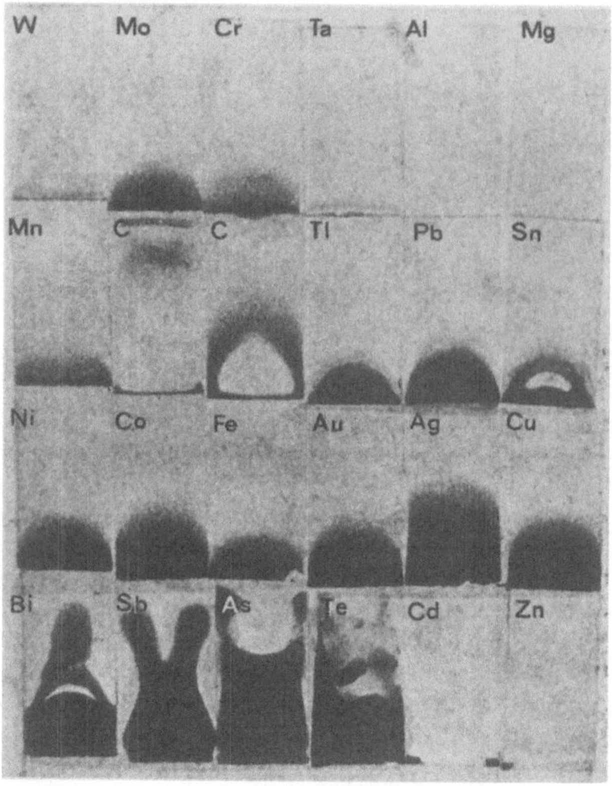

Fig. 3.1. Deposition of material sputtered from different metal cathode plates in a hydrogen discharge on a glass plate perpendicular to the cathode surface. The distributions for C, Bi, Sb, As, and Te are attributed to cracking of volatile hydrides by electron impact all over the discharge, whereas for the other metals the deposited amount decreases with increasing distance from the cathode

action of the bombarding ions with the target atoms. This model was, however, soon abandoned as cathode erosion was also found in noble gas discharges. The observation of deposition patterns of sputtered material for some combination of discharge gas and cathode materials, such as H^+ on C, Bi, Sb, As, and Te which were very different from the ones found for nonreactive materials (Fig. 3.1), again led *Güntherschulze* (1926) [3.3] to the conclusion that chemical reactions must have contributed to the erosion. Since then the expression "chemical sputtering" has been used in the literature for all the different aspects of the chemical interactions of the projectile and target atoms during sputtering. Though sputtering has meanwhile been largely investigated and has found important technical applications, the chemical contributions to the sputtering process are still little investigated and no attempt has been made to give a clear identification and definition for these processes.

Sputtering has been defined by *Sigmund* in [Ref. 3.4, Chap. 2] as the erosion at the surface of a solid as a consequence of energetic particle bombardment that is observable in the limit of small particle currents and fluences. This means that processes appearing due to ion bombardment such as macroscopic beam heating causing evaporation as well as erosion due structural modifications of a surface layer (which require a threshold fluence like blistering and flaking) are not included in the sputtering process. But, in principle, the definition includes the chemical effects in sputtering because it does not depend on any details of the atomistic process like collision cascades, and chemical interactions between energetic ions and surface atoms of a solid may also take place in the limit of low fluences. However, by increasing the fluence, ions generally accumulate in the surface layer and the influence of chemical interactions on the sputtering process becomes more pronounced.

3.1.1 Formation of an Altered Surface Layer

Energetic ions bombarding a solid partly penetrate the surface. They are slowed down and may come to rest in the surface layer of the material. Rare gas ions can generally be trapped only at damage sites in the target material or precipitate as separate phase like gas bubbles. The enthalpy of solution is so small and the binding energy to defects so large that the interstitial solution of rare gases has not been observed (Chap. 7). In addition to these processes, reactive ions can be dissolved or may form a compound phase in the target material [3.5, 6]. At high fluences, generally an equilibrium concentration of trapped ions will be reached which is determined by a balance among the bombarding ion flux, the diffusion of the implanted ions into the bulk and back to the surface, and surface removal due to sputtering [3.7–9]. Here, preferential sputtering of the resulting compound surface layer, i.e., sputtering of the different components not proportional to their surface concentrations, must be taken into account (Chap. 2). Finally, surface diffusion or segregation of implanted ions and target atoms may lead to the formation of surface molecules with different binding energies to the surface.

The development of an altered surface layer has also been found for bombardment by nonreactive ions in a reactive gas atmosphere. If the number of reactive atoms arriving at the surface is comparable or larger than the number of atoms removed by sputtering, a compound layer can be formed which was found to spread over a thickness equivalent to the range of the bombarding ions. This may be formed by recoil implantation and cascade mixing [3.10–12] or diffusion of the atoms originally absorbed on the surface. For sputtering of a solid with a reactive ion species in a reactive residual gas, two-fold chemical interactions and synergistic effects may occur.

The build-up of the altered surface layer generally leads to a different sputtering behaviour than for the original surface. The sputtering yield can be decreased or increased and the composition and the distributions of the sputtered species can be different. The effects leading to these changes may be divided into two groups:

i) Effects due to the presence of trapped ions: The implanted and chemically bound atoms at the surface will also be sputtered. They will take up momentum and energy from the cascade and thus decrease the sputtering yield of the original target atoms. If implanted atoms segregate or build up on the surface, the sputtering yield for atoms of the original solid may even decrease to zero [3.13].

Trapped ions in the altered surface layer will modify the spread of the collision cascade, especially if the mass of the trapped ions is very different from the mass of the target atoms.

ii) Effects due to changes of the binding energies: The compound formed in the surface layer and on the surface will generally lead to a binding energy of the surface which is different from the original solid. A lower binding energy will result in an increase, a higher binding energy, in a decrease of the sputtering yield. Compounds with low binding energies equivalent to high vapor pressures may evaporate due to spike effects in dense collision cascades [3.14]. For sufficiently low binding energies, thermal desorption of compound molecules can lead to additional erosion. This thermal release will be best observable if the yields from collisional effects are low

3.1.2 Definition

Traditionally, the separation between physical and chemical sputtering has been the ejection mechanism of the particles.

In *physical sputtering*, also named knock-on sputtering [3.4], the sputtered particles receive sufficient energy to overcome the surface binding by atomic collisions from the incident particle. This sputtering process may be altered by implantation with subsequent changes of the binding energy, but as long as the atoms are ejected via a collision cascade or a local spike, the process will be called *chemically enhanced* or *chemically reduced physical sputtering*. These effects have occasionally been called physichemical sputtering [3.15, 16].

In *chemical sputtering*, molecules are formed on the surface due to a chemical reaction between the incident particles and the target atoms which have a binding energy low enough to desorb at the temperature of the solid under investigation. Molecules or atoms with low binding energy or in antibonding states at the surface can also be produced for a short time by electronic excitation during bombardment with ions as well as with electrons or photons. This excitation process, which especially in insulators can last long enough to lead to erosion, has also recently been named chemical sputtering [3.17]. But as no compound formation between incident ions and target atoms occurs like in the case of high-temperature erosion of graphite by He and Ar

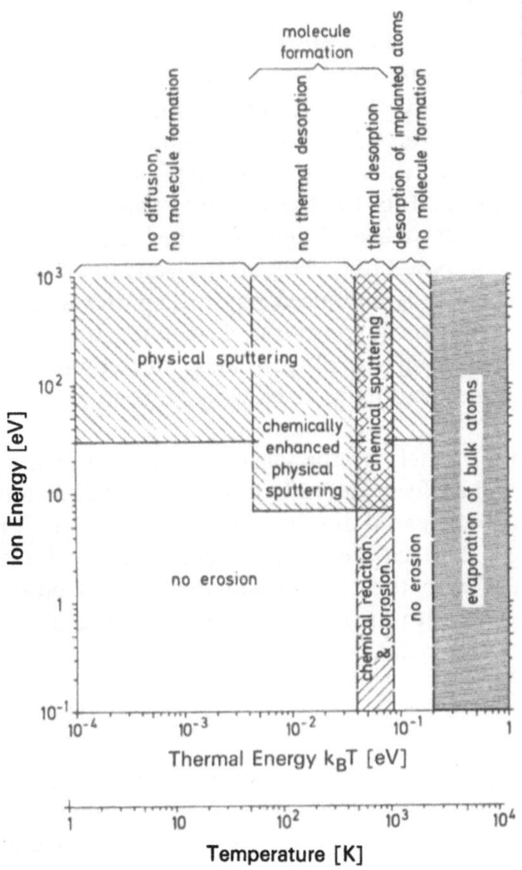

Fig. 3.2. Limits in ion energy and surface temperature for erosion processes like physical sputtering, chemical sputtering and evaporation

bombardment [3.144, 145], this phenomenon may rather be called radiation-enhanced evaporation. Some of these effects are treated in Chap. 4.

The border line between chemical and physical sputtering is not sharp. Compound surface layers are formed in most ion-solid interactions and the formation of compound molecules are often observed in sputtering. For small binding energies, the distinction whether it is the kinetic energy from the incident particle leading to the ejection or just the surface temperature of the target, possibly increased by the heat of formation of the compound which leads to thermal desorption, is often difficult.

In Fig. 3.2, an attempt is made to schematically indicate the conditions for the particle energy E and the surface temperature T for the different sputtering and erosion processes.

The limits which divide the parameter ranges are not sharp and are determined by the binding energy of surface atoms or molecules formed in chemical reactions. Surface atoms or molecules to be released must receive these energies either from the bombarding ion or from thermal energy. In the latter case a temperature equivalent to about 1/30 of the binding energy will lead to measurable molecule desorption or bulk atom evaporation, whereas in the former case, the threshold energy for physical sputtering can be estimated to about four times the surface binding energy [3.18]. Additional limits may occur at temperatures, firstly, where the onset of surface diffusion enables the formation of molecules and secondly, where the implanted ions become thermally desorbed. In the example in Fig. 3.2, the absolute values for the sputtering of carbon by hydrogen ions have been taken. Here the lower binding energy of hydrocarbon molecules compared to carbon atoms leads to a lowering of the threshold energies, to chemically enhanced physical sputtering and at sufficiently high temperatures to chemical sputtering. For other ion target combinations, e.g., aluminium sputtering by oxygen ions, the compound molecules have higher binding energies leading to chemically reduced physical sputtering.

All temperature limits depend ultimately on the sensitivity of the measurement. They may be influenced by spike effects and radiation enhanced diffusion as well as by the energy which may be released during molecule formation. For the considerations shown in Fig. 3.2 it has been assumed that the volatile molecule is in thermal equilibrium with the surface and the recombination energy is not immediately transferred into kinetic motion of the molecule; a point which will be discussed later (Sect. 3.5).

3.1.3 Summary

The first experimental indication of chemical sputtering will arise from an observation of the total sputtering yield, which is higher than for other ion-target combinations of similar mass. The sputtering yield should show a strong temperature dependence and no sharp threshold energy.

Following the above definition the occurrence of chemical sputtering may then be determined by two experimental observations

— The material is removed as compound molecules. This can be observed by mass analysis of the sputtered species.
— The compound molecules are released by thermal desorption. This can be investigated by an energy analysis of the sputtered molecules.

Only in recent years have adequate techniques been developed for observation of these requirements. Section 3.2 contains a short description of some possible *experimental set-ups.*

The *formation of a compound* between incident particles and target atoms requires the slowing down of the incident particles in the target material to energies low enough and the trapping for times long enough for a chemical bond to occur. The field of chemical trapping of the injected particles in materials and the formation of an altered layer has been the topic of recent reviews [3.19–21] and will shortly be presented in Sect. 3.3.

Section 3.4 will emphasize that chemical effects occur for most ion-target interactions, mostly as *chemically reduced* or *chemically enhanced physical sputtering*. Even for self-ion bombardment or bombardment with inert gas ions, chemical effects may occur with reactive atoms from the residual gas.

Examples of *chemical sputtering* will be presented in Sect. 3.5. The complexity of the chemical sputtering mechanism will be demonstrated by the example of hydrocarbon formation during the bombardment of graphite with hydrogen.

Apart from the topics mentioned, several further aspects of chemical processes between impinging energetic atoms and the solid surface have been investigated and described elsewhere, e.g., surface catalysis [3.22] and plasma chemistry [3.23].

3.2 Experimental Methods

Experimental methods investigating chemical effects during sputtering must also consider the fate of the injected reactive ions in contrast to inert gas sputtering. Their fate must be considered in the determination of the sputtering yield and requires the mass and energy analysis of the molecular composition of sputtered particles and the determination of the surface composition of the target. As reactive atoms from the residual gas may also lead to chemical surface modifications, UHV conditions are a basic requirement for measurements of chemical effects in sputtering.

3.2.1 Measurements of the Partial Sputtering Yield

In general, the sputtering yield is determined using the equation

$$Y=\frac{(\text{signal})\cdot(\text{sensitivity factor})}{(\text{ion dose})}. \quad (3.1)$$

If the signal only refers to one sputtered species, one obtains the partial sputtering yield Y_i, while for all species the total sputtering yield Y is obtained. In order to understand the chemical effects in reactive ion sputtering, the partial sputtering yields for the target atoms Y_T, for the implanted ions Y_I and for all possible compound molecules $Y_{\mathrm{T}_n\mathrm{I}_m}$ must be determined. This requires that

i) the observed signal must be selective to detect the different sputtered particles, for example, by mass analysis or characteristic optical lines;

ii) the sensitivity factor being, for example, the ionization efficiency and transmission of a mass analyzer or the oscillator strength for optical lines, must be known or calibrated for each sputtered species;

iii) the incident beam of reactive ions should be mass analysed. The ion dose can be measured within $\pm 2\%$ from the collected charge. For beams of neutral reactive atoms, possibly containing both atomic and molecular species, the arrival rate generally has to be measured separately by special techniques [3.24, 25]. The error introduced by the dose measurements may in this case be large.

By observing all sputtered species T_nI_m, the total sputtering yield may then be divided into partial sputtering yields for target atoms T and implanted ions I

$$\left.\begin{aligned} Y_{\mathrm{T}} &= \sum_{n,m} n\cdot Y_{\mathrm{T}_n\mathrm{I}_m} \\ &\text{and} \\ Y_{\mathrm{I}} &= \sum_{n,m} m\cdot Y_{\mathrm{T}_n\mathrm{I}_m}, \end{aligned}\right\} \quad \text{for} \quad n,m=0,1,2\ldots, \tag{3.2}$$

where n is the number of target atoms and m the number of reactive ions per molecule.

The partial sputtering yield for implanted ions Y_{I} is generally less than one for low fluences and increases with the concentration of implanted ions. If Y_{I} stays below 1, a build-up of surface layers from the bombarding ions occurs. Steady-state sputtering is only possible for $Y_{\mathrm{I}}=1$.

a) Ionization and Mass Analysis

As the sputtered atoms and molecules are predominantly neutral, one way of detection is ionization and mass analysis in a quadrupole mass analyser. However, the ionization probability of energetic sputtered particles is generally low and decreases with increasing energy. Therefore, in some of the first experimental apparatus, the sputtered particles were allowed to thermalise in many collisions with the wall of the vacuum chamber before ionisation in an electron impact ioniser. The gas analysis is done inside a relatively small volume around the target and allows fast scans over the interesting parts of the mass spectrum.

Such apparatus has been used to investigate the formation of hydrocarbons in bombarding graphite or carbides with hydrogen ions [3.26]. This set-up has the disadvantage that during collisions with the vacuum wall all information about the state as well as the energy and angular distributions of the sputtered particles is lost.

In a later and more refined set-up (Fig. 3.3 [3.27]), the analyzing quadrupole mass spectrometer is directed toward the beam spot. This allows analysis of the reaction products without any collision with inner surfaces. To increase the sensitivity, the impinging particle beam is mechanically chopped and the mass spectrometer signal is processed by a lock-in amplifier. In addition, the amplitude and the phase shift of the modulated signal allows the determination of the reaction probability as well as the reaction time constants for the selected reaction products. In the system, shown in Fig. 3.3, the target can be heated by an electron beam and the temperature is controlled by optical and infrared pyrometers.

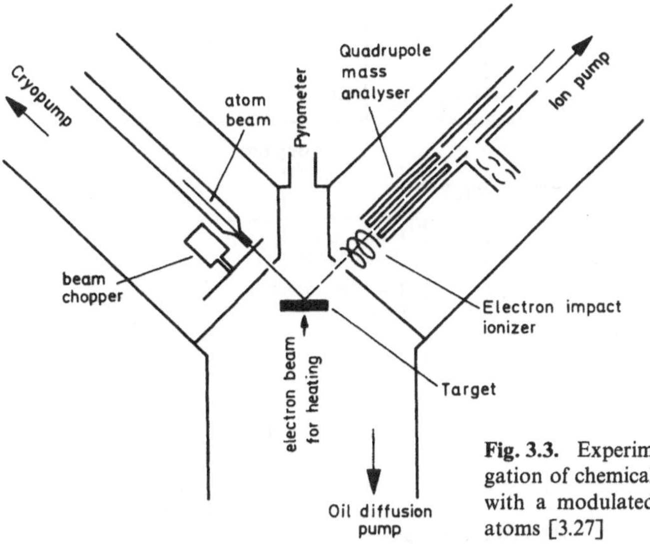

Fig. 3.3. Experimental set-up for investigation of chemical reactions of solid surfaces with a modulated atomic beam of reactive atoms [3.27]

Remaining disadvantages are the very low ionisation probability and the modification of sputtered species upon ionisation. During ionisation, molecular species may be cracked and only identified by their cracking products. Further energetic sputtered atoms have a much lower detection probability than thermally desorbed molecules. In order to obtain quantitative results, the apparatus should be calibrated separately for the possible reaction products, e.g., using variable leaks of known conductance.

Another possible set-up [3.28, 29] makes use of a low pressure plasma discharge (Fig. 3.4). The plasma serves not only as an ion source for sputtering, but also to ionize the neutral sputtered particle. The ionisation process of the sputtered neutral species is different in different pressure regimes in the discharge.

At relatively high pressures [3.28], the Penning-ionisation dominates, i.e., ionisation due to a collision with an excited atom of the discharge. At low pressures [3.29], the neutrals are predominantly ionised by collisions with electrons. These differences in the ionisation processes must be taken into account for the interpretation of the data. After ionisation, the sputtered particles are energy analysed and finally mass analysed in a quadrupole mass analyser. The neutral sputtered particles can, in principle, also be ionised by the electron impact ioniser behind the sampling orifice. However, the spectrum of ions extracted from the plasma is far richer in observable species relative to the postionised neutrals and has provided more insight into the discharge chemistry [3.28]. Energy analysis was included to separate sputtered neutrals ionised in the discharge from negative sputtered ions, which are accelerated in the Langmuir-sheath in front of the target and from neutrals ionised in the sheath in front of the sampling orifice. The total sputtering yield is measured by

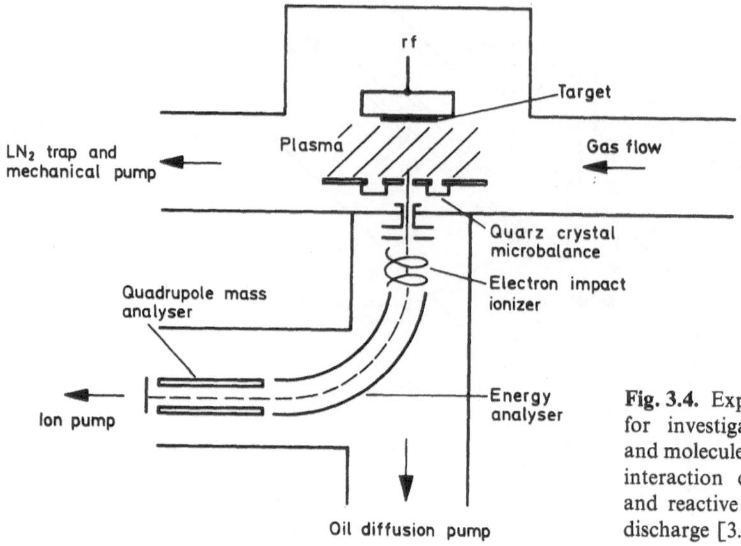

Fig. 3.4. Experimental set-up for investigations of atoms and molecules sputtered in the interaction of solid surfaces and reactive ions from a gas discharge [3.28]

condensation of all sputtered species on a quarz crystal microbalance in this set-up.

A disadvantage of this technique is that charged sputtered species are significantly influenced by the sheath potential in front of the target and substrate plane. This might be of importance for the sputtering of alkali compounds which results in predominantly charged species.

b) Matrix Isolation

A different approach to determine the sputtered species in sputtering experiments is matrix isolation which has been successfully applied by *Gruen* et al. [3.30, 31]. The ion bombarded target is surrounded by a liquid nitrogen and a liquid helium cooled shield. Sputtered species like ions and neutral atoms, stable molecules and unstable chemical radicals are continuously condensed on the potassium chloride plate, which is held at 14 K together with an inert matrix gas. The matrix flow is adjusted to yield a 10^3–10^4 : 1 ratio of matrix atoms to sputtered species on the plate to separate sputtered atoms or molecules from one another and prevent reactions in the matrix. After sputtering the refrigerated matrix assembly is spectroscopically analysed in situ through the Dewar windows. Absorbance versus frequency measurements are performed, showing characteristic absorption lines from the sputtered species. The advantage of this measuring technique is that besides the ions, also neutral atoms and molecules and even unstable radicals can be detected in the trapping matrix.

The matrix isolation technique has been successfully applied in sputtering of Au and Mo [3.30]. It has allowed identification of metal oxides and nitrides in sputtering experiments [3.32–36] and gives information about the thermodynamics of the sputtering of binary targets [3.34, 35]. For absolute measurements, the collection efficiency on the matrix and the oscillator strength for the absorption transition for the investigated species in the matrix must be known. In the application to chemical sputtering only little is known about the oscillator strengths for molecular sputtering products. So far, only relative measurements have been performed which compared sputtering with inert and reactive ions, but the method is still awaiting its application to quantitative sputtering measurements.

3.2.2 Surface Analysis Combined with Mass Spectrometry

A variety of measurements have been performed to characterise the chemical composition of the target surface after ion bombardment. The techniques are listed in Table 3.1 together with the extracted information on the implanted surface layer.

In most earlier measurements, these techniques were not applied in situ in the sputtering chamber. The in situ analysis, however, is essential for identification of chemical surface reactions. Recently, several systems especially designed for ion-surface interaction measurements in chemical sputtering have been used. One example is shown in Fig. 3.5 which was applied to investigate the reaction occurring in reactive sputtering of Si with CF_4 [3.42]. The target material, in this case Si, is evaporated on a quartz crystal microbalance which was mounted on a rotatable sample holder. The reaction products were

Table 3.1

Technique	Analysed depth	Extracted information	Ref.
Colour change	5000 Å	Thickness of implanted layer	[3.37]
Ion beam analysis: Rutherford-backscattering, Nucl. reaction techniques	100–5000 Å	Depth distribution of average composition	[3.38] [3.39]
Electron diffraction:		Phase formation	[3.6]
RHEED	1000 Å–1 μm	Lattice structure	
LEED	3 Å	Surface structure	[3.42]
Laser induced Raman spectroscopy	1000–2000 Å	Phonon modes Lattice structure localized chemical bonds	[3.40]
Infrared absorption spectroscopy	Transmission	Localised chemical bonds	[3.41]
Auger spectroscopy	3 Å	Chemical composition Chemical state from line shifts	[3.42]
Desorption spectroscopy	–	Binding energies	[3.43]

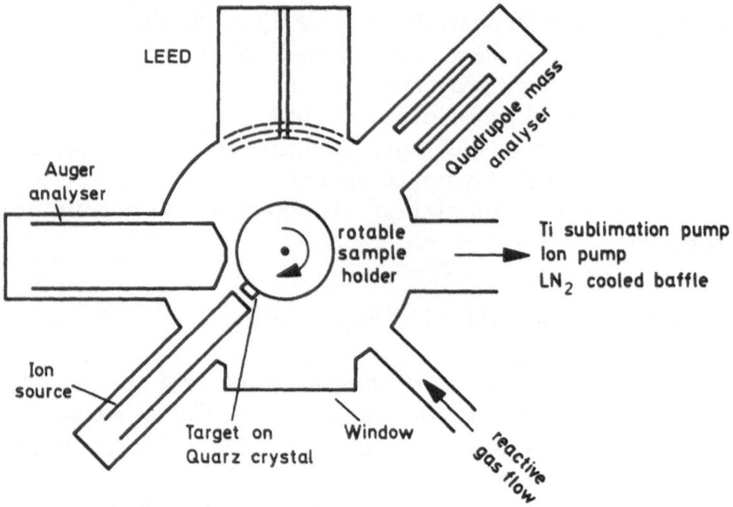

Fig. 3.5. Experimental set-up for investigations of chemical reaction products and the surface composition during the interaction of reactive atoms and ions with solid surfaces [3.42]

analysed by an off-sight quadrupole mass spectrometer. For surface composition measurements, the sample could be moved into an Auger electron spectrometer or, in a similar set-up, into an x-ray photoelectron spectrometer. By this means, the dependence of the reaction yield on the different bonding conditions of C, F, and Si could be evaluated.

In a similar instrument [3.44], an additional retarding dispersive energy analyser coupled to the quadrupole mass analyser allowed the extraction of the energy distribution of selected charged sputtered species during sputtering of different metals with Ar and O_2.

3.2.3 Measurement of the Energy Distribution of Sputtered Particles

The energy distribution of the sputtered particles which are mostly neutral can give valuable information about the ejection mechanism. Ions and postionised neutrals can be analysed with an apparatus as presented in Fig. 3.4. The energy of ions leaving the plasma may, however, additionally depend on the position, where the neutral atom has been ionized. Therefore, the energy analysis is used in this set-up only to separate directly sputtered ions from postionised sputtered neutrals. The energy distribution of the sputtered postionised species has also been measured from the current on an ion collector opposite the target using a retarding field [3.45, 46]. With this set-up, however, a mass analysis of sputtered particles is not possible.

The best experimental results for the energy distribution of sputtered particles have been obtained by time-of-flight methods [3.47]. One possible experimental set-up is shown in Fig. 3.6 [3.48].

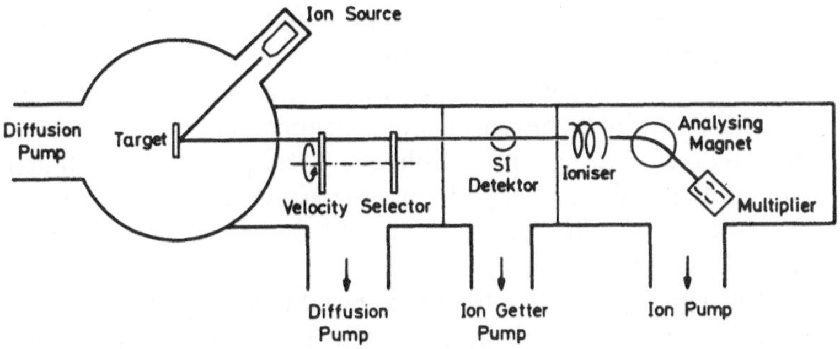

Fig. 3.6. Experimental set-up for the velocity analysis of sputtered atoms and molecules [3.48]. SI stands for surface ionization

The time-of-flight technique is especially suited for the investigation of alkali metals and compounds where the efficient surface ionisation technique can be used for detection (Sect. 3.5.1d).

For atoms other than alkali metal atoms, the detection efficiency is much lower. To improve the sensitivity and the signal-to-noise ratio, a correlation technique can be used in which a cross correlation is performed between the detected signal and the randomly pulsed initial particle beam [3.49]. This method has been used to investigate the chemical sputtering of Si under Ar bombardment and simultaneous exposure to XeF_2 gas [3.50].

Another very promising method for energy analysis of sputtered neutral particles is based on measurements of the Doppler shift of laser resonance excitation [3.51, 52]. For the case of Ar sputtering of Ti in an oxygen atmosphere, changes in the energy distribution for Ti atoms have been observed for different degrees of surface oxidation [3.53].

3.3 Trapping and Compound Formation of Implanted Ions

The first stage in the possible chemical interaction of the sputtering ions and the target material is that part of the ions are implanted and slowed down in the material. At temperatures low enough to suppress diffusion, the implanted atoms accumulate in the implanted layer until saturation occurs. This is determined by surface removal due to sputtering or by radiation induced and/or thermal out-diffusion through the surface (Chap. 7). At higher temperatures, diffusion into the bulk and through the surface alters the implantation profile.

The final physical state of implanted material depends on the interplay of the gradual addition of impurity atoms and radiation damage resulting from atomic collisions. The implanted atoms may form an interstitial solution within the target material or may be retained at substitutional or damage positions.

The formation of chemical bonds between the ions and the target atoms is frequently observed. If the concentration for solubility is exceeded, domains of well-defined stoichiometric compound phases may form surrounded by the bulk material [3.5, 6, 54]. At high concentrations of trapped inert gas ions, they may agglomerate and precipitate as separate phases. These processes combined with the radiation damage will generally change the structure of the material. Such changes in surface structure and topography due to implantation of gaseous ions are dealt with in Chaps. 6, 7.

Virtually any atom can be injected into any material to large concentrations by ion implantation. At low temperatures, metastable alloys may form [3.55, 56] which are not accessible by conventional thermal methods. Upon annealing the stoichiometry of the alloys approaches phases expected from equilibrium phase diagrams [3.5], i.e., chemical implantation has taken place. These effects concerned with metallurgy by ion implantation are closely related to sputtering by reactive ions. In this section we will concentrate on H, O, N, C, F implantation where most investigations on chemical sputtering have been performed. The solid-state chemistry of implanted atoms in general has recently been treated in several reviews [3.19, 57].

3.3.1 Chemical Trapping of Hydrogen

Trapping, re-emission and thermal desorption of hydrogen implanted into several materials has been investigated by different techniques [3.58, 59], especially in respect to fuel recycling problems in fusion reactors [3.60].

a) Metals

For several metals, hydrogen has a positive heat of solution and phase diagrams predict the formation of hydrides in certain temperature ranges. The detailed electronic structure of these alloys is not yet fully resolved; they are normally metallic and show at best only partially ionised bonds. The formation of these phases is generally accompanied by a change in the crystal structure [3.62, 63].

The formation of hydride phases has been found, for example, for hydrogen and deuterium implantation at energies of 4–25 keV into Ti [3.38, 64–66] and Zr [3.67] at room temperature. The observation of these phases is rather indirect in the work by *Yonts* and *Strehlow* [3.68] and *Bohdansky* et al. [3.64, 65], showing large volume swelling of the implanted layer by optical and SEM microscopy. With x-ray diffraction [3.66, 67], scanning Auger spectroscopy [3.69] as well as transmission electron microscopy [3.70], it could be shown that initially micron-sized reaction zones are formed which finally grow into continuous hydride layers.

In the metals Ti, Zr, Nb, and Er, hydrogen trapping was found to be strongly dependent on the implantation temperature [3.38, 64, 71] (Fig. 3.7). At low temperatures, the trapped amount saturates at low fluences, around room

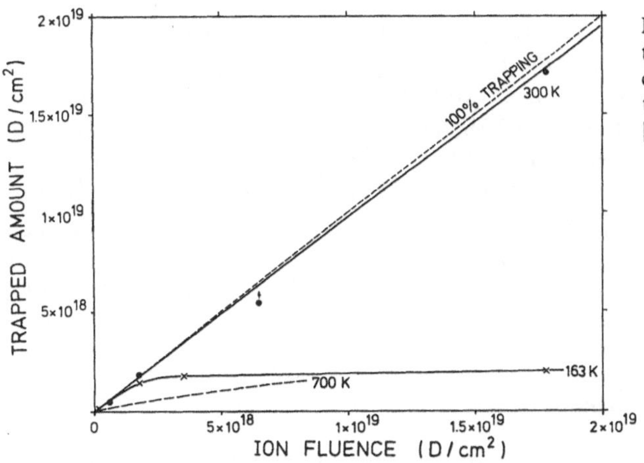

Fig. 3.7. Dependence of the trapped amount on ion fluence for D in Ti at different temperatures: (——) 6.7 keV [3.38] (- - -) 18 keV [3.61]

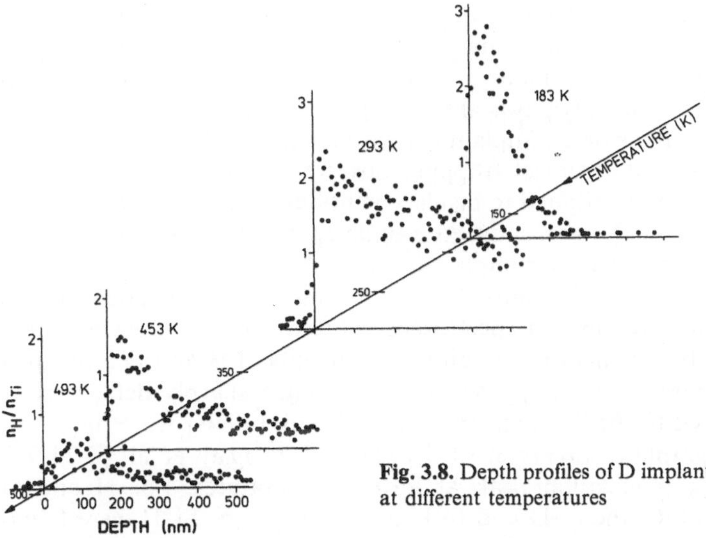

Fig. 3.8. Depth profiles of D implanted at 6.7 keV into Ti at different temperatures

temperature trapping reaches about 100% and does not saturate even at very high fluences, while at temperatures above 500 K, trapping decreases below 100%. This behaviour can be understood from the depth distributions of the implanted hydrogen atoms.

Figure 3.8 shows the depth profiles for 6.7 keV D implanted into Ti at a fluence of $6 \times 10^{18}\,cm^{-2}$ for different temperatures. At low temperatures, the diffusion of D is low and a surface layer comparable to the ion range saturates with D/Ti ratios exceeding even the stoichiometric hydride of TiD_2. At room temperature the diffusion of deuterium in the hydride phase formed initially within the implantation range is fast enough to transport the deuterium to the hydride-metal interface where hydride builds up continuously.

The surface concentration generally stays below the stoichiometric value and a surface barrier prevents the release of hydrogen. No saturation occurs even at fluences up to $10^{20}\,D\,cm^{-2}$ and the thickness of the hydride layer increases with the ion fluence [3.38, 67]. Up to 470 K, weight increase measurements still show close to 100% trapping [3.65], and concentration profiles reveal diffusion into the bulk with near-surface concentrations depending on implantation flux [3.38]. At still higher temperatures the hydrogen atoms can become thermally re-emitted at the surface. Very similar behaviour of hydrogen in Ti has also been found in other hydride-forming materials like Zr [3.67].

b) Semiconductors and Carbon

In semiconductor materials such as Si, Ge and in carbon, the formation of localised chemical bonds is expected. For hydrogen implantation into Si and Ge, *Gruen* et al. [3.72] have found the Si–H, Si–D, Ge–H bonds by infrared spectroscopy. Absorption bands are found which can be attributed to the hydride and deuteride local-mode frequencies. It could be shown that at fluences of $10^{16}\,cm^{-2}$, mainly physical trapping occurs, while at higher fluences an increasing percentage of the implanted hydrogen is chemically trapped. The strong enhancement of chemical trapping observed on going from H to D implantation has been attributed to displacement damage as a prerequisite for chemical trapping [3.72]. Laser-induced Raman-spectroscopy revealed a large degree of a lattice disorder after implantations.

For H and D implantation into C [3.40], chemical trapping could not be observed so readily by infrared spectroscopy due to interference with other absorption lines. In thermal desorption measurements of H and D implanted electro-carbon graphite, *Erents* [3.43] found a single sharply defined peak which he attributed to the breaking of the C–H bond. In later work, however, using pyrolytic graphite, *Erents* et al. [3.73] and *Langley* et al. [3.74, 75] observed a broad re-emission peak at lower temperatures. For H and D implantation into SiC, the C–D and C–H bonds as well as Si–H were found using infrared spectroscopy [3.40].

Trapping measurements of H and D implanted in carbon at room temperature gave a linear increase of the trapped amount with fluence eventually reaching a saturation (Fig. 3.9) [3.74–77]. This indicates that at room temperature, the trapped hydrogen is bound to its trapping sites and diffusion is negligible. The lattice structure of pyrolytic graphite is completely destroyed upon D implantation [3.40], as measured by laser induced Raman scattering. Rupturing and reformation of C–C bonds led to the occurrence of acetylenic $(-C{\equiv}C-)_n$ bonds. Depth profiles of implanted D [3.74, 75, 78, 79] indicate that D is trapped within the damaged surface layer. The trapping of D in pyrolytic graphite has been studied for different temperatures by *Doyle* et al. [3.80] and *Scherzer* et al. [3.78]. For temperatures between 300 K and 900 K, it was found that the number of D atoms retained in saturation decreases

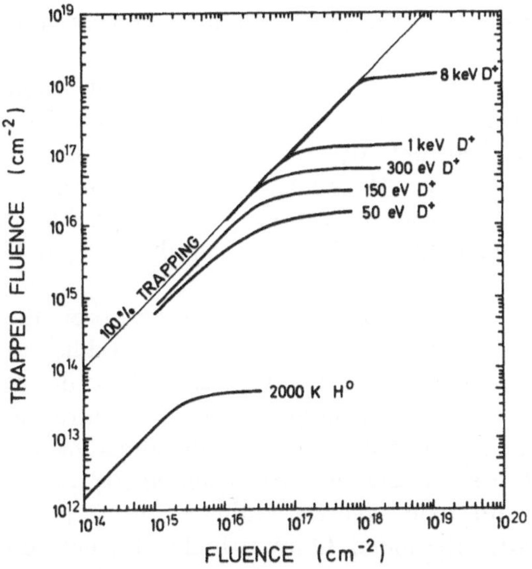

Fig. 3.9. Trapped amount of deuterium in graphite as a function of implantation fluence for different implantation energies [3.24, 74–77]

monotonically with increasing implantation temperature. This was attributed to the annealing of radiation-induced defects such as broken bonds between carbon atoms which are responsible for the trapping of H [3.78]. The post irradiation annealing of radiation damage, however, has been observed only at temperatures around 1300 K [3.40].

c) Oxides

Similar to the trapping of hydrogen in semiconductors, the trapping in oxides leads to the formation of localised chemical bonds. The chemical implantation reaction for the case of Al_2O_3 is [3.41]

$$Al_2O_3 + H \longrightarrow Al_2O_2(OH). \tag{3.3}$$

Such trapping reactions have been observed by infrared adsorption bands for SiO_2 [3.81] and TiO_2 [3.82, 83]. In Al_2O_3 the amount of chemically trapped D is larger than H, indicating the contribution of radiation damage which is larger for D implantation [3.41]. This has not been observed for TiO_2 [3.82]. *Gruen* et al. [3.41] observed a saturation of chemically trapped D in α-saphire. At higher implantation fluences an increasing portion of the ions is physically trapped with higher mobility and released through the surface. Blister formation after high fluence bombardment has also been found (Chap. 7).

3.3.2 Chemical Trapping of C, O, N

For a large number of metals and for silicon, the formation of carbides, oxides, nitrides has been reported upon C,O, and N implantation (see, e.g., [3.84]). The implantation of C and N into materials, in particular, is an important technique for surface refinement [3.20, 21, 85].

The experimental methods for observing the compound formation range from colour changes to electron diffraction in a transmission microscope. With electron diffraction it has been observed that Al_4C_3 and AlN are formed for 40 keV C and N implantation in crystalline Al, respectively, [3.86]. In many cases the produced compound is amorphous and can only be identified after subsequent annealing at elevated temperatures [3.86] as in the case of C implantation into Fe.

Oxide formation during oxygen implantation into metals has been the subject of many studies. Recently, *Giani* et al. [3.6] have reviewed 24 different systems and come to the following conclusion: at low fluences of the order of 10^{15} cm^{-2}, ion beam analysis indicates a low average oxygen concentration whereas electron diffraction patterns on selected areas [3.87, 88] already show oxide phases, which are dispersed in small islands. The stoichiometry of the oxides formed generally agrees with oxide phases evaluated from phase diagrams in thermal equilibrium with the metal. Only at higher fluences do oxygen-rich phases form. The average oxygen concentration in saturation may be limited by oxygen sputtering and may never reach values found in bulk oxides.

For O, N, and C implantation into Si, the formation of continuous compound films has been observed. SiO_2 layer formation has been reported for bombardment with 60 keV O^+-ions at fluences to 10^{18} ions cm^{-2} and temperatures to 600 K [3.89]. The oxide thickness depends on ion energy only and did not increase after prolonged bombardment. For 40 and 60 keV oxygen bombardment, stoichiometric SiO_2 layers were observed extending up to the range of the implanted ions [3.90]. When the ion fluence was not sufficient, the oxidation was incomplete. Oxide and nitride layers have also been produced by 35 keV implantation up to 2×10^{18} cm^{-2} [3.91]. SiC crystallites embedded in bulk silicon could only be observed after annealing to temperatures $\gtrsim 1100$ K following a 200 keV implantation of C at room temperature [3.92]. SiC was recently synthesized by the implantation of 4×10^{17} Si cm^{-2} at 40 keV into diamond at RT as well as 4×10^{17} C cm^{-2} into Si [3.93], where the kinetics of SiC growth depended on ion energy [3.94]. In all cases, carbide, oxide and nitride formation have been demonstrated by characteristic infrared absorption bands and x-ray diffraction.

3.3.3 Implantation of F and Cl

The formation of chemical compounds due to F and Cl implantation into materials is strongly anticipated in view of the chemical reactivity of these ions. Reactions of fluorine with metals and semiconductors have been the subject of many studies and very often yield volatile products, as will be seen in Sect. 3.5. Surface adsorbed F has been studied using x-ray photoelectron spectroscopy.

In the case of F and CF_3 implantation into Si and SiO_2 [3.42], a chemical shift of the Si line has been observed indicating the formation of an Si–F bond at the surface. Detailed investigations using the same method showed a large coverage of a silicon surface with SiF_2 complexes after exposure to F_2 and XeF_2 gases. No investigations of the depth distribution have been performed [3.95].

3.4 Chemically Enhanced and Reduced Physical Sputtering

The trapping of implanted ions in the solid and the formation of a compound layer influence the sputtering yield by several effects:

i) the target surface atoms are partly shielded by the implanted ions;
ii) the surface binding energy changes compared to the original material;
iii) the development of the cascade which finally leads to sputtering is changed.

Furthermore, changes in the composition of sputtered species, in the energy and angular distributions as well as charge states may occur. These effects are similar to the effects observed in the sputtering of compound materials (Chap. 2).

3.4.1 Sputtering Yield

Both decreases and increases of the sputtering yield have been observed during sputtering of elements with reactive ions as well as sputtering in a residual gas containing reactive atoms.

a) Sputtering with Reactive Ions

In the first systematic study, *Almén* and *Bruce* [3.13] investigated the sputtering yield of Ta, Cu, and Ag with a large variety of elemental ions of 45 keV. The sputtering yield was determined by measuring the weight loss of the samples after high dose bombardment.

The amount of collected ions was measured separately and the corresponding weight increase due to the implanted ions was taken into account [3.96]. For many projectile ions, the sputtering yield was found to depend on ion fluence. For inert gas sputtering, a lower sputtering yield was initially found but after saturation at higher fluences, the sputtering yield increased, while the sputtering yield for reactive ions generally tended to decrease with fluence [3.13]. In saturation, lower sputtering yields were usually measured while in some cases, even a weight increase due to the build-up of surface layers was observed. Build-up of surface layers only occurs for projectiles with self-sputtering yields lower than 1 and is dependent on target temperature.

The sputtering yield data obtained for Cu-targets at room temperature for a large variety of primary ions measured for low [3.97] and high fluences [3.13] are shown in Fig. 3.10.

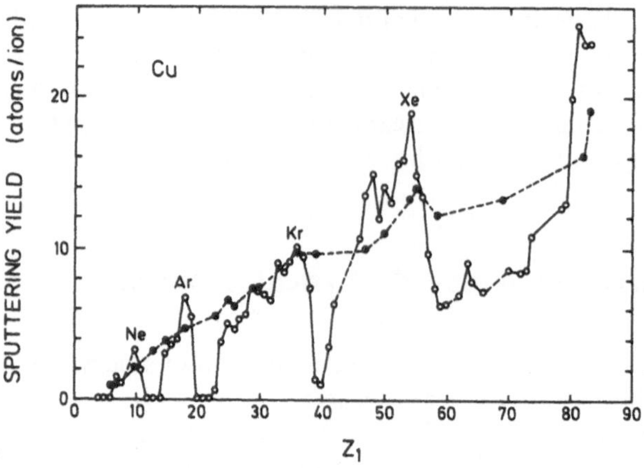

Fig. 3.10. Sputtering yield of Cu with 45 keV ions of various atomic number Z_1; (—○—) high fluence measurements [3.13]; (--●--) low fluence measurements [3.97]

The yields show a general increase with increasing ion mass. The sputtering yield for high fluences measured by *Almén* and *Bruce* show superimposed a periodicity comparable to the periodic systems of elements. For inert gases, a pronounced maximum always occurs, whereas for other ions, the sputtering yield is lower and sometimes decreases to zero. Similar data have been obtained for Ag and Ta target materials. The more recent results for much lower fluences from *Andersen* and *Bay* [3.97] are determined for thin films evaporated onto a quartz-oscillator microbalance. With this technique, yields for Si and Au are also measured [3.98]. Generally yields were determined relative to self-sputtering yields. *Andersen* and *Bay* also investigated the dependence of the sputtering yield on the ion dose and found strong dose effects. For Sn^+ bombardment the yield decreases with fluence, while for Bi^+, Tm^+ ions, a strong increase of the yield is observed with fluence. The ratio of the sputtering yield to the Cu self-sputtering yield, however, did not change with ion fluence.

The observed periodicity in the sputtering yields measured at high fluences are due to the change of the surface layer by implantation. In the cases where the yield decreases to zero, e.g., for C and Ca, the build-up of a solid layer of collected ions was observed preventing any sputtering from the backing material at high fluences [3.96]. The slight increase of the sputtering yield for inert gas ions for high ion fluences has been explained by a change in the structure of the surface layer by gas agglomeration and bubble formation (Chap. 7). For the case of low but nonzero yields, e.g., for V, Y, and Zr, the sputtering yield was representative for the alloy obtained by implantation. From the observations that the sputtering yield relative to self-sputtering before and after reactive ion sputtering remained unchanged, *Andersen* and *Bay* conclude that the chemical surface condition, represented by the surface binding energy U_0, is responsible for the change of the absolute sputtering yield. A comparison of the yield changes with the surface binding energy (Fig. 3.11) of the projectile material reveals, indeed, striking similarities. For sputtering with ions from materials with high surface binding energy, the yields

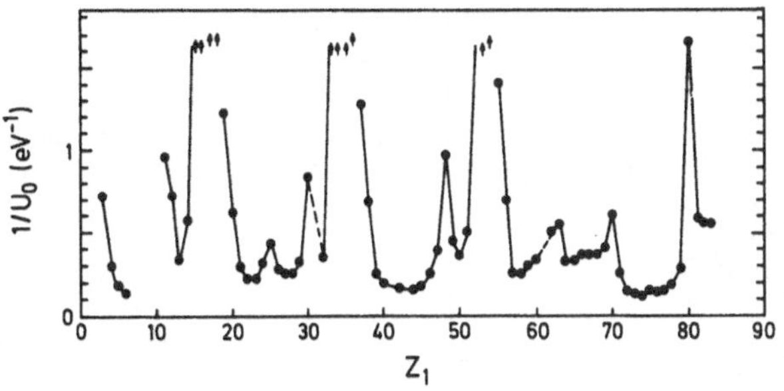

Fig. 3.11. The reciprocal surface binding energy U_0 of elements with atomic number Z_1

decrease while for ions with low surface binding energy, the yields increase with ion fluences.

A comparison of the sputtering yields of C, Ni, Fe, TiC, SiC, and W with oxygen, carbon, and neon ions in the energy range of 100 eV to 10 keV has been performed by *Hechtl* et al. [3.99–101]. It could be shown that sputtering with carbon on all samples investigated leads to build-up of carbon layers, as the self-sputtering yield of carbon does not exceed unity in this energy range. Oxygen sputtering generally gives sputtering yields close to the neon values above 10 keV decreasing much faster toward lower energies. This was attributed to the formation of a surface oxide layer and a higher surface binding energy which leads to an increase in the threshold energy for sputtering. For oxygen sputtering of graphite and tungsten, chemical sputtering and chemical enhanced physical sputtering occur and are described later (Figs. 3.12, 3.18).

b) Sputtering in a Reactive Atmosphere

The influence of surface compound formation on the sputtering yield can also be observed for sputtering with nonreactive ions in a reactive gas atmosphere. In an uncontrolled way this occurred in nearly all early sputtering measurements in poor vacuum and has been a major concern in all well-defined sputtering yield measurements. Due to the deposition of hydrocarbon molecules and subsequent cracking by the beam, a decrease of the sputtering yield with increasing residual gas pressure has usually been observed [3.13, 102, 103].

Systematic investigations in a well-defined residual gas atmosphere have been conducted in recent years. If the arrival rate for reactive gas atoms is higher than the sputtering rate, a stable compound phase can be formed at the target surface which may extend over many monolayers due to diffusion or recoil implantation of chemisorbed atoms [3.104]. In a plot of the sputtering yield *vs* the ratio of the arrival rate of reactive gas atoms to sputtering ions, a transition of the sputtering yield is expected. In Fig. 3.12, three examples are given for sputtering in a reactive gas atmosphere. Figure 3.12a shows the sputtering of Ti with 11 keV Ar ions in a residual gas containing various amounts of N_2, O_2 or H_2 atoms [3.105, 106]. The sputtering yield was measured by the time needed for sputtering through thin Ti films. It can be seen that at distinct ratios of the arrival rate of reactive atoms to Ar-ions, the

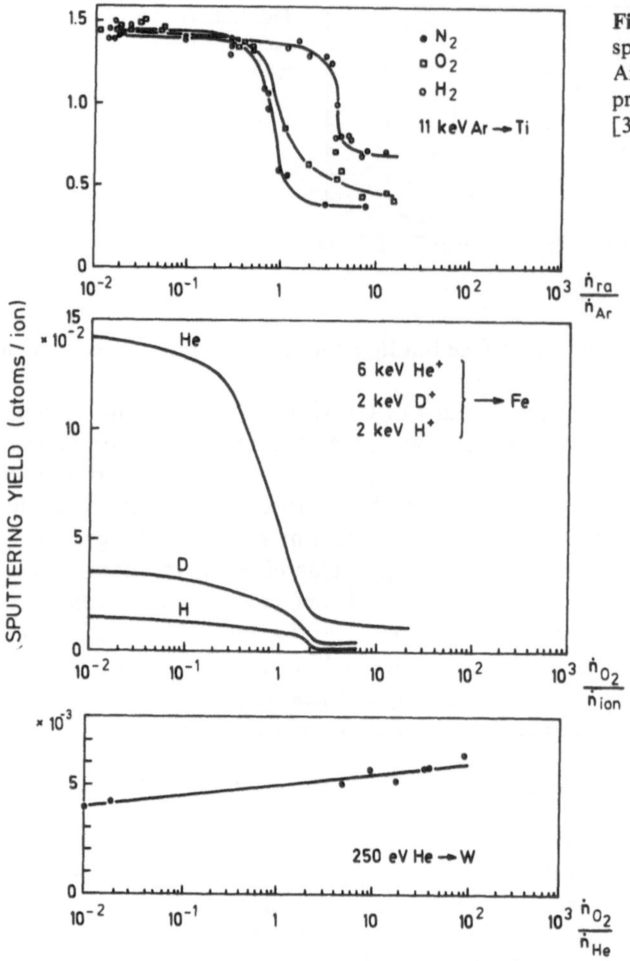

Fig. 3.12. Dependence of the sputtering of Ti, Fe, and W by Ar, He, H, and D on the partial pressure of reactive atoms [3.105, 106, 108, 109]

sputtering of Ti atoms decreases sharply, the decrease amounting to 50–70%. Qualitatively similar results have been obtained using 2.3 keV Ar sputtering [3.107]. Figure 3.12b shows the sputtering yield of Fe as measured by the characteristic light emission after laser-excitation of neutral Fe atoms in front of the target for He, D, and H sputtering and various O_2 residual gas pressures [3.108]. Again a drastic reduction in the Fe sputtering yield is observed, amounting in this case to about a factor of 8–10 for neutral atoms and 5 for the total yield measured by weight loss. In contrast to these observations, Fig. 3.12c shows for the case of W bombarded with 250 eV He ions a slight, steady increase of the sputtering yield with increasing O_2 partial pressure [3.109].

For these observations different explanations have been proposed. In the case of Ti, stable compound formation is expected at the surface. Although the surface composition has not been determined, TiO or TiO_2, Ti_2N and TiH_2 formation can be assumed during bombardment in O_2, N_2, and H_2 residual gas, respectively. These stable phases may have total sputtering yields comparable to pure Ti, only modified by a slightly different surface binding energy [3.53]. The partial sputtering yield of Ti, however, which is in fact measured here, will be lowered by 50–60% due to the dissolution of Ti atoms in the compound phase [3.106].

This effect can only partly account for the drastic reduction of the Fe partial sputtering yield with increasing O_2 partial pressure. In this case it was assumed that oxygen atoms, adsorbing at the surface, shield the Fe atoms from being sputtered. As sputtered oxygen atoms are continuously renewed from the residual gas at high partial pressures, the partial sputtering yield for Fe atoms is strongly reduced [3.108].

In the case of W sputtering with He ions in an O_2 residual gas, an oxide layer will also be formed. In this case, measurements of the energy dependence of W sputtering yield show as well as for oxygen bombardment at ion energies close to the sputtering threshold, an additional branch for sputtering corresponding to a reduced surface binding energy [3.111]. This contribution increases with increasing oxygen partial pressure (Fig. 3.12) and indicates that oxide molecules are sputtered, which are known to have a much lower surface binding energy than W atoms [3.110]. A similar additional sputtering mechanism has been found for O^+ ion sputtering of W below 600 eV [3.101].

Only three examples have been given to illustrate the influence of the residual gas effects but many observations are reported in the literature. Changes in sputtering yields are well known for experiments on the production of thin films sputter deposited in a glow discharge containing reactive ions [3.112–114] and must be considered in sputtering experiments in poor vacuum [3.1].

3.4.2 Sputtered Species

Together with the change in sputtering yield, the nature of the sputtered particles has been found to change due to chemical trapping. Besides metal atoms and molecules [3.115–117], molecules of compound masses have also been found among the sputtered species [3.47].

a) Sputtered Molecular Ions

A large amount of mass spectrometric studies of the sputtered atomic and molecular ions have been performed in SIMS (Secondary Ion Mass Spectroscopy) [3.47] investigations [3.116 and 3.118, Chap. 2]. For sputtering with reactive ions or with inert gas ions in a residual reactive gas atmosphere, the fraction of sputtered ions is substantially increased. Thus, in respect to a quantitative and sensitive analysis of surface layers by SIMS, the secondary ions formed, for example, in sputtering with O^+, N^+, and Cs^- ions have been investigated in some detail [3.118, 119].

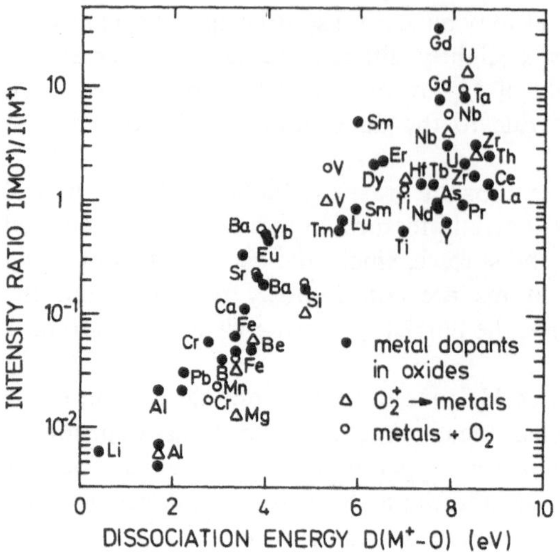

Fig. 3.13. Relative intensity of oxide ions $I(MO^+)/I(M^+)$ versus bond dissociation energy $D(M^+\text{–}O)$ for oxygen sputtering of metals and inert gas sputtering of oxides or metals in an oxygen atmosphere [3.119]

In the following a few examples are presented which illustrate molecule formation. During sputtering of Fe with O_2^+, iron oxide ions up to Fe_3O_2 have been observed [3.120]. Sputtering of oxygen-covered W by Ar ions produced predominantly WO^+, but also WO_2^+ and WO_3^- ions [3.121]. In general, the measurements on a variety of metals containing various amounts of oxygen in the surface layer showed that the oxygen-to-metal ratio in the sputtered molecules increases with the oxygen concentration in the surface layer [3.122].

For inert gas sputtering of bulk oxides and metals covered with surface oxygen as well as oxygen bombardment of metals, the intensity of sputtered oxide molecules to sputtered metal atoms increases with increasing bond dissociation energy of the oxide molecule ion. It ranges from 10^{-2} at a dissociation energy of 2 eV to 10 at dissociation energies of 8 eV [3.119] (Fig. 3.13). The dissociation energies are taken from the literature and have been collected in [3.123].

b) Sputtered Neutral Molecules

The technique of matrix isolation already mentioned in Sect. 3.2, allows detection of all sputtered particles. This technique has been applied to identify the particle removed during sputtering of metals with reactive ions. For sputtering of Ti with H and D ions, *Krauss* and *Gruen* [3.124] reported the observation of TiH and TiD. For hollow-cathode sputtering of Ti and Al in a mixture of Ar and O_2, the molecular state of the sputtered species was investigated for different O_2 concentrations [3.35, 125]. The results are summarized in Table 3.2.

For oxygen concentrations up to 10^{-2}%, atomic Ti sputtering could be observed. At 0.1% O_2, the sputtered particles were only TiO. At higher O_2 concentrations, TiO_2 molecules were also found and dominated at 5.5% O_2. In

Table 3.2. Amount of sputtered oxide molecules for different percentages of oxygen in the sputtering gas

Material / Sputtering gas	Ti [3.124] band intensity for Ti	TiO	TiO_2	Al [3.35] ratio of metal atoms to oxide molecules n_{Al}/n_{AlO}	Relative ratio of oxide molecules I_{Al_2}/I_{AlO}
99.995% Ar				10^4	20
0.01% O_2	Strong	Weak	No		
0.015% O_2				10^2	10
0.1% O_2	No	Only	No		
0.15% O_2				25–29	2
0.5% O_2				4.4	0.6
0.5% O_2	No	Weak	Strong		
1% O_2				5.5	0.3
5.5% O_2	No	No	Only		

the same experimental set-up Al was also sputtered and similar results have been obtained [3.35] (Table 3.2). Oxygen additions to the sputtering gas of more than 0.5% lead to oxide-molecule sputtering comparable to Ar sputtering of bulk oxides. For sputtering of Nb with O_2 and N_2, the observation of NbO and NbN molecules in the matrix is reported [3.126].

The influence of chemical reactions can be very pronounced if a compound is sputtered with ions which react with only one of the components. Preferential sputtering of one component of the compound, which is generally observed for sputtering with nonreactive ions (Chap. 2), may then be significantly enlarged. The difference in the composition of sputtered particles for sputtering of Al_2O_3 with 50 keV Ar and 15 keV H ions has been investigated by *Gruen* et al. [3.34, 35]. With the matrix isolation technique, only Al, Al_2, AlO, AlO_2 molecules, which were previously found to dominate the secondary ion mass spectrum [3.140], were followed. Striking differences in the sputter products for Ar and H bombardment were observed. The results are shown in Fig. 3.14 where the visible absorption spectra of sputtered products from single crystal saphire isolated in Ar-matrices are plotted. For 50 keV Ar bombardment of a fluence of $2\times10^{17}\,cm^{-2}$ where about 100 monolayers are sputtered off, mostly AlO molecules are detected in the matrix, the ratio of n_{Al}/n_{AlO} being 0.7 (Curve *C*). For a proton fluence of $2\times10^{19}\,cm^{-2}$, where only 30 monolayers are sputtered, roughly the same ratio of Al atoms to AlO molecules is found, but already at this low fluence a broad shoulder indicating sputtering of Al_2O appears (Curve *A*). At higher proton fluences the ratio of n_{Al}/n_{AlO} increases to about 11–14 and the intensity of the Al_2O peak is 12 times higher than the AlO peak (Curve *B*).

The ratios of Al atoms sputtered in the form of Al-oxide molecules have been explained by *Gruen* et al. by considering the composition of the vapour

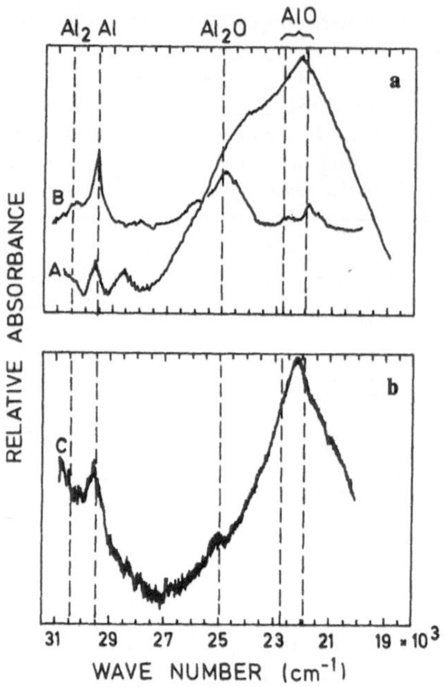

Fig. 3.14a, b. Absorption spectra of particles sputtered from saphire and isolated in an Ar-matrix. Absorption bands due to Al, Al_2, AlO, and Al_2O are observed [3.35]

(a) 15 keV $H \rightarrow Al_2O_3$
A 6×10^{16} atoms cm^{-2} removed
B 3.6×10^{17} atoms cm^{-2} removed
(b) 50 keV $Ar \rightarrow Al_2O_3$
2×10^{17} atoms cm^{-2} removed

pressures above an $Al_{2+x}O_{3-x}$ compound for various values of x, assuming an effective surface temperature of 1800 K. For small values of x, i.e. depletion in oxygen, the ratio of the vapour pressure of Al_2O to AlO increases drastically. The differences between Ar and H sputtering are explained by the strong reducing action of protons on oxides, to form volatile H_2O molecules leading to a much stronger depletion of oxygen in the surface layer than found in physical sputtering with Ar (Chap. 2). The formation of H_2O, which has a high vapour pressure and desorbs quickly at room temperature, indeed fulfills the definition of chemical sputtering and will be further mentioned in Sect. 3.5.

3.5 Chemical Sputtering

The effects discussed in the previous paragraph, where the chemical interaction between the implanted ions and the atoms of the solid has led to a reduction and in a few cases also to an increase of the sputtering yield, were named chemically reduced or enhanced physical sputtering. In this section we shall deal with chemical sputtering defined as the process where the molecules which are formed between the target atoms and the bombarding ions are volatile, i.e., they have a surface binding energy low enough to desorb at the temperature where sputtering occurs.

According to this definition, the appearance of chemical sputtering may be inferred from different experimental observations:

i) The sputtering yield should show strong variations with surface temperature;
ii) molecules formed between the bombarding ions and target atoms should be observable;
iii) compared to physical sputtering, no sharp threshold energy should be observable;
iv) the energy distribution of the sputtered molecules should be close or equivalent to the target surface temperature;
v) chemical sputtering should be strongly selective for different combinations of target atoms and sputtering ions;
vi) the activation and inhibition of the sputtering process by the state of the bombarding reactive ions and the target atoms should be pronounced.

Cases of chemical sputtering reported in the literature will be reviewed according to these experimental observations.

In all cases the following processes will occur:

i) implantation or adsorption of the reactive ions bombarding the surface;
ii) diffusion of implanted ions back to the surface;
iii) diffusion on the surface and the formation of molecules;
iv) desorption of the molecules.

Each of these processes may again consist of various steps which are very often not sufficiently experimentally investigated. For the case of the hydrogen reaction with carbon, much work has been done and is presented in Sect. 3.5.3 as an example for the kinetics of volatile molecule formation.

3.5.1 Experimental Observations

a) Temperature Dependence of the Sputtering Yield

Chemical sputtering may only be observable in a certain temperature interval (Fig. 3.2). Not only is the desorption of the compound molecule generally a thermally activated process, but each step in the chemical reaction is temperature dependent, leading in most cases to strong variations of the chemical sputtering yield. This is in contrast to physical sputtering of elemental materials, where the yield has been found to be nearly independent of target temperature until the temperature approaches the melting point [3.141, 142].

Chemical sputtering and its variation with temperature is best observable if the physical sputtering yield is small compared to the maximum chemical sputtering yield. As an example, the sputtering yields of carbon for bombardment with hydrogen and oxygen ions as a function of the target temperature are shown in Fig. 3.15 as obtained by several independent groups. Figure 3.15a shows the temperature dependence of the sputtering yield for hydrogen bombardment with energies between 0.1 and 30 keV. In cases (a–c) [3.127–129], the total sputtering yield was measured by weight loss or volume loss, while in cases (j) and (k) [3.130, 131] only the chemical sputtering yield is determined by calibrated measurements of the CH_4 partial pressures. Additionally, data for

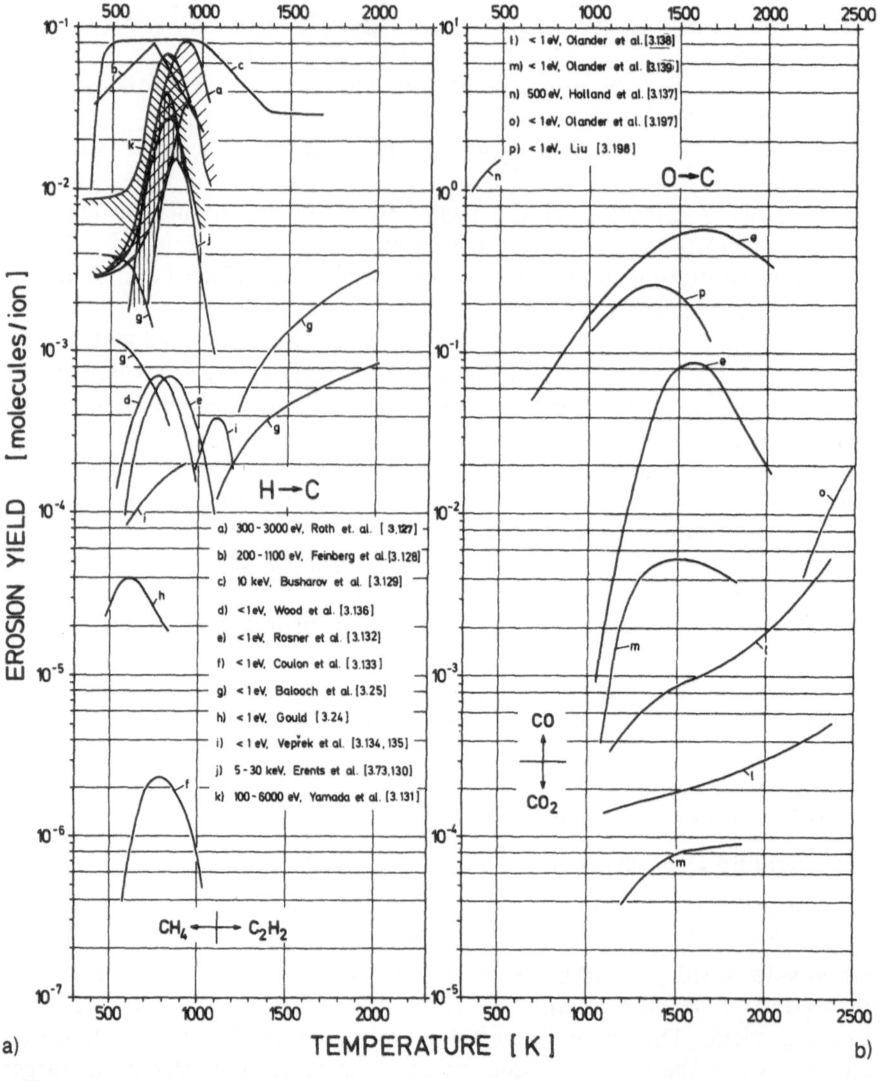

Fig. 3.15. Erosion yield of graphites due to interaction with hydrogen and oxygen as a function of temperature. (*a–f, i, n*) measured by change in weight or volume, (*g, k, j–m, o, p*) measured by emission of CH_4, C_2H_2, CO or CO_2. The shaded areas between curves *a, h*, and *j* indicate the range of reaction yields for different incident ion energies. Curves *l* and the higher curves *g* are obtained on prism planes, curves *m* and the lower curves *g* on basal planes of pyrolytic graphite. The lower curve *e* is obtained for O_2, curve *o* for H_2O attack of graphite

atomic hydrogen with energies below 1 eV are included (d–i) [3.24, 25, 132–136].

Though various types of graphite have been employed by the different groups [3.127–132] for chemical sputtering by energetic hydrogen ions, the measured yields differ by less than a factor of 3. Radiation induced amorphisation may lead at high fluences to similar surface structures for different types of graphite [3.40]. For thermal atomic hydrogen, however, the yields obtained by different investigators [3.24, 25, 132–136] spread from 10^{-6} to 10^{-3} atoms per hydrogen atom. This may be due to the different types of graphite or to the different sources of atomic hydrogen used. Further, the experimental difficulties in determining the atomic neutral hydrogen currents may also contribute to the differences.

In nearly all cases the sputtering yield exhibits a maximum at temperatures between 720 K and 920 K. The physical sputtering yield of carbon by energetic hydrogen is of the order of 10^{-2} atoms per ion between 100 eV and 1 keV, while the maximum of the chemical sputtering yield reaches values close to 10^{-1} atoms per ion. In one case [3.25] the chemical reaction yield does not show a maximum, but decreases monotonically with temperature. In this case surface coverage effects may change the temperature dependence, as discussed in (Sect. 3.5.2b). The same investigation reports a high temperature branch, showing increasing C_2H_2-formation with increasing temperature.

The chemical sputtering of graphite with oxygen also shows a strong temperature dependence (Fig. 3.15b). *Holland* and *Ojha* [3.137] found monotonically increasing yields for 500 eV oxygen ions at temperatures between RT and 500 °C. At higher temperatures, *Olander* et al. [3.138, 139] observed CO and CO_2 formation with yields increasing with temperature. Similar measurements have been reported by *Rosner* and *Allendorf* [3.132] for oxygen and OH radicals impinging on carbon samples.

Often hysteresis effects in the temperature dependence are found, with lower reaction probabilities reported when going from high temperatures to lower than when going from low to higher temperatures. This effect has been observed for the hydrogen-carbon and oxygen-carbon reaction by different authors [3.25, 139, 143]. It is explained by the different concentrations of the

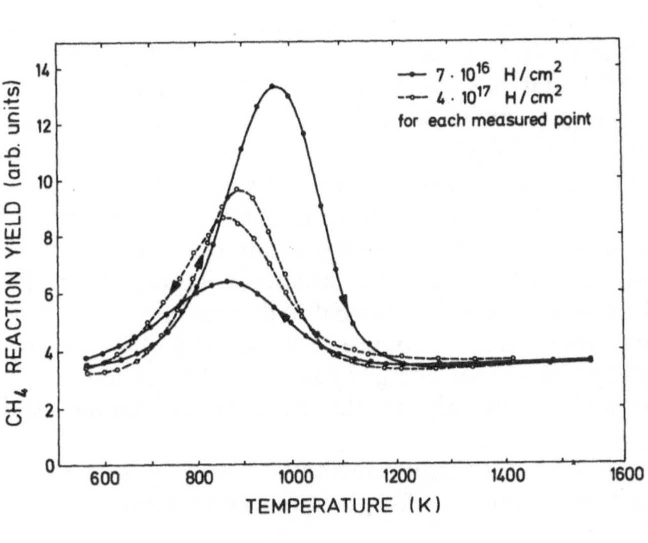

Fig. 3.16. Variation of the CH_4 reaction yield with target temperature for 2 keV H^+ bombardment of graphite. Arrows on the curves through the data indicate the direction of temperature change while the data were obtained

reactive atoms in the material at the start of the measurements and may lead to a drastic reduction of the chemical erosion after high-temperature [3.143] annealing of the sample. For energetic hydrogen bombardment of pyrolytic graphite, Fig. 3.16 shows hysteresis effects in the methane production yield. The difference in yield for increasing and decreasing temperature depends on the accumulated fluence of reactive ions before measuring the yield at a given temperature, and is the largest for fast changing temperatures. As the saturation concentration of hydrogen in graphite decreases with increasing temperature [3.78, 80], the sample will be oversaturated if the temperature is rapidly increased and depleted if the temperature is decreased and saturation will only be reached after a certain time. As the methane production is a function of hydrogen surface concentration (Sect. 3.5.2), different methane yields will result. It could be found [3.144] that the fluence needed to establish equilibrium, amounts to $1.8 \times 10^{18}\,\mathrm{H\,cm^{-2}}$ at 2 keV, both for increasing and decreasing the temperature. A value of $1 \times 10^{18}\,\mathrm{H\,cm^{-2}}$ has been determined by *Yamada* et al. [3.131] for 1 keV hydrogen bombardment.

The temperature variation of the sputtering yield is a dominant feature in chemical sputtering but it has been examined in some detail only for hydrogen sputtering of carbon and silicon and some carbides [3.146–148]. Chemical sputtering of silicon by fluorine and chlorine ions, for example, should also show strong temperature effects. Their investigation may contribute to a better understanding of the mechanisms.

b) Observation of Volatile Sputtered Molecules

In the first observation of chemical sputtering (Fig. 3.1), the nature of the volatile products has not been determined and only inferred from thermodynamical data, while in more recent experiments they have been analysed by residual gas analysis.

During bombardment of carbon samples with hydrogen ions at energies between 100 eV and 30 keV, *Roth* et al. [3.127], *Erents* et al. [3.73, 130], *Busharov* et al. [3.129, 149], and *Yamada* et al. [3.131, 151, 152] showed CH_4 formation depending on target temperature. CH_4 molecules were also found for bombardment of SiC, TiC, and B_4C with hydrogen at elevated temperatures [3.146–148, 152, 153]. For the reaction of hydrogen atoms with carbon films, *Rye* [3.154] could see, in addition to the dominant CH_4 peak, up to 10 additional masses formed during the interaction with atomic hydrogen, these being cyclic compounds consistent with the structure of graphite. C_2H_2 formation as expected from the results of thermal hydrogen interaction with graphite at temperatures above 1000 K, could not be found [3.144]. At low temperatures, however, especially at ion energies below 1 keV, C_2H_4 molecules were found to contribute significantly to the chemical sputtering yield [3.144, 155].

During bombardment of Si at different temperatures with hydrogen ions, *Roth* [3.109] observed SiH_4 using an off-sight quadrupole mass analyser.

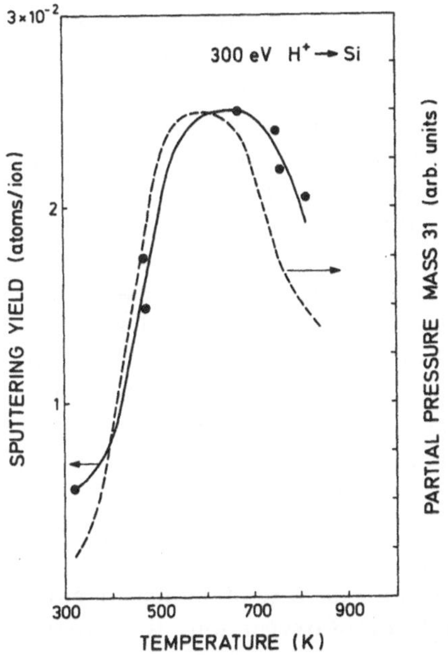

Fig. 3.17. Correlation of the temperature dependence of the total sputtering yield with the observed partial pressure of SiH_4 during sputtering of Si with 500 eV H^+ ions

Figure 3.17 shows for varying temperature the correlation of the total sputtering yield as measured by the weight loss method with the observed mass signal 30 and 31, cracking products from the parent mass 32. It can be seen that the increase in sputtering yield around 600 K measured by weight-loss correlates well with the observation of SiH_4.

During bombardment of Si with CF_3^+ ions [3.42], as well as in a CF_4 discharge [3.156], the formation of SiF_4 molecules is reported at room temperature. Also in this case the SiF_4 molecules in the residual gas spectrum have been inferred from the cracking product SiF_3^+. In more recent studies that use on-sight quadrupole mass analysis (Fig. 3.6) less fluorinated Si radicals were more abundant than SiF_4 [3.50].

For the chemical reaction of solid metals, carbides, nitrides and C with atomic and molecular hydrogen, oxygen, nitrogen, fluorine and chlorine gases, OH^- radicals and H_2O vapor at elevated temperatures, there are many results on the molecular reaction products [3.132, 138, 157, 158]. These erosion rates are not reviewed in detail and are only reported in the following as far as they correlate with energetic ion sputtering.

c) Threshold Energy for Sputtering

Chemical sputtering occurs if the binding energy of the molecules formed between the incident ion and the target atom on the surface is comparable to the thermal energy. This means that chemical sputtering should be observable down to very low bombarding energies and a threshold energy should not appear. For high bombarding energies in the keV range, the chemical sputtering yield is ultimately limited by the stoichiometric ratio of target atoms to reactive ions in the volatile molecule. Towards lower energy the yield may decrease as the energy deposited in the surface layer decreases which contributes to the chemical reaction yield. At still lower energies an increasing fraction of the bombarding ions is reflected, thus additionally reducing the chemical sputtering yield. At energies below 1 eV chemical sputtering should merge gradually into chemical gas-solid reactions.

As physical sputtering decreases steeply at low energies towards a threshold energy [3.18], chemical sputtering will dominate the erosion at low enough energies. *Güntherschulze* [3.3] has already reported that chemical processes dominate the erosion mechanism in a hydrogen discharge at energies below 600–1200 eV and he attributed the increase of the sputtering yield towards higher energies to additional physical sputtering. Figure 3.18 shows his original results for sputtering of As, Bi, and Sb. It should be mentioned, however, that the experimental conditions of this early work [3.3] as well as the high observed yields indicate that impurity ions such as oxygen rather than protons may be responsible for the physical sputtering.

In Fig. 3.19 a more recent example is shown for the case of bombardment of pyrolytic graphite with a mass analysed ion beam of oxygen ions as compared to neon and carbon ions [3.100].

While the sputtering yields for neon and carbon ion bombardment decrease to zero at about 80 eV, the yield for oxygen ion bombardment, being close to 1, decreases only slowly at low energies. Indeed, carbon shows erosion by chemical reactions with oxygen even at thermal ion energies, especially at

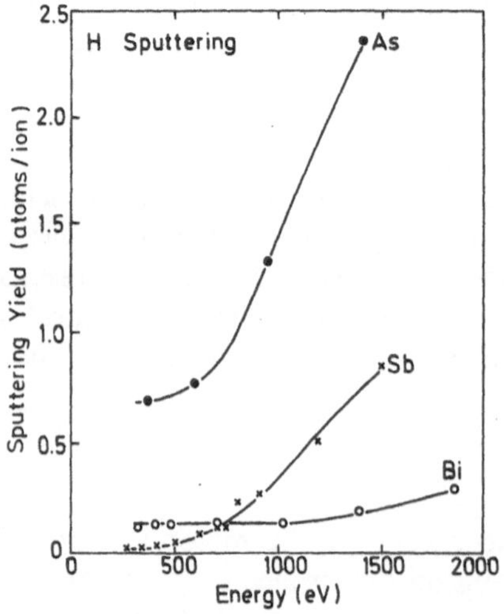

Fig. 3.18. Energy dependence of sputtering yield of As, Sb, B with H^+ ions at room temperature [3.3]

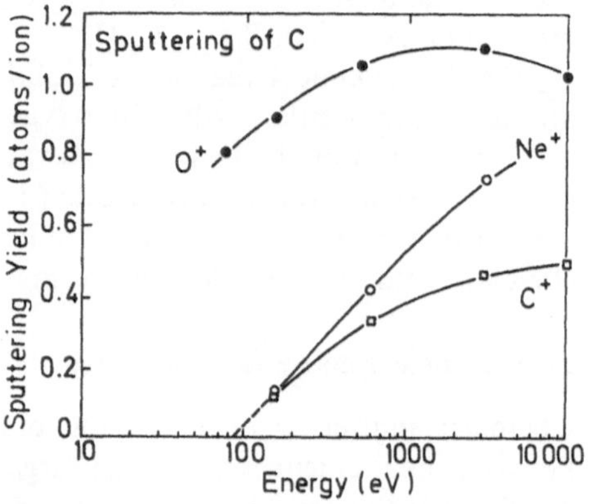

Fig. 3.19. Energy dependence of carbon sputtering with O^+, Ne^+, and C^+ ions at room temperature [3.101]

surface temperatures above 800 K where CO is the most abundant volatile molecule observed (see Fig. 3.15b).

Finally, all known data for the attack of materials with thermal atomic and molecular gases can serve as examples for chemical sputtering taking place at energies below the threshold for physical sputtering. There are data on the attack of Mo and W by O, Cl, and F [3.132], of Pt and W by F [3.159], of Ta and UO_2 by F [3.157, 158], of SiC, B, and BN by O and Cl [3.158], of C by H [3.25] and O [3.138], as well as coupled reactions of O and N with SiC and B [3.132].

d) Energy Distribution of Sputtered Particles

The energy distribution of sputtered particles allows, in principle, the distinction between physical and chemical sputtering. From the different mechanisms for particle ejection in physical and chemical sputtering, differences in the energy distributions of sputtered particles are expected. In chemical sputtering the thermal desorption of molecules should lead to a distribution close to a Maxwellian energy distribution representing the surface temperature T of the solid if the emitted particles are thermally accomodated to the surface. This yields

$$\frac{dY(E_1)}{dE_1} \propto E_1 \, e^{-E_1/k_B T}, \tag{3.4}$$

where k_B is Boltzmann's constant and E_1 the energy of the sputtered atoms. For physical sputtering in the linear collision regime ($E_1 \ll \gamma E$), the energy distribution is given by [3.4, 160]

$$\frac{dY(E_1)}{dE_1} \propto \frac{E_1}{(E_1 + U_0)^{3-2m}}. \tag{3.5}$$

Here, E is the incident ion energy, U_0 the surface binding energy and m the exponent for the interatomic potential being generally $m \approx 0.2$ to 0. For $m=0$ this distribution has a maximum at $1/2\, U_0$ and decreases as $(1/E_1)^2$ for high energies. Unfortunately, there are few measurements of the energy distribution of sputtered neutral particles for the case of chemical sputtering. For deuterium bombardment of SiC [3.161] the mean energy of sputtered hydride molecule ions has been found much lower than sputtered C, Si or SiC ions.

For the case of chemical sputtering of Si with Ar^+ ions in a XeF_2 atmosphere energy distributions have been measured for SiF_4 molecules using the apparatus described in Fig. 3.6 [3.50]. Apart from the reaction product SiF_4, as determined by its cracking product SiF_3^+, also similar quantities of SiF_2 and SiF radicals were observed. The energy distribution of SiF_3^+ and SiF_2^+ is shown in Fig. 3.20 compared to the distributions expected for thermal evaporation of SiF_4 molecules and for physical sputtering of Si atoms. The decrease of the sputtered flux of SiF_2^+ and SiF_3^+ towards higher energies proportional to E_1^{-2}

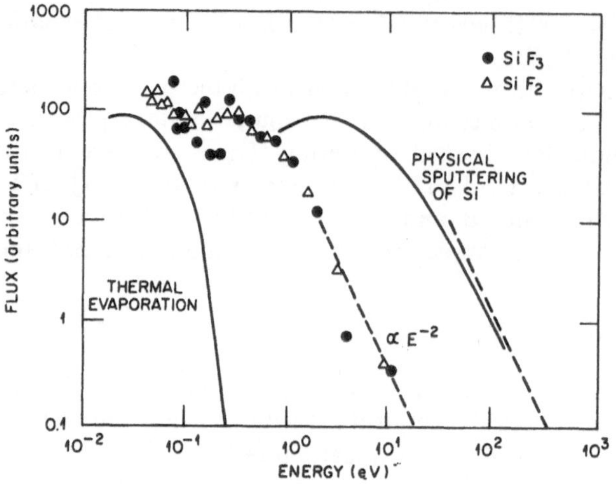

Fig. 3.20. Energy distribution of molecules sputtered by 3 keV Ar^+ ions from Si at XeF_2 flux of 7×10^{16} cm^{-2}. The Maxwell-Boltzmann distribution for the target temperature (300 K) and the distribution from a collision cascade are indicated [3.50]

indicates physical sputtering of molecules, however with a surface binding energy as low as 0.3 eV. At low energies there is an indication for additional thermal desorption of reaction products accommodated to the surface at 300 K i.e. chemical sputtering. Thus, both processes chemically enhanced physical sputtering and chemical sputtering could be distinguished by the energy distribution of sputtered particles.

A review on energy distributions of sputtered atoms and molecules from ionic crystals and condensed gases is given in [3.162] and shows the strength of these measurements in explaining the sputtering mechanisms.

e) Selectivity

One important application of chemical sputtering, sputter etching with reactive ions in semiconductor fabrication, will be reported in detail in [3.118]. This application makes use of the high selectivity of chemical sputtering, e.g., for the etching of Si compared to SiO_2.

The selectivity in physical sputtering, i.e., the ratio of the sputtering yield of materials with similar mass by the same ions, is mainly determined by the binding energy of the surface atoms. The yield ratio, therefore, may be of the order of 2–3.

In chemical sputtering the selectivity may reach much larger values. The chemical reactivity of the bombarding ions with different target materials of similar masses may be orders of magnitude different.

This selectivity is, for example, used in sputtering surface films without attacking the bulk material. To clean the inner wall of large fusion devices [3.163, 164] or ion accelerators [3.165] from surface carbon and oxide layers, their reactivity towards oxygen and hydrogen is used. During a glow discharge in a reactive gas, the wall is bombarded with large fluxes of the reactive ions which remove the surface layer by forming volatile CO or H_2O molecules especially at elevated temperatures of 400–500 K. If the reactive ion energy is low enough, the underlying stainless steel will not be eroded by physical sputtering. The cleaning of vacuum walls of carbon surface layers of several thousand Å has been reported [3.166].

Reactive sputter etching is used, for example, in the removal of surface layers of SiO_2 on Si without attacking the bulk silicon. In special reactive gas mixtures selectivity values up to 30 are reported for the yield ratio of SiO_2 to Si [3.167]. This strong selectivity is obtained, for example, in high frequency discharges in a CF_4–H_2 etch gas. The targets are mounted on the RF cathode where they are subject to energetic ion bombardment from the plasma (Fig. 3.4). For bombardment with CF_4 ions, both SiO_2 and Si react with the F atoms and SiF_4 molecules are formed. These molecules are volatile and desorb rapidly. The residual carbon atoms now cause a difference in chemical sputtering of SiO_2 and pure Si.

— For pure Si the C atoms can be removed only in the form of C_2F_6 molecules which has indeed been observed in the effluent gas of the discharge.
— For SiO_2, however, the residual carbon atoms can form volatile CO and CO_2 molecules and thus can be removed from the surface without consuming additional fluorine, which is then available to remove more Si atoms.

This leads to higher chemical sputtering yields for Si in the presence of oxygen as can be seen in Fig. 3.21a, where the increase of the silicon etch rate with increasing oxygen up to 10% in the F_4 etch gas is shown.

This higher chemical sputtering of SiO_2 than of Si can further be increased by adding hydrogen to the discharge gas. If H_2 is added, the CF_4 etch gas is made deficient in fluorine and for both Si and SiO_2, the chemical sputtering yield decreases but much more strongly for Si than for SiO_2, as can be seen from Fig. 3.21b. At 40% H_2 the selectivity, i.e., the ratio of SiO_2 sputtering to Si sputtering, increases to values close to 30 [3.167, 168], for more than 40% H_2 polymerisation of C on the silicon surface has been found [3.167]. The details

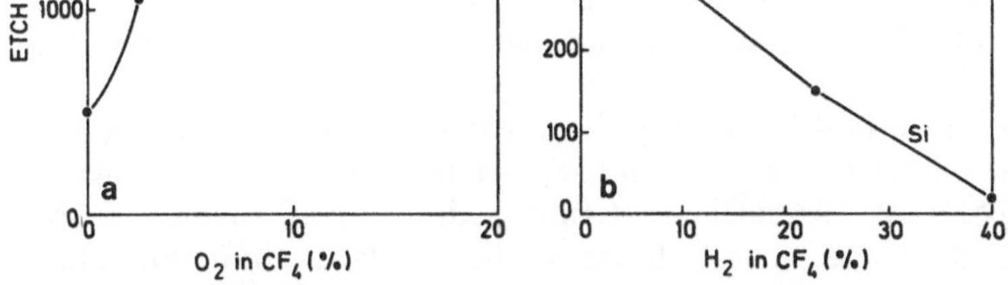

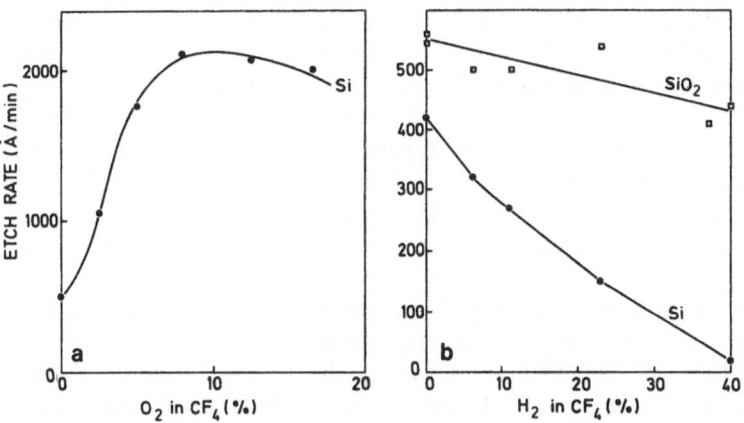

Fig. 3.21. (**a**) Dependence of Si etch rate on % O_2 in CF_4 etching gas [3.149]. (**b**) Dependence of SiO_2 and Si etch rates on % H_2 in CF_4 etching gas [3.167]

of the chemical reactions involved are still under active discussion and remain unresolved. It appears, however, that the above model developed by *Coburn* and *Winters* [3.156] describes the experimental results at least qualitatively well.

Another example of the selectivity of chemical sputtering is the preferred reaction of the bombarding ions with only one component of a compound. Depletion of this component in a surface layer may appear much more pronounced than observed in collisional sputtering ([3.11] and Chap. 2).

The interaction of energetic hydrogen ions with carbides has been extensively studied. *Braganza* et al. [3.146] found an increase of the sputtering yield of B_4C and SiC at 450 and 800 K, respectively, for 20 keV H^+ bombardment together with the observation of CH_4. In the case of SiC, the increased sputtering yield together with CH_4 formation could only be observed until a surface layer comparable to the thickness of the implanted layer was depleted of carbon. Subsequent annealing to 1200 K reproduced the initial surface composition, either due to diffusion of C from the bulk to the surface or due to evaporation of Si, and the effect could be repeated [3.146]. Indeed, it was shown by *Mohri* et al. [3.148] that annealing to 1200 K and above could lead to surface segregation of carbon layers. The same authors [3.169] observed at temperatures close to room temperature preferential loss of Si in a surface under deuterium bombardment, whereas He and Ar sputtering left the surface composition unchanged. This may be due to the different temperature regimes of chemical sputtering of Si and C, as can be seen from Figs. 3.15, and 17. For the case of TiC sputtering with protons it has been shown that carbon atoms are removed in form of CH_4 molecules, while physical sputtering of Ti atoms limits the overall erosion [3.152].

A strong interaction of hydrogen ions with oxygen in oxide surfaces is often reported. A reduction of the oxide in the surface layer due to H_2O formation and subsequent desorption is expected, as in a reaction of atomic and molecular hydrogen with oxides at elevated temperatures [3.170].

The production of H_2O by an rf discharge of H_2 in a pyrex bottle was demonstrated by *Blauth* et al. [3.171, 172]. *McCracken* and *Patridge* observed predominantly H_2O ions in the mass spectrum of secondary ions extracted from a hydrogen plasma interacting with glass [3.173]. The dependence of H_2O formation during exposure of oxidised stainless steel to atomic H atoms on the fluence has been studied by *Ishibe* and *Oyama* [3.174]. They found the H_2O production initially proportional to the H-fluence being roughly 10^{-2} H_2O/H atoms. With increasing fluence, i.e., with increasing depletion of the oxide layer, the reaction probability decreased until values of the order of 3×10^{-3} were observed.

The studies mentioned above present cases of chemical reactions of hydrogen with oxides activated by pre-dissociation of hydrogen molecules, rather than cases of chemical sputtering with energetic hydrogen ions. Unfortunately, there are no ion beam studies of H_2O formation during bombardment of oxides with hydrogen ion beams. The finding of a strong surface reduction of Ta_2O_5 during proton bombardment [3.39, 175], though being currently explained with collisional sputtering models, may well be partly due to chemical sputtering.

3.5.2 Activation and Inhibition of Chemical Sputtering

The reactions involved in chemical sputtering can be enhanced or decreased by a change in the state of the reacting atoms or by the addition of small quantities of impurity atoms. For different cases of chemical sputtering a different stage, i.e., adsorption, diffusion, molecule formation or desorption, may be the rate-limiting step. Each of these stages can be activated or inhibited by a change in the state of the reacting atoms. Thus, the observation of variations of the sputtering yield upon the state of the surface atoms or the molecular state of the bombarding particles may also be an indication for chemical sputtering. The following sections are illustrated by specific examples of each type of process. Although the examples are representative, they are only a few of the examples which could have been chosen.

a) State of the Bombarded Surface

The structure and topography of the bombarded surface can have a considerable influence on the chemical reaction rate. This is already well known from catalysis where kinks and steps on the surface show a much higher reaction probability than the atoms in a flat surface [3.22].

For thermal hydrogen atoms, the large variations of chemical erosion of different sorts of graphite (Fig. 3.15) as well as the dependence of the erosion yield on the pre-treatment of the carbon sample found by *Coulon* and *Bonnetain* [3.133] is probably due to the different crystalline structure at the surface. For pyrolytic graphite a higher reactivity of the surface perpendicular to the hexogonal basal planes (prism plane surfaces) compared to the surface parallel to the basal planes has been observed by different authors [3.25, 134]. *Gould* [3.24], and *Wood* and *Wise* [3.136] assume that the chemical reaction takes place at specific reaction sites. The density of these sites depends on surface orientation as well as on surface damage.

In order to show the dependence of reaction probability on the crystal damage, *Veprek* et al. [3.176] implanted 2 MeV He ions into pyrolytic graphite cut parallel to the basal planes. The carbon was subsequently eroded by a chemical reaction with hydrogen atoms in a high frequency low pressure discharge at 920 K. The erosion yield as a function of eroded depth was determined by the weight loss of the samples (Fig. 3.22). The erosion of the damaged graphite was found to be considerably higher and remained higher than undamaged graphite until a depth of about 20 μm was eroded, which is far more than the projected range of the 2 MeV He ions. The authors assume that lattice stress may lead to a much deeper damaged surface layer. The examination of different kinds of graphite

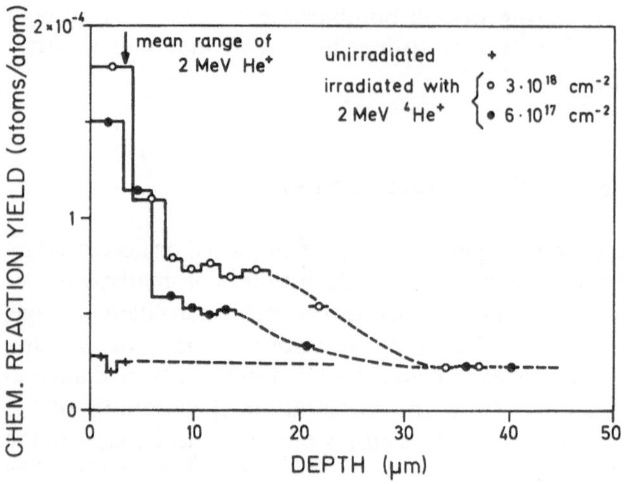

Fig. 3.22. Chemical erosion yield of graphite in a hydrogen plasma at 920 K as a function of eroded depth after prebombardment with different fluences of 2 MeV He^+ [3.176]

showed that the erosion yield in a surface layer comparable to the range of the damaging ions was almost equal though the erosion yield of the undamaged graphite showed strong differences for the different graphites used.

For the bombardment of graphite with energetic hydrogen ions no strong influence of the surface structure on the chemical sputtering yield was found. This may be due to the formation of an amorphised surface layer by implantation, as was observed by *Wright* et al. [3.40] by Raman spectroscopy, where many broken carbon bonds are reaction sites for hydrogen trapping. *Roth* et al. [3.127, 155] could show that the maximum chemical sputtering yield of graphite at 920 K has the same energy dependence as the energy deposited into nuclear collisions in the surface (Fig. 3.30). The reaction of pyrolytic graphite with atomic hydrogen increases by more than a factor of 40 due to simultaneous 5 KeV Ar bombardment [3.177]. The finding that thermal atomic hydrogen together with electron bombardment above 40 eV of the graphite surface yields enhanced methane production rates similar to the rates found with energetic hydrogen [3.178, 179] may be an indication that bonds may be broken by electronic excitation also. Recently, also UV irradiation has been shown to enhance the reaction yield with atomic hydrogen [3.180]. In all cases it has been assumed that nuclear or electronic energy transfer changes the nature of binding of surface atoms thus creating reaction sites for methane formation. The enhancement with electrons and photons is, however, very small and could not been reproduced by other researchers within the sensitivity of their equipment [3.177, 181]. Thermal atomic hydrogen in addition to energetic hydrogen bombardment, however, does not lead to further increased methane yields [3.151].

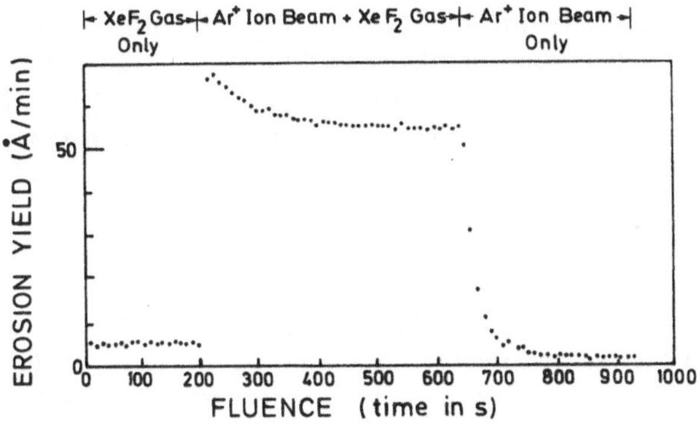

Fig. 3.23. Ion-assisted gas-surface chemistry using 450 eV Ar^+ ions and XeF_2 on silicon [3.182]

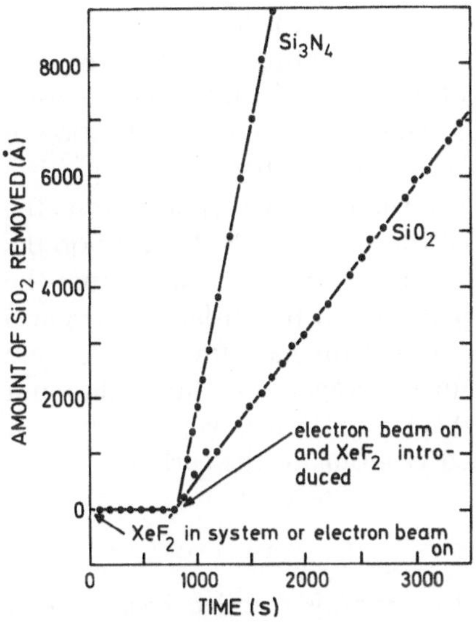

Fig. 3.24. Electron-assisted gas-surface chemistry using 1500 eV electrons and XeF_2 on SiO_2. Neither exposure to XeF_2 nor an electron beam produces etching by itself. Simultaneous exposure produces an etch rate of 200 Å/min [3.182]. The effect is even more pronounced in Si_3N_4

A similar dependence on the surface state, as was observed for graphite, has been observed in investigations of the chemical erosion of Si and SiO_2. *Coburn* and *Winters* [3.182] have found a strong increase in the etch rate of silicon by XeF_2 of F_2 molecules if the surface is simultaneously bombarded with Ar ions (Fig. 3.23).

Similarly, the etching of SiO_2 and Si_3N_4 in XeF_2 was only possible when the surface was simultaneously bombarded by electrons (Fig. 3.24 [3.182]).

It is known from Auger measurements that XeF_2 dissociatively chemisorbs on the surface of SiO_2, Si_3N_4, and SiC [3.183]. In the case of silicon, which is etched in XeF_2-gas in the absence of radiation and where the measurements indicate that the surface adsorption-dissociation process is the rate-limiting step, the strong increase in erosion yield was attributed to an increase of surface sites where XeF_2 can be dissociatively chemisorbed [3.182, 50]. SiO_2 and Si_3N_4

are not etched in XeF_2 alone. Therefore, the rate-limiting step must be the formation of the SiF_4 molecule between the adsorbed fluorine and the silicon atoms. The electron bombardment of SiO_2 may break the Si–O bond [3.184] and the oxygen could evaporate as O_2. The elemental Si left can then react readily with the fluorine and be released as SiF_4.

b) Influence of Small Impurity Concentrations

Chemical erosion can be influenced by small amounts of molecules or atoms on a surface which may act as a catalyst. *Vossen* [3.185] observed that in rf sputter etching of organic material in an oxygen glow discharge, the presence of 5–6% Cu on the surface decreased the etch rate by several orders of magnitude. Simultaneous cone formation on the eroded surface has been observed. The reduction in the etch rate was explained by local electric fields around the tips of the cones which can deviate incoming O^+ ions to the tips so that less oxygen ions can strike the exposed organic regions of the surface [3.185]. On the other hand, small amounts of copper have been shown to enhance the gasification of graphite in molecular oxygen, hydrogen or H_2O above 1100 K [3.186].

Balooch et al. [3.187] have studied the formation of methane from the reaction of H_2 molecules with graphite and its dependence on Pt traces on the surface. Without Pt no methane formation was found, as expected from the low sticking probability of H_2 on C. With platinum at the surface, which acts as catalyst to dissociate hydrogen molecules [3.188], methane formation occurs with a maximum at intermediate platinum coverage. Too little platinum provides insufficient dissociative area for the H_2 molecules while too much platinum prevents the access of the adsorbed H atoms to the carbon.

c) Molecular State of Bombarding Ions

Only in the case of bombardment with low energy reactive gases can a difference between molecular and atomic species be expected for the first stage of the chemical interaction, i.e., the adsorption of the reactive particles.

For the reaction of hydrogen at energies up to some eV with graphite there are various reported experiments showing an increase in the erosion yield by atomic compared to molecular hydrogen. *Gould* [3.24] observed an increase in erosion yield by orders of magnitude upon dissociating the introduced hydrogen molecules at a hot tungsten filament in front of the carbon sample. *Balooch* and *Olander* [3.25] could not observe any reaction of molecular hydrogen with graphite within their detection limit. Upon dissociation of the molecular hydrogen in an effusion oven heated to 2200 K they observed reaction yields up to 5×10^{-3} (Fig. 3.15a). This difference was attributed to the difference in sticking probability between atomic and molecular hydrogen. *Wood* and *Wise* [3.136], *Rosner* and *Allendorf* [3.132] and *Coulon* and *Bonnetain* [3.133] extracted atomic hydrogen from an rf discharge to observe measureable reaction yields.

For the reaction of graphite with oxygen, *Rosner* and *Allendorf* observed a similar difference between atomic and molecular oxygen [3.132]. Figure 3.15b shows the temperature dependence of the erosion yield for O atoms and O_2 molecules (Curve *e*). Similar differences were found by *Nordine* for the attack of metals with fluorine [3.159].

Recently chemical erosion of Si by SF_6 molecules could be stimulated by simultaneous laser irradiation [3.95], whereas Si is not eroded by laser irradiation or in SF_6 gas alone. From the dependence of the erosion yield on gas pressure, laser pulse energy as well as from surface analytic measurements, it could be concluded that the SF_6 molecules are dissociated by the laser irradiation into SF_5^+ and F^- ions. In addition, the collision between excited SF_6 atoms and the Si surface results in dissociative chemisorption. The fluorine ions react readily with Si surface atoms to form volatile SiF_4 molecules.

3.5.3 Mechanism of Chemical Sputtering

Chemical sputtering is a multi-step process which finally leads to the formation of the volatile molecule. Out of the limited number of observed cases of chemical sputtering, the reaction of H with graphite has received most interest and most work has been done to investigate the kinetics of CH_4 and C_2H_2 formation, covering the energy range from thermal energies to 30 keV H^+. There are other systems where reaction kinetics have been proposed such as, for example, the reaction of F with silicon [3.156] and refractory metals [3.159] or the reduction of oxides by H [3.189], but as an example only the H–C reaction will be treated here in more detail.

The reaction of hydrogen with carbon in thermodynamic equilibrium gives a qualitative temperature dependence for methane and acetylene formation. A detailed kinetic model has been developed for the surface reaction of atomic hydrogen with pyrolytic graphite. Subsequently, the application of these results to the case of energetic hydrogen ions will be discussed.

a) Thermodynamic Equilibrium Reaction of Hydrogen and Carbon

Before analysing the data shown in Fig. 3.15a in detail, an approach to the hydrogen-carbon interaction is to calculate the equilibrium partial pressure of different gaseous hydrocarbons above a carbon surface heated in a hydrogen atmosphere. In Fig. 3.25 the results for the partial pressures are shown for H_2, H, CH_4, and C_2H_2. Only these gases have been considered as experiments show that other hydrocarbons can be neglected. At temperatures above 3000 K similar calculations show the importance of other hydrocarbons and carbon sublimation [3.190].

The calculations have been performed using the following set of equations

$$nC + \frac{m}{2}H_2 \rightleftarrows C_nH_m \quad n=0,1,2, \quad m=1,4,2, \tag{3.6}$$

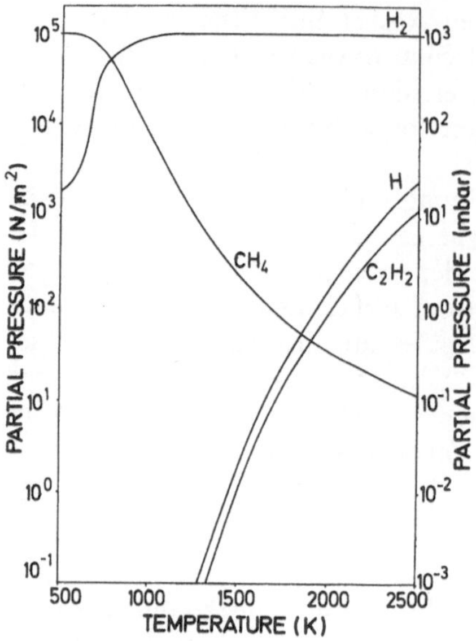

Fig. 3.25. Equilibrium partial pressure of hydrogen and hydrocarbon gases above carbon. The total pressure is constant at 1 atm

where the reaction constants are taken from the entropies and enthalpies of formation listed in the JANAF tables [3.191]. These equations are coupled with the assumption of a constant total pressure:

$$p_{H_2} + p_H + p_{CH_4} + p_{C_2H_2} = 1 \text{ atm}. \tag{3.7}$$

A comparison with molecular beam experiments by *Balooch* and *Olander* [3.25] (Fig. 3.15a, Curve *g*) already shows qualitative agreement: the clear separation of reaction products is predicted by thermodynamic equilibrium. Below about 800 K, CH_4 is the dominant reaction product, whereas C_2H_2 evolves at temperatures above 1200 K. Between 1000 and 2000 K the main residual gas is molecular hydrogen.

In order to obtain a more quantitative agreement, *Balooch* [3.192] modified the thermal equilibrium calculations assuming an atomic hydrogen gas in thermal and chemical equilibrium with the solid surface which produced the same number of adsorbed hydrogen atoms as the impinging atomic hydrogen beam J_H with a sticking coefficient η_s for hydrogen atoms [3.193, 195]. He introduced the production rates for hydrocarbons

$$R_{C_nH_m} \propto \eta_{C_nH_m} p_{C_nH_m}, \tag{3.8}$$

the $\eta_{C_nH_m}$ being the equilibration probability of the molecules on the surface. Additionally, he replaced (3.7) by a relation for hydrogen conservation

$$\eta_s J_H = \sum_{n,m} m R_{C_nH_m}. \tag{3.9}$$

An agreement could only be obtained assuming that atomic hydrogen and methane molecules only partly reach equilibrium at the carbon surface, the equilibration probability decreasing strongly with increasing temperature.

The need for these assumptions show that molecular and ion beam experiments represent nonequilibrium conditions. The sticking coefficient of impinging atoms, the diffusion from the end of range back to the surface, surface diffusion and the equilibration probability on the surface enter into the consideration, being functions of particle energy, surface coverage and/or target temperature.

Although the thermodynamic equilibrium considerations may give qualitative indications for the reaction products and temperature dependence, kinetic models should be better suited for molecular and ion beam techniques.

b) Kinetic Reaction Model for Thermal Atomic Hydrogen

For the reaction of a beam of atomic hydrogen with a carbon surface, different kinetic models have been proposed [3.24, 25, 136]. The most detailed one by *Balooch* and *Olander* [3.25] explains the results of their experiments using a modulated atomic hydrogen beam in the apparatus described in Sect. 3.2.1a. The parameters entering in the model are derived by comparing the predicted dependence of values such as the reaction probability and the phase lag on temperature and beam intensity with the measured dependencies.

The basic idea for the kinetic models is the adsorption of hydrogen on a carbon surface, the formation of C–H bond and the successive addition of hydrogen atoms until CH_4 is desorbed. The equilibrium distances and potential energies of different hydrocarbons on a carbon surface are shown schematically in Fig. 3.26.

The chemical reactions on the surface assumed in the kinetic model are summarised schematically in Table 3.3 together with the reaction constants in the notation of [3.25] with $R = 1.99 \times 10^{-3}$ kcal/mole K being the gas constant.

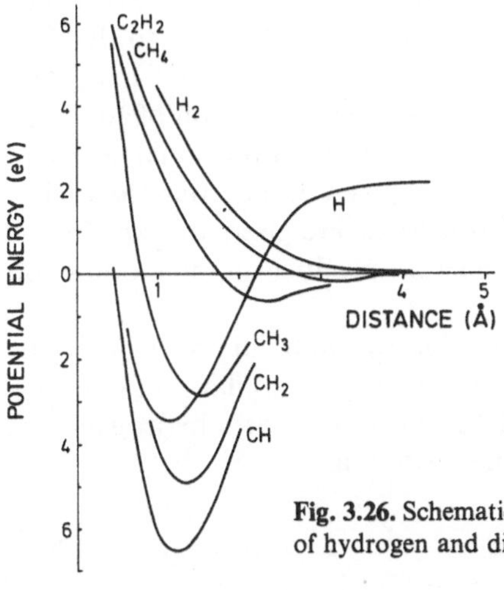

Fig. 3.26. Schematic diagram of the potential energies for the adsorption of hydrogen and different hydrocarbons on a carbon surface

Table 3.3

		Prism plane	Basal plane
H(g) $\xrightarrow{\eta_s}$ H(ads)	η_s	0.02	0.006
H(ads) $\overset{H\sqrt{D}}{\rightleftharpoons}$ H(sol'n)	$H\sqrt{D}$	$2.7\times10^{6}\,e^{\frac{-9.6}{RT}}$	$3.85\times10^{4}\,e^{\frac{-5.4}{RT}}$
H(ads)+H(ads) $\xrightarrow{k_2^e}$ H_2 (g)	k_2^e	$1.06\times10^{-2}\,e^{\frac{-15.9}{RT}}$	$1.30\times10^{-4}\,e^{\frac{-18.5}{RT}}$
H(ads)+C $\overset{K_1}{\rightleftharpoons}$ CH(ads)			
CH(ads)+H(ads) $\xrightarrow{K_2}$ CH_2(ads) CH_2(ads)+H(ads) $\xrightarrow{k_3}$ CH_3(ads) CH_3(ads)+H(ads) $\xrightarrow{\text{fast}}$ CH_4(g) methane branch low temperature	$K_1\,K_2\,k_3$	$1.27\times10^{-18}\,e^{\frac{-3.3}{RT}}$	$2.17\times10^{-21}\,e^{\frac{0.9}{RT}}$
CH(ads)+CH(ads) $\xrightarrow{k_2}$ C_2H_2(g) acetylene branch high temperature	$K_1^2\,k_2$	$1.59\,e^{\frac{-32.5}{RT}}$	$1.82\times10^{-3}\,e^{\frac{-26.8}{RT}}$

Molecular hydrogen does not adsorb on a carbon surface [3.25], whereas atomic hydrogen is strongly bound. The hydrogen atoms adsorbed may then react with surface carbon atoms leading to the formation of CH, CH_2, CH_3 which have gradually lower binding energies and a slightly larger equilibrium distance [3.195]. Finally, methane is formed which is physically adsorbed with a low binding energy of about 0.2 eV and at an equilibrium distance of 3.2 Å [3.196]. It will desorb spontaneously at temperatures above 50 K. Acetylene is formed on the surface by a combination of two CH groups.

The elementary steps for both products are combined in a way which produces an overall reaction order greater than unity. The best fit to the data has been achieved assuming the methane formation proportional to c_H^3 and acetylene proportional to c_H^2, where c_H is the surface concentration of adsorbed hydrogen atoms. Recombination of atomic hydrogen proportional to c_H^2 to form molecular hydrogen, which then desorbs, is taken into account as well as hydrogen diffusion into the bulk. These processes are used to explain the response of the phase lag of the desorbed reaction product on the modulation frequency.

During bombardment a surface concentration c_H will build up which can be calculated from a mass balance equation for hydrogen similar to (3.9). In a steady-state experiment where the solid is saturated with hydrogen and diffusion into the bulk is absent, this can be written as

$$\eta_s' J_H = 2k_2^e c_H^2 + 2K_1^2 k_2 c_H^2 + 4K_1 K_2 k_3 c_H^3 , \qquad (3.10)$$

where $\eta_s' = \eta_s\,(1 - c_H/c_0)$ is the sticking coefficient on a surface covered with hydrogen.

The supply term on the left-hand side takes surface saturation into account as the surface concentration c_H approaches the saturation concentration c_0. The saturation concentration depends on temperature [3.78, 80] and has been determined experimentally to $4.5 \times 10^{13}\,\mathrm{H\,cm^{-2}}$ thermal atomic hydrogen at room temperature (Fig. 3.9). Trapping to much higher concentrations has recently been reported by *Flaskamp* et al. [3.199, 200] but in this case, absorption in the bulk rather than adsorption on the surface may have accounted for the high values.

The loss terms on the right-hand side describe the surface recombination and desorption of molecular hydrogen, the loss of hydrogen due to acetylene and methane formation, respectively.

Solving (3.10) for the surface concentration c_H, the carbon erosion yield is given by

$$Y = \frac{1}{J_H}(K_1 K_2 k_3 c_H^3 + 2K_1^2 k_2 c_H^2), \tag{3.11}$$

where the first term in the brackets on the right-hand side represents methane formation while the second term, acetylene formation.

Low Surface Coverage. In the atomic beam experiments mentioned above [3.25], the hydrogen flux and the sticking probability were low enough to prevent effects of surface saturation ($c_H \ll c_0$). A comparison of (3.11) with the measured yields allows the values for the combinations of the reaction constants $K_1 K_2 k_3$ for the methane and $2K_1^2 k_2$ for the acetylene formation to be determined. The derived values [3.25] are also given in Table 3.3.

The hydrogen-sticking coefficient η_s extracted from the data is in good agreement with values in the literature [3.24, 199–202] which range between 0.015 and 0.004. In this experiment, even at the highest beam intensities of $6 \times 10^{16}\,\mathrm{cm^{-2}\,s^{-1}}$ the average surface concentration was still less than 3×10^{11} atoms $\mathrm{cm^{-2}}$ resulting in a coverage of less than 10^{-3} H/C atoms. In this reaction model a substantial mobility of hydrogen atoms on the graphite surface must be assumed at all temperatures. Above 1000 K the C_2H_2 formation also implies a large surface mobility of CH species. No detectable amount of the free radicals CH, CH_2, CH_3 were observed to be emitted from the surface in these experiments which confirms that these compounds are strongly bound to the graphite surface (Fig. 3.26). Only when a stable gaseous hydrocarbon CH_4 or C_2H_2 is formed can it leave the surface immediately.

The temperature dependence of the reaction probability for C_2H_2 and CH_4 as calculated from the model for pyrolytic graphite cut parallel to the prism and basal plane are shown in Fig. 3.27 together with the experimental values.

The results from quasi-equilibrium calculations [3.192] mentioned in Sect. 5.3.1 are also included in Fig. 3.27. For the dependence of the reaction

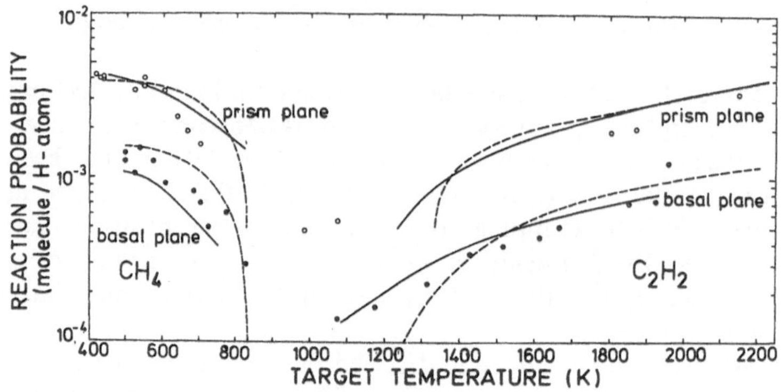

Fig. 3.27. Comparison of the temperature dependence of measured reaction probabilities [3.25] for CH_4 and C_2H_2 formation with the kinetic model (——) and results from a quasi-equilibrium calculation (---) [3.192]

probability on beam intensity and modulation frequency as well as for the dependence of the phase lag on temperature and modulation frequency, the agreements between the measured and calculated values are also good.

The kinetic model has been extended by *Balooch* and *Olander* [3.25] to high pressure gasification data of carbon in a hydrogen gas. The model fits the high pressure data surprisingly well using the kinetic constants obtained at a pressure about 6 orders of magnitude lower. The best fit was again obtained setting the sticking coefficient of molecular hydrogen to zero and regarding only atomic hydrogen in equilibrium with the H_2 atmosphere. Therefore, the surface coverage with hydrogen still remains below 0.1% at $P_{H_2}=1$ atm.

High Surface Coverage. For higher atomic hydrogen fluxes, as used in most experiments which gave the results shown in Fig. 3.15, surface coverage saturation effects have to be taken into account. Especially at low temperatures, where the recombination of hydrogen is low and only methane formation reduces the surface concentration, the saturation value c_0 is reached. At low temperatures the hydrogen concentration remains constant and an increase in methane formation with increasing temperature is expected reflecting an increase in the reaction constant $K_1K_2k_3$. At higher temperatures surface recombination reduces the hydrogen concentration and the methane formation decreases.

Indeed, most of the experimental data show a maximum in methane production at a temperature T_{max} around 800–900 K. In these experiments hydrogen fluxes up to $10^{21}\,H^+\,cm^{-2}\,s^{-1}$ [3.132, 135, 136] were used.

Solving (3.11) for c_H and taking the surface saturation value c_0 into account, the rate for CH_4 formation can be shown to have a maximum. The temperature T_{max} of the maximum depends on the hydrogen current density used and on the saturation surface coverage c_0. Using the same kinetic constants as obtained from the atomic beam experiments [3.25], the dependence of the reaction rate

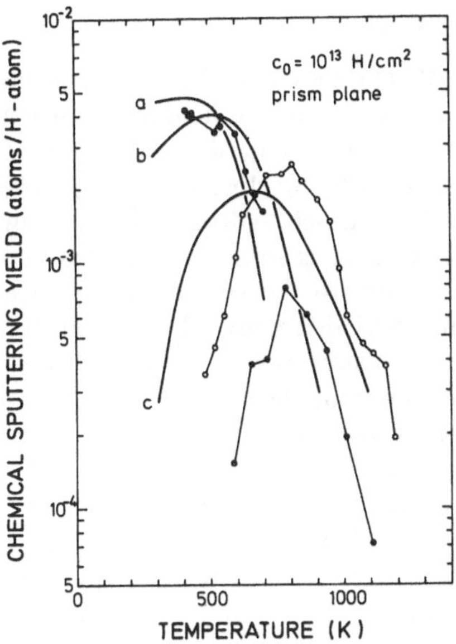

Fig. 3.28. Comparison of the temperature dependence of the reaction probability for CH_4 from experiment using different hydrogen fluxes J_H [3.25, 132, 136] with the kinetic model, assuming a surface saturation c_0 of 10^{13} H cm^{-2} [(*a*) $J_H = 10^{15}$ H cm^{-2} s^{-1}, (*b*) $J_H = 10^{17}$ H cm^{-2} s^{-1}, (*c*) $J_H = 10^{19}$ H cm^{-2} s^{-1}]

on temperature has been calculated for fluxes between 10^{15} and 10^{21} H cm^{-2} s^{-1}. The results of these calculations are shown in Fig. 3.28 for a prism plane surface together with experimental results [3.25, 132, 136].

For atomic hydrogen fluxes larger than 10^{17} H cm^{-2} s^{-1}, this model gives a maximum in the CH_4 formation appearing at temperatures T_{max} above room temperature as observed experimentally. At the fluxes used by *Wood* and *Wise* [3.136] and *Rosner* and *Allendorf* [3.132], i.e., 5×10^{19} H cm^{-2} s^{-1}, T_{max} is of the order of 550–650 K as compared to experimental values of 770–850 K. One possible reason for this slight disagreement might be that the saturation coverage c_0 determined by *Gould* for room temperature is substantially lower at elevated temperatures, as has been found for the bulk saturation concentration [3.78, 80] of implanted hydrogen. To explain still higher T_{max}, as reported recently by *Brewer* et al. [3.135], this model fails. Additional effects such as, for example, desorption of hydrogen due to ion, electron or photon impact from the glow discharge, then have to be taken into account.

The kinetic model by *Balooch* and *Olander* [3.25], while describing adequately a large number of experimental results, fails to describe the absolute chemical erosion yield observed in some other experiments [3.24, 133]. The uniqueness of the model, in principle, thus remains doubtful. Other kinetic models, however, cannot satisfactorily describe the set of kinetic data obtained by [3.25], e.g., the kinetic model proposed by *Wood* and *Wise* [3.136] which assumes the addition of H to carbon atoms until a CH_3 radical is reached, which is then converted to CH_4 by an impinging H_2 molecule. Moreover, the model does not allow methane formation without the presence of molecular hydrogen.

c) Erosion of Graphite by Energetic Hydrogen Ions

For hydrogen bombardment of carbon with energies much above thermal energies, part of the ions are kinetically reflected while the remainder are buried beneath the surface within the implantation range. Generally, reflection is neglected, i.e., the sticking probability is assumed to be close to unity [3.76] independent of flux and fluence. In steady state, as much hydrogen diffuses out from the bulk through the surface as implanted. It is reasonable to assume [3.73] that the hydrogen diffusing from the bulk to the surface reacts in a similar way with surface atoms as hydrogen atoms thermally adsorbed on the surface. However, to explain the decrease in erosion at lower temperatures, mechanisms other than the coverage dependence of the sticking coefficient (3.11) must limit the hydrogen concentration at low temperature. *Erents* et al. [3.73, 130] proposed for this mechanism an ion induced desorption process. They further neglected the depletion of hydrogen due to methane and acetylene formation and accounted for thermal desorption of hydrogen by a term $(c_H)^n/\tau$ corresponding to a mean surface residence time $\tau=\tau_0\exp(+Q_2/RT)$. The hydrogen balance equation can be written as

$$J_H=\frac{(c_H)^n}{\tau}+J_H\sigma c_H\,. \tag{3.12}$$

Here σ is the cross section for ion bombardment desorption. Equation (3.12) is equivalent to (3.10) for $\eta=1$, if $\sigma=1/c_0$ and $\tau=1/2k_2^e$.

$$J_H(1-\sigma c_H)=(1/\tau)(c_H)^n\,. \tag{3.13}$$

Equation (3.12) can be solved for c_H and *Erents* et al. [3.73, 131] assumed the rate of methane formation to be proportional to $(c_H)^m$ with a thermal activation energy Q_1. Contrary to the kinetic model, the best empirical fit to the measured data (Fig. 3.15) could be obtained for $m=n=1$. For this case, they get for the chemical sputtering yield

$$Y\propto\frac{e^{-Q_1/RT}}{J_H\sigma+\tau_0^{-1}e^{-Q_2/RT}} \tag{3.14}$$

and the maximum sputtering yield occurs at a temperature T_{max}:

$$T_{max}=\frac{Q_2}{R}\ln\left[\frac{(Q_2-Q_1)}{J_H\tau_0\sigma Q_1}\right]^{-1}. \tag{3.15}$$

Their experimental values for 20 keV H^+ erosion of pyrolytic graphite could be fitted to (3.14) using $\sigma=10^{-16}\,cm^2$, $Q_1=25$ kcal/mole and $Q_2=47$ kcal/mole and $\tau_0=10^{-13}$ s [3.73]. They predicted a change of T_{max} with energy as observed by *Roth* et al. [3.127] as σ is supposed to vary with energy

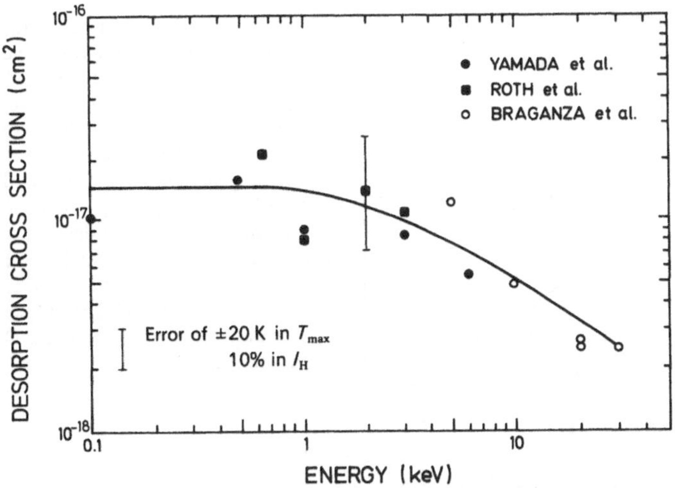

Fig. 3.29. Calculated desorption cross sections from measurement of T_{max} [3.127, 136, 131] using (3.15)

[3.203, 204], whereas *Smith* and *Meyer* [3.205] pointed out that the change in J_H has a strong influence on T_{max} [3.130, 144]. Taking both dependencies together, the ion-induced desorption cross section is calculated for energies between 100 eV and 30 keV from the data of *Braganza* et al. [3.130], *Roth* et al. [3.127], and *Yamada* et al. [3.131], and is shown in Fig. 3.29. The values for σ are lower than ion-induced desorption cross-sections for H on W, Mo, and SS [3.206], which is not surprising in view of the strong chemical binding to the graphite surface (Fig. 3.26). In view of (3.13) the cross section σ is related to the surface-saturation concentration $c_0 = 1/\sigma$. The resulting large values for c_0 of about 10^{17} H cm^{-2} indicate that within a surface layer of about 200 Å all trapped hydrogen is involved in the chemical reaction. Recently, measured values for the surface roughness factor for graphite, being for the densest pyrolytic graphite as high as 40 [3.207, 208] would, however, reduce this value to a surface layer of 5 Å.

Once the energy dependence of the desorption cross section is known, the model of *Erents* et al. [3.73] also gives an energy dependence of the maximum chemical sputtering yield at T_{max} (3.11). This dependence is shown in Fig. 3.30 and has been fitted to *Braganza* et al.'s data at 10–30 keV [3.130]. Also shown in this figure are the experimental measured chemical sputtering yields, measured at a hydrogen or deuterium flux of $J_H = 10^{15}$ H cm^{-2} s^{-1} [3.155] using the weight-loss method. Determining the chemical sputtering yield from the CH_4 formation alone yields a slightly different energy dependence [3.131, 209, 155]. Furthermore, C_2H_4 hydrocarbon formation becomes important at ion energies below 1 keV [3.155].

In contradiction to Erents model, (3.14), at intermediate energies between 0.3 and 0.6 keV, a maximum is found. This was already realized by *Roth* et al.

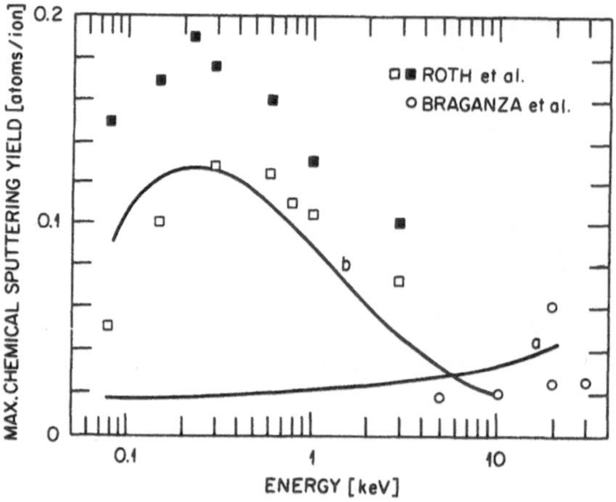

Fig. 3.30. Energy dependence of the maximum chemical sputtering yield at T_{max}. Experimental data [3.130, 155] have been measured or reduced to a hydrogen flux of $J_H = 10^{15}$ H cm^{-2} s^{-1} using (3.14). Curve *a* gives the dependence of (3.14). For comparison, Curve *b* shows the dependence of the energy deposited in the carbon surface for energy transfers larger than 8 eV as calculated by TRIM [3.210]. Open symbols are for H, full symbols for D bombardment

[3.127] and *Yamada* et al. [3.131]. It has been argued that radiation damage introduced into the surface layer by nuclear collisions is responsible for the increase in yield. This is emphasized by the higher yields for deuterium compared to hydrogen ions. For comparison, the energy deposited into nuclear collisions which transfer more than 8 eV, as calculated by TRIM computer simulation [3.210], is also shown in Fig. 3.30. The influence of radiation damage on chemical reactions has already been emphasised in Sect. 3.5.2a.

Erents et al.'s model [3.73] is based on *Balooch* and *Olander's* [3.25] equations but has been strongly simplified. The thermal desorption term is assumed to be proportional to c_H instead of c_H^2 as expected during recombination. The reaction depends only linearly on the surface concentration, while for methane formation it should be proportional to c_H^3 as in the kinetic model. An inclusion of a desorption term proportional to c_H^2 and a methane formation probability proportional to c_H^3 could not yield overall fitting of the data [3.130]. The model for high ion energies, which describes the experimental results reasonably, must, therefore, be considered as an empirical description of the surface interactions. It gives a good description of the maximum of the sputtering yield at T_{max} and its dependence on energy and ion flux, but to calculate absolute values, the energy deposited into surface damage has to be taken into account. Values extracted from the model, as the ion-induced desorption cross section σ or the activation energies for the chemical reaction, are primarily empirically-fitting parameters.

In addition to the maximum in temperature, *Gould* [3.24] as well as *Busharov* [3.129, 149] have observed a maximum of the reaction rate with ion fluence for hydrogen atoms at thermal energies and for energetic hydrogen ions, respectively. This has been tentatively explained by *Gould* with an interaction of adjacent CH_3 complexes which blocks the CH_4 formation because the surface coverage increases with ion dose. *Yamada* [3.131] could show for energetic hydrogen ions that this initial maximum of the chemical sputtering yield is strongly dependent on the hydrogen concentration in the target prior to bombardment. If the target was prebombarded at room temperature, a strong maximum in the chemical erosion rate at 900 K occurred at 1×10^{17} H cm^{-2}, while without prebombardment at room temperature, no maximum could be observed.

In summary, the models presented can reasonably describe the experimental results for the erosion of carbon with atomic thermal hydrogen. Thermodynamic equilibrium already predicts CH_4 and C_2H_2 to be the dominant reaction products, CH_4 being formed at low temperatures and C_2H_2 prevailing above 1000 K. The kinetic models developed for molecular beam experiments must also take into account surface coverage effects. At high hydrogen energies, ion-induced desorption may limit the surface coverage and radiation damage may be responsible for the energy dependence of the reaction yield.

Similar models have been proposed for the system O on C and F on UO_2 by *Olander* et al. [3.138, 139, 197]. For most other known systems of chemical sputtering, similar models have not yet been developed. The hydrogen-carbon kinetic model has been presented here in some detail to elucidate the difficulties involved in determining the detailed mechanism of chemical sputtering. But even in this case, only the formation of methane at low temperatures and acetylene at high temperatures are studied in detail. The observations by *Rye* [3.154] of a vast variety of other hydrocarbons formed in small quantities in the interaction of H with C at elevated temperatures has not yet been satisfactorily understood.

Acknowledgement. In preparing this contribution during the years 1977 to 1981, I had many clarifying discussions with H. Winters, S. Veprek, K. Erents, and D. Gruen. Their suggestions as well as the invaluable assistance of R. Behrisch and J. Bohdansky are gratefully acknowledged.

References

3.1 W.R.Grove: Philos. Trans. R. Soc. London **11**, 87 (1852)
3.2 V.Kohlschütter: Jahrb. Radioakt. **9**, 355 (1912)
3.3 A.Güntherschulze: Z. Phys. **36**, 563 (1926)
3.4 R.Behrisch (ed.): *Sputtering by Particle Bombardment* I, Topics Appl. Phys., Vol. 47 (Springer, Berlin, Heidelberg, New York 1981)
3.5 S.M.Myers: Radiat. Eff. **49**, 95 (1980)
3.6 E.Giani, D.K.Murti, R.Kelly: Proc. Symp. on Thin Film Phenomena – Interfaces and Interactions, The Electrochem. Soc. Proc. Vol. **78-2**, 443 (1978)
3.7 J.P.Biersack: Radiat. Eff. **19**, 249 (1973)
3.8 F.Schulz, K.Wittmaack: Radiat. Eff. **29**, 31 (1976)
3.9 Z.L.Liau, J.W.Mayer: J. Vac. Sci. Technol. **15**, 1629 (1978)
3.10 P.Sigmund, A.Gras-Marti: Nucl. Instrum. Methods **168**, 389 (1980)
3.11 A.Gras-Marti, P.Sigmund: In *Proc. Symp. on Sputtering* (SOS), eds. by P.Varga, G.Betz, F.P.Viehböck (Techn. Univ. Wien, 1980) p. 512
3.12 U.Littmark, W.O.Hofer: Nucl. Instrum. Methods **168**, 329 (1980)
3.13 O.Almén, G.Bruce: Nucl. Instrum. Methods **11**, 279 (1961)

3.14 P.Sigmund: In *Inelastic Ion-Surface Collisions*, eds. by N.H. Tolk, J.C. Tully, W. Heiland, C.W. White (Academic Press, New York 1977) p. 121
3.15 ERDA Task group (1978)
3.16 D.M.Gruen, S.Veprek, R.B.Wright: In *Plasma Chemistry* I, Topics Current Chem. (Springer, Berlin, Heidelberg, New York 1980)
3.17 Y.-Y.Tu, T.J.Chuang, H.F.Winters: In *Proc. Symp. on Sputtering* (SOS), eds. by P. Varga, G. Betz, F.P. Viehböck (Techn. Univ. Wien, 1980) p. 337
3.18 J.Bohdansky, J.Roth, H.L.Bay: J. Appl. Phys. **51**, 2861 (1980)
3.19 G.K.Wolf: *Topics in Current Chem.* **85**, 1 (1979)
3.20 Proc. 1st Intern. Conf. Ion Beam Modification of Materials, eds. by J. Gyulai, P. Lohner, E. Pasztor. Radiat. Eff. **47–49** (1980)
3.21 Proc. 2nd Intern. Conf. Ion Beam Modification of Materials, eds. by R.E. Benenson, E.N. Kaufmann, G.L. Miller, W.W. Scholz. Nucl. Instrum. Methods **182/183** (1981)
3.22 G.A.Somorjai: Surf. Sci. **89**, 496 (1979)
3.23 Proc. of 4th Intern. Symp. on Plasma Chemistry, eds. by S. Veprek, J. Hertz, Univ. Zürich, Switzerland (Aug. 1979)
3.24 R.K.Gould: J. Chem. Phys. **63**, 1825 (1975)
3.25 M.Balooch, D.R.Olander: J. Chem. Phys. **63**, 4772 (1975)
3.26 G.M.McCracken, J.H.C.Maple, H.H.H.Watson: Rev. Sci. Instr. **37**, 860 (1966)
3.27 R.H.Jones, D.R.Olander, W.J.Siekhaus, J.A.Schwarz: J. Vac. Sci. Technol. **9**, 1429 (1972)
3.28 J.W.Coburn, E.Kay: In *Proc. 7th Intern. Vacuum Congr. and 3rd Intern. Conf. on Solid Surf.*, eds. by R. Dobrozemsky, F. Rüdenauer, F.P. Viehböck, A. Breth, Vienna (1977) p. 1257
3.29 H.Oechsner, W.Gerhard: Phys. Lett. **40**A, 211 (1972)
3.30 J.Bates, D.M.Gruen, R.Varma: Rev. Sci. Instrum. **47**, 1506 (1976)
3.31 D.M.Gruen, S.L.Gaudioso, R.L.McBeth, J.L.Lerner: J. Chem. Phys. **60**, 89 (1974)
3.32 J.S.Shirk, A.M.Bass: J. Chem. Phys. **52**, 1894 (1970)
3.33 D.H.Carstens, D.M.Gruen, J.F.Kotzlowski: High Temp. Sci. **4**, 436 (1972)
3.34 D.M.Gruen, P.A.Finn, D.L.Page: Nucl. Technol. **29**, 309 (1976)
3.35 P.A.Finn, D.M.Gruen, D.L.Page: In *Radiation Effects on Solid Surfaces*, Adv. in Chem. Series **158**, 30 (1976)
3.36 D.W.Green, W.Korfmacher, D.M.Gruen: J. Chem. Phys. **58**, 404 (1973)
3.37 T.E.Parker, R.Kelly: J. Phys. Chem. Solids **36**, 377 (1975)
3.38 J.Roth, W.Eckstein, J.Bohdansky: Radiat. Eff. **48**, 231 (1980)
3.39 H. v. Seefeld, R.Behrisch, B.M.U.Scherzer, Ph.Staib, H.Schmidl: Proc. 7th Intern. Conf. Atomic Collisions in Solids (Moscow State University Publishing House 1981) p. 327
3.40 R.B.Wright, R.Varma, D.M.Gruen: J. Nucl. Mater. **63**, 415 (1976)
3.41 D.M.Gruen, B.Siskind, R.B.Wright: J. Chem. Phys. **65**, 363 (1976)
3.42 J.W.Coburn, H.F.Winters, T.J.Chuang: J. Appl. Phys. **48**, 3532 (1977)
3.43 S.K.Erents: Inst. Phys. Conf. Ser. **28**, 318 (1976)
3.44 A.R.Krauss, D.M.Gruen: Appl. Phys. **14**, 89 (1977)
3.45 H.Oechsner, L.Reichert: Phys. Lett. **23**, 90 (1966)
3.46 H.Oechsner: Z. Phys. **238**, 433 (1970)
3.47 A.Benninghoven, C.A.Evans, Jr., R.A.Powell, R.Shimizu, H.A.Storms (eds.): *SIMS* II, Springer Ser. Chem. Phys., Vol. 9 (Springer, Berlin, Heidelberg, New York 1979)
A.Benninghoven, J.Giber, J.László, M.Riedel, H.W.Werner (eds.): *SIMS* III, Springer Ser. Chem. Phys., Vol. 19 (Springer, Berlin, Heidelberg, New York 1982)
3.48 G.P.Können, J.Grosser, A.Haring, A.E. de Vries, J.Kistemaker: Radiat. Eff. **21**, 171 (1974)
3.49 M.Szymonski, H.Overeijnder, A.E. de Vries: Radiat. Eff. **36**, 189 (1978)
3.50 R.A.Haring, A.Haring, F.W.Saris, A.E. de Vries: Appl. Phys. Lett. **41**, 174 (1982)
3.51 D.Hammer, E.Benes, P.Blum, W.Husinsky: Rev. Sci. Instrum. **47**, 1178 (1976)
3.52 A.Elbern, E.Hintz, B.Schweer: J. Nucl. Mater. **76/77**, 143 (1978)
3.53 E.Dullni, E.Hintz: Verhandl. DPG (VI) **16**, 908 (1981)
3.54 R.S.Nelson: In *Applications of Ion Beams to Metals*, eds. by S.T.Picraux, E.P. EerNisse, F.L. Vook (Plenum Press, New York 1974) p. 221
3.55 J.M.Poate: Radiat. Eff. **49**, 81 (1980)
3.56 J.M.Poate: J. Vac. Sci. Technol. **15**, 1636 (1978)

3.57 E.N.Kaufmann, L.Buene: Nucl. Instrum. Methods **182/183**, 327 (1981)

3.58 J.F.Ziegler, C.P.Wu, P.Williams, C.W.White, B.Terreault, B.M.U.Scherzer, R.L.Schulte, E.J.Schneid, C.W.Magee, E.Ligeon, J.L.Écuyer, W.A.Lanford, F.J.Kuehne, E.A.Kamykowski, W.O.Hofer, A.Guivarch, C.H.Filleux, V.R.Deline, C.A.Evans, Jr., B.L.Cohen, G.J.Clark, W.K.Chu, C.Brassard, R.S.Blewer, R.Behrisch, B.R.Appleton, D.D.Allred: Nucl. Instrum. Methods **149**, 19 (1978)

3.59 J.Bøttiger: J. Nucl. Mater. **78**, 161 (1978)

3.60 Proc. 3rd Intern. Conf. on Plasma Surface Interaction in Controlled Fusion Devices. J. Nucl. Mater. **76/77** (1978)
Proc. 4th Intern. Conf. on Plasma Surface Interaction in Controlled Fusion Devices. J. Nucl. Mater. **94/95** (1980)
Proc. 5th Intern. Conf. on Plasma Surface Interaction in Controlled Fusion Devices. J. Nucl. Mater. **111/112** (1982)

3.61 E.S.Hotston, G.M.McCracken: J. Nucl. Mater. **68**, 277 (1977)

3.62 W.M.Müller, J.P.Blackledge, G.G.Libowitz (eds.): *Metal Hydrides* (Academic Press, New York 1968)

3.63 *Gase und Kohlenstoff in Metallen*, eds. by E.Fromm, E.Gebhardt (Springer, Berlin, Heidelberg, New York 1976)

3.64 M.K.Sinha, J.Roth, J.Bohdansky: *Proc. 9th Symp. on Fusion Technology* (Garmisch-Partenkirchen, FR Germany 1976) p. 41

3.65 J.Bohdansky, J.Roth, M.K.Sinha, W.Ottenberger: J. Nucl. Mater. **63**, 115 (1976)

3.66 I.Sheft, A.M.Reis, Jr., D.M.Gruen, S.W.Paterson: J. Nucl. Mater. **59**, 1 (1976)

3.67 W.Möller, J.Bøttiger: J. Nucl. Mater. **88**, 95 (1980)

3.68 O.C.Yonts, R.A.Strehlow: J. Appl. Phys. **32**, 192 (1961)

3.69 R.H.Stulen: Appl. Surf. Sci. **5**, 212 (1980)

3.70 A.Pontau, K.L.Wilson, F.Greulich, L.G.Haggmark: J. Nucl. Mater. **91**, 343 (1980)

3.71 G.M.McCracken, J.H.Maple: Brit. J. Appl. Phys. **18**, 919 (1967)

3.72 D.M.Gruen, R.Varma, R.B.Wright: J. Chem. Phys. **64**, 5000 (1976)

3.73 S.K.Erents, C.M.Braganza, G.M.McCracken: J. Nucl. Mater. **63**, 399 (1976)

3.74 R.A.Langley, R.S.Blewer, J.Roth: J. Nucl. Mater. **76/77**, 313 (1978)

3.75 B.M.U.Scherzer, R.A.Langley, W.Möller, J.Roth, R.Schulz: Nucl. Instrum. Methods **194**, 497 (1982)

3.76 G.Staudenmaier, J.Roth, R.Behrisch, J.Bohdansky, W.Eckstein, P.Staib, S.Matteson, S.K.Erents: J. Nucl. Mater. **84**, 149 (1979)

3.77 W.R.Wampler, D.K.Brice, C.W.Magee: J. Nucl. Mater. **102**, 304 (1981)

3.78 B.M.U.Scherzer, R.Behrisch, W.Eckstein, U.Littmark, J.Roth, M.K.Sinha: J. Nucl. Mater. **63**, 100 (1976)

3.79 J.Roth, B.M.U.Scherzer, R.S.Blewer, D.K.Brice, S.T.Picraux, W.R.Wampler: J. Nucl. Mater. **94/95**, 601 (1980)

3.80 B.L.Doyle, W.R.Wampler, D.K.Brice: J. Nucl. Mater. **103/104**, 513 (1981)

3.81 P.L.Mattern, G.J.Thomas, W.Bauer: J. Vac. Sci. Technol. **13**, 430 (1976)

3.82 B.Siskind, D.M.Gruen, R.Varma: J. Vac. Sci. Technol. **14**, 537 (1977)

3.83 M.Guermazi, P.Thevenard, P.Faisant, M.G.Blanchin, C.H.S.Dupuy: Radiat. Eff. **37**, 99 (1978)

3.84 G.Dearnaley, J.H.Freeman, R.S.Nehon, J.Stephen: *Ion Implantation* (North-Holland, Amsterdam 1973)

3.85 G.Dearnaley: In *Nucl. Physics Methods in Material Research*, eds. by K.Bethge, H.Baumann, H.Jex, F. Rauch (Vieweg, Braunschweig 1980) p. 56

3.86 P.V.Parlow, E.I.Zonin, D.I.Tetelbaum, V.P.Lesnikov, G.M.Ryzhkov, A.V.Parlow: Phys. Status Solidi A: **19**, 373 (1973)

3.87 J.H.Wilson, K.H.Goh, K.G.Stephens: Thin Solid Films **33**, 205 (1976)

3.88 D.K.Murti, E.Giani, M.Arora, R.Kelly: To be published

3.89 M.Watanabe, A.Tooi: Jpn. J. Appl. Phys. **5**, 737 (1966)

3.90 S.S.Gill, I.H.Wilson: In *Proc. 1st Conf. on Ion Beam Modification of Materials*, eds. by J.Gyulai, T.Lohner, E.Pasztor (Budapest, Hungary 1978) p. 1231

3.91 J.H.Freeman, G.A.Gard, D.J.Mazey, J.H.Stephen, F.B.Whiting: In *Proc. Conf. Ion Implantation* (Reading, England 1970) p. 73
3.92 J.A.Borders, W.Beezhold: In *Ion Implantation in Semiconductors*, eds. by I. Ruge, J. Graul (Springer, Berlin, Heidelberg, New York 1971) p. 241
3.93 I.P.Akimchenko, K.V.Kisseleva, V.V.Krasnopevtsev, Yu.V.Milyutin, A.G.Touryanski, V.S.Vavilov: Radiat. Eff. **33**, 75 (1977)
3.94 I.P.Akimchenko, K.V.Kisseleva, V.V.Krasnopevtsev, A.G.Touryanski, V.S.Vaivlov: Radiat. Eff. **48**, 7 (1980)
3.95 T.J.Chuang: J. Chem. Phys. **72**, 6303 (1980)
3.96 O.Almén, G.Bruce: Nucl. Instrum. Methods **11**, 257 (1961)
3.97 H.H.Andersen, H.Bay: Radiat. Eff. **13**, 67 (1972)
3.98 H.H.Andersen, H.Bay: Radiat. Eff. **19**, 139 (1973)
3.99 H.L.Bay, J.Bohdansky, E.Hechtl: Radiat. Eff. **41**, 77 (1979)
3.100 E.Hechtl, J.Bohdansky, J.Roth: In *Proc. Symp. on Sputtering*, eds. by P.Varga, G.Betz, F.P.Viehböck (Techn. Univ. Wien, 1980) p. 834
3.101 E.Hechtl, J.Bohdansky, J. Roth: J. Nucl. Mater. **103/104**, 333 (1981)
3.102 R.Behrisch: Erg. Exakt. Naturw. **35**, 295 (1964)
3.103 O.C.Yonts, D.E.Harrison: J. Appl. Phys. **31**, 447 (1960)
3.104 W.Wach, K.Wittmaack: Nucl. Instrum. Methods **149**, 259 (1978)
3.105 W.O.Hofer, H.Liebl: In *Ion Beam Surface Layer Analysis*, eds. by O.Meyer, G.Linker, F.Käppeler (Plenum Press, New York 1976) p. 659
3.106 W.O.Hofer, H.L.Bay, P.J.Martin: J. Nucl. Mater. **76/77**, 156 (1978)
3.107 J.Hrbek: Thin Solid Films **42**, 185 (1977)
3.108 R.Behrisch, J.Roth, J.Bohdansky, A.P.Martinelli, B.Schweer, D.Rusbüldt, E.Hintz: J. Nucl. Mater. **94/95**, 645 (1980)
3.109 J.Roth: Unpublished data
3.110 R.Kelly, N.Q.Lam: Radiat. Eff. **19**, 39 (1973)
3.111 J.Roth, J.Bohdansky, W.Ottenberger: Max-Planck-Institut für Plasmaphysik, Garching, FRG, Rpt. IPP 9/26 (1979)
3.112 T.Abe, T.Yamashina: Thin Solid Films **30**, 19 (1975)
3.113 W.D.Westwood: Progr. Surf. Sci. **7**, 71 (1976)
3.114 J.Heller: Thin Solid Films **17**, 163 (1973)
3.115 R.F.K.Herzog, W.P.Poschenrieder, F.G.Satkiewicz: Radiat. Eff. **18**, 199 (1973)
3.116 G.Staudenmaier: Radiat. Eff. **13**, 87 (1972)
3.117 F.Shinoki, A.Itoh: Jpn. J. Appl. Phys. *Suppl.* 2, *Pt.* 1, 505 (1974)
3.118 R.Behrisch (ed.): *Sputtering by Particle Bombardment* III, Topics Appl. Phys. (Springer, Berlin, Heidelberg, New York) to be published
3.119 K.Wittmaack: Surf. Sci. **89**, 668 (1979)
3.120 A.E.Morgan, H.W.Werner: Appl. Phys. **11**, 193 (1976)
3.121 A.Benninghoven, C.Plog, N.Treitz: Int. J. Mass Spectrom. Ion Phys. **13**, 415 (1974)
3.122 C.Plog, L.Wiedmann, A.Benninghoven: Surf. Sci. **67**, 565 (1977)
3.123 A.E.Morgan, H.W.Werner: J. Chem. Phys. **68**, 3900 (1978)
3.124 A.R.Krauss, D.M.Gruen: J. Nucl. Mater. **63**, 380 (1976)
3.125 D.H.W.Carstens, J.F.Kozlowski, D.M.Gruen: High Temp. Sci. **4**, 301 (1972)
3.126 D.M.Gruen: In *The Chemistry of Fusion Technology*, ed. by D.M.Gruen (Plenum Press, New York 1972) p. 215
D.M.Gruen: J. Nucl. Mater. **53**, 220 (1974)
3.127 J.Roth, J.Bohdansky, W.Poschenrieder, M.K.Sinha: J. Nucl. Mater. **63**, 222 (1976)
3.128 B.Feinberg, R.S.Post: J. Vac. Sci. Technol. **13**, 443 (1976)
3.129 N.P.Busharov, E.A.Gorbatov, V.M.Gusev, M.I.Guseva, Yu.V.Martynenko: Sov. J. Plasma Phys. **2**, 321 (1976)
3.130 C.M.Braganza, S.K.Erents, G.M.McCracken: J. Nucl. Mater. **75**, 220 (1978)
3.131 R.Yamada, K.Nakamura, K.Sone, M.Saidoh: J. Nucl. Mater. **95**, 278 (1980)
3.132 D.E.Rosner, H.D.Allendorf: Proc. Int. Conf. Heter. Kinetics at Elevated Temp. (Univ. Pennsylvania 1969) p. 231 and references therein

3.133 M.Coulon, L.Bonnetain: J. Chim. Phys. **71**, 711, 717, 725 (1974)
3.134 S.Veprek, M.R.Haque, H.R.Oswald: J. Nucl. Mater. **63**, 405 (1976)
3.135 R.Brewer, H.Stuessi, S.Veprek, A.P.Webb: J. Nucl. Mater. **93/94**, 634 (1980)
3.136 B.J.Wood, H.Wise: J. Phys. Chem. **73**, 1348 (1969)
3.137 L.Holland, S.M.Ojha: Vacuum **26**, 53 (1976)
3.138 D.R.Olander, R.H.Jones, J.A.Schwarz, W.J.Siekhaus: J. Chem. Phys. **57**, 421 (1972)
3.139 D.R.Olander, W.Siekhaus, R.Jones, J.A.Schwarz: J. Chem. Phys. **57**, 408 (1972)
3.140 R.Castaing, O.Hennequin: Adv. Mass Spectrom. **5**, 419 (1971)
3.141 R.S.Nelson: Philos. Mag. **11**, 291 (1965)
3.142 J.Bohdansky, J.Roth, A.P.Martinelli: Presented at the 4th Intern. Conf. on Solid Surfaces, Cannes (1980) to be published
3.143 T.Abe, K.Obara, Y.Murakami: J. Nucl. Mater. **91**, 233 (1980)
3.144 J.Roth, J.Bohdansky, K.L.Wilson: J. Nucl. Mater. **111/112**, 775 (1982)
3.145 V.Philipps, K.Flaskamp, E.Vietzke: J. Nucl. Mater. **111/112**, 781 (1982)
3.146 C.Braganza, G.M.McCracken, S.K.Erents: Proc. Intern. Symp. Plasma Wall Interaction (Jülich 1976) p. 257
3.147 M.Mohri, K.Watanabe, T.Yamashina: J. Nucl. Mater. **75**, 7 (1978)
3.148 M.Mohri, K.Watanabe, T.Yamashina, H.Doi, K.Hyakawa: J. Nucl. Mater. **85/86**, 1185 (1979)
3.149 N.P.Busharov, E.A.Gorbatov, V.M.Gusev, M.I.Guseva, Yu.V.Martynenko: J. Nucl. Mater. **63**, 230 (1976)
3.150 K.Sone, H.Ohtsuka, T.Abe, R.Yamada, K.Obara, O.Tsukakoshi, T.Narusawa, T.Satake, M.Mizuno, S.Komiya: Proc. Intern. Symp. Plasma Wall Interaction (Jülich 1976) p. 323
3.151 R.Yamada, K.Nakamura, M.Saidoh: J. Nucl. Mater. **98**, 167 (1981)
3.152 R.Yamada, K.Nakamura, M.Saidoh: J. Nucl. Mater. **111/112**, 744 (1982)
3.153 A.E.Pontau, K.L.Wilson: J. Vac. Sci. Technol. **20**, 1322 (1982)
3.154 R.R.Rye: Surf. Sci. **69**, 653 (1977)
3.155 J.Roth, J.Bohdansky: IEA Workshop on Graphite, Rigi-Kaltbad, Swizerland (Oct. 1982) to be published
3.156 J.W.Coburn, H.F.Winters: J. Vac. Sci. Technol. **16**, 391 (1979)
3.157 A.J.Machiels, D.R.Olander: Surf. Sci. **65**, 325 (1977)
3.158 A.J.Machiels, D.R.Olander: High Temp. Sci. **9**, 3 (1977)
3.159 P.C.Nordine: J. Electrochem. Soc. **125**, 498 (1978)
3.160 M.W.Thompson: Philos. Mag. **18**, 377 (1968)
3.161 S.Kato, T.Satake, M.Mohri, T.Yamashina: J. Nucl. Mater. **103/104**, 351 (1981)
3.162 M.Szymonski, A.E.deVries: *Proc. Symp. on Sputtering*, ed. by P.Varga, G.Betz, F.P. Viehböck (Techn. Univ. Wien, 1980) p. 320
3.163 W.Poschenrieder, G.Staudenmaier, Ph.Staib: J. Nucl. Mater. **93/94**, 322 (1980)
3.164 H.F.Dylla: J. Nucl. Mater. **93/94**, 61 (1980)
3.165 A.G.Mathewson: Proc. Intern. Symp. Plasma Wall Interaction, Jülich (1976) p. 517
3.166 W.Poschenrieder: Max-Planck-Institut für Plasmaphysik, Garching, FRG (private communication)
3.167 L.M.Ephrath: J. Electrochem. Soc. **126**, 1419 (1979)
3.168 R.A.H.Heinecke: Solid State Electron. **18**, 1146 (1975)
3.169 T.Yamashina, M.Mori, K.Watanabe, H.Doi, K.Hayakawa: J. Nucl. Mater. **76/77** (1978)
3.170 A.A.Bergh: Bell Syst. Tech. J. **154**, 261 (1965)
3.171 E.Blauth, E.Mayer, F.Schwirzke: Proc. 5th Intern. Conf. Ionization Phenomena in Gases, Munich (1961) p. 545
3.172 E.W.Blauth, E.H.Meyer: Z. Angew. Phys. **19**, 549 (1965)
3.173 G.M.McCracken, J.W.Patridge: J. Nucl. Mater. **63**, 773 (1976)
3.174 Y.Ishibe, H.Oyama: J. Nucl. Mater. **85/86**, 1191 (1979)
3.175 E.Taglauer, W.Heiland: Appl. Phys. Lett. **23**, 950 (1978)
3.176 S.Veprek, A.P.Webb, H.R.Oswald, H.Stuessi: J. Nucl. Mater. **68**, 32 (1977)
3.177 E.Vietzke, K.Flaskamp, V.Philipps: J. Nucl. Mater. **111/112**, 763 (1982)
3.178 C.I.H.Ashby, R.R.Rye: J. Nucl. Mater. **92**, 141 (1980)

3.179 C.I.H. Ashby, R.R. Rye: J. Nucl. Mater. **103/104**, 489 (1981)
3.180 C.I.H. Ashby: 29th Symp. Am. Vac. Soc., J. Vac. Sci. Technol. (to be published)
3.181 A.A. Haasz, P.C. Stangeby, O. Anciello: J. Nucl. Mater. **111/112**, 757 (1982)
3.182 J.W. Coburn, H.F. Winters: J. Appl. Phys. **50**, 3189 (1979)
3.183 H.F. Winters, J.W. Coburn: Appl. Phys. Lett. **34**, 70 (1979)
3.184 B. Carriere, B. Lang: Surf. Sci. **64**, 209 (1977)
3.185 J.L. Vossen: J. Appl. Phys. **47**, 544 (1976)
3.186 R.T.K. Baker, J.J. Chludzinski, Jr.: Carbon **19**, 75 (1981)
3.187 M. Balooch, R. Behrens, D.R. Olander: Private communication
3.188 S.L. Bernasek, G.A. Somorjai: J. Chem. Phys. **62**, 3149 (1975)
3.189 K.J. Dietz, F. Waelbroeck, P. Wienhold: Jülich-IPP-Report 1448 (1977)
3.190 B. Lersmacher, H. Lydtin, W.F. Knippenberg, A.W. Moore: Carbon **5**, 205 (1967)
3.191 JANAF Thermo-Chemical Tables, eds. by D.R. Stull, H. Prophet, NSRDS-NBS 37 (1971)
3.192 M. Balooch: Jpn. Appl. Phys. **16**, 1557 (1977)
3.193 J.C. Batty, R.E. Stickney: J. Chem. Phys. **51**, 4475 (1969)
3.194 J.C. Batty, R.E. Stickney: Research Lab. of Electronics. MIT Tech. Rpt. **473** (1969)
3.195 *Handbook of Physics*, eds. by E.V. London, H. Ochshaw (McGraw-Hill, New York 1958)
3.196 C. Pisani, R. Ricca, R. Dovesi: Proc. 2nd Intern. Conf. Solid Surf. Jpn. J. Appl. Phys. Suppl. 2, Pt. 2, 269 (1974)
3.197 D.R. Olander, T.R. Acharya, A.Z. Ullman: J. Chem. Phys. **67**, 3549 (1977)
3.198 G.N.-K. Liu: MIT Tech. Rpt. **186** (1973)
3.199 K. Flaskamp, H.R. Ihle, G. Stöcklin, E. Vietzke, K. Vogelbruch, C.H. Wu: Proc. Intern. Symp. Plasma Wall Interaction, Jülich (1976) p. 285
3.200 P. Hucks, K. Flaskamp, E. Vietzke: J. Nucl. Mater. **93/94**, 558 (1980)
3.201 G. Beitel: J. Vac. Sci. Technol. **8**, 647 (1971)
3.202 G. Beitel: J. Vac. Sci. Technol. **6**, 224 (1968)
3.203 E. Taglauer, W. Heiland: J. Nucl. Mater. **76/77**, 328 (1978)
3.204 S.K. Erents: Nucl. Instrum. Methods **170**, 455 (1980)
3.205 J.N. Smith, Jr., C.H. Meyer, Jr.: J. Nucl. Mater. **76/77**, 193 (1978)
3.206 A. Sagara, K. Kamada: J. Nucl. Mater. **111/112**, 812 (1982)
3.207 K. Watanabe, K. Nakamura, S. Maeda, Y. Hirohata, M. Mohri, T. Yamashina: J. Nucl. Mater. **85/86**, 1081 (1979)
3.208 R. Brewer, S. Veprek, T. Yamashina, S. Maeda, M. Mohri: IEA Workshop on Graphite, Rigi-Kaltbad, Swizerland (Oct. 1982) to be published
3.209 R. Yamada, K. Nakamura, K. Sone, M. Saidoh: J. Nucl. Mater. (to be published)
3.210 W. Eckstein: Private communication (1981)

4. Sputtering by Electrons and Photons

Peter C. Townsend

With 18 Figures

Surfaces of insulators and semiconductors can be eroded by bombardment with low energy (5–100 eV) electrons and photons. The mechanisms causing sputtering under these conditions operate via a localised excitation or ionisation followed by a relaxation of the lattice and initiation of a short replacement collision sequence. These processes which have been studied in most detail in respect to radiation damage in insulators can be directly related to the few sputtering investigations performed, especially with alkali halides. Generally, halide atoms are sputtered preferentially in close-packed crystal directions, while alkali atoms are emitted isotropically, probably by evaporation after enrichment on the surface. For laser-irradiation, besides evaporation, sputtering by multiphoton absorption is also possible.

Sputtering by electrons and photons can be observed only at atomically clean surfaces. The surface topography which develops shows cones presumably caused by contaminants and pits probably due to higher sputtering around dislocations.

For bombardment with electrons at energies above a few 100 keV, knock-on sputtering caused by a direct energy transfer to lattice atoms has also been observed.

4.1 Ion and Electron Sputtering Mechanisms

Sputtering of atoms from the surface of a solid during ion beam irradiation is a general phenomena and it is well known that for metals, energetic ions transfer energy in binary collision events to remove lattice atoms. Similarly, for electron bombardment with very high energy electrons (e.g., $E > 200$ keV), direct collisions with nuclei cause atomic displacements. Such atoms may be released from the solid and hence even an electron beam can cause sputtering. A high energy electron experiment of this type will be described in Sect. 4.9. However, for low energy electrons or photons (e.g., $E \sim 5$ to 100 eV), it is less obvious that they can cause decomposition or sputtering from a crystal lattice. For these events we must seek fairly subtle ways of using the available energy to trigger mechanisms which will abstract momentum [e.g., 4.1–3], and perhaps additional energy, from the crystal.

The examples available to us in the literature include many works on photon-induced decomposition (termed photolysis) which will be mentioned in

Sect. 4.7.2 as well as more extreme examples in which the photon absorption triggers exothermic reactions and explosions.

High power laser beams can generate surface erosion (Sect. 4.7.1) but in many instances this is merely an example of thermal evaporation and will not be considered here. However, there are definite examples of sputtering which result from photon absorption and these will be discussed with particular reference to the alkali halides. The low energy mechanisms which lead to sputtering can operate either by excitation or ionisation processes and are confined to insulators and semiconductors.

Such mechanisms must also be considered during ion beam sputtering events as a considerable proportion of the ion energy can be dissipated via electronic interactions. This may lead directly to sputtering. Further, the combination of nuclear and ionisation events can noticeably modify the rate of damage production, e.g., [4.4], annealing [4.5] and sputtering [4.6–8].

To be consistent with ion beam terminology, we will define a sputtering yield Y as

$$Y = \frac{\text{atoms removed}}{\text{incident photon (or electron)}}.$$

Whilst this expression is of the form used for ion beam experiments, it is worth noting that the electronically induced sputtering may only be effective for one element of a compound (e.g., I from KI). Thus, the measured sputtering yield will be strongly influenced by the rate of removal from the surface of the secondary species. For example, the accretion of a monolayer of nonvolatile metal on the surface may totally inhibit sputtering. Some early data for low energy electron sputtering of KI is shown in Fig. 4.1 [4.9]. In the conditions of this experiment the sputtering yield has a maximum value of 6. We shall see that the threshold energy for both defect formation and sputtering in alkali halides is that of an exciton (i.e., 5–10 eV).

The number of successful searches for photon or low energy electron sputtering is limited and there are several reasons why attempts may not have

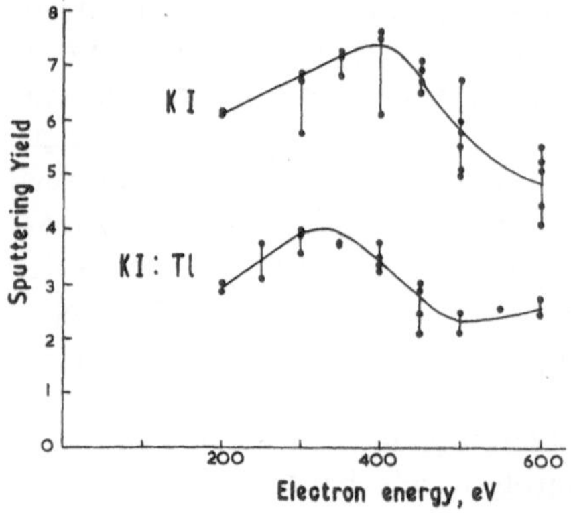

Fig. 4.1. The sputtering yield of KI and KI : Tl for electron irradiation at 220 °C [4.9]. (Note that Tl acts as an exciton trap thus, as will be apparent from Sects. 4.3.4, the net sputtering yield is reduced

succeeded. Firstly, the intense optical sources, namely lasers, produce photons of only a few eV and multiphoton or x-ray sources are relatively weak. Similarly, except for Auger or x-ray inner shell excitations, the rate of energy deposition into the lattice is only a few eV/Å which may be insufficient to stimulate displacements. Secondly, photons or low energy electrons only transfer energy to the *electronic* system within the crystal and negligible momentum is available. This is in strong contrast with the situation for ion beam induced sputtering. During a sequence of binary collisions within a collision cascade, large amounts of energy and momentum pass between the atoms so that displacements and sputtering readily occur. For an ion beam, directed normal to a surface which sputters, it is still necessary to conserve momentum for the system and since the sputtered particles are directed away from the surface we see that in the final accounting for their momentum they must have taken momentum from the crystal lattice.

By contrast, the low energy mechanisms operate by localising energy at a single lattice site with only minor disturbance to the neighbouring atoms. Therefore, the process of phonon coupling to the atom which is to be displaced is quite critical as there is an immediate need for the momentum. Further, the small amounts of energy available to start the displacement and sputtering events reduce the range of the collision events. In particular, we can see from Fig. 4.2 that if the mechanism is selective in the compound AB and only atoms of type A are sputtered, then for a clean surface and/or one in which B type atoms are subsequently removed (e.g., by evaporation), this is not a problem. Due to their low energies the binary collisions of type A atoms with other atoms are unlikely to cause further displacements or sputtering. Therefore, a surface coating of impurities will inhibit the sputtering process as shown in Fig. 4.2. Consequently, observations of photon or electron induced sputtering are rare and are only likely to be recorded if one deliberately ensures that the surface is clean.

Ion beam sputtering will actively clean surfaces by nuclear collisions and in such a situation the role of the electronically driven sputtering can contribute to the total sputtering yield.

The importance of low-energy mechanisms which can cause sputtering has only become apparent over the last ten years but there will be much further

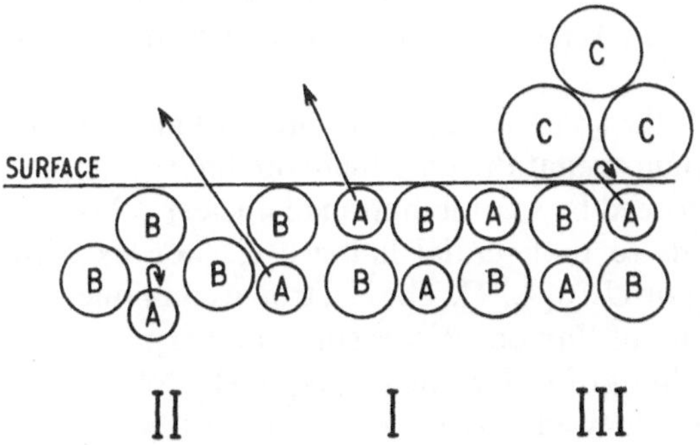

Fig. 4.2. Sputtering from a compound AB, the surface layer is nonstoichiometric in region II, perfect in region I and contaminated in region III. If only atoms of type A are ejected from lattice sites and possess small amounts of energy, then only in region II will A atoms be sputtered

interest in the subject. Not only for the understanding of the physical principles but also because the effects will, or are being, revealed for many insulating materials. For example, Auger electron spectroscopy is not a passive measurement for insulators which undergo electronically stimulated sputtering. Thus the recorded Auger analysis is of a surface layer perturbed by sputtering. The problem has now been identified in many systems, notably oxides and halides. A review by *Pantano* and *Madey* [4.10] gives examples of some 20 materials which undergo surface decompositions during Auger electron spectroscopy. The effect is apparent in both amorphous and crystalline systems including materials of technological interest such as silicon dioxide [4.11] and $LiNbO_3$ and $LiTaO_3$ [4.12]. Other oxides showing electronically induced instability include SnO_2 [4.13], ZnO [4.11], PbO, and molybdenum oxides [4.14]. Of the halides the mechanism for decomposition is well established for the alkali halides and will be discussed in this chapter but many other halides show halogen loss during excitation including $ZnBr_2$, AgBr, CaI_2, PbI_2 [4.15], CaF_2 [4.16] or UF_4 [4.17]. In these cases a detailed mechanism is still the subject of speculation. The models include a stage of energy localisation followed by an electronic relaxation and particle rearrangement. In general, the models require ionisation or multiple-ion excitation (e.g., by Auger decay of deeply bonded electrons) and may closely resemble those models proposed for desorption of absorbates from the surface. Various aspects of the topic were reviewed by authors in a recent conference on electronically induced desorption [4.18]. The desorption via electronic changes was described in the Menzel-Gomer-Redhead model [4.18–20] and modified by *Brenig* [4.17], *Antoniewicz* [4.21], and by *Knotek* and *Fiebleman* [4.18, 22].

Lower energy processes can sputter a solid and in general the lower limit seems to be an energy comparable with the energy gap or exciton transitions. For alkali halides we shall see the process is quite well understood and it has been briefly reviewed elsewhere [4.18, 23]. Although for alkali halides the sputtering is a linear function of the exciton production there are numerous examples of materials which show a nonlinear dependence on the sputtering yield with light intensity [4.24]. Some examples are CdSe, GaP, ZnO, and TiO_2.

We have already mentioned that during ion bombardment the electronic contribution to sputtering is effective so it must be considered in ion beam cleaning of glass or crystalline insulators, in particular noting that the electronic mechanisms frequently only remove a single element from a compound.

In high-energy ion implantation the sputtering yield is directly linked to the rate of energy deposition in electronic excitation. One therefore finds examples of sputtering in insulators which proceed by a combination of nuclear collisions and electronic excitation, this leads to preferential halogen ejection as with alkali halides [4.8, 25] CaF_2 [4.26] or UF_4 [4.27]. The electronic component is very effective in causing the break-up of Van der Waals solids and experiments on the sputtering of condensed gas films of A, Kr, Xe, O_2, N_2, CH_4, NH_3, SO_2, CO_2, and H_2O have recently been reviewed by *Brown* et al. [4.28]. Ion-beam induced desorption of absorbate molecules may also proceed via an electronic excitation. The literature includes results for desorption of very large absorbate molecules of several thousand atomic mass units, and this feature is being used in the development of a powerful analytical technique for the study of bio-organic molecules. Recent conference proceedings [4.29] give results and applications but offer little agreement as to the detailed mechanism of the sputtering.

High-power lasers are now in common use and by multi-photon processes photon energies sufficient to cause sputtering are achieved. A look much further into the future suggests that if laser driven fusion systems provide significant quantities of neutrons, then they could, in turn, cause ionisation damage and sputtering in the optical components of the laser amplifiers. This would, in turn, increase light scattering and, in extreme cases, would cause self destruction of the amplifier glass by scattered laser power.

In this chapter we shall show the general principles by which photons or electrons can stimulate electronic transitions that are precursors of atomic displacements. We then follow the ways in which the excited states relax to eject atoms from the crystal. There are close parallels between the studies of bulk defects (i.e., as exemplified by optical absorption) and the sputtering. In the case of the alkali halides the relevant theory and experiments are well developed and one now has a consistent model for the entire process.

The only reason that the alkali halide case is understood is that considerable effort has been devoted to it. Continued study of other systems can and should draw on this earlier experience and the field is ideally positioned for rapid expansion.

4.2 Defect Formation and Sputtering of Insulators

It has been known for nearly 50 years that alkali halides form colour centres when exposed to ionising radiation. At first it was thought that the vacancies generated in the halogen sub-lattice were just a result of special sites in the crystal as the rate of colouration is strongly influenced by impurities and mechanical deformation. Several models for the mechanism of defect formation were proposed and rejected. However, even if these early ideas are inappropriate for the alkali halides, they should not be forgotten as they may have relevance for the damage mechanisms in other materials.

In all cases, the major steps are the absorption and localisation of energy from the photon or electron beam followed by an ejection of the atom from the lattice site. For alkali halides we shall see in Sect. 4.4 that a more detailed analysis requires

i) the production of an excition;
ii) diffusion of the exciton until it is trapped at a lattice site (perhaps with a preference for impurity or surface sites),

iii) a long-lived excited state whilst the lattice relaxes, and
iv) decay of the excited state and a displacement of the halogen ion.

For alkali halides, the separation of the halogen interstitial away from the vacancy is by a replacement collision sequence along a row of halogen ions. This is clearly shown by the directional character of the halogen ejection.

The sputtering is apparent in the case of alkali halides with clean surfaces and one can record the sputtering yield, directionality, threshold energy, temperature dependence and lifetime effects for the process. For alkali halides at low temperature or with contaminated surfaces, sputtering is not detected although defect formation in the bulk material is shown by the optical absorption. Photon (or electron) induced sputtering may be a fairly general phenomenon for insulators and a reasonable basis for extending the range of materials is to commence with those insulators which show colour centre formation under ionising radiation.

Within a solid, stable defects only exist if the displaced atoms and vacant lattice sites are separated so that lattice vibrations do not cause annealing. Indeed, short pulse irradiations frequently reveal optical absorption bands from transient defects which exist for only a few nanoseconds, even at 4 K. For long term stability the separation of the vacancy-interstitial pair should be at least 4 or 5 lattice sites at ambient temperature. However, for sputtering from the surface layers, defect stability is much less important and one may observe sputtering from processes which do not generate stable defects in the interior of the material. Again this may suggest that a more general search for photon sputtering of insulators might be fruitful.

The first step for defect production is to localise energy at a single atomic site within the lattice. This may be a direct event such as when photon energy is absorbed by stimulating an electronic transition at a defect site in the crystal, or it may be a two part process in which free electrons and holes, or bound electron-hole pairs (excitons), are formed but diffuse before being trapped at a specific site in the crystal.

To summarise the methods of coupling energy into the lattice for the first step, we have:

a) direct collision events with particles of high energy;
b) ionisation of regions of the lattice (with a corresponding reduction in the displacement energy and the surface binding energy);
c) individual atomic sites where the chemical bonds are destroyed (i.e., either partial or complete ionisation of the bonds to neighbouring atoms);
d) excitation to bound states.

For the secondary step of sputtering or defect formation we may follow with:

i) ejection into an unrelaxed lattice;
ii) assisted ejection by lattice relaxation;
iii) thermal diffusion of ions in the excited state;

iv) thermal diffusion of ions in different charge states;
v) interchange of excited ions on lattice sites with interstitials;
vi) electrostatic repulsion of multiply charged ions;
vii) relaxation into new chemical species with an associated change in lattice energy and momentum
viii) initiation of replacement collision sequences.

In principle, any combination of the two steps can cause sputtering. We may also emphasise that whilst the initial excitation may occur in some 10^{-15} s, as with ion bombardment, the time scale of the secondary steps may be very long.

Delaying features can occur whilst an exciton or free charge is mobile in the lattice, for this a time scale of 10^{-9} s is not unreasonable. Further, after localisation of the energy, relaxation of the excited region takes place and since this involves many phonons, the shortest time is some 10^{-12} s. In some excited states, for example, where adjacent ions reform into a molecular state, the perturbation can easily exist for 10^{-8} s.

One should note that these various processes introduce considerable time delays between the initial absorption of the energy and the sputtering. Experimentally this is important because for measurements of the velocity spectrum of the ejected particles one cannot use a time-of-flight technique linked to a pulsed irradiation source. A post-sputtering velocity filter or Doppler measurement must be chosen as alternatives (Sect. 4.4.4).

In the following examples we shall note the well documented case of the alkali halides where the key steps are of type d) and vii) or viii). For silver halides, steps d) plus iv) seem appropriate but for other insulators there is great uncertainty in the details of the mechanisms.

We have been pedantic in listing the possible primary and secondary stages of the sputtering process and it is not possible to cite examples of all of them. However, in many systems where ionisation alone causes sputtering or it assists nuclear collision sputtering, we do not yet know the details of the process and for future model building these lists may be stimulating.

Sections 4.3–5 will take a rather detailed look at the work in the alkali halides. In these materials the state of knowledge is quite far advanced and as such one may use these studies as an example of what is possible. It is unfortunate that no other system can be quoted in such depth.

Stages a) and i) are really only appropriate for ion beam sputtering but they will be briefly treated in Sect. 4.8.

4.3 Defects in Alkali Halides

A multitude of stable defects exist in the halogen sublattice of the alkali halides. For the purposes of our discussions the relevant ones are those which are first formed during irradiation, namely, the vacancy (F) and interstitial centre (H).

Table 4.1. Major defects in alkali halides

F	An electron trapped by a halogen ion vacancy.
H	An interstitial halogen atom; the defect is stabilised by a hole which is trapped by a linear array of four halogens spread over three lattice sites. The axis of this interstitial is ⟨110⟩.
H_2	A di-interstitial.
H_A	An H centre next to an alkali impurity.
V_K	A hole trapped by a pair of halogen ions which relax along a ⟨110⟩ direction.

Related defects are the impurity modification of H, i.e., the H_A, and a self-trapped hole defect V_K. Definitions are given in Table 4.1. Within the literature other centres and their structure are documented [e.g. 4.1–3, 30]; the list of these is formidable and includes $F^+(\alpha)$, F^-, $F_2(M)$, $F_3(R)$, $F_4(N)$, F_A, F_Z, F_2^-, F_3^-, I, I_A, H_Z as variations of the basic structures listed in Table 4.1.

Alkali halides are important because not only have many defect centres been identified but also optical absorption experiments with pulsed electron [4.11–37] or photon beams [4.38] identify that at all temperatures the initial defects are the F and H pair. F centres are formed within some 11 ps and the H centres develop within nsecs in many alkali halides. Other defects are the result of secondary processes such as diffusion, aggregation or annealing.

For sputtering it is the initial defect formation which is most interesting. Once thermal equilibrium is established within the crystal, energetic ejection of particles is unlikely although the diffusion of defects (e.g., interstitials) to the surface and their evaporation can contribute to the total sputtering yield.

4.3.1 The Original Excitonic Mechanism

For alkali halides an excitonic mechanism for vacancy formation was proposed independently in rather similar models by *Pooley* [4.39, 40] and *Hersh* [4.41]. The idea had also been appreciated by *Luschick* et al. [4.42, 43]. They based their respective models, which differ only in detail, on essentially the same experimental results observed with crystals of KI.

The measured F centre production yield and the efficiency of the luminescence in KI in the temperature region between 80 and 200 K show that F centre production increases with temperature whereas the luminescence decreases. To account for this anticorrelation it was proposed that the original excitation formed an exciton state which could relax and decay by alternative paths of luminescence or defect formation. For alkali halides the relaxed exciton seemed particularly reasonable as an $(e^- + h^-)$ trapped on a pair of ⟨110⟩ adjacent I^- ions could also be viewed as a $(V_K + e^-)^*$ state, i.e., an excited state of an electron trapped near a V_K centre. The difference in temperature dependence between the radiative and nonradiative decay paths is then explained by a small energy barrier which must be overcome for the defect-forming path.

Figure 4.3 represents *Pooley's* original configurational co-ordinate diagram of the ground and excited states for an $(e^- + h^+)$ pair in KI. The presentation

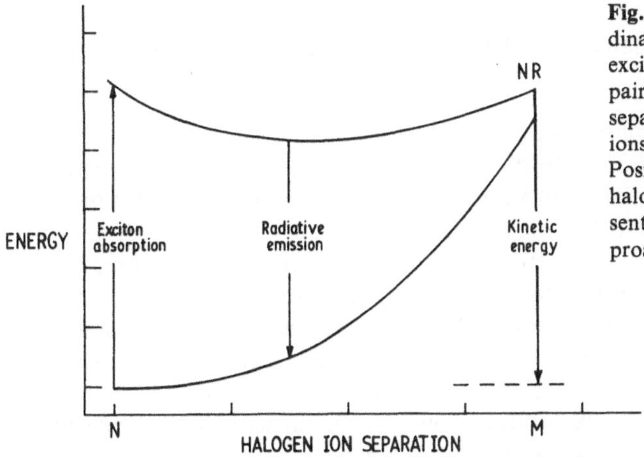

Fig. 4.3. The configuration coordinate diagram of the ground and excited states for an electron-hole pair in KI as a function of the separation of the two halogen ions which have trapped the hole. Position *N* corresponds to normal halogen ion separation, *M* represents formation of a close approach in a molecular-like state

gives the energy levels of the system as a function of the separation of adjacent iodine ions along the ⟨110⟩ axis. The point labelled *N* is for normal separation and *M* corresponds to a close approach and the formation of a molecular ion. This diagram has the essential features that ion movement is thermally activated. For low temperatures the relaxation is via the minimum of the excited state so there is luminescence. At high temperatures there is a nonradiative recombination of the electron and hole at *NR*. The kinetic energy is available for the initiation of a displacement event and since the system forms two negatively charged halogens in close contact, there will obviously be a separation of the component ions.

For the secondary step of the displacement *Pooley* suggested that the energy is unequally shared between two ions. He argued that because of the symmetry of the molecule in the lattice the ions separate with momentum vectors along ⟨110⟩ directions and that there is a subsequent replacement collision sequence along a ⟨110⟩ direction due to this momentum. Since the separation of the interstitial halogen from its original site is expected to form a stable Frenkel pair, one should consider the range of the replacement sequence. The range of the replacement collision sequence would be dependent on both the initial energy of the replacement collision sequence and on the energy loss in each replacement. *Pooley* [4.44] had demonstrated the replacement collision sequence in various alkali halides with a computer simulation. This was encouraging in that it allowed replacement collisions to take place for most alkali halides with the energy available from the exciton, although the range of the replacement sequence would have been insufficient to achieve a stable separation of the vacancy and interstitials. In the Pooley model it is of interest to note that:

i) The threshold energy is less than the valence-conduction band gap;
ii) the electron-hole recombination of a relaxed exciton $(V_K+e)^*$ has two competing processes, one of which gives luminescence and the other produces displacements;

iii) it is necessary for the subsequent ⟨110⟩ replacement collision sequence that the energy is unequally shared between two halogen ions of the relaxed exciton.

According to the original version of the Pooley and Hersh model, the final defect formation is delayed until the $(e^- + h^+)$ pair have relaxed to the ground state, i.e., after the decay of a singlet exciton or a triplet exciton. But it has been shown that the lifetime of a singlet exciton is about 10 ns and for a triplet exciton is between 10^{-6} and 10^{-3} s for most of the alkali halides [4.45]. However, the results of pulse experiments indicate that the defect formation does not take place after the electron-hole recombination of either singlet or triplet excitons. Further, the original Pooley model requires that energy be shared unequally between the excited ions of the molecular ion pair. This asymmetry does not generally produce one component with enough energy for damage to follow and so the overall damage efficiency is expected to be low. In fact, the observed rate is high and approaches 80% efficiency per absorbed photon in some cases. One therefore must consider alternative exciton relaxation processes which result in the formation of a Frenkel defect.

4.3.2 Modifications to the Basic Model

Many experiments have been made to clarify the correlation between the luminescence and defect formation and to include data from a wide range of alkali halides [4.31–38, 45, 46]. The results suggest that the basic excitonic mechanism is reasonable but more subtlety is required to improve the efficiency of vacancy-interstitial separation. For example, *Smoluchowski* et al. [4.47] extended the Pooley model by considering that in the collision sequence of halogens along the ⟨110⟩ row, the moving entity is effectively a halogen neutral rather than the larger ion. Thus, the momentum and replacement events are driven by the heavy neutral and the hole associated with the H centre is transported by electronic rearrangement. The net effect is to reduce the losses in each replacement event and so achieve stable F and H separations.

Variations on this theme involve consideration of the electronic states; these are most conveniently phrased in terms of the Σ_u, Π_g, Π_u, etc., states of a V_K centre for which experimental data are available [4.48–50].

Itoh and *Saidoh* [4.51] proposed an energetically more efficient mechanism for maintaining a replacement collision sequence in which the moving entity is a polarized V_K centre.

The advantage of such a system is that the momentum is still directed along ⟨110⟩ directions but the energy losses of each replacement step are minimized as the molecule is effectively reduced in cross section in the perpendicular plane. *Itoh* and his co-workers [4.52] originally suggested the model as a result of pulse-irradiation experiments in which they studied the formation of F, H, H_A, and H_2 (i.e., V_4) centres. Figure 4.4 shows a comparison of the "normal"

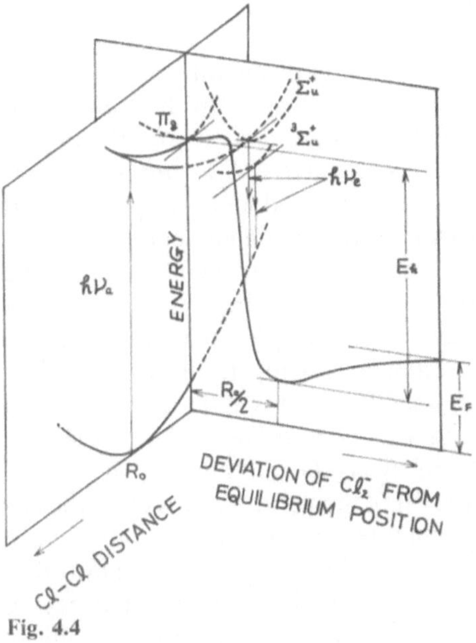

Fig. 4.4

Fig. 4.4. Configuration coordinates curves for the relaxation of an exciton. Along the coordinates of Cl–Cl distance, the exciton is relaxed into $^1\Sigma_u^+$, $^3\Sigma_u^+$ or Π_g states. The potential for the deviation of the Cl_2^- molecule ions as a whole is shown in the other plane. $^1\Sigma_u^+$ and $^3\Sigma_u^+$ states are stable and recombination luminescence is emitted. The Π_g state would result in a replacement sequence with the kinetic energy E_k and final products would be a pair of well-separated F and H centres with formation energy E_F

Fig. 4.5a–c. Excitations of the NaCl lattice are indicated in (**a**) by the symmetric exciton state σ and the directional features Π which causes relaxation to a V_k like defect. (**b**) shows the separation of F and H centres along the ⟨110⟩ halogen axis. An extension of the replacement collision sequence across the surface produces directional halogen ejection (**c**)

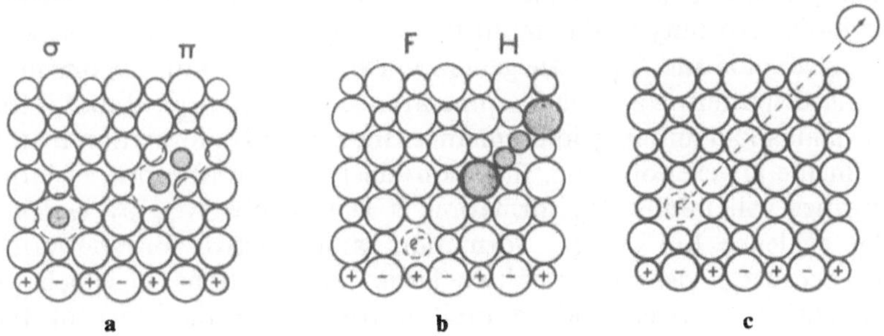

Fig. 4.5

configuration energy diagram with that of the modification via plots presented in two different directions in the Itoh and Saidoh model.

Here the X_2^- molecule (e.g., Cl_2^-) is in an excited state which must be one of the Σ_u, Π_u or Π_g states. After the molecule has relaxed it is in a double potential well formed by the two halogen vacancies with effectively $X^{-1/2}$ in each. However, the symmetric configuration is not stable against movement in the ⟨110⟩ direction, nor against distortions induced by impurity ions if the molecule is in the Π_g state. Hence, the replacement sequence will commence since the Π_g state already implies a p orbital directed along the ⟨110⟩ axis. Alternatively, a V_K centre which is in the Σ_u state will remain in the symmetric configuration and relax radiatively through singlet or triplet states.

Figure 4.5a shows two possible ways in which the exciton can excite the lattice. The symmetric excitation σ cannot lead to defect formation whereas the π excitation relaxed to a V_K structure may. Figure 4.5b gives an indication of the defect products with the vacancy and interstitial centre separated along the ⟨110⟩ axis.

Itoh et al. [4.51] also introduced the concept of an interaction volume between newly formed defects and the existing impurities. In their model this interaction occurs between the replacement collision sequence and the impurities. They observed both a higher fraction of $H_A/(H+H_A)$ centres in doped crystals and more efficient H_2 production than H production in pure crystals immediately after the pulsed electron irradiation at higher temperatures. In their interpretation, this is the result of a longer replacement collision sequence with an activation energy term of 0.03 eV in KBr, which is less than the activation energy for H-centre diffusion. It is possible to offer alternative models for the increase in interaction volume with temperature [4.53–55].

In one of the variations of the basic model [4.55] it is assumed the excitons are generated uniformly in the crystal and are free to diffuse before initiating a collision sequence. Therefore, the volume of crystal, or number of excitons reaching the surface, is an increasing function of temperature. It is only after the diffusion stage that the defect formation commences. Sites which trap excitons may be either impurities, dislocations or surface sites. Since such sites introduce a distortion into the lattice, they provide not only a region for exciton decay but also asymmetry in the lattice which favours the start of a replacement collision sequence in close proximity to the impurity and could explain the reason why the growth curves of H and H_A centre generation are parallel, and why one may expect enhanced sputtering at higher temperatures. It is also argued that if the presence of small alkali impurity ions enhances the rate of H_A formation by the asymmetry in the Π_g exciton state, then conversely, the substitution of large alkali impurities will inhibit H_A formation, a feature observed by *Guiliani* [4.56]. The evidence for the H_A formation is instructive for sputtering experiments because it demonstrates that lattice imperfections, in which we include the surface, actively induce exciton decay and the start of the replacement collision sequences. In this chapter, with the emphasis on sputtering, this point is of particular reference as it follows that exciton generation within several hundred nanometres of the surface will contribute to the sputtering yield, not by long range collision events but by exciton transport to the surface [4.9, 55].

4.3.3 Recent Views on the Excitonic Mechanism

Among the more recent review papers which discuss photon induced damage in alkali halides [4.2, 3, 27, 57], the review by *Williams* [4.2] considers in detail a variety of exciton states and whether or not they can be the precursor for defect formation. In Fig. 4.6 the state labelled *P* for the ions relaxing along the ⟨110⟩ row of halogen ions can generate the F and H centres in the correct state. Alternative levels may exist which result in alternative excited states of the F and H centres. For several alkali halides the secondary paths may contribute to

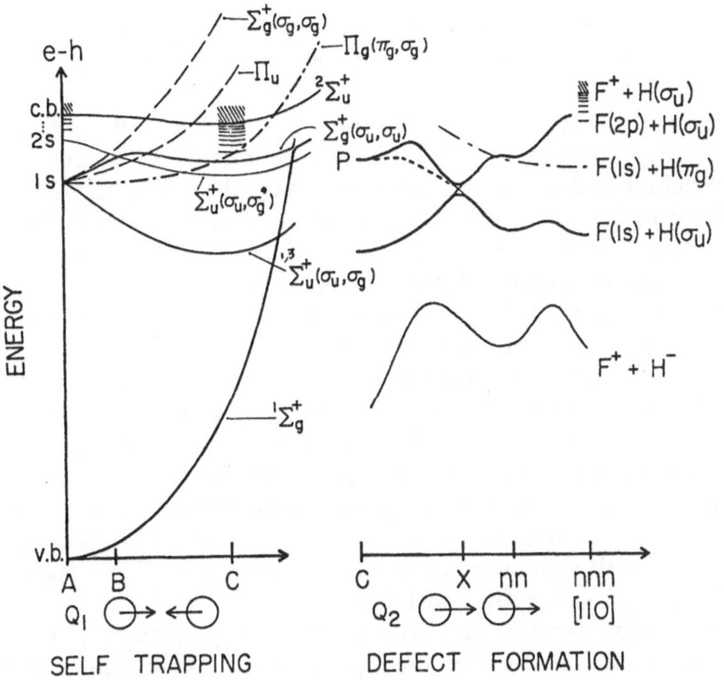

Fig. 4.6. A possible energy scheme for the movement of two halide ions after the capture of an exciton. Relaxation to the state labelled *P* can act as a precursor to the direct formation of a vacancy-interstitial pair (F + H), [4.2]

defect formation. Similarly, the equivalent diagrams for other systems such as MgF_2 or CaF_2 can successfully predict how relaxation could induce defects in these materials. However, without an efficient mechanism for separation of the components the defects may only appear as transient features. Nevertheless, if one's primary interest is in sputtering, then such a relaxation mode in the lattice may be significant. For our purposes we should note that the theory indicated by the energy levels of Fig. 4.6 is now quite complicated in detail but the broad features remain the same for an exciton induced relaxation.

4.4 Sputtering from Alkali Halides

4.4.1 Halogen Sputtering from Alkali Halides

In Sect. 4.3 we concentrated on processes which cause halogen vacancies and interstitials and only mentioned in passing that such displacements may be sensed by halogen ejection from perfect alkali halide surfaces. The consequence of a ⟨110⟩ replacement collision sequence is shown in Fig. 4.5c. When the sequence intersects the surface the final halogen atom is ejected. Thus, we expect halogen atoms to be preferentially ejected along ⟨110⟩ and to have energies less than the maximum available in the excitonic process (i.e., about 5–10 eV).

Whilst we have predicted that the same mechanism can produce either defect formation or sputtering, we should be aware that there are also differences between bulk and surface experiments. Firstly, the bulk defect model required a mechanism for the F and H centres to reach a stable separation. Therefore, alternative directions such as the ⟨211⟩ were ignored, even though there is the possibility of forming V_K-like halogen molecular ions, as the replacement collision sequence might be only a single step before the energy is dissipated to the lattice. For sputtering this is not a limitation as a sub-surface event can impart ⟨211⟩ type directionality and eject a halogen even if the "sequence" is only one step long. Therefore, in the pattern of the sputtered halogen we may expect features such as ⟨211⟩ in addition to the ⟨110⟩ events.

Secondly, the efficiency for sputtering is probably similar for all the alkali halides because the important collision sequences involve excitons trapped near the surface, hence the ejection range is very short. This was observed experimentally [4.58] with sputtering yields ranging from 3 to 17 for 500 eV electrons in NaBr, NaCl, NaF, KBr, and KCl.

The colour centre data also provides evidence for the replacement collision chains, together with the efficiency of their production. Alkali halides are particularly interesting as they form a set of some 13 compounds with the same crystal structure but different ionic sizes and interionic spaces. For our collision sequence to form a crowdion interstitial, we see that key factors are the size of the additional neutral atom, diameter d, and the amount of free space s between halogen ions in the ⟨110⟩ halogen direction (Fig. 4.7). Figure 4.8 shows two possibilities of F, H pairs. At low temperatures, very close pairs may be stable so one interstitial *B* can be squeezed into the row by occupying space available in just two interhalogen regions (i.e., the minimum need is $s/d \sim 1/2$). At high temperatures, only distant F, H pairs will exist so the argument changes to the use of three spaces hence the threshold space condition is $s/d \sim 1/3$. If the collision chain is well spaced then it will cause defocussing so the efficiency of propagation will decrease; this happens when $s \geqq d$.

The range of alkali halides gives a wide range of s/d values to test this consequence of a halogen replacement collision sequence [4.58]. Figure 4.9 shows data of the efficiency of forming F centres as a function of s/d. Low and high temperature results show the expected knees at $s/d \sim 1/2$; 1/3 and 1. It is clear that beyond the critical lower value all the alkali halides damage with similar efficiency whereas for the more tightly-packed systems, defect formation is inhibited. Since the efficiency changes by some 10^4, one suspects that in crystals such as LiI, NaI, LiCl, NaBr, the defects only form near surfaces, dislocations or other defects.

For the sputtering yield the s/d value is of minor importance as the ejection involves very short range events. However, as we shall see in Sect. 4.4.5, it does influence the energy distribution of the ejected particles.

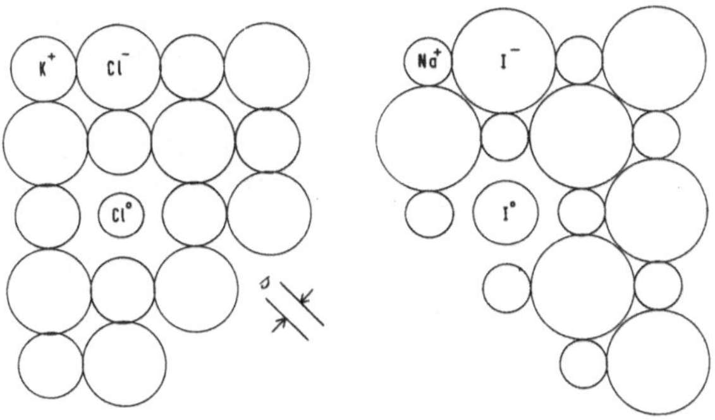

Fig. 4.7. The relative sizes of ions in the alkali halide lattices of NaI and KCl. The parameter s, the space between anions and the size of the neutral halogen d are also shown

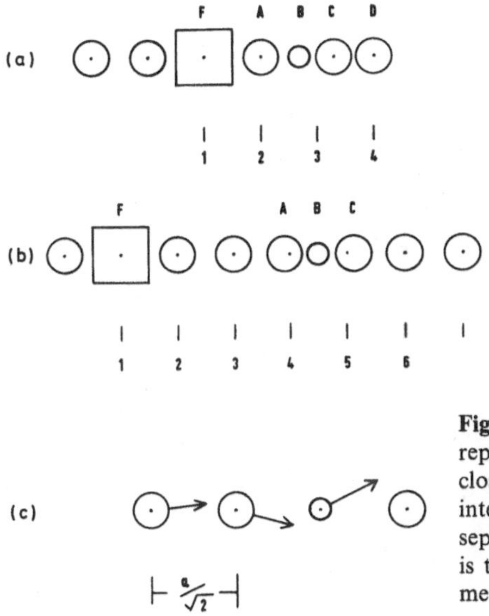

Fig. 4.8. (**a**) The space required for a very short replacement collision sequence. The FH pair are close and the crowdion B is squeezed in by use of two inter ionic gaps (i.e., $d \sim 2s$ is a limit). (**b**) Well separated FH pairs, three gaps used therefore $d \sim 3s$ is the limit. (**c**) Defocussing of a very open replacement chain; for alkali halides this occurs for $s > d$

4.4.2 Alkali Atom Sputtering

After a series of halogen atoms have been sputtered from the surface there will have been a change in the crystal stoichiometry where the near surface region is halogen deficient [4.59–62]. Such an imbalance in the alkali/halogen ratio has several alternative consequences. Firstly, the system may remain as a pseudo-stable material of (alkali-F centres) in which the halogen sites are occupied by only an electron. A second alternative is that metal ions may evaporate so that one would observe sputtering plus evaporation of the entire material. Thirdly, the surface layers may collapse into a new stable phase. The alkali halide literature is sufficiently extensive that there is evidence for each of these alternatives after different conditions for damage. However, in more general terms one might consider similar results for other insulators as an indication that low energy sputtering is a possibility in them.

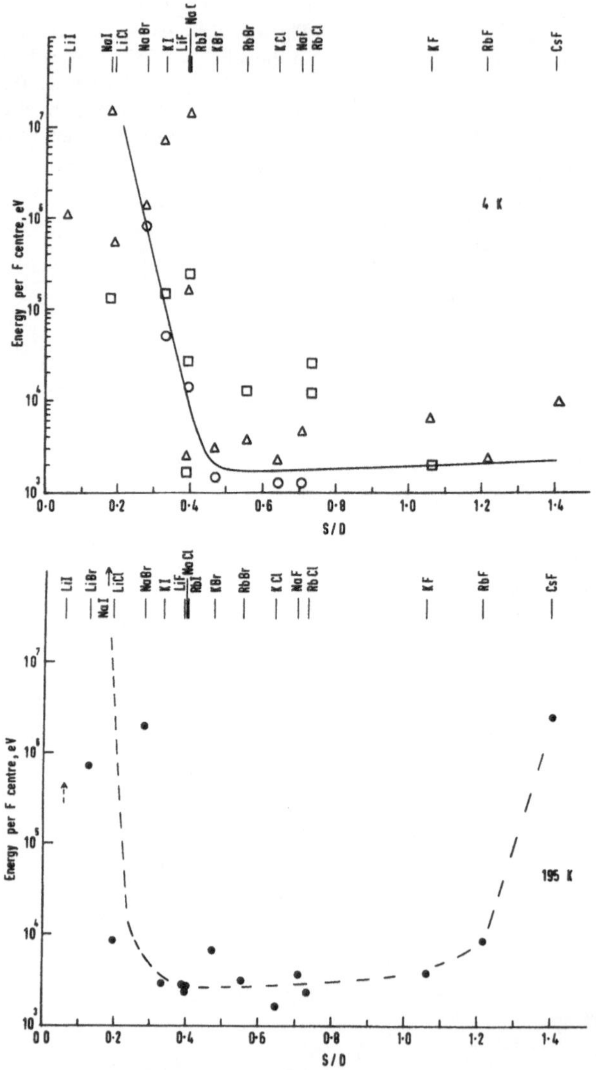

Fig. 4.9. The average energy required to form F centres in alkali halides at 4 and 195 K. (○) and (□) are from x-ray data, (△) and (○) are from 400 keV electron irradiation

Evidence for the pseudo-stable alkali-F centre crystal had been observed prior to the exciton model of defect formation. *Carroll* and *Birnbaum* [4.59] bombarded LiF in an electron microscope with oxygen ions to make in situ thinning and produced a sample which appeared to be only lithium but arranged in the fcc structure of LiF.

Later, Auger electron spectroscopy [4.60–62] measurements of the surface structure of alkali halides all noted a change in the alkali/halogen ratio during the electron irradiation. The rate of loss of halogen ions was a function of temperature and for KI this was consistent with the exciton model of sputtering in that at low temperature, halogen atoms were ejected but the metal excess remained frozen on the surface, whereas at high temperatures, the halogen ejection was followed by metal evaporation thereby returning the surface layer to a normal ratio of halogen to metal atoms.

Phase changes have not been reported as such but, in electron microscope studies of alkali halides with very high defect concentrations, the collapse of the interstitials into platelets has been observed [4.63]. Such platelets could change the sputtering yield. Sodium clusters have also been reported to occur after electron irradiation of NaCl [4.64].

The evidence for directional halogen ejection combined with thermal evaporation of metal atoms will now be presented in more detail.

4.4.3 Alkali Halide Sputtering Patterns

The first measurements of alkali halide sputtering patterns caused by low energy electrons were simply recorded visually on cooled silica plates. The quality of the patterns was very variable and the most clearly defined ones occurred when the sputtered region of the crystal was uniformly eroded and free of cones or other topographical features. The major features of the electron induced sputtering patterns are shown in Fig. 4.10 in which the silica plate shows a white deposit where there is deposition of both sodium and chlorine [4.64]. Patterns produced by photon excitation in the exciton absorption band using a laser and multiphoton absorption [4.65, 66] were recorded for the halogen component of the emission using a channel plate and mass spectrometer (Fig. 4.11). There are differences between the two types of pattern particularly around the ⟨100⟩ normal but in both cases the ⟨110⟩ and ⟨211⟩ features are clearly resolved. The ⟨110⟩ and ⟨211⟩ features of the electron beam sputtering patterns were also identified with halogens by a separate determination of the metal emission [4.64]. Alkali atoms were detected by a surface ionisation detector which was only sensitive to sodium atoms. The observed cosine distribution of intensity about the electron beam axis confirms the idea that the alkali atoms are randomly emitted from the alkali halide surface.

Both types of excitation give the ⟨110⟩ halogen features suggested by the early theory of bulk defect formation. Additionally, ⟨211⟩ spots are major features of the sputtering pattern. This, as outlined in Sect. 4.4.1, is reasonable if we assume that ejection in specific directions is dominated by very short replacement collision chains. Recent attempts [4.66] to find ⟨211⟩ ejection further from the [100] surface normal using laser irradiation have been unsuccessful. Also more accurate measurements suggest all the spots have a centre of mass away from the crystallographic directions.

The additional central features seen in electron sputtering experiments are not predicted by the present theory and have not been seen in the photo-induced decomposition. We should note that the pulsed optical stimulation was made via a four-photon absorption process to the exciton band in crystals which were at room temperature. The high laser powers generate fields up to 10^6 V cm^{-1} and these may influence, or distort, the sputtering pattern. Electron experiments were made at high temperature and in a steady-state (i.e., not

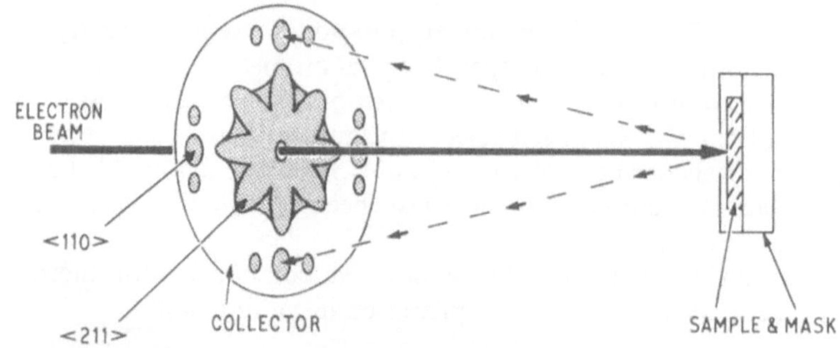

▲ **Fig. 4.10.** A schematic view of the collection system for viewing the sputter pattern from alkali halides. The silica collector plate is cooled. Major directional features of ⟨110⟩ and ⟨211⟩ are labelled. Other spots and maxima are noted

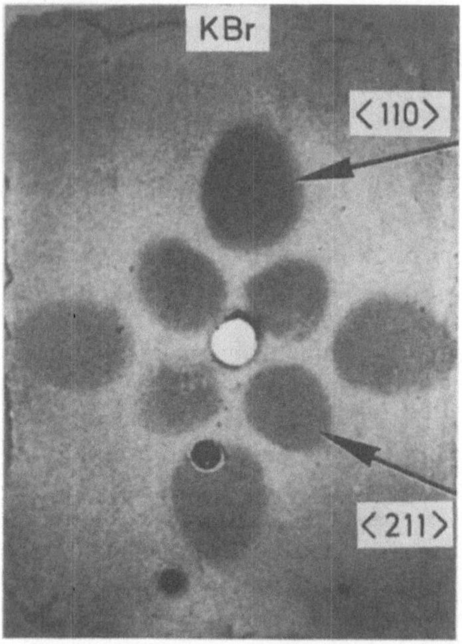

◀ **Fig. 4.11.** Bromine emission pattern obtained from a room temperature (100) KBr crystal face after exposure to 15 pulses of a ruby laser. The major ⟨110⟩ and ⟨211⟩ features are clearly shown

pulsed) mode. Therefore, there are differences between these two types of experiment and at this stage we lack the data to speculate on the physical reasons for them.

The major conclusion from the similarities of the results are that alkali halides can sputter halogen atoms and do so via a replacement collision sequence which is triggered by absorption of only 5–10 eV from exciting or ionising radiation.

4.4.4 Correlation with Luminescence

The original model of bulk defect production in alkali halides was based on the observation that the absorbed energy produced either luminescence or F centres (vacancies). Likewise, we may test that the sputtering yield and

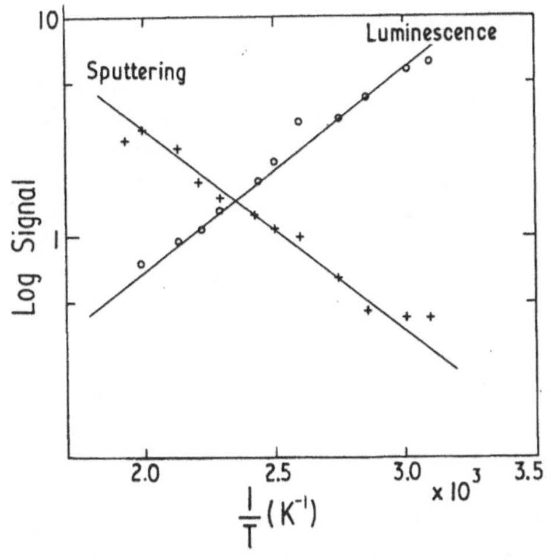

Fig. 4.12. Simultaneous measurements of the luminescence and sputtering yield from electron irradiated NaCl

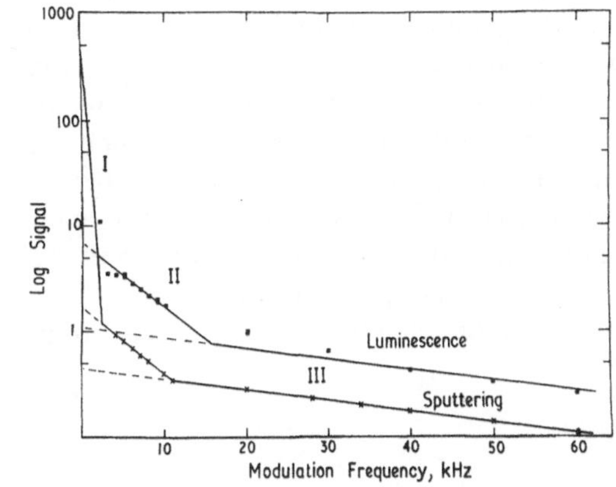

Fig. 4.13. Separation of frequency dependent components of the luminescence and sputtering for NaCl at 220 °C during irradiation with 1700 eV electrons. Data points are only shown for stages II and III

luminescence are alternative uses of the absorbed energy. In simultaneous measurements with NaCl at "high" temperatures from 20 to 200 °C, the anticorrelation is evident and the processes have the same activation energy. Figure 4.12 shows the results [4.64] from this electron irradiation experiment.

Measurements at a single temperature also indicate that neither luminescence nor sputtering is a simple process but that there are components in the yield intensities which show different lifetime effects. For NaCl, more detailed studies [4.67] made by modulating the primary electron beam reveal long lifetime components in the range 10^{-3} to 10^{-5} s for both the luminescence and the sputtering (Fig. 4.13). In this figure we see that at one temperature the lifetimes of both the sputtering and luminescence components follow the same pattern. The measured lifetimes probably indicate a time delay whilst the system decays through higher ionisation states. It is possible to influence the population of the energy states by selecting resonance energies of the electron beam via the inner shell transitions of electrons in the atoms. To change the fraction of excitons in high and low states the primary electron beam was swept

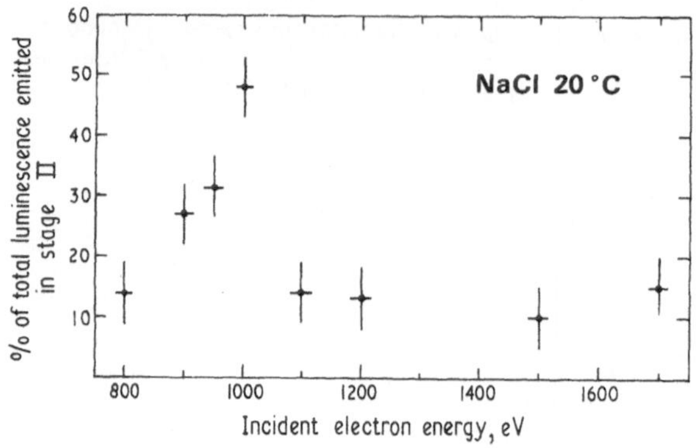

Fig. 4.14. A luminescence measurement of NaCl at 20 °C. The 430 nm emission in the stage II component (Fig. 4.13) changes as a function of the incident electron energy

through the energy range 800–1500 eV and a strong resonance was observed at 1000 eV (Fig. 4.14). Since the excitation energy for the sodium K_α line is 1040 eV, it is reasonable to suggest that the promotion of an inner shell electron has generated a higher state exciton. On this basis the three regions, I, II, and III were attributed to sputtering and luminescence from a first exciton state, for the major component of the signal (I), with additions from excitons which are diffusing in a higher state. These higher state systems then exhibit different lifetimes in the surface (II) and bulk regions (III). The suggestion is tentative for the NaCl data but from the more general viewpoint of sputtering we see that if the process involves an indirect step where energy is stored or diffused, as here by the excitons, there may be an appreciable time lag between the energy input and the sputtering.

4.4.5 Velocity of Ejected Particles

The major effort to measure the velocity spectra of energetically ejected material from alkali halides has been made by the group at FOM [1] [4.15, 18, 69, 70]. They used a velocity selector in their analysis to circumvent the delay problems which occur between the electron beam pulse and the final ejection event [4.70]. It is interesting to note that thermal ejection events can be significantly delayed and a comparison with the NaCl lifetime measurements, shown in Fig. 4.13, is in good agreement with the halogen sputtering component with a lifetime of $\sim 6 \times 10^{-4}$ s being thermal. In all cases the metal emission follows a thermal energy distribution.

Not all alkali halides show the hyperthermal component and the data so far can be separated into two types by reference to the (S/D) parameter of Fig. 4.9. Only those alkali halides with $S/D > 1/3$ have hyperthermal emission, as expected if we consider these to be the candidates for focussed replacement collision sequences.

1 FOM Institute for Atomic and Molecular Physics.

A large fraction of the halogen yield (up to 50%) is thermal in character so one may interpret the total sputtering as a mixture of ejected halogens and those which diffuse to the surface.

Molecular (X_2) emission is also detected. A summary of the present state of the velocity measurements is as follows:

i) energetic halogens are emitted from crystals at temperatures below 200 °C;
ii) only halides with $s/d > 1/3$ show energetic emission;
iii) the ratio of molecular to atomic emission X_2/X_1 decreases with temperature for halides with $s/d < 1/3$ but increases if s/d is $> 1/3$;
iv) lifetime measurements show the same features as the luminescence but the delayed emission is predominantly thermal.

All these observations fit into the broad outline of defect mechanisms proposed for the alkali halides but detailed interpretations are still subjective [4.71, 72]. One now suspects, however, that the halogen ejection contains component features from short replacement collision sequences which intersect the surface, long range diffusion from within the crystal, and metastable molecular structures at the surface. Further experiments are needed to clarify the ideas and the interpretation would benefit from a knowledge of the energy spectra of directional emission.

One already knows that the photon induced sputtering patterns look simpler than those formed during electron bombardment and it is also likely that one excites fewer of the exciton states, therefore there is a very strong need to extend the work to laser induced sputtering.

4.5 Surface Topography

The alkali halide experiments reported here have all had problems of reproducibility because of surface contaminants which did not sputter during electron irradiation. It has, therefore, only been possible to secure the electron induced sputtering in UHV and clean systems where the surface layer is an alkali halide one. As with ion beam sputtering experiments of rough and contaminated surfaces, the angular dependence of the sputtering yield and differences in sputtering yield for different materials leads to the usual formation of cones, mesas, pits and other surface features [4.62]. Photon sputtering problems are particularly severe because the sputtering yield of many surface contaminants is zero and the surfaces develop the same artistically interesting shapes as for the electron beam experiments.

Once beyond the surface problem there are still other features characteristic of the material. The regions around emergent dislocations sputter more rapidly than the plane areas and so develop pits. Additionally, many impurities in alkali halides decorate the dislocation lines, thus, associated with the rapidly sputtering pits is a core of nonsputtered material which produces a cone. A

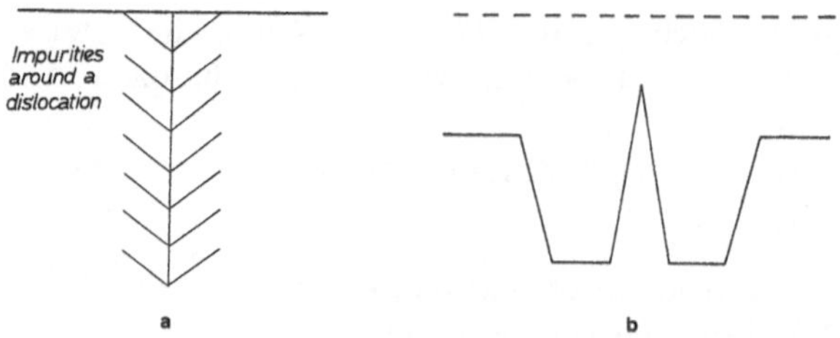

Fig. 4.15a, b. A model to explain features such as shown in Fig. 4.16d. Sputtering of a region (a) which contains an impurity decorated dislocation leads to (b) where the reduced sputtering coefficient of the impurities generates cone formation. In addition, the surrounds sputter more rapidly because of flux enhancement and the higher sputtering rate of the strained lattice

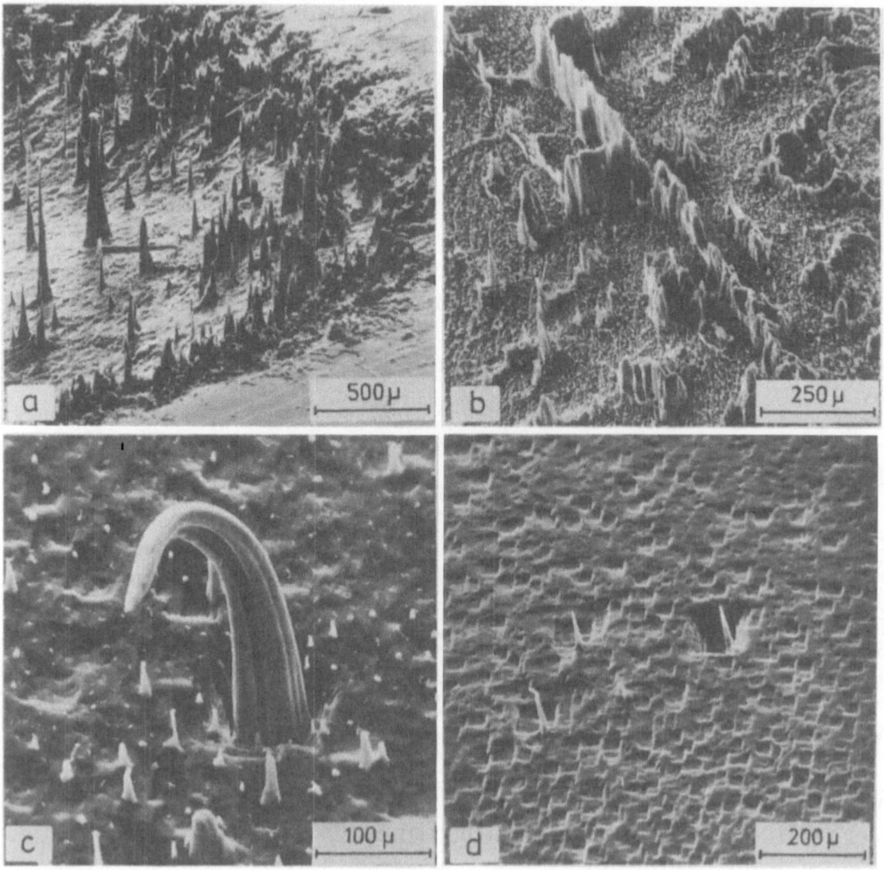

Fig. 4.16a–d. Surface features which developed on NaCl during irradiation with electrons. (**a**) Cones formed on air-cleaved NaCl by an electron irradiation. (**b**) A range of cones formed on air-cleaved NaCl by electron irradiation. The line of the cones corresponds to the position of a surface cleavage step. (**c**) An unusual cone formed during 1 kV electron irradiation of NaCl. (**d**) A surface of NaCl after prolonged electron irradiation. The level of the surface is 1 mm below the original crystal surface

schematic view of the surface profile is shown in Fig. 4.15 and examples may also be seen in Fig. 4.16.

A brief comment on Fig. 4.16 exemplifies these points. In Fig. 4.16a one sees cones, with at least one example (centre left) decorated by a large block of contaminant which may have acted as a shield. Figure 4.16b shows a continuous line of cones which are the extension of a cleavage step on the pre-irradiated crystal. Figure 4.16c shows a rather large bent cone feature that demonstrates the size of cones that may form. Small "square" sputtered etch pits are also visible. These are seen more clearly in Fig. 4.16d. It is interesting to note that in this example the larger cones are associated with the larger pits. A model for this type of pit formation is sketched in Fig. 4.15.

4.6 Silver Halides

Electron beam sputtering experiments are not very numerous and it is therefore useful to cite a second class of material that does erode during bombardment and for which we have a moderate understanding of the mechanism of dissociation.

The silver halides dissociate and sputter with a very low threshold energy, namely, by absorption of a photon which generates an exciton. Because of their importance in photography [4.73–75], we shall present one model for the dissociation. The precise details of the steps are unknown but the energy is directly absorbed in the exciton absorption band in pure material or, alternatively, the exciton is generated in a dye coupled to the lattice. In the latter case the photons of lower energy are efficient and the film may respond to visible or near infrared wavelengths.

In all cases it seems necessary to dissociate the electron and hole pair of the free exciton and separate the charges. The actual details of the model are in some doubt but two alternatives are (i) that the electron is trapped at a silver ion which is still on a lattice site and (ii) that the ion is a silver interstitial.

The process in either case is to form a neutral silver atom ($Ag^+ + e^- \rightarrow Ag^0$). The silver is then mobile in the neutral charge state. The hole is removed to a trap on the surface of the silver halide grain. Impurity states on the surface serve a two-fold purpose: firstly, they produce an electric field extending into the grain which separates the electron and hole of the exciton and secondly, the impurity states trap the hole once it has migrated to the surface. The argument is phrased in terms of the small grains used in photography but a similar argument would hold for large single crystals. However, in this case one should realize that the impurity atoms tend to precipitate in the region of dislocation lines. Such lines are therefore electrically charged with respect to the perfect lattice [4.76] and so are equivalent to the surface impurity states of crystallites.

The second stage of defect formation is by diffusion of the neutral silver atom from a lattice site to an interstitial site. Such sites are unstable but if several silver atoms (three or four) aggregate, they form a colloidal speck of silver which is stable (the latent image); all these states can be separately identified by optical absorption [4.77]. The details of colloidal speck formation may involve special surface sites but once again the details are uncertain.

Negative halogen ions may be neutralised by the hole and are then able to thermally escape from the surface, or react with the surrounding matrix in the case of photographic emulsions. The loss of the halogen and precipitation of the silver into clusters has the necessary characteristics to demonstrate a second case of photon sputtering. It differs from the alkali halide model in that the movement of the interstitials is by diffusion and therefore release from the surface is likely to be by thermal evaporation and without any directional features. No experimental data on the directionality is available.

4.7 Surface Erosion by Photons

4.7.1 Laser Damage and Sputtering

In the earlier discussion of alkali halide sputtering, we mentioned one example of laser induced sputtering [4.65, 66]. For the particular example one required a high power laser as the excitation was not direct but by a multiphoton process. For the alkali halides, the high power light source is not essential as one may directly sputter them with light absorbed in the exciton energy levels [4.62, 78], but since one is dealing with wide band gap materials, such experiments are difficult as there is a shortage of intense light sources at the high photon energies.

High power optical illumination can produce both damage and sputtering and there are now many examples of the problems that result from laser induced damage. From the viewpoint of this chapter, the destruction of surfaces by the dissipation of large quantities of power is not particularly relevant as in most cases the materials are thought to vaporise and/or undergo dielectric breakdown rather than damage by the more subtle methods discussed so far. To appreciate the large quantities of power that are available, and needed, for dielectric breakdown, one should recall that electric field strengths of 1–5 $\mathrm{MV\,cm^{-1}}$ are required but optical illuminations in the 50–100 $\mathrm{MW\,mm^{-2}}$ produce such fields and this is quite an accessible power range for modern laser systems.

Damage and sputtering which result solely from power absorption are simple to distinguish from the indirect processes as they are generally a function of the power absorbed above some average threshold level, and, unless confused by a strong absorption feature, are insensitive to the wavelength used.

We may list some of the more obvious mechanisms for power absorption as follows.

i) Optical transitions produced by light absorption at colour centres.
ii) Excitation and acceleration of conduction band electrons.

iii) Interband excitation by multiphoton processes.

iv) Avalanche effects which result from the energy imparted to the conduction band electron causing ionisation of lattice ions. In insulators the initial concentration of conduction band electrons may be negligible so the onset of avalanche effects will require primary steps from processes such as i) or iii). Thus, we see that the avalanche damage is likely to be a function of the total damage which is retained in an illuminated area as well as the power level used. In applications where one is using pulsed laser beams, the duration and repetition rate of the pulses, as well as the peak power, will control the probability of erosion. For such reasons the quoted damage threshhold powers for different materials may vary over orders of magnitude (for example, see the review paper of *Wood* et al. [4.79]).

v) In some materials the optical beam changes the refractive index of the solid so that even if one had achieved a uniform irradiation of a homogeneous material, power gradients from self focussing of the beam would develop.

Before rejecting further discussion of laser induced erosion as merely the result of thermal evaporation or shock, we should note that the apparent damage threshold power levels show very wide variations for the same material and we may need to look for more subtle processes which assist the sputtering (the problem is also of interest to ion beam research where ion beam polishing of laser materials is used).

For optical components, one may try to reduce power absorption or scattering from inhomogeneities, colour centres, refractive index gradients and surface features but in compounds, one may have the further problem that the stoichiometry of the surface layers is changed more readily than in the interior of the solid. The reasons are diverse and may range from preferential ejection of one element (as in the alkali halides) to outdiffusion from a multiphase system. For example, one may speculate that the reason why $LiNbO_3$ has such a wide and low range of quoted damage threshold [4.79] is that the surface layers have a different stoichiometry from the bulk material and the exact composition is a function of the method of preparation of the surface. Many articles on $LiNbO_3$ discuss this problem [e.g., 4.80–82]. Not only is there general agreement that the surface of the material contains many defect levels but also it is well established that lithium atoms will outdiffuse from the surface at quite modest temperatures [4.83]. Thus, one has a situation where the surface absorbs power because it is not perfect and the consequent rise in temperature can lead to further changes in composition; a situation which is unstable and will lead to breakdown.

It is interesting to compare this high power system, where the rate of breakdown accelerates with changes in surface composition, with the earlier, low energy and low power mechanisms that were discussed. In the alkali halides it was clear that the photon absorption ejected halogen atoms but unless one could displace the metal ions by a secondary process such as thermal evaporation, then the sputtering was inhibited. The low vapour pressure of the alkali metals ensured that for the alkali halides at temperatures not much above room temperature, the sputtering yield remained finite. A more extensive search for low energy electron sputtering [4.15] showed that $ZnBr_2$, AgBr, CaI_2, and PbI_2 all emitted halogen atoms but the rate of emission was retarded or even stopped as the surface layers built up an excess of the metal. At the time of writing it is not possible to offer more specific interpretations of this data.

To conclude we see that whilst high power laser damage and sputtering is basically a simple matter of energy absorption and independent of wavelength, any type of inhomogeneity which can exist in the material can cause the onset of the damage. In the early stages of the breakdown point defects do have a role and one should not be surprised that any material which potentially has low energy defect mechanisms may also show low thresholds for laser damage.

4.7.2 Photon Assisted Thermal Decomposition

The energy required for dissociation of ions from a lattice is a function of the electronic state of the bonds, thus, one may reasonably suppose that the rate of thermal decomposition can be enhanced if the ions are raised to an excited state. One may also suppose that materials that thermally decompose are potentially likely to show photolytic decomposition. Similarly, if the kinetic energy is provided by ion bombardment, rather than by photons, one might expect to see a higher sputtering yield in such materials if they are illuminated during ion beam sputtering. Such experiments may not yet exist but the parallel effects of higher damage rates within insulators (e.g., SiO_2, MgO, Al_2O_3, $LiNbO_3$) caused by simultaneous ionisation and displacement events are known. The effects are most apparent with light ions where the large electronic energy losses ionise the track of the ion path and hence reduce the displacement energy for the nuclear collisions.

We shall now consider the trivial case of photon induced release of surface adsorbed molecules. Here, ionisation assisted release is equally applicable to metallic or insulating materials. For example, in studies of gas release from stainless steel and Al_2O_3 surfaces, both low [4.84, 85] and high energy photons [4.86] have proved effective. The data is sparse and sometimes conflicting as the released species are sensitive to the range of surface states, and hence other adsorbed species. Consequently, measurements of ion species show different fractions of CO_2, CO, O_2, CH_3, H_2, etc., for thermal and photon stimulated desorption.

Photon assisted decomposition is apparent at low levels of optical illumination. For example, in CdS [4.87] it has been seen for levels as low as 10^3 to $10^5\,\mu W\,cm^{-2}$. For these experiments, single crystals of CdS were held at temperatures between 680 and 760 °C and the rate of decomposition of the stable *C* face was observed. Illumination with photons at the energy of the band gap enhanced the rate of decomposition but the response to changes in light intensity was slow. One, therefore, excludes a mechanism of direct ionisation of surface atoms and instead suggests that the light produces a change in the density of charge in the surface states or, alternatively, a change in the charge state of ions which diffuse to the surface before evaporation. In nonstoichiometric CdS, the addition of excess *S* ions enhanced the optical effect but excess Cd did not.

One should also note that illumination of CdS with high intensity laser light (10^8 W cm^{-2}) changes the luminescence spectra produced at 77 K [4.88]. The modifications of the band edge absorption and luminescence were interpreted in terms of exciton states bound to donor centres and the changes produced by the laser reflect the formation and movement of sulphur vacancies and interstitials. If this interpretation is correct then one should suspect the presence of a low energy defect formation process in CdS.

For the purpose of this chapter the CdS example is suitable because firstly it demonstrates that one may optically enhance the rate of thermal decomposition. Secondly, it is an effect which is sensitive to excitation wavelengths and impurities so it is electronic in origin. The low temperature changes suggest an intrinsic mechanism for defect formation so a search for direct photon induced sputtering might be profitable.

4.8 Near Threshold Events with Relativistic Electrons

In the preceding sections sputtering by direct interactions with the primary low energy electron beam has not been possible. At higher electron energies above some 0.5 MeV, the relativistic electrons posses sufficient energy and momentum to directly displace lattice atoms by a collision. Indeed, by a comparison of calculations and experimental data one can estimate the displacement energy from the energy dependent cross section for the displacement process, e.g., [4.89]. Atomic defects formed at the surface can, of course, cause sputtering and we shall now extend the discussion of electron induced sputtering to this same range as one may use the energy dependence to resolve the low threshold energy for surface sputtering and the higher energy effects originating in the interior. A particularly clear series of experiments has been made by *Cherns* et al. [4.90–92] who viewed transmission sputtering from thin films of gold in a high-voltage electron microscope (HVEM). Their experimental arrangements, shown in Fig. 4.17, enabled them to view sputtering from a single grain and although the gold to collector separation is only some 5 μm, they could resolve the details of the three $\langle 110 \rangle$ ejection spots in subsequent analyses.

To explain the angular widths and intensity variations with the $\langle 110 \rangle$ spots, they first used a [4.90] computer simulation using the Marlowe code [Ref. 4.93, Chap. 3] developed by *Robinson* and *Torrens* [4.74] to predict how the primary recoils were focussed. They assumed that with a low electron energy of 0.5 MeV, where the maximum energy transferred (T_m) to a gold atom is only 8.3 eV, it was not possible to form directional displacements or

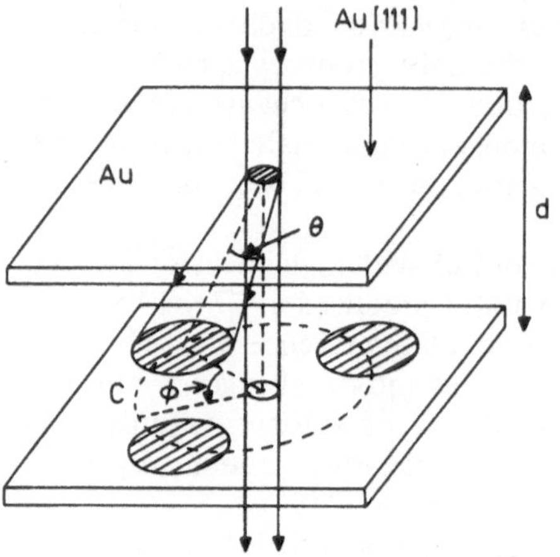

Fig. 4.17. Diagram of experimental arrangement for investigating sputtering from gold foils in the HVEM. Gold atoms transmission-sputtered by electrons in the direction (θ, ϕ) were collected on a carbon foil situated at a distance d below the gold. At energies >600 keV, gold deposit thicknesses showed maxima at angles $\phi = 120°$ apart, as illustrated schematically

sputtering. Those atoms which leave the (111) surface show a distribution which resembles a thermal emission pattern about the axis of the electron beam. However, on increasing the energy to 0.6 MeV ($T_m \sim 10.6$ eV), one can generate the dynamic events in which surface atoms are ejected. The directions are determined by the repulsive and attractive potentials with the neighbours and hence one sees spots centred around $\langle 110 \rangle$ directions.

The dependence of the measured spot patterns on the primary energy is shown in Fig. 4.18 where we see that on further increasing the primary energy, the spot pattern reduced in width. The explanation of this is that at the higher electron energies (e.g., $E_{\text{electron}} = 1.1$ MeV, $T_m = 25.4$ eV), the energy imparted to an atom several layers beneath the surface can still lead to ejection of a surface atom if sufficient energy reaches the surface. This is only possible if one has directed the energy so that there is a focused collision sequence from the struck atom up to the surface atom that is sputtered. Since the energy loss is a function of the interatomic potential, the angle of the primary recoil with the crystal axis and the number of layers in the sequence, only well directed sequences will cause ejection and the resulting sputtering patterns have a narrow angular width. *Cherns* et al. [4.90–92] assumed only recoils within 25° of the $\langle 110 \rangle$ axis could lead to a focused sequence. If the energy loss per layer is some 3 eV, then for the 1 MeV electron irradiation they estimate that they obtain equal contributions from direct surface events and long range focused events. In this case, long range is some 4 layers. For collision sequences which deliver too little energy for ejection, the surface atoms are displaced to leave a surface vacancy and the atom moves to an adatom position on the gold surface. Thus, not only is there a threshold energy for escape but there is at the same time a discontinuity in the angular distribution of the movement of the displaced atoms. This bending of the trajectories was expected on the later computer simulation of *Cherns* et al. [4.91] where they used a molecular dynamics model [Ref. 4.93, Chap. 3] to perform the calculation of the atom trajectories.

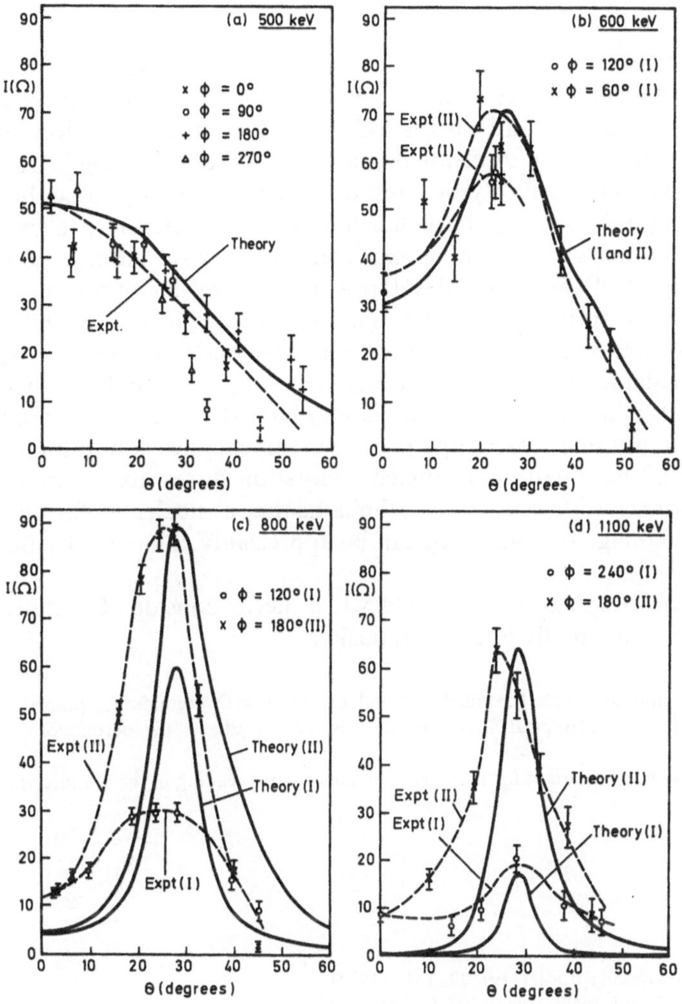

Fig. 4.18a–d. The angular distribution function of sputtered atoms $I(\Omega)$. The points joined by broken curves show experimental points. The solid lines are the theory of *Cherns* et al. [4.90]

Finally, *Cherns* [4.92] has considered how the surface migration of the surface vacancies can explain the development of pits and roughening of the (111) gold surface.

This work with the HVEM forms a link between the low energy sputtering events and the normal ion beam sputtering. For the low energy experiments it demonstrates that there is a real possibility of separating the surface and near surface events by measurements of the angular distribution of the sputtering patterns. At present this is a refinement which has not yet been experimentally achieved.

4.9 Conclusion

The sputtering mechanism by electrons and photons for the alkali halides is specific to these compounds and successfully described the formation of radiation damage as well as the directional character of the sputtering.

A much more general possibility of photon induced sputtering exists but experimentally, the search for such effects has not been very intensive. In part because it is readily masked by either extrinsic or intrinsic surface contaminants. The concept of sputtering, developed from the effects of ion beam bombardment, carries with it the idea that many of the ejected particles have energies in excess of the thermal spectrum of ion energies. If one widens the concept to include enhancement of thermal decomposition or photolysis, then the range of materials which might "sputter" is very large. Further, one should not ignore the possibility that low energy processes can enhance sputtering or radiation damage and in situations of mixed irradiation fluxes, for example, photon or electronic excitation occurring at the same time as nuclear collisions, the overall rates of damage or sputtering can be appreciably enhanced by the low energy mechanisms.

In conclusion one suggests this is a field which needs considerably more exploration, both experimentally and theoretically.

Acknowledgements. I am grateful to Drs. Braunlich, Schmid, Cherns, and Overeijnder for preprints of their work and to The Philosophical Magazine and Radiation Effects for permission to reproduce diagrams.

My thanks also go to Dr. R. Behrisch for a critical and constructive reading of the manuscript.

References

4.1 M. Saidoh, P. D. Townsend: Radiat. Eff. **27**, 1–12 (1975)
4.2 R. T. Williams: Semicond. Insulators **3**, 251 (1978)
4.3 P. D. Townsend, F. Agullo-Lopez: J. Phys. (Paris) Colloq. C**6** (Suppl. 7, Vol. 41) 279 (1980)
4.4 D. W. Palmer: Inst. Phys. Conf. Ser. **31**, 144 (1977)
4.5 F. Jaque, P. D. Townsend: Nucl. Instrum. Methods **182/183**, 781 (1981)
4.6 M. Szymonski: Radiat. Eff. **52**, 9 (1980)
4.7 L. E. Seiberling, J. E. Griffith, J. A. Tombrello: Radiat. Eff. **52**, 201 (1980)
4.8 J. P. Biersack, E. Santner: Nucl. Instrum. Methods **132**, 229 (1976)
4.9 Y. Al Jammal, D. Pooley, P. D. Townsend: J. Phys. C **6**, 247 (1973)
4.10 C. G. Pantano, T. E. Madey: Appl. Surf. Sci. **7**, 115 (1981)
4.11 B. Carriere, B. Lang: Surf. Sci. **64**, 209 (1977)
4.12 V. H. Ritz, V. M. Bermudez: Phys. Rev. B**24**, 5559 (1981)
4.13 Y. Shapira: J. Appl. Phys. **52**, 5696 (1981)
4.14 T. T. Lin, D. Lichtman: J. Vac. Sci. Technol. **15**, 1689 (1978)
4.15 H. Overeijnder, M. Szymonski, A. Haring, A. D. de Vries: Radiat. Eff. **36**, 63 (1978)
4.16 C. L. Strecker, W. E. Moddeman, J. T. Grant: J. Appl. Phys. **52**, 6921 (1981)
4.17 W. P. Ellis: J. Vac. Sci. Technol. **14**, 1316 (1977)

4.18 *Desorption Induced by Electronic Transitions*, DIET I, ed. by N.H.Tolk, M.M.Traum, J.C.Tully, and T.E.Madey, Springer Ser. Chem. Phys., Vol. 24 (Springer, Berlin, Heidelberg, New York 1983)
4.19 D.Menzel, R.Gomer: J. Chem. Phys. **41**, 3311 (1964)
4.20 P.A.Redhead: Can. J. Phys. **43**, 886 (1964)
4.21 P.R.Antoniewicz: Phys. Rev. B**21**, 3811 (1980)
4.22 M.L.Knotek, P.J.Feibelman: Phys. Rev. Lett. **40**, 964 (1978)
4.23 P.D.Townsend: Nucl. Instrum. Methods **198**, 9 (1982)
4.24 T.Nakayama, N.Itoh: Radiat. Eff. **67**, 129 (1982)
4.25 U.Bangert, D.E.Arafah, U.Sassmannshausen, K.Thiel, P.D.Townsend: Nucl. Instrum. Methods **209/210**, 1111 (1983)
4.26 D.E.Arafah, U.Bangert, K.Thiel, P.D.Townsend: Nucl. Instrum. Methods **209/210**, 1105 (1983)
4.27 L.E.Seiberling, C.K.Meins, B.H.Cooper, J.E.Griffith, M.H.Mendenhall, T.A.Tombrello: Nucl. Instrum. Methods **198**, 17 (1982)
4.28 W.L. Brown, W.M. Augustyniak, E. Simmons, K.J. Marcantonio, L.J. Lanzerotti, R.E.Johnson, J.W.Boring, C.T.Reiman, G.Foti, V.Pirronello: Nucl. Instrum. Methods **198**, 1 (1982)
4.29 Ion induced desorption of molecules from bio-organic solids. Nucl. Instrum. Methods **198** (1982)
4.30 A.M.Stoneham: *Theory of Defects in Solids* University Press (Oxford 1975)
4.31 M.Ueta: J. Phys. Soc. Jpn. **23**, 1265 (1967)
4.32 Y.Kondo, M.Hirai, T.Yoshinari, M. Ueta: J. Phys. Soc. Jpn. **26**, 1553 (1969)
4.33 M.Hirai, Y.Kondo, T.Yoshinari, M.Ueta: J. Phys. Soc. Jpn. **30**, 440 (1971)
4.34 Y.Kondo, M.Hirai: J. Phys. Soc. Jpn. **30**, 1765 (1971)
4.35 Y.Kondo, M.Hirai, M.Ueta: J. Phys. Soc. Jpn. **33**, 151 (1972)
4.36 T.Karasawa, M.Hirai: J. Phys. Soc. Jpn. **33**, 1728 (1972)
4.37 T.Karasawa, M.Hirai: J. Phys. Soc. Jpn. **34**, 276 (1973)
4.38 I.N.Bradford, R.T.Williams, W.L.Faust: Phys. Lett. **35**, 300 (1975)
4.39 D.Pooley: Solid State Commun. **3**, 241 (1965)
4.40 D.Pooley: Proc. Phys. Soc. **87**, 245 (1966)
4.41 H.N.Hersh: Phys. Rev. **148**, 928 (1966)
4.42 Ch.B.Lushchik, G.G.Liid'ya, M.A.Elango: Sov. Phys.—Solid State **6**, 1789 (1965)
4.43 Ch.B.Lushchik, I.K.Vitol, M.A.Elango: Sov. Phys.—Solid State **10**, 2166 (1969)
4.44 D.Pooley: Proc. Phys. Soc. **87**, 257 (1966)
4.45 D.Pooley, W.A.Runciman: J. Phys. C: **3**, 1815 (1970)
4.46 A.E.Hughes, D.Pooley, H.U.Rahman, W.A.Runciman: AERE Rpt. R. 5604 (1967)
4.47 R.Smoluchowski, O.W.Lazareth, R.D.Hatcher, G.J.Dienes: Phys. Rev. Lett. **27**, 1288 (1971)
4.48 M.N.Kabler: In *Point Defects in Solids*, Vol. 1, ed. by J.H.Crawford, Jr., L.M.Slifkin (Plenum Press, New York 1972) p. 327
4.49 R.T.Williams, M.N.Kabler: Solid State Commun. **10**, 49 (1972)
4.50 R.T.Williams, M.N.Kabler: Phys. Rev. B**9**, 1897 (1975)
4.51 N.Itoh, M.Saidoh: J. Phys. (Paris) **34**, C–9, 101 (1973)
4.52 M.Saidoh, J.Hoshi, N.Itoh: J. Phys. Soc. Jpn. **39**, 155 (1975)
4.53 M.N.Kabler: Lecture Notes for NATO Advance Study Institute on Radiation Damage Processes in Materials, Corsica (1973)
4.54 G.J.Dienes, R.Smoluchowski: J. Phys. Chem. Solids **37**, 95 (1976)
4.55 P.D.Townsend: J. Phys. C **9**, 1871 (1976)
4.56 G.Giuliani: Phys. Rev. B**2**, 464 (1970)
4.57 P.D.Townsend: Surf. Sci. **90**, 256 (1979)
4.58 P.D.Townsend: J. Phys. C **6**, 961 (1973)
4.59 D.B.Carroll, H.K.Birnbaum: J. Appl. Phys. **36**, 2658 (1965)
4.60 P.W.Palmberg, T.N.Rhodin: J. Phys. Chem. Solids **29**, 1917 (1968)
4.61 T.E.Gallon, I.G.Higginbotham, M.Prutton, H.Tokutaka: Surf. Sci. **21**, 224 (1970)

4.62 D.J.Elliott, P.D.Townsend: Philos. Mag. **23**, 249 (1971)
4.63 G.Chassagne, D.Durand, J.Serughetti, L.W.Hobbs: Phys. Status Solidi A: **40**, 629 (1977); **41**, 183 (1977)
4.64 A.Friedenberg, Y.Shapira: Surf. Sci. **87**, 581 (1979)
4.65 P.D.Townsend, R.Browning, D.J.Garlant, J.C.Kelly, A.Mahjoobi, A.J.Michael, M.Saidoh: Radiat. Eff. **30**, 55 (1976)
4.66 A.Schmid, P.Braunlich, P.K.Rol: Phys. Rev. Lett. **35**, 1382 (1975)
4.67 A.Schmid, P.Braunlich, P.K.Rol: To be published
4.68 P.D.Townsend, A.Mahjoobi, A.J.Michael, M.Saidoh: J. Phys. C: **9**, 4203 (1976)
4.69 H.Overeijnder: Thesis, Amsterdam (1978)
4.70 H.Overeijnder, R.R.Tol, A.D.de Vries: Surf. Sci. **90**, 265 (1979)
4.71 F.Agullo-Lopez, P.D.Townsend: Phys. Status Solidi (b) **97**, 575 (1980)
4.72 M.Szymonski: Proc. Symp. on Sputtering (SOS), Techn. Univ., Wien 1980) p. 761
4.73 C.E.K.Mees, T.H.James: *The Theory of the Photographic Process* (Macmillan, London 1966)
4.74 L.Slifkin: In *Solid State Dosimetry* ed. by S.Amelinckx (Gordon and Breach, London 1969) Chap. 2
4.75 J.F.Hamilton, L.Slifkin: Solid State Phys. (to be published)
4.76 K.L.Kliewer: J. Phys. Chem. Solids **27**, 705 (1966)
4.77 H.Kanzaki, S.Sakuragi: Photogr. Eng. **17**, 69 (1973)
4.78 P.D.Townsend, D.J.Elliot: Phys. Lett. **28** A, 587 (1969)
4.79 R.M.Wood, R.T.Taylor, R.L.Rouse: Opt. Laser Techn. **7**, 105 (1975)
4.80 J.J.Amodei, D.L.Staebler: RCA Rev. **33**, 71 (1972)
4.81 D.Redfield, W.J.Burke: J. Appl. Phys. **45**, 4566 (1974)
4.82 W.Bollmann: Phys. Status Solidi (a) **40**, 83 (1977)
4.83 I.P.Kaminow, J.R.Carruthers: Appl. Phys. Lett. **22**, 326 (1973)
4.84 G.W.Fable, S.M.Cox, D.Lichtman: Surf. Sci **40**, 571 (1973)
4.85 M.J.Drinkwine, V.Shapira, D.Lichtman: Adv. Chem. Ser. **158**, 171 (1976)
4.86 S.Brumbach, M.Kaminsky: Adv. Chem. Ser. **158**, 183 (1976)
4.87 G.A.Somorjai: Surf. Sci. **2**, 298 (1964)
4.88 M.S.Brodin, N.A.Davydova, I.Yu Shablii: Sov. Phys.– Semicond. **10**, 375 (1976)
4.89 J.W.Corbett: Solid State Phys. Suppl. **7** (1966)
4.90 D.Cherns, F.J.Minter, R.S.Nelson: Nucl. Instrum. Methods **132**, 369 (1976)
4.91 D.Cherns, M.W.Finnis, M.D.Mathews: Philos. Mag. **35**, 593 (1977)
4.92 D.Cherns: Surf. Sci. **90**, 339 (1979)
4.93 R.Behrisch (ed.): *Sputtering by Particle Bombardment* I, Topics Appl. Phys., Vol. 47 (Springer, Berlin, Heidelberg, New York 1981)
4.94 M.T.Robinson, I.M.Torrens: Phys. Rev. B**9**, 5008 (1974)

5. Sputtering of Solids with Neutrons

Rainer Behrisch

With 25 Figures

With the application of intense neutron fluxes in science, technology and power generation, the effects of neutron bombardment, not only in the bulk material, but also at the surface become important. One of these surface effects is neutron sputtering, i.e., the removal of atoms from the bombarded surfaces. Generally neutron sputtering yields are small, typically 10^{-5} to 10^{-4} atoms per incident neutron, of which 10^{-8} to 10^{-6} atoms per neutron are radioactive-primary-knockon atoms. Only for fissile materials sputtering yields of the order of unity are found.

The few sputtering yields measured with reactor neutrons and 14 MeV neutrons for Au, Nb, V, and Fe have large uncertainties due to the limits of sensitivity in detecting the material removed. The values tend to be lower but still agree within the experimental uncertainties with sputtering yields calculated from linear transport theory. Radioactive primary knockon emission yields have been measured with high sensitivity. They are related to mean projected ranges of these atoms in the solid.

The emission of microparticles as observed for UO_2 and by some investigators for Nb, V, and Si can only be understood if these are pre-formed on the target surface, either by previous intense bombardment or by special mechanical treatment. One release mechanism may then be an electrostatic charging of the particles and of the partly insulating surface and subsequent electrostatic repulsion.

5.1 Neutron Irradiation Effects

When energetic neutrons bombard a solid, the major effects observed are the creation of radioactivity and changes in the composition and structure of the material. The radioactivity is induced by nuclear reactions between the neutrons and the atomic nuclei of the solid, while the changes in the composition and structure are due to the production of transmuted atoms and due to radiation damage produced in the crystal lattice [5.1]. The primary process causing radiation damage is the transfer of kinetic energy to a target atom, which becomes a primary knockon atom, during the interaction with a neutron. The mean transferred energy depends on the neutron energy and target mass. It is of the order of 100 keV for 14 MeV neutrons and proportionally lower for lower neutron energies. Thermal neutrons can create primary knockon atoms with energies of 50–500 eV due to the recoil energy received in (n, γ) processes [5.2], or in the form of fission products with energies of about 60 MeV [5.3]. The primary knockon atoms distribute their energy to other lattice atoms and to electrons via collision cascades [5.4]. When a cascade has come to rest, most of the energy has been dissipated to phonons and electrons. In addition, Frenkel defects, i.e., vacancies and interstitial atoms, and larger disordered regions are left in the lattice [5.1, 4]. In metals, the energy transferred to electrons does not contribute to damage, while in insulating materials this energy may lead to local excitations, which can transfer energy to lattice atoms. The latter mechanism can be more effective in creating radiation damage than the direct energy transfer to a target atom (Chap. 4).

When energy is transferred to a surface or a near-surface atom, sputtering can occur. Thus, sputtering is a manifestation of radiation damage in the surface layers of a solid and is expected to occur for any radiation causing radiation damage in the bulk of a solid. The emission of radioactive primary knock-on atoms (or radioactive recoils) represents a fraction of the sputtered atoms, but can be measured separately with high sensitivity. Such an emission named "Emanationsvermögen" for the case of gaseous atoms has already been a topic of many early investigations [5.5, 6].

The cross section for the interaction between a neutron and an atom of the solid is mostly of the order of 10^{-8} Å^2 and thus much smaller than the area of an elementary unit lattice cell (about 5 Å^2). The mean free path of neutrons is large, typically a few cm, and primary knockon atoms are created uniformly in the solid. Collisions with near-surface atoms causing sputtering are rare and the surface composition and structure, as well as the development of the cascade started by the primary knockon atoms, is not modified until very high neutron fluences.

In a crystalline solid the trajectories of energetic atoms, and the spread of a collision cascade may be influenced by the lattice structure due to channeling, blocking and focusing [Ref. 5.7, Chap. 3]. In neutron sputtering experiments channeling of the incident neutrons cannot be observed. Similar to ion sputtering the primary knockon atoms generally start from lattice sites. Their trajectories are influenced by the blocking effect and will be close to those in a random material. The contribution of the focusing effect is mostly small due to the small acceptable solid angle and also due to thermal lattice vibrations and damage in the crystal [Ref. 5.7, Chaps. 2, 3].

The amount of sputtering is measured by the sputtering yield Y which is defined as the mean number of atoms removed from the surface of a solid per neutron passing through this surface [5.7]. Because neutrons can penetrate through a solid with a thickness of the order of centimeters, we have to distinguish [5.8] between *backward sputtering* occurring at the side where the neutrons enter the solid (Fig. 5.1a), which is similar to normal ion-beam sputtering, and *forward sputtering* occurring at the side where the neutrons emerge from the solid (Fig. 5.1a), which can be compared to transmission sputtering of very thin films in

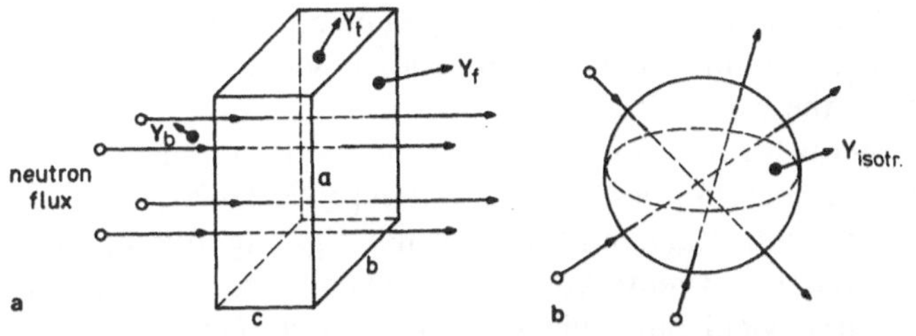

Fig. 5.1. (**a**) Sputtering from different surface areas of a cube subjected to a directed flux of neutrons (Y_f: forward sputtering yield, Y_b: backward sputtering yield, Y_t: transverse sputtering yield) (**b**) Sputtering from a sphere subjected to an isotropic flux of neutrons

ion bombardment [Ref. 5.7, Chaps. 2, 4]. We further deal with *transverse sputtering* at those areas of a solid which are parallel to the neutron flux (Fig. 5.1a), and *isotropic sputtering* if a probe is subjected to an isotropic flux of neutrons such as that existing close to a reactor core (Fig. 5.1b).

Forward sputtering yields are generally of the order of 10^{-5} to 10^{-4} atoms per neutron, except for fissile materials where yields of the order of 1 have been found. The transverse sputtering yields should have a similar magnitude to the forward sputtering yields and they depend on the detailed shape of the probe. Backward sputtering yields are a factor of 10 to 20 lower than forward sputtering yields. Isotropic sputtering yields are equal to the sum of the backward sputtering yield at the surface where the neutrons enter the solid, the transverse sputtering yield and the forward sputtering yield at the exit of the neutrons from the solid, averaged over all directions of the incident neutrons. Finally, the yields Y^p for emitting radioactive primary knockon atoms are two to three orders of magnitude lower than the sputtering yields.

There is a general physical interest in neutron sputtering because of its relation to radiation damage [5.8, 9] and to ranges of energetic ions in solids [5.6, 10–14]. Further, neutron sputtering turned out to be an important process for emitting radioactive wall atoms into the coolant channels in nuclear reactors [5.14–17].

In fusion research, neutron sputtering is of interest as being one of the mechanisms which introduce atoms from the solid walls of the plasma chamber into the hot deuterium-tritium plasma [5.18–22]. Here they represent impurities which cool the plasma by radiation and decrease the number of fusion reactions due to a depletion of deuterium and tritium at fixed electron density. In early fusion reactor design studies, the first wall was proposed to be made of niobium [5.23]. Therefore, for this material neutron sputtering yields were investigated in some detail [5.22]. However, due to the high radioactivity produced by 14 MeV neutrons in Nb [5.24] and the high cost, other materials were envisaged in subsequent fusion reactor design studies. These were vanadium [5.25], the nickel alloy PE 16 [5.26], 316 stainless steel [5.27] and materials with low atomic number Z such as carbon or carbides [5.28–31]. In any case it turned out that impurity production by neutron sputtering at the first wall is not expected to be a limiting factor in fusion devices if compared to other impurity sources at the walls [5.22, 32].

Sputtering measurements for neutrons are still very few in number and the results obtained show large uncertainties, as reviewed in [5.22, 33–36]. Thus, in the following, firstly the experimental conditions for investigating total neutron sputtering yields and radioactive primary knockon atom emission yields will be described and the measured yield data will be summarized. The theoretical treatment of neutron sputtering yields, first derived by *Sigmund* [5.8], will be outlined including some remarks about spikes and mechanisms for the emission of microparticles and chunks.

5.2 Sputtering Experiments with Neutrons

The experimental techniques for measuring neutron sputtering yields are, in principle, similar to those applied in ion beam sputtering [Ref. 5.7, Chap. 4] and [5.33–37]. A given sample is exposed in high vacuum to a known neutron dose and the amount of material removed from the surface is measured. However, the neutron fluxes available are much lower than the ion fluxes used for sputtering experiments. Furthermore, the sputtering yields expected are much smaller than ion sputtering yields. Thus, only much less than the equivalent of one monolayer could be sputtered away in the experiments performed up to now. The most sensitive techniques available had to be applied to detect the material removed and deposited on a collector.

5.2.1 Neutron Sources

Sources for neutrons can be fission reactors or nuclear reactions with energetic ion beams or photons [5.38–43]. Most sources do not give single energy neutrons but a broad energy spectrum which is further modified by the absorption and scattering properties of the materials surrounding the source. As the neutron energy is one parameter determining the sputtering yield, sputtering measurements should be performed preferably with single energy neutrons or with neutrons where the energy distribution is known. The energy distributions have been measured for most neutron sources by time-of-flight techniques and/or by the activation of different elements having different threshold energies [5.38, 44–52]. Representative distributions for different neutron sources are shown schematically in Fig. 5.2.

For fission neutrons the energy spectrum generally shows a peak slightly below 1 MeV and a mean energy of about 2 MeV, while the highest energy is about 8 MeV [5.38, 46, 49, 50]. For a description of the neutron irradiation

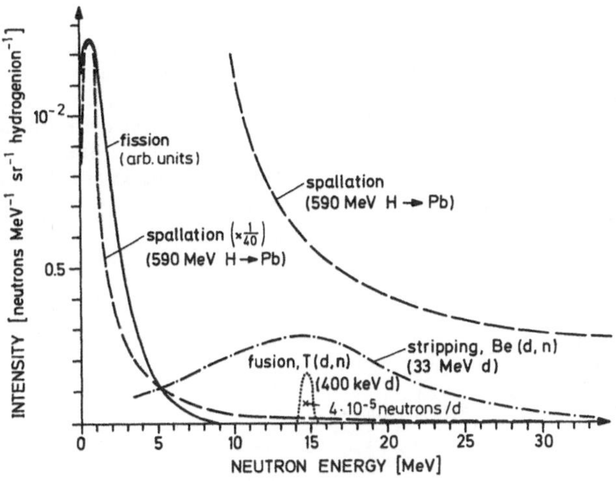

Fig. 5.2. Energy spectra of neutrons produced by different sources [5.41–50]

effects of such a continuous neutron spectrum, generally, different energy regions are regarded:

— the thermal neutron flux $J_{n,\text{therm}}$, i.e., neutrons with energies $E < 1$ eV;
— the epithermal neutron flux $J_{n,\text{epith}}$, i.e., neutrons with energies $1\,\text{eV} \lesssim E \lesssim 100\,\text{keV}$;
— the fast neutron flux $J_{n,\text{fast}}$, i.e., neutrons with energies $100\,\text{keV} \lesssim E \lesssim E_{\text{max}}$.

In most sputtering as well as radiation damage experiments, only the fast neutron flux is regarded because these neutrons can transfer sufficient energy so that larger cascades develop. However, epithermal and thermal neutrons also contribute some sputtering and radiation damage via primary knockon atoms (Sect. 5.3). The fast neutron fluxes which can be obtained near the core of today's high flux reactors are of the order of $10^{15}\,\text{cm}^{-2}\,\text{s}^{-1}$ [5.53, 54]. Up to now, however, neutron sputtering experiments have been performed only at reactors with much lower fast neutron fluxes of up to a few times 10^{12} fast neutrons $\text{cm}^{-2}\,\text{s}^{-1}$ [5.38, 50].

The neutron sources giving a relatively peaked energy distribution make use of the fusion reaction *T(d, n)*. A beam of 160–400 keV deuterium ions is directed onto a target containing tritium mostly in the form of a TiT_2 film on the surface, being thicker than the range of the deuterons [5.41, 42, 55, 56]. The energy of the neutrons in the laboratory system depends slightly on the energy of the incident deuterium ions and the emission angle relative to the direction of the ion beam, as shown in Fig. 5.3. Because they slow down in the target, the energies of the deuterium ion and thereby also of the neutrons produced in the forward direction decrease. Thus, for 400 keV deuterons, the mean neutron energy in the forward direction is about 14.8 MeV [5.55]. The diameter of the deuterium beam is typically 1 cm. At larger distances from the TiT_2 target the source can be regarded as a point source emitting neutrons nearly isotropically

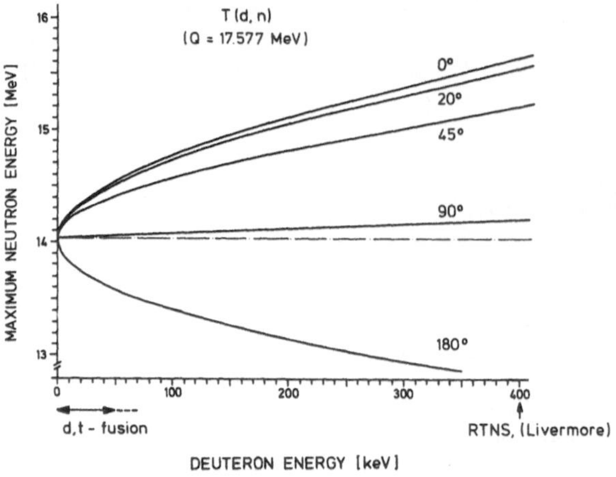

Fig. 5.3. Energy of neutrons produced with the $T(d, n)$ reaction by an energetic deuterium beam directed onto a tritium target in the forward (0°, 20°, 45°), transverse (90°), and backward (180°) directions

in all directions but with slightly different energies. High neutron fluxes can be obtained only very close to the target where, however, the flux gradient is largest. Earlier facilities had fluxes of $\lesssim 10^{10}$ neutrons $cm^{-2}\,s^{-1}$ [5.57, 58], while more recent sources with rotating targets produced fluxes of $\simeq 10^{11}$ neutrons $cm^{-2}\,s^{-1}$ [5.56] and up to 10^{13} neutrons $cm^{-2}\,s^{-1}$ [5.55] at a distance of about 1 cm from the TiT_2 target. The operation time of these sources is, however, limited to times between 5 and 50 h due to a depletion of the tritium in the TiT_2 target caused by the implanted deuterium and the α-particles produced [5.55, 56]. The efficiency of such sources is about 4×10^{-5} neutrons per deuteron.

A more intense neutron source with an efficiency of about 4×10^{-2} neutrons per sr. and deuteron in the forward direction is the stripping source using the Be(d, n) or Li(d, n) reactions [5.43–45, 48, 49, 59, 60]. It was used especially to study the neutron radiation damage which will occur with 14.1 MeV neutrons in future fusion reactors. The neutrons produced by stripping reactions of about 30 MeV deuterons in Be or Li leave the target mostly in the forward direction and have a broad energy distribution (Fig. 5.2) [5.44, 45, 48, 49]. The advantage of this source is that the neutrons are produced in a beam with a relatively small angular spread, giving lower activation of the surrounding structure. However, due to the broad energy distribution, a measurement of the flux to better than 10% is difficult. With this source, the most accurate measurements of neutron sputtering yields have been performed at energies around 15 MeV [5.61].

The most intense neutron fluxes per hydrogen ion, both at low and high energies (Fig. 5.2), can be obtained from a spallation source [5.43, 51, 62]. Neutrons are produced by bombardment of heavy nuclei (W, Pb, ^{238}U) with protons of about 600 MeV. The emission of the neutrons can be explained partly in analogy to the sputtering of solids by knocking out the neutrons from the nucleus and partly by neutron evaporation from the nucleus [5.62].

Further neutron sources are photo neutron sources using the bremsstrahlung of $\simeq 100$ MeV electron beams. These sources as well as the spallation sources are especially developed for pulsed operations [5.62] and have not been used for sputtering experiments.

5.2.2 Experimental Conditions

Some of the experimental set-ups used for neutron sputtering measurements are shown schematically in Fig. 5.4 [5.22, 57, 58, 63–70]. Generally, the samples to be investigated are surrounded by collector foils made of extremely clean Al, Si or C. The samples together with the collectors are placed in a capsule or a chamber, pumped to high vacuum and then exposed to the neutron flux. The neutron fluence is measured by the activation of monitor foils. After irradiation, the collector foils are removed and analysed for measuring the amount of material deposited.

a) Surface and Vacuum Conditions

In order to perform well-defined sputtering measurements, the surfaces of the solids to be investigated should be atomically clean at the beginning. During bombardment the mean number of atoms released per unit surface area of the target by sputtering should be higher than the number of atoms and molecules from the residual gas hitting and sticking on the surface ([Ref. 5.7, Chap. 4] and [5.37]). Neither of these conditions has been fulfilled in most neutron sputtering experiments. In contrast, several experiments were performed with samples having "technical" surfaces, i.e., a surface layer of oxides, carbides, water or other residuals from, e.g., chemical or electrolytic polishing [5.69]. Some of the apparatus used are bakeable so that a vacuum in the 10^{-9} mbar range could be obtained [5.22, 26, 35, 57, 58, 61, 66–69]. However, during neutron bombardment the residual gas pressure increased to the 10^{-8} or 10^{-6} mbar region due to desorption from the walls of the chamber. Because the sputtering yields are very low, even clean surfaces become covered with residual gas layers during the experiments at these gas pressures. Such contaminated surfaces generally give lower sputtering yields for the atoms of the bulk material than clean surfaces, as demonstrated in ion-beam experiments [5.71]. In neutron sputtering experiments with energies above a few MeV, however, the mean energy transferred to a target atom is of the order of 100 keV. Such a primary knockon atom can initiate a large collision cascade and cause sputtering of 10^2 to 10^3 atoms at the surface, forming a small crater [5.34, 72]. If such a large number of atoms is removed in one sputtering event, the influence of a surface layer on the yield will be smaller than for the case of low energy ion sputtering where only one or a few atoms are removed in one event.

Finally, the surface topography has to be known. For very rough surfaces, a correlation was found between surface microfinish and the size of microparticles which were occasionally observed on the collectors in neutron sputtering

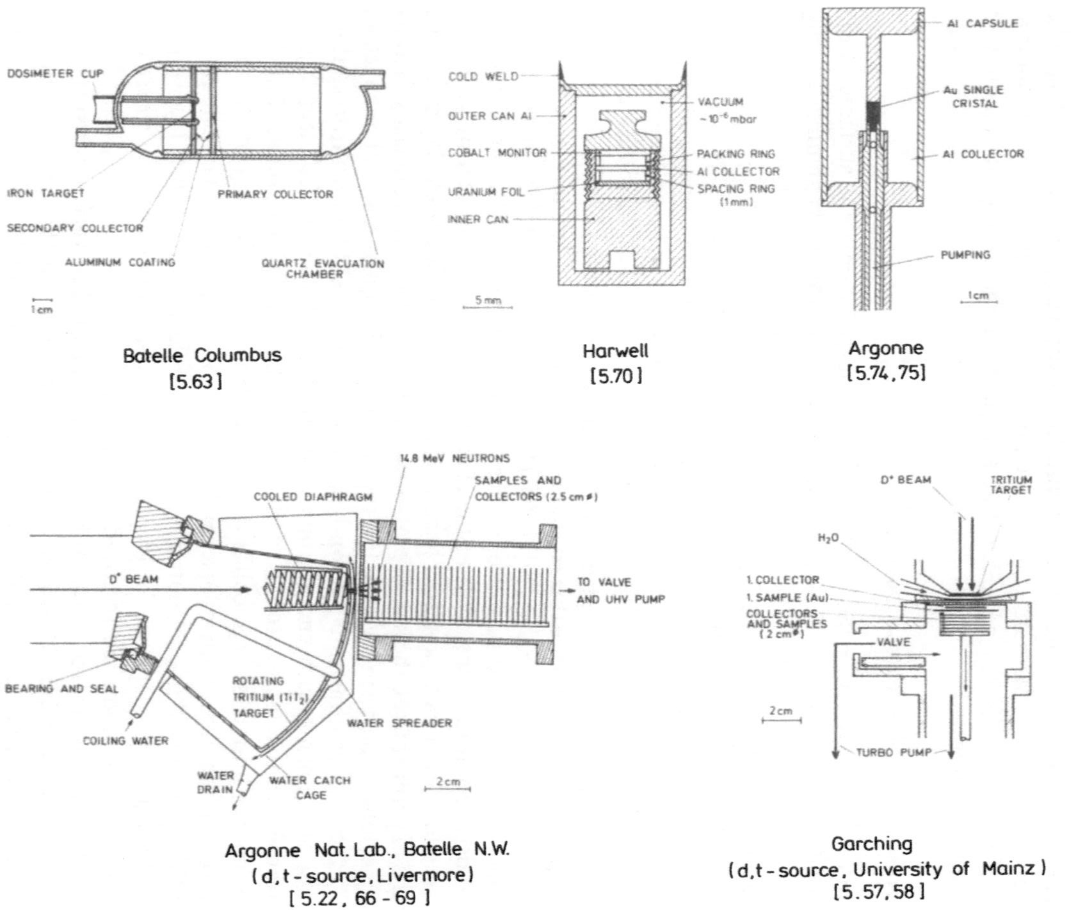

Fig. 5.4a, b. Schematic diagram of the apparatus used for sputtering measurements **(a)** with reactor neutrons and **(b)** with ≈ 14 MeV neutrons

experiments [5.33, 73]. An influence of mechanical surface treatment, especially mechanical stress in the surface layers on the sputtering yield, could not be proved with the techniques and the sensitivities of today's experiments [5.22, 33, 66–69].

b) Limitations of Collector Measurements

The technique of measuring neutron sputtering yields by condensation of the removed atoms on collectors has several limitations:

The sticking probability of the target atoms on the collectors should be close to one. For sputtered atoms a sticking probability of 0.5–0.9 has been reported [5.74, 75]. It depends on the surface conditions, the temperature and the amount of adsorbed surface layers [5.76], but in nearly all neutron sputtering measurements, a sticking probability of one was assumed.

It is difficult to clean the target surface in situ before the neutron bombardment is started, for example, in a glow discharge and/or by annealing. Thus, the assembly and handling of the targets and collectors has to be performed extremely carefully to avoid any surface debris or loosely bound microparticles being knocked off from the target and reaching the collectors. A UHV manipulator to separate the targets from the collectors for ion sputter cleaning in an adjacent chamber immediately before neutron irradiation has been applied only in one case by *Harling* et al. [5.61].

As targets and collectors are both subjected to neutron bombardment, material from the collector is also sputtered onto the target. Resputtering of target material deposited on the collector can only be neglected if the deposits are small, i.e., much below one monolayer. This is fulfilled in most neutron sputtering measurements except for some sputtering experiments of uranium [5.77, 78].

The sputtered atoms can have energies above the threshold energy for ion sputtering at the surface where they are deposited. Especially primary knockon atoms can be emitted in the forward direction with the maximum energy T_m which can be transferred from a neutron to an atom of the solid (Sect. 5.3.1). For 14 MeV neutrons, T_m lies between a few 100 keV and a few 1 MeV, depending on the mass M_2 of the target atoms. If such energetic primary knockon atoms emitted from the collector foil hit the target, they can cause more backsputtering than the neutron bombardment. This effect has not been considered in most experiments and is presumably the reason that measured backward sputtering yields are too high relative to measured forward sputtering yields when compared with theoretical predictions [5.8, 79–83].

c) Measurement of the Deposits on the Collectors

The material sputtered and deposited on the collectors is typically much less than 1/100 of a monolayer, i.e., less than 2×10^{13} atoms cm^{-2} and the most sensitive techniques had to be applied for a quantitative analysis. These are

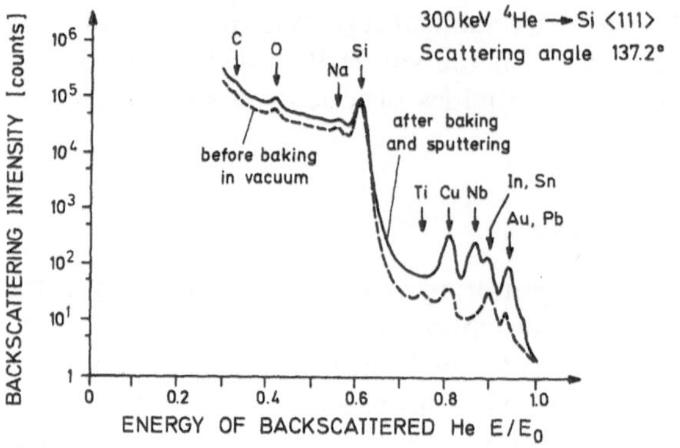

Fig. 5.5. Rutherford backscattering spectrum for 300 keV ^{4}He from a collector consisting of a pure Si single crystal. The channeling effect was applied to suppress backscattering from the Si [5.82]

neutron activation analysis, nuclear track detectors and surface analysis techniques.

Firstly, it turned out to be a major problem finding sufficiently clean collector foils. The contamination of a material with foreign atoms is typically in the ppb to ppm range, corresponding to 5×10^{11} to 5×10^{14} atoms cm^{-2} in a 0.1 mm thick collector foil. If just the surface or a thin surface layer is analysed, the bulk contamination can be neglected. However, even extremely clean bulk materials can get a large degree of surface contamination from handling or from segregation, especially after baking in vacuum (Fig. 5.5).

In more relyable experiments the target had been activated with thermal neutrons before the sputtering with fast neutrons. If the activated atoms are sputtered proportional to their concentration in the bulk the yield can be measured directly from the activation on the collector. Also the angular distribution of the sputtered material can be obtained, for example, by autoradiography [5.65, 85]. The activation of the target has the advantage that any contamination on the collectors does not effect the measurements.

In most experiments the collectors have been analysed after the sputtering experiment by activation with thermal neutrons. In order to avoid the loss of activated atoms by thermal neutron sputtering, i.e. due to the recoil energy in the (n, γ) process (Sect. 5.2.5a) [5.86–89], the collector foils have to be covered during activation by another foil [5.35, 57, 58, 68, 69]. This technique has the disadvantage that all impurities present in the collector material are also measured. Therefore contamination of the collector material has to be checked carefully in blank experiments. The neutron activation technique was especially applied for gold, by detecting the ^{198}Au with a half-life of 2.7 days and a γ energy of 4.11.8 keV [5.18, 39, 57, 58, 61, 64, 68, 69, 90–93]. The sensitivity for measuring Au is about 5×10^8 atoms; however, most collector materials are contaminated with at least 10^{10} Au atoms cm^{-2} [5.57, 58, 64]. The neutron activation technique was also applied to measure sputtered Niobium by detecting the ^{94m}Nb atoms having a half-life of 6.29 min. By additionally applying a fast chemical separation, a sensitivity of 2×10^{11} Nb atoms on a 5 cm^2 collector area could be obtained [5.35, 68, 69].

If sputtering is performed by irradiation with a spectrum of fast and thermal neutrons, activation takes place simultaneously [5.38, 63]. However, it is generally not possible to separate those target atoms which have been activated on the target and then became sputtered from those atoms which are emitted by thermal neutron sputtering. Thermal neutron sputtering yields must be investigated separately and substracted.

In all experiments the radioactive primary knockon atoms produced by energetic neutrons in $(n, 2n)$, (n, p), (n, α)... nuclear reactions can again be detected separately due to different γ-ray energies [5.10, 16, 84].

For sputtering of fissile materials such as ^{233}U, ^{235}U or ^{239}Pu with neutrons, the amount of sputtered fissile material deposited on the collectors could be measured with a sensitivity of $\simeq 1.5 \times 10^{13}$ atoms cm^{-2} by counting the α-particles from radioactive decay [5.94, 95], or with a sensitivity of 10^7 atoms cm^{-2} by detecting the fission events on the collector if introduced in a reactor using a special fission counter [5.77, 95] or with nuclear track detectors [5.78]. In the latter case, the collectors have been placed on mica foils and irradiated by thermal neutrons. Some of the U-atoms undergo fission (cross section: $\simeq 500$ barn) and the fission tracks produced on the mica can be made visible in the microscope by etching with HF and can be counted. For a thermal neutron fluence of 5×10^{15} neutrons cm^{-2}, one fission track in the mica corresponds to about 10^6 U-atoms on the collector. Thus this technique is even more sensitive than the neutron activation analysis for Au. It has also been applied in ion sputtering and for other fissile materials [5.96–98].

Finally, several surface analysis techniques have been used to measure the amount of material deposited on the collectors. These are ion beam analysis techniques such as RBS and SIMS [5.99–102] and electron beam analysis such as SEM and Auger Spectroscopy [5.102–104].

The **R**utherford **b**ack**s**cattering analysis (RBS), provides absolute areal densities in atoms cm^{-2}. With a single crystalline collector such as silicon the channeling effect can be used to suppress the signal from the bulk material, a sensitivity of $\simeq 10^{12}$ atoms cm^{-2} can be achieved for heavy atoms [5.82, 99]. A depth resolution below 10 nm can be obtained in order to discriminate against impurities in the collector material and a lateral resolution of 0.1 mm is possible [5.99]. An example of a RBS spectrum for the analysis of a Si single crystal collector is shown in Fig. 5.5. This collector was used in a 14 MeV proton sputtering experiment of Nb as a simulation of neutron sputtering [5.80–82]. In the figure, the RBS spectrum before baking in vacuum is also shown. Impurity traces of C, O, Na, Cu, In, Sn, and Au are found on the surface, which increased after baking. In neutron sputtering experiments RBS analysis has been applied in [5.33, 66, 67, 73, 80–82].

The surface analysis technique with the highest sensitivity of $\simeq 10^8$ atoms cm^2 and a lateral resolution of about 2 μm is the **s**econdary **i**on **m**ass **s**pectroscopy (SIMS) microprobe [5.100, 102]. This technique gives, however, only semi-quantitative results. It was applied in neutron sputtering to investigate the surfaces of the targets [5.22] and to measure the lateral distribution of the sputtered material on the collectors [5.33, 66, 67, 73].

The most widely used instrument to image the surface topography and composition is the scanning electron microscope (SEM) [5.103, 104]. An image of the elements on the surface with a lateral resolution of $\simeq 10$ nm and a sensitivity of about 10^{13} atoms cm^{-2} can be obtained. In neutron sputtering, SEM has been applied for investigating the target surfaces and the collectors with respect to the possible emission of microparticles and chunks [5.22, 33–35, 57, 58, 66–69, 92, 93, 105].

5.2.3 Measured Sputtering Yields

Neutron sputtering yields have been measured with fast neutrons for Au, Nb, Fe, V, and Co, and with thermal neutrons for $U(UO_2)$ and Pu. Except for Au, U, and Pu, all measured yields represent only an order of magnitude estimate of the upper limits due to ill-defined experimental conditions. For sputtering with thermal neutrons by (n, γ) processes no definite data have been reported in the literature. Some reliable data are available for the emission yields of radioactive primary knockon atoms produced with thermal neutrons due to the recoil energy from (n, γ) reactions and fission processes or with energetic neutrons in $(n, 2n)$, (n, p), (n, α) nuclear reactions.

a) Sputtering Yields for Au and Nb

The most extensive sputtering yield measurements have been performed for Au [5.18, 38, 39, 57, 58, 64, 65, 68, 85, 90, 91, 106] and for Nb [5.22, 33, 66–69, 73, 92, 93, 105, 106], both with reactor neutrons as well as with neutrons from a $T(d, n)$ and $Be(d, n)$ neutron source. Au was investigated because of its high detection sensitivity by neutron activation analysis and because a relatively clean surface could be obtained, while Nb had been of some interest as a first wall material in fusion devices [5.23, 24].

The results of neutron sputtering yield measurements with reactor neutrons and with about 14.8 MeV neutrons from $T(d, n)$ and $Be(d, n)$ sources incident on Au and Nb foils are summarised in Figs. 5.6, 5.7. Here the measured amount of material, or an upper limit of it, found on the collectors in forward, backward and random directions, is plotted as a function of the neutron dose used in the different experiments. This way of plotting the results gives better information about the different measurements than just giving the sputtering yields [5.69]. Lines representing different sputtering yields are also included. The measurements have been performed mostly at room temperature by measuring the amount of deposition on a collector foil assuming a sticking probability equal to 1. The targets typically had a diameter of about 1–2 cm so that the neutron fluence and total neutron dose had similar values. In most cases polycrystalline targets were used at room temperature. The few experiments with single crystals [5.33, 65, 69, 73] did not give different sputtering yields within the experimental error.

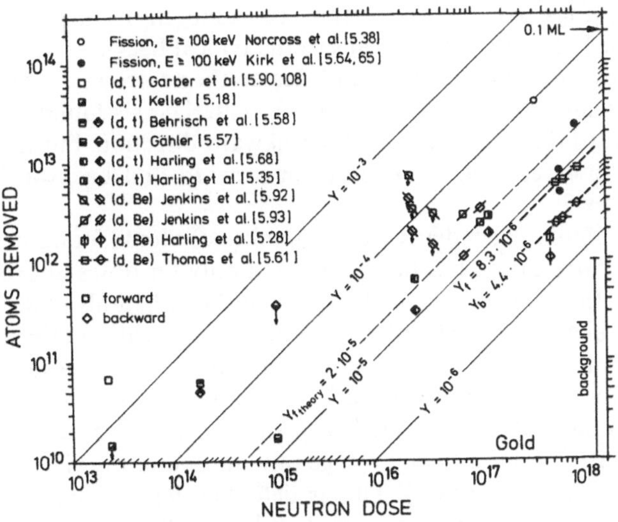

Fig. 5.6. Measured erosion at the entrance (backward) and exit (forward) side of Au foils bombarded with different doses of neutrons. A typical Au contamination on the collectors is indicated on the right by "background". The diagonal lines correspond to different sputtering yields

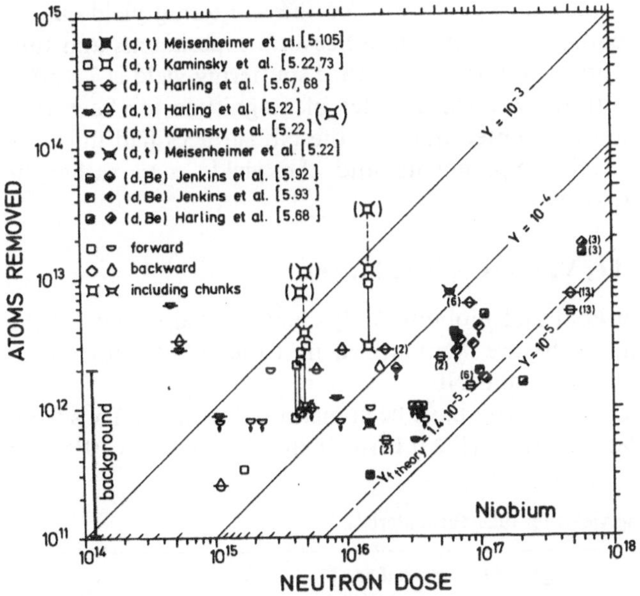

Fig. 5.7. Measured erosion at the entrance (backward) and exit (forward) side of polycrystalline Nb foils bombarded with different fluences of neutrons. A typical Nb contamination on the collectors is indicated on the left by "background". The diagonal lines correspond to different sputtering yields. The points ⌘ give the numbers, where chunks are included

In the figures, no differentiation has been made between targets with different surface treatments such as mechanical polish, light or heavy chemical etch or electrochemical polish or just optically ground [5.22, 33, 69, 73]. Microparticles or chunks which have been found on the collectors by two experimental groups (Sect. 5.2.4) in sputtering rough surfaces contribute less than 10% to the total sputtering yield [5.73]. These sputtering yield measurements are also included in Fig. 5.7. For the measured points in parenthesis, the statistics concerning the number of chunks are uncertain [5.33, 66, 67, 73, 107].

As a general result, Figs. 5.6, 5.7 show that the amount of material measured on the collector foils is just slightly above the typical background of 10^{12} atoms cm^{-2} and shows no significant dependence on the neutron dose for doses smaller than 10^{17} neutrons. In most cases, more material was found on the collectors in the forward direction than in the backward direction, but the reverse result was also observed. The ratio between forward and backward emission is generally below 3.

For *gold* (Fig. 5.6) a definite dependence of the number of atoms found on the collectors upon the neutron dose has been measured only for doses above 6×10^{17} neutrons [5.61]. In this experiment the target surfaces had also been cleaned by ion sputtering in a separate chamber. The sputtering yields obtained were 8.3×10^{-6} atoms/neutron in the forward direction and 4.4×10^{-6} atoms/neutron in the backward direction. For neutrons from a fission reactor, the most refined experiments have been performed with single crystals at $\simeq 40$ K with neutron doses in the 10^{18} range and an average random sputtering yield of 8×10^{-6} atoms/neutron is reported [5.65]. This is definitely higher than the lowest value of $\simeq 2\times 10^{-6}$ atoms/neutron reported earlier [5.39].

For *niobium* (Fig. 5.7) only an upper limit for the sputtering yield can be given. The points in parenthesis give the sum of several experiments and thus cannot be counted. The upper limit for the forward sputtering yield is 7×10^{-6} to 5×10^{-5} atoms/neutron and the same order of magnitude for backward sputtering yields. It should be pointed out that the Nb surfaces had not been cleaned in nearly all of the experiments and the yields may rather be representative for Nb_2O_3 or NbC.

b) Sputtering Yields for Fe, V, Co (Cu, Mo, In, W)

For materials other than gold and niobium, only single measurements of the yields are available. The results are summarized in Table 5.1, together with neutron sputtering yields for Au and Nb.

Random sputtering yields for iron have been measured with neutrons from a fission reactor at a fast neutron ($E \geqq 0.1$ MeV) fluence of $\approx 2\times 10^{18}$ cm^2

Table 5.1. Measured sputtering yields for different materials

		Sputtering yield [atoms/neutron]				
		Au	Nb	Fe	Co	V
14 MeV neutrons	forward	8×10^{-6}	$\lesssim 10^{-5}$			$<5\times 10^{-5}$ [5.109]
	backward	$\lesssim 4\times 10^{-6}$	$\lesssim 10^{-5}$		$<1.6\times 10^{-5}$ [5.93]	
Reactor neutrons $E>0.1$ MeV	random	$<8\times 10^{-6}$		$<5\times 10^{-3}$ [5.63]		

[5.63]. In the experiment the samples had been exposed to both the fluxes of fast and thermal neutrons. The sputtering yields due to the fast neutrons were calculated from the number of radioactive ^{59}Fe atoms found on the collector, which were produced by (n, γ)-reactions from ^{58}Fe, contained with 0.3% in natural Fe. The relatively high yields reported may be due to thermal neutron sputtering and/or contamination of the collectors.

The neutron sputtering yields for Co were measured with neutrons from a Be(d, n) source at a fluence of $\simeq 1.4 \times 10^{17}$ neutrons [5.93]. The amount of sputtered Co collected on graphite was measured by neutron activation analysis. Two measurements have been performed with a factor of 2.6 difference in the yields. The mean value is given in Table 5.1.

The sputtering yield measurement on V have been performed with neutrons from a T(d, n) source at a dose of $(4\text{–}7) \times 10^{15}$ neutrons [5.109]. It was reported that samples with a very rough surface finish can emit chunks, leading to an increase of the sputtering yield by a factor of about 5.

Finally, neutron sputtering has been investigated for Cu, Mo, In, and W with a neutron dose of $\simeq 2.4 \times 10^{13}$ neutrons from a T(d, n) source. The sputtered material deposited on the collectors was investigated by neutron activation analysis, but no deposits have been found. Taking the detection sensitivity for the different elements, only an upper limit for the sputtering yield of $Y < 4 \times 10^{-2}$ for Cu, $Y < 10^{-4}$ for Mo, $Y < 3.6 \times 10^{-3}$ for In, and $Y < 1.1 \times 10^{-2}$ for W could be given [5.18].

c) Sputtering Yield for Solids Containing Fissile Atoms

If a solid contains fissile atoms, the fission products produced, predominantly at thermal neutron bombardment, can be regarded as primary knockon atoms which cause sputtering from the surface layers. The fission products have atomic masses of about 95 and 140 and an energy of 60–70 MeV [5.3], which is considerably higher than generally applied in sputtering [Ref. 5.7, Chaps. 2–5]. The energy loss of these energetic primary knockon atoms in the solid occurs predominantly to electrons and only a small fraction is transferred into kinetic energy of target atoms. The deposited energy density is very large and high ionization will occur along the trajectory of the fission product ([Ref. 5.7, Chap. 2] and [5.110–113]).

Sputtering experiments with thermal neutrons on solids consisting of mostly fissile material or containing some fissile atoms have been performed for ^{233}U and $^{233}UO_2$ [5.94], ^{235}U and $^{235}UO_2$ [5.70, 78, 95, 114–116], and for ^{239}Pu [5.95]. As the sputtering yield is directly proportional to the number of fissile atoms in the solid, mostly relative yields, i.e., the mean number of sputtered atoms per fission event occurring in the bulk [5.77] or the mean number of sputtered atoms per fission product emitted at the surface, i.e., Y/Y^{p} [5.70, 78, 94, 95, 114, 115], are given. The fission product or primary knockon emission yield Y^{p} has been calculated (Sect. 5.3.2) or measured separately (Sect. 5.2.5).

Table 5.2. Neutron sputtering yields for solids containing fissile atoms

Material	Thermal neutron fluence [neutrons cm^{-2}]	Y/Y_p	Y atoms/neutron	Ref.
$^{233}UO_2$ disc	$5\text{–}10 \times 10^{16}$	24 ± 15	0.08 ± 0.04	*Laptev* and *Ersher* [5.94]
^{233}U discs, mech. pol.	$1\text{–}2 \times 10^{16}$	1200 ± 600	4.2 ± 2	
^{235}U (enriched) ^{234}U (enriched) 12 μm foil oxidised or mech. polished	$< 2 \times 10^{14}$	≈ 2000	6.5 ± 1.5	*Rogers* and *Adam* [5.95]
	$> 10^{16}$	< 30	< 0.1	
^{235}U (93%) ^{234}U (enriched) polished foils	6.5×10^{14} (6 measurements)	≈ 460	2.2	*Rogers* [5.70]
	3×10^{14}	≈ 2000	~ 10	
^{235}U (93%) ^{234}U (enriched) sputter deposit (~ 1 nm)	5×10^{16}	$\sim 10^4$		
^{235}U (90%) evaporated films 0.1 μm – 10 μm	$< 5 \times 10^{14}$	1500 ± 500 (0.2 μm foil)	0.4 ± 0.1	*Biersack* et al. [5.78]
		760 ± 400 (10 μm foil)	5 ± 2	
	Predamage 1.5×10^{17}	2600 ± 1500 (including microparticles)	17 ± 10	
$^{238}UO_2$, ^{235}U(0.715%, natural composition) disc	$\lesssim 7 \times 10^{17}$	9	2.7×10^{-4}	*Nilsson* [5.77]
^{238}U, ^{235}U (0.715%, natural composition) disc, electropolish	$\lesssim 2.3 \times 10^{17}$	4.3	1.1×10^{-3}	
Al, Cu, Nb, ^{235}U(1%) disc	4×10^{17}	No microparticles		*Blewitt* et al. [5.116]
^{239}U	$0.2\text{–}2 \times 10^{16}$	3500 ± 100	6.0 ± 1.5	*Laptev* and *Ersher* [5.94]

The experiments have been performed on thin films or bulk material using the collector technique (see Fig. 5.4). In one case a dynamically pumped vacuum system was also used [5.77], but the vacuum achieved was only in the 10^{-5} mbar range. Because uranium and plutonium are chemically very reactive, the surfaces were covered with oxide layers in all experiments. As expected with the collector technique, for high neutron fluences ($>2\times10^{14}$ neutrons cm^{-2}) the measured sputtering yields decrease with increasing neutron fluence due to resputtering from the collector. This resputtering was found to be caused predominantly by energetic fission products emitted from the target and hitting the collector foil and only to a small extent by fission processes of the collected material. At a mean collected amount corresponding to a layer of about 1 nm, an equilibrium between deposition and removal at the collectors was reported [5.70, 95].

The measured sputtering yields for the materials investigated are summarized in Table 5.2. Generally, on the average about 1000–2000 atoms are sputtered per emitted fission product. For highly enriched material this gives a sputtering yield of about 5 atoms/neutron, while it is correspondingly lower for solids containing only a small concentration of fissile atoms. For thin films with a thickness smaller than the mean range of the fission fragments, a higher number of atoms emitted per fission fragments but a lower sputtering yield are reported [5.70, 78]. Bulk UO_2 gives a sputtering yield a factor of 5–50 lower than uranium metal which also has, however, an oxide layer on the surface [5.77, 94].

A detailed comparison of the yields published by different authors is difficult because different numbers have been used in the data evaluation. The mean range of the fission fragments were reported to be 4.5–5.1 μm in U metal [5.77], 7 μm in U metal [5.95], 7.3 μm in U metal [5.94], and 10 μm in UO_2 [5.78, 115]. The cross section for fission by thermal neutrons is of the order of 300–600 barn. It depends on the neutron energy [5.117] and thus the exact energy distribution of the neutrons in each of the reactors used.

5.2.4 Sputtered Material

a) Sputtered Atoms, Angular Distribution

The material released at a surface in sputtering of solids by ion bombardment is mostly neutral atoms in the ground state. Only less than 5% are ions and less than 1% is emitted in the form of small atom clusters containing between two and about 10 atoms. The energy distribution of the sputtered atoms has a maximum at a few eV and decreases about $\sim 1/E^2$ toward higher energies for the linear cascade regime ([Ref. 5.7, Chap. 2] and [5.118]). The angular distribution of the emitted atoms depends on the surface structure and in the case of single crystals on the orientation [5.7, 118]. However, the majority of atoms are generally emitted with an angular distribution which is close to a cosine distribution.

For neutron sputtering, the material released has been only little investigated. The angular distribution was measured from the distribution of the deposition on a collector. For sputtering of an Au single crystal with neutrons from a fission reactor, *Kirk* et al. [5.65] report an approximately cosine distribution of the sputtered material. No preferred emission in the close-packed crystal directions was found. For sputtering of polycrystalline Au with 14 MeV neutrons, *Gähler* [5.57] found a distribution of the atoms sputtered and condensed on a collector which is also in agreement with a cosine distribution.

In one experiment [5.108], electrical currents were measured between the target and the collectors which were both enclosed in an evacuated capsule and exposed to the neutron flux of a fission reactor. For different target materials, these currents show a dependence on target mass similar to sputtering yields for ion bombardment. For a fast ($E \geqq 1$ MeV) neutron flux of 2×10^{12} neutron $cm^{-2}\,s^{-1}$, the currents from a 1 cm^2 size target are reported to lie between 10^{-7} and 10^{-9} A. If these measured currents were mainly due to sputtered ions and only partly caused by secondary electrons, the measured currents would correspond to sputtering yields of 3×10^{-1} to 3×10^{-3} ions/neutron. Such yields are, however, orders of magnitude larger than the neutron sputtering yields measured by deposition of the sputtered material on collectors. Thus, secondary electrons may make a major contribution to the measured currents. The angular distribution of the sputtered ions as deduced in these investigations from the electrical currents between different polycrystalline and single crystalline targets and an array of collectors and the energy distribution deduced from retarding field measurements are questionable.

b) Microparticle Emission

The nature and size of the material released in neutron sputtering has been investigated in a few experiments both for sputtering with 14 MeV neutrons and for sputtering fissile material with thermal neutrons. Similar to sputtering by ion bombardment, the majority of the material released should be single atoms, probably in the ground state. But additionally in some neutron sputtering investigations, larger clusters containing about $\simeq 10^4$ atoms [5.114, 115], up to about 5×10^7 atoms [5.78] or even up to 5×10^{12} atoms [5.33, 66, 67, 73, 98] have been found, the size of which was correlated to the surface topography of the target.

In sputtering of highly enriched U and UO_2 films with thermal neutrons, the material released was collected on carbon foils positioned 1 mm in front of the target and subsequently analysed by transmission electron microscopy (TEM) [5.114, 115]. The published pictures taken at neutron fluences of 10^{14} to 10^{17} neutrons cm^{-2} do not differ much for sputtering of U and UO_2 films. But *Rogers* [5.114, 115] concluded from these investigations that different mechanisms contribute to sputtering for the metal and the oxide.

Uranium metal is sputtered via a collision cascade to a major part in the form of microparticles having a diameter of about 7.3 nm and containing about 10^4 atoms. On the average, every 16th fission product emits such a microparticle. This was concluded from the size and number of black spots on the collectors which increase with neutron fluence [5.114].

Vacuum deposited UO_2-films with a grain size of ≈ 5 nm are sputtered mostly via single atoms, such as evaporation in a spike. Initially, on the average,

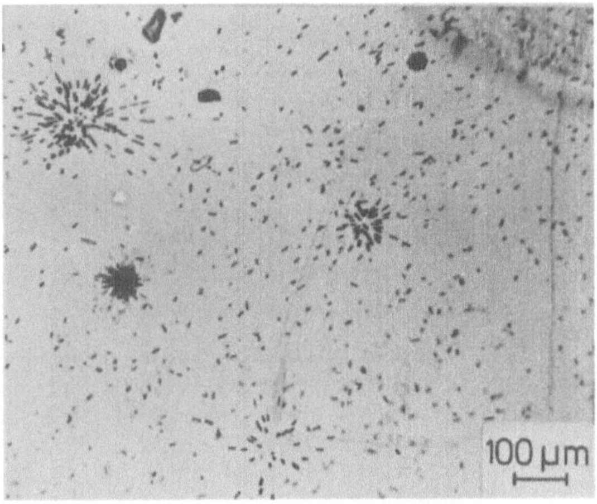

Fig. 5.8. Fission tracks in mica from uranium deposits on a collector after irradiation with thermal neutrons and chemical etching, as used to measure neutron sputtering yields of uranium. The star patterns indicate microparticles of $\sim 10^7$ atoms on the collector. The larger star pattern are due to the fact that the collector and the mica were not in close contact [5.78]

one fission fragment leaving the surface ejects about 5×10^4 atoms. This was concluded because for increasing neutron fluence, the black spots found on the collectors tend to increase more in size than in number. After bombardment with a fluence of about 5×10^{16} neutrons cm^{-2}, the emission yield decreases to about 30 uranium atoms per emitted fission products. This decrease is explained by growth of the grains in the target, i.e., the UO_2-film to a size of ≈ 10 nm giving better cooling of the spikes [5.115].

Biersack et al. [5.78] sputtered 1–10 μm thick polycrystalline and amorphous UO_2 films (^{235}U, 90%) which had been vapour deposited on Al, quartz or CaF_2. After the neutron irradiation for sputtering, the collectors were removed, covered with mica films and irradiated for a second time with thermal neutrons. The mica films were subsequently etched in HF and the fission tracks became visible in an optical microscope (Fig. 5.8). Besides many isolated tracks, each corresponding to $\approx 10^6$ ^{235}U-atoms on the collector, some star-like pictures were found which are explained as being due to microparticles on the collector containing up to 5×10^7 atoms. Microparticles smaller than $\approx 10^6$ atoms cannot be seen with the thermal neutron dose used here. Such star-like pictures are not found for sputtering with neutron fluences below 5×10^{14} cm^{-2} corresponding to 6×10^{15} fission events per cm^3. For higher fluences the probability of finding star-like pictures increases. A scanning electron microscopy picture of the UO_2 target surface showed that the initially flat surface had microcraters at the neutron fluence where star-like pictures indicate the removal of microparticles. Inclusion of the microparticles in the total sputtering yield gives an increase of the yield maximum by a factor of about 3 (Table 5.2) [5.78].

In sputtering experiments with 14 MeV-neutrons on Nb, Si, and V, it was reported that micron-size particles named chunks can be observed on the collectors. The chunks had linear dimensions of 0.5–5 μm containing 10^{10} to

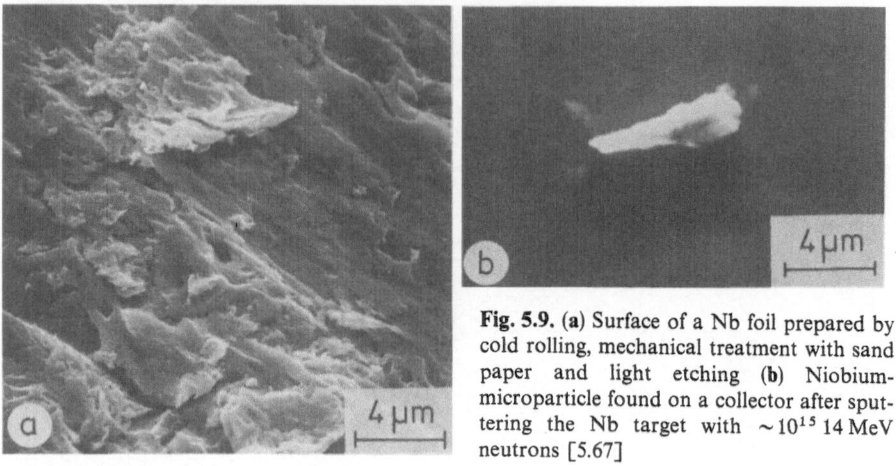

Fig. 5.9. (**a**) Surface of a Nb foil prepared by cold rolling, mechanical treatment with sand paper and light etching (**b**) Niobium-microparticle found on a collector after sputtering the Nb target with $\sim 10^{15}$ 14 MeV neutrons [5.67]

10^{15} particles. They were reported to be emitted with a velocity of about $10\,\mathrm{cm\,s^{-1}}$ corresponding to a mechanical momentum of up to $10^{-9}\,\mathrm{g\,cm\,s^{-1}}$. The chunks were found only for sputtering of targets with a rough "technical" surface which was preparted by "optical grinding" with sand paper and a subsequent short electrolytical polish. Chunk emission of these surfaces was observed especially at low neutron fluences, but only some investigators could find chunks for such surfaces [5.12, 33, 66, 67, 73, 105, 107, 109], while others did not [5.16, 22, 35, 43, 50, 51, 57, 58, 61, 64, 68, 69, 80–82, 92, 93].

Figure 5.9a shows a scanning electron micrograph of a Nb surface from which chunk emission is expected [5.22]; a picture of a Nb-chunk found on a Si collector [5.67] is shown in Fig. 5.9b. Generally, the size of such microparticles correlates with the roughness of the target surface which was sputtered [5.33, 73].

5.2.5 Radioactive Primary Knockon Emission Yield

At neutron bombardment the primary knockon atoms created in a nuclear reaction can be radioactive and those emitted at the surface can be detected with high sensitivity. These atoms start with higher energies and can be emitted from larger depth than the majority of sputtered atoms. Thus for the atoms released at the surface the fraction of radioactive atoms is larger than produced in the bulk [5.6, 17]. This is important for the radioactive contamination of cooling channels in nuclear reactors [5.6, 16, 17, 35]. The measurements of the number of primary knockon atoms released per incident neutron, called the primary knockon emission yield (Y^{P}) gives information about the mean projected ranges of these energetic atoms in the solid [5.6, 13–17]. As primary knock-on atoms start from lattice sites, the angular distributions of atoms emitted from single crystals is influenced by the blocking effect [5.119].

The radioactive primary knockon emission yield (Y^{p}) depends in two ways on the energy E of the bombarding neutrons. Firstly, the reaction cross section $\sigma(E)$ is a strong function of neutron energy (Sect. 5.3) and secondly, the average energy of the primary knockon atoms and thus their emission probability at the surface is about proportional to the neutron energy. Similarly to sputtering, radioactive primary knockon atoms may be emitted for thermal, slow, and fast neutron bombardment, thus the energy distribution of the neutron flux $J_{n}(E)$ at the measuring position has to be known in detail to determine the yield.

a) (n, γ) Emission Yields, Thermal Neutron Sputtering

After the capture of a thermal neutron and subsequent γ emission the recoil energy T transferred to the atom is of the order of 100 eV and thus higher than surface and bulk binding energies in a solid (Sect. 5.3.1) [5.86–89, 120, 121]. Recoil atoms created near the surface can thus be directly emitted, but they may also create small cascades and cause sputtering of other surface atoms. Typically, about one half of the recoils created in the upper atomic layer are emitted [5.86], while the total thermal neutron sputtering yield, i.e., including other cascade atoms, is expected to be about a factor of 2–3 higher than the recoil emission yield.

Measurements of primary knockon emission yields at thermal neutron bombardment have been performed only for Au [5.39, 86], In [5.86], V [5.112], Nb [5.122, 123], Al [5.13], and Cu [5.13]. The results are summarised in Table 5.3. In order to calculate Y^{p} from the work of *Yosim* and *Davis* [5.86], a mean cross section had to be assumed [5.117] which is also given in the table.

Table 5.3. (n, γ)-Radioactive primary knockon emission yields

Material	Y^{p} [atoms/neutron]	Ref.
^{116}In	7.4×10^{-8} ($\sigma\simeq200\,b$)	*Yosim* and *Davies* [5.86]
^{198}Au	10^{-8} ($\sigma\simeq50\,b$)	
	3.3×10^{-10}	*Verghese* [5.39]
^{52}V	2.3×10^{-9}	*Matsuura* et al. [5.122, 123]
^{94m}Nb	1.6×10^{-10}	
^{28}Al	6.1×10^{-10} ($\sigma\simeq3.8\,b$)	*Biersack* [5.13]
^{64}Cu	9.7×10^{-9} ($\sigma\simeq0.23\,b$)	

By applying electrical potentials between target and collector plates in these measurements and in further investigations [5.86, 124], it was shown that about half of the recoils are emitted as positive ions, the remainder are neutrals. No negative ions are observed.

b) Primary Knockon Emission by Fast Neutrons

The cross section for $(n, 2n)$, (n, p) and (n, α) reactions becomes significant only at neutron energies above a few MeV [5.117]. At these energies the neutron flux from fission reactors is very low, but it is significant for fusion neutrons and neutrons from accelerator sources.

As will be shown in Sect. 5.3, the primary knockon emission yield Y^{p} is directly proportional to the first moment of the range distribution of the knockon atoms in the solid projected onto the surface normal which is, in most cases, equal to the mean projected range $\langle R_{\perp}(T)\rangle$. Therefore, primary knockon emission measurements have mostly been evaluated with respect to mean ranges of the primaries in the target material. For a directed neutron flux at normal incidence we get for the yield per unit surface area (Sect. 5.3.3)

$$Y^{\mathrm{p}}(E)=N\left\langle \sigma(E)\left\langle R_{\perp}\left(\frac{E}{1+A}\right)\right\rangle\right\rangle, \tag{5.1}$$

where N is the atomic density in the solid, $\sigma(E)$ the cross section for neutrons of energy E and $T=E(1+A)^{-1}$ (Sect. 5.3.3). Averaging has to be performed according to the energy distribution $J_{\mathrm{n}}(E)$ of the neutrons.

The radioactive primary knockon emission yield Y^{p} has mostly been measured with the collector technique (Sect. 5.2.2) by dividing the number of radioactive atoms condensed on a collector A_{c} by the fluence $f_{\mathrm{n,tot}}$ of neutrons in the energy range of interest. Generally, $f_{\mathrm{n,tot}}$ has to be determined separately, for example, from the activation of a test sample. For bombardment with single-energy neutrons the mean projected range $\langle R_{\perp}(T)\rangle$ can then be obtained from (5.1). For a neutron spectrum an average range $\tilde{R}_{\perp}$ may be obtained from (5.1), which is defined by $\tilde{R}_{\perp}=\langle\sigma(E)\langle R_{\perp}(T)\rangle\rangle/\langle\sigma(E)\rangle$ and is specific for $f_{\mathrm{n}}(E)$ of the neutron source.

It is possible to obtain $\langle R_{\perp}(T)\rangle$ or $\tilde{R}_{\perp}$ also directly without knowing $f_{\mathrm{n,tot}}$ or $\sigma(E)$ and $\langle\sigma(E)\rangle$, when the target itself is taken as the test sample. The number of radioactive atoms A_{t}, produced per unit area in the target of thickness c by a directed neutron fluence $f_{\mathrm{n,tot}}$ is given by $A_{\mathrm{t}}=Nc\langle\sigma(E)\rangle f_{\mathrm{n,tot}}$. The number of radioactive primary knockon atoms emitted and condensed on the collector A_{c} can be written as $A_{\mathrm{c}}=Y^{\mathrm{p}}f_{\mathrm{n,tot}}$. Together with (5.1), we obtain for a directed neutron flux [5.6, 13, 14, 125, 126]

$$\langle R_{\perp}(T)\rangle \quad \text{or} \quad \tilde{R}_{\perp}=\frac{A_{\mathrm{c}}}{A_{\mathrm{t}}}c\,. \tag{5.2}$$

Table 5.4. Radioactive primary knockon emission yields for reactor neutrons

Material and Reaction	Reactor	Y^p [atoms/neutron]	$\tilde{R}$ [nm]	Ref.
$^{27}Al(np)^{27}Mg$	BER (Berlin Experimental Reactor)	3.4×10^{-8} ($\langle\sigma\rangle=50$ mb)	455	*Zimen* and *Ertel* [5.127]
	HTR (Hitachi Training Reactor)		418	*Mitsui* [5.15]
$^{27}Al(n,\alpha)^{24}Na$	HTR		1370	
	BER	8.22×10^{-10} ($\langle\sigma\rangle=0.6$ mb)	910	*Ertel* and *Zimen* [5.128]
$^{56}Fe(n,p)^{56}Mn$	Ber	2.1×10^{-10} ($\langle\sigma\rangle=0.87$ mb)	113	*Wiechmann* et al. [5.129]
$^{58}Ni(n,p)^{58}Co$	BER	8×10^{-9} ($\langle\sigma\rangle=89$ mb)	38.5	*Oberhauser* and *Wiechmann* [5.130]

For a random neutron flux, $\langle R_\perp(T)\rangle$ or $\tilde{R}_\perp$ is a factor of 4 smaller than the mean projected range in the forward direction $\langle R_p\rangle$ ([5.6] and Sect. 5.3.2). Having determined $\langle R_\perp(T)\rangle$ or $\tilde{R}_\perp$ it is now possible to calculate Y^p with (5.1), which will depend on the energy interval taken to calculate $\langle\sigma\rangle$, see (5.2).

For a *random flux of fission neutrons*, radioactive primary knockon emission yields Y^p and average ranges $\tilde{R}_\perp$, measured with the collector technique for Al, [5.15, 127, 128], Fe [5.129] and Ni [5.130] are summarized in Table 5.4. For the measurements of *Mitsui* [5.15], the fission neutron spectrum and thus the mean cross section is not given so that the values for the average ranges could not be converted into emission yields.

For a *directed flux of* 14 MeV *neutrons*, both from a T(d,n) and a Be(d,n) source (Fig. 5.2), investigations about radioactive primary knockon yields have been performed for emission in the forward, backward and transverse directions [5.16, 57, 58, 125, 126, 131, 132]. The corresponding first moments of the projected ranges in the different directions have been measured directly, without knowing the cross section by using (5.2), or have been calculated from the measured recoil emission yields [5.126] with the known cross sections according to (5.1) [5.16, 117].

For gold the measured radioactivity on the collectors in the forward and backward directions, obtained with a T(d,n) neutron source, are plotted in Fig. 5.10 as a function of the neutron dose [5.16]. The linear increase with neutron dose demonstrates that the results are not influenced by any background on the Si-collectors, as was obviously the case in the measurements of the total sputtering yields.

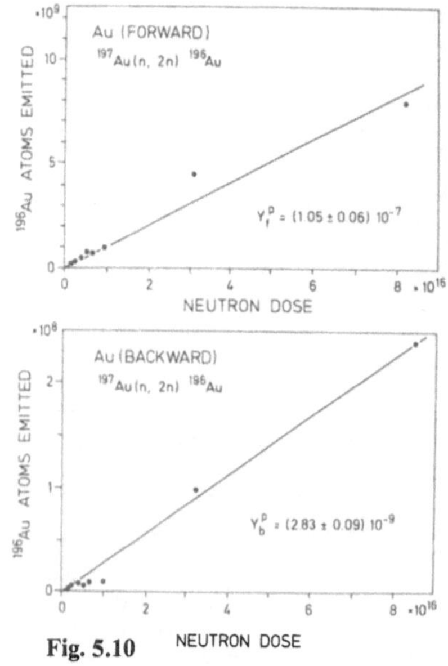

Fig. 5.10

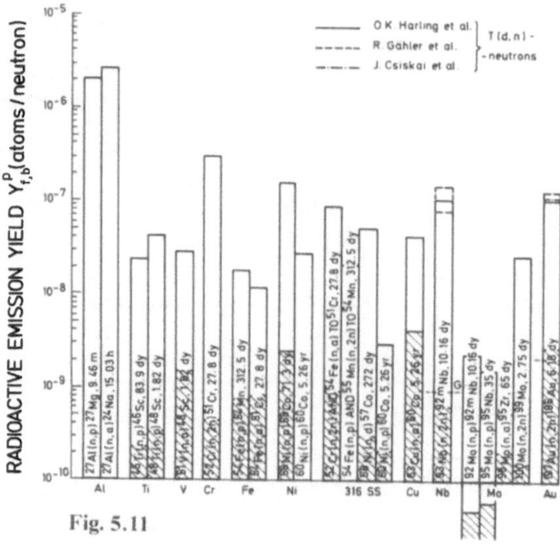

Fig. 5.11

Fig. 5.10. Forward and backward emission of ^{196}Au atoms from a gold foil as a function of the 14 MeV neutron dose [5.16]

Fig. 5.11. Radioactive primary knockon emission yields in forward (white areas) and backward (dashed areas) directions for different materials and different reactions at 14-MeV neutron bombardment [5.16, 57, 58, 125, 132]

The results for forward and backward yields measured for different materials are summarized in Fig. 5.11. The yields are typically in the range of 10^{-8} to 10^{-7} atoms/neutron; the values obtained in investigations with different neutron sources agree well within the experimental uncertainties. Backward emission yields are mostly a factor of 10–100 smaller than forward emission yields (Figs. 5.10, 11). A table with the measured values can be found in [5.16]. For a Be(*d, n*) neutron source with the much broader neutron spectrum (Fig. 5.2), the primary knockon emission yields are somewhat lower; for Au and Nb about a factor of 2 because for the neutrons with lower energy, the cross section for the nuclear reaction is smaller [5.16, 131].

In one investigation the transverse primary knockon emission yield of Au was measured with a Be(*d, n*) neutron source [5.131]. The polished and cleaned gold targets of thickness $a = 0.01$ cm were packed between high purity carbon catcher foils and exposed to the neutron flux parallel to the target surfaces. After neutron irradiation the activity of ^{196}Au was counted on both the target and the collectors, giving a value of $A_c/A_t \simeq 1.6 \times 10^{-5}$. In order to obtain the transverse radioactive emission yield for one side of a cube, we have to multiply this value by $N\langle\sigma\rangle a$ and we obtain[1] $Y_t^p = 8 \times 10^{-9}$ atoms/neutron. For a T(*d, n*)

1 $N_{Au} = 5 \times 10^{22}$ cm^{-3}, $a = 0.01$ cm; for Be(*d, n*) neutrons on Au $\langle\sigma\rangle \simeq 10^{-24}$ cm^2 [5.16], while for T(*d, n*) neutrons, $\sigma = 2.2 \times 10^{-24}$ cm^2.

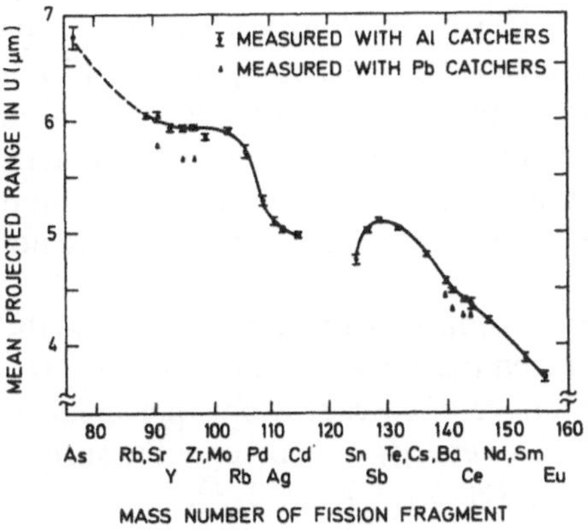

Fig. 5.12. Mean projected ranges of several fission products in uranium, as obtained from primary knockon emission yields. The lower mass fission fragments started with energies of ~70 MeV, while the higher mass fission fragments started with ~60 MeV [5.84]

neutron source we would obtain a value by a factor of about two higher, still being a factor of 5–6 lower than the forward emission yields [5.16]. These values give a mean transverse range of 5.1 nm, and for the mean range projected on the surface normal, a value of 1.6 nm. These values can be compared to the mean projected forward range of 9.2 nm.

c) Emission of Fission Fragments

The determination of the emission yields for different fission fragments at the surface of fissile material is a straightforward method of measuring the mean projected range of the different fission fragments in the material [5.84]. As the fission fragments are emitted from lattice sites, the blocking effect causes that they cannot be channeled and their trojectories are close to those in a random lattice. For uranium the mean projected ranges have been determined according to (5.2) for 28 fission fragments from the emission yields [5.84]. Films of 25 µm thickness of enriched U have been used and the collector technique with subsequent chemical disolution was applied. The results for the mean ranges are summarised in Fig. 5.12. If we take a mean range of 5 µm and a fission cross section of 500 *b*, we get from (5.1) a mean primary knockon or fission fragment emission yield for highly enriched uranium of

$$Y^{\mathrm{p}} = 3 \times 10^{-3} \left(\frac{\text{fission fragments}}{\text{neutron}}\right).$$

This value has been used in Table 5.2 for converting the reported values of Y/Y^{p} for uranium into sputtering yields Y.

5.3 Theoretical Considerations

Sputtering processes can generally be described using the concepts developed for describing radiation damage in the bulk of a material [5.1, 4, 7, 8, 133–150]. An incident neutron interacts with an atom of the solid producing an energetic primary knockon atom with energy T. These primary knockon atoms collide further with lattice atoms and with electrons and resulting in a collision cascade. In the following we shall concentrate on metals where only the portion $\nu(T)$ of the primary knockon energy T, which is deposited into the kinetic energy of atoms, contributes to the cascade, leading to displacements and sputtering [Ref. 5.7, Chaps. 2, 3] and [5.8]. Further, single crystal effects are not regarded (Sect. 5.1).

If a primary knockon atom and the collision cascade are initiated in the near-surface layers, energetic atoms may reach the surface and be released. The sputtering yield is derived as the mean number of cascade atoms per incident neutron which reach the surface with sufficient energy to overcome the surface potential, which is of the order of 2–10 eV [Ref. 5.7, Chaps. 2, 3].

5.3.1 Primary Knockon Atoms and Cascades in an Infinite Solid

The creation of primary knockon atoms by the incident neutrons can occur by elastic and inelastic collisions with energetic neutrons, by nuclear reactions such as (n, γ), $(n, 2n)$, (n, p), and (n, α), or by fission processes. In fissile material the fission products are regarded as primary knockon atoms.

The cross section for the interaction of neutrons with atomic nuclei has been investigated in some detail for all elements as a function of neutron energy [5.117, 151, 152]. The *differential cross section* $d\sigma(E, T)/dT$ for transferring an energy T to a target atom of mass M_2 generally shows oscillations as a function of T [5.149, 151, 152]. As a first approximation, however, sometimes a hard sphere cross section corresponding to isotropic scattering is taken [5.1]. This is reasonably accurate only for low energies, i.e., $E \lesssim 1$ MeV [5.151–153]. The values for the *total cross section*, $\sigma(E)$, for gold are shown in Fig. 5.13 [5.117, 147]. They have a similar magnitude for most elements, giving a mean free path for neutrons in a solid of centimeters.

As most irradiations are performed with neutrons having a broad energy distribution $J_n(E)$, we define an average cross section $\langle\sigma\rangle$ given by

$$\langle\sigma\rangle = \frac{1}{J_{n,\mathrm{tot}}} \int_0^\infty J_n(E) dE \tag{5.3}$$

with $J_{n,\mathrm{tot}} = \int_{E_1}^{E_2} J_n(E) dE$ being the total neutron flux. The integration is generally performed only for the energy range of interest, $E_1 \leqq E \leqq E_2$, such as the thermal, the epithermal or the fast neutron flux, where $\sigma(E)$ is large.

The maximum energy T_m, which can be transferred in an elastic collision from a neutron of energy E to a target atom is given by

$$T_m = \frac{4A}{(1+A)^2} E, \tag{5.4}$$

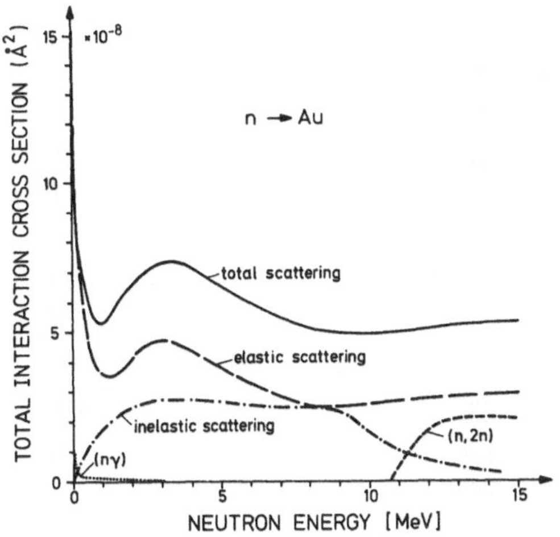

Fig. 5.13. Total neutron cross section for gold as a function of neutron energy

where $A = M_2/M_n$ is the ratio between the mass M_2 of the target atom and the mass M_n of the neutron. Except for very light target atoms we have $A \gg 1$ and (5.4) reduces to $T_m \simeq (4/A)E$. The mean transferred energy depends on the differential cross section $d(E, T)/dT$. For hard-core collisions the mean transferred energy is $T_m/2$, but for neutron energies above $\simeq 1$ MeV it is much lower [5.147, 153–155]. In inelastic collisions the transferred energy is generally peaked at approximately $T_m/4$ for $A \gg 1$ (Sect. 5.3.3). For fast neutrons ($E \geqq 0.1$ MeV) these transferred energies lie in the range of a few keV to a few 100 keV.

At thermal neutron bombardment in the (n, γ) process generally several γ-quanta with energies of 1–5 MeV are emitted [5.156, 157]. The angular correlation and time delays between the γ-emission processes can mostly be neglected in calculating the mean energy transferred to a target atom [5.120, 121]. This is given by [5.87]

$$T = \frac{\sum E_\gamma^2 w(E_\gamma)}{2M_2 c^2} \tag{5.5}$$

or

$$T = \frac{537}{A} \sum E_\gamma^2 w(E_\gamma) \, [\mathrm{eV}] \, .$$

Here $c \simeq 3 \times 10^8 \, \mathrm{m\,s^{-1}}$ is the velocity of light, E_γ is the energy of the emitted γ-radiation in units of MeV, $w(E_\gamma)$ is the probability for emitting a γ-quantum with energy E_γ, and $\sum$ means the sum over all emitted γ-quanta [5.156, 157]. Average recoil energies from (n, γ) processes have been calculated from (5.5) for several elements, the results are shown in Table 5.5 [5.87, 89, 120, 163].

Table 5.5. Cross sections and calculated average recoil energies for different elements for (n, γ)-reactions with thermal neutrons [5.52, 163]

Element	$\sigma_{\text{therm}}\,(b)$	$\langle T \rangle$ [eV]
Be	0.01	1912
C	0.003	891
Na	0.53	566
Mg	0.063	745
Al	0.23	812, 771[a], 801[b]
Si	0.16	730
K	2.1	413
Ca	0.43	468
Ti	6.1	481
V	5.05	426
Cr	3.1	670
Mu	13.3	385
Fe	2.56	473
Co	37.2, 37[a]	362, 362[a]
Ni	4.43, 4.8[a]	592, 525[a]
Cu	3.8, 3.8[a]	413, 374[a], 425[b]
Zn	1.1[a]	299[a]
Zr	0.2	162
Nb	1.15	125
Mo	2.65, 2.7[a]	119, 160[a], 169[b]
Rh	156[a]	113[a]
Pd	8[a]	125[a]
Ag	63.45, 63[a]	136, 136[a], 125[b]
Cd	2450[a]	129[a]
In	196[a]	86[a]
Ta	31.3	3
W	18.5, 19.2[a]	14, 51[a], 61[b]
Re	86[a]	39[a]
Pt	8.8[a]	70[a]
Au	98.8, 89.8[a]	68, 68[a]
Pb	0.17	140

[a] [5.87] [b] [5.120]

For a description of the development of the collision cascade started by the primary knockon atom, several quantities have been introduced, as shown in Figs. 5.14 and 15 ([Ref. 5.7, Chap. 2] and [5.138, 158]):

(i) The trajectory of a primary knockon atom in the solid can be described by *different ranges*, as shown in Fig. 5.14. and the corresponding range distributions, $F_R(T, \theta, x)dx$ (Fig. 5.15).

(ii) Kinetic energy will be deposited at different depths x with a distribution of the *deposited energy* $F_D(T, \theta, x)dx$ and

$$\int_{-\infty}^{+\infty} F_D(T, \theta, x)dx = v(T)\,. \tag{5.6}$$

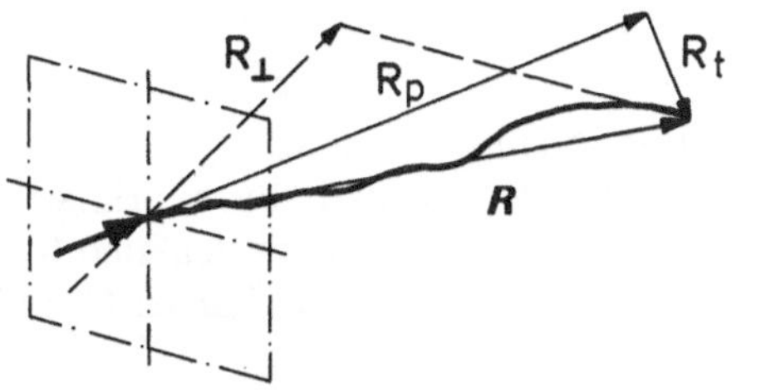

Fig. 5.14. Different components of the range of an energetic atom in a solid. $\boldsymbol{R}$ = vector range, R_p = projected range, R_t = transverse range and $R_\perp$ = range projected on the direction of the surface normal

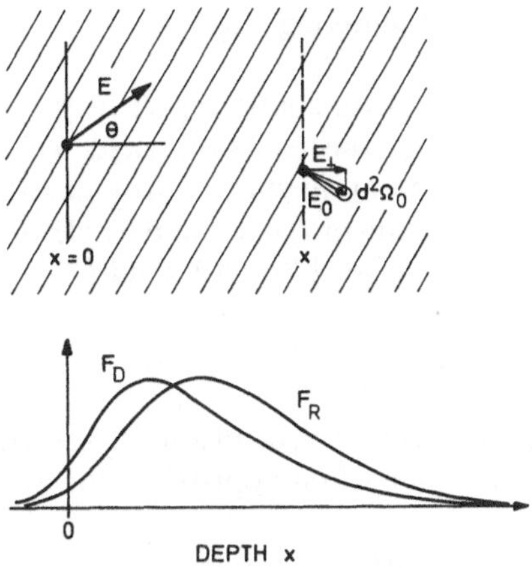

Fig. 5.15. Range distribution and deposited energy density for ions starting at $x=0$ with energy E and an angle θ relative to the starting plane in an infinite target [5.138, 158]

(iii) For describing the sputtering process, the *average number of cascade atoms*, $j(T, \theta, x, E_0, \theta_0)dE_0 d^2\Omega_0$, was introduced, which penetrate a plane x from one side with energy E_0, dE_0 in the direction $\theta_0, d^2\Omega_0$ for each primary knockon atom starting at $x=0$ with energy T, in the direction θ, averaged over a large number of primary knockon atoms [Ref. 5.7, Chap. 2].

These distributions and their interrelations can be calculated with Boltzmann transport equations [Ref. 5.7, Chaps. 2, 8] or with computer simulation codes [Ref. 5.7, Chap. 3] and [5.82]) for single crystals, for polycrystalline material and for amorphous solids [5.159].

For an amorphous solid, *Sigmund* ([Ref. 5.7, Chap. 2] and [5.8, 144]) obtained the following relationship between the desposited energy density F_D and the average number of atoms j penetrating a plane x in the direction $\theta_0, d^2\Omega_0$ with energy E_0, dE_0 for each primary knockon atom starting at $x=0$ with energy T in the direction θ [Ref. 5.7, Eq. (2.3.6) with $j=J/\psi$]:

$$j(T, \theta, x, E_0, \theta_0)dE_0 d^2\Omega_0 = F_D(T, \theta, x)\frac{\Gamma_m dE_0}{E_0|dE_0/dx|}\cos\theta_0 \frac{d^2\Omega_0}{4\pi}, \tag{5.7}$$

where $\Gamma_m = m/[\psi(1) - \psi(m-1)]$ and $\psi(x) = d[\log\Gamma(x)]/dx, \Gamma(x)$ being the Γ-function, dE_0/dx is the stopping power for atoms with energy E_0 in the solid and m a parameter describing the interatomic potential with $m=1$ at very high recoil energies and $m \approx 0$ at low energies.

Equation (5.7) relates the sputtering yields to the deposited energy density. It represents an asymptotic solution of the Boltzmann transport equation for the linear cascade regime, i.e., at keV energies and for $E_0 \ll T$. At such E_0, the atoms in the cascades are shown to move isotropically in all directions.

5.3.2 Sputtering Yields

The neutron sputtering yield is obtained from (5.7) in two steps:

First, the emission yield $\tilde{Y}(T,\theta,x)$ at a surface x caused by a primary knockon atom starting from $x=0$ with energy T and angle θ is calculated by integrating (5.7) over all particles which may overcome the surface barrier U_0 at x. This gives [Ref. 5.7, Chap. 2]

$$\tilde{Y}(T,\theta,x)=\Lambda F_{\mathrm{D}}(T,\theta,x). \tag{5.8}$$

The factor Λ depends on the properties of the solid only. For a planar surface potential a first approximation is [Ref. 5.7, Eqs. (2.3.8–11)]

$$\Lambda=\frac{0.042}{NU_0} \tag{5.9}$$

with the dimension Å-eV^{-1} where the atomic density N is measured in Å^{-3} and the surface potential U_0 in eV. For U_0 the value of the heat of sublimation is a good approximation [Ref. 5.7, Chaps. 2, 3].

The sputtering yield is then obtained from (5.8) by integration over all depth x where primary knockon atoms start. The mean number, energy and starting directions of the primary knockon atoms created by one incident neutron per unit of depth depend on the directions of the neutron flux relative to the orientation of the surface.

a) Forward and Backward Sputtering

For a directed neutron flux hitting a surface at an angle ϑ relative to the surface normal, the mean number of primary knockon atoms N^{p} with energy T, dT created for one incident neutron per unit depth x, normal to the surface, is given by

$$N^{\mathrm{p}}(E,T)dT=\frac{N}{\cos\vartheta}\frac{d\sigma(E,T)}{dT}dT. \tag{5.10}$$

The neutron sputtering yield is then given by

$$Y(E)=\int_{U_0}^{T_{\mathrm{m}}}\int N^{\mathrm{p}}(E,T)\tilde{Y}(T,\theta,x)dx\,dT. \tag{5.11}$$

Here the starting directions θ of the primaries with respect to the neutron direction ϑ are determined by the neutron energy E, the transferred energy T and for inelastic collisions the energy Q lost or gained [5.160]. This means that the integration over T in (5.11) includes the integration over θ. The integral limits in (5.11) depend on the geometry. If the target is much thicker than the range of the primary knockon atoms ($c \gg R_p$), we obtain from (5.11) with (5.8, 9) and for the forward (and the backward) sputtering yield:

$$Y_{f,(b)}(E,\vartheta) = \frac{NA}{\cos\vartheta} \int_{U_0}^{T} \frac{d\sigma(E,T)}{dT} dT \int_{0,(-\infty)}^{\infty,(0)} F_D(T,\theta,x)dx\,. \tag{5.12}$$

An integral of the form as in (5.12) also occurs in the evalution of neutron radiation damage in the bulk of a material [5.147, 153, 154, 161–163]. It has first been named *specific damage energy* [5.147] and later *damage energy cross section D(E)*, and is defined by

$$D(E) = \int_{E_{th}}^{T_m} \frac{d\sigma(E,T)}{dT} v(T)dT. \tag{5.13}$$

For low neutron energies, corresponding to primary knockon energies T in the range of 100 eV, the damage energy $v(T)$ is close to T[5.137, 138, 147]. Here the damage-energy cross section approaches the nuclear stopping cross section $S_n(E)$ for the neutrons in the solid [5.137, 138]. At higher neutron energies $D(E)$ is always less than $S_n(E)$.

The damage energy cross section $D(E)$, also written as $\langle\sigma(E,T)v(T)\rangle$ has been investigated in some detail with respect to bulk radiation damage [5.89, 147, 153, 154, 161–163]. For the lower limit in the integral mostly $E_{th}=0$ is taken which may not be correct, especially for low-energy primary knockon atoms. Generally, the damage energy $v(T)$ should become zero for $T<E_{th}$, i.e. for energies below a threshold energy [5.164]. For comparing the effects measured at neutron sources having different energy distributions, similarly to (5.3) average-damage-energy cross-sections $\langle D\rangle$ have been calculated for each source for different elements [5.147, 163].

For the calculation of sputtering yields with (5.12), the integrals over the damage energy function have to be calculated. In the case of forward sputtering, the integration in depth is performed over the main part of the damage function (Fig. 5.15). Thus, for angles of incidence not too large we may put [5.8]:

$$\int_0^{\infty} F_D(T,\theta,x)dx \simeq v(T). \tag{5.14}$$

For angles of incidence close to $\pi/2$ only about half of the energy is deposited in the solid and we get

$$\int_0^{\infty} F_D(T,\theta,x)dx \simeq \frac{v(T)}{2}. \tag{5.15}$$

By introducing (5.14, 15) into (5.12) and using (5.13), we get for the *forward sputtering yield*

$$Y_{\mathrm{f}}(E,\vartheta)=\frac{N\Lambda}{\cos\theta}K(\vartheta)D(E). \tag{5.16}$$

Here $K(\vartheta)=1$ for $\vartheta=0°$ and $K(\vartheta)=1/2$ for ϑ close to $\pi/2$ with values in between for other ϑ.

For an evaluation of the backward sputtering yield, we regard the *energy reflection coefficient* (also called *sputtering efficience*) for ions bombarding a solid with energy T at an angle of incidence θ, which is defined by [Ref. 5.7, Chap. 4] and [5.8, 165, 167])

$$\gamma(\theta, T)=\frac{1}{T}\int_{-\infty}^{0} F_{\mathrm{D}}(T,\theta,x)dx. \tag{5.17}$$

This definition assumes that the reflected energy is carried predominantly by sputtered atoms, while back-scattered atoms make only a minor contribution. This should apply for self ion bombardment.

The energy reflection coefficient has been investigated by calorimetric methods as a function of energy and angle of incidence θ, with ion beams and for $\theta \lesssim 45°$ the experimental results agreed reasonably with the formula ([Ref. 5.7, Eq. (4.2.8)] and [5.167])

$$\gamma(\theta, T)=\gamma(0, T)+[\tfrac{1}{2}-\gamma(0, T)](1-\cos\theta)^2. \tag{5.18}$$

Here $\gamma(0, T)$ is the energy reflection coefficient for normal incidence, which is typically of the order of the 0.02–0.03. Recently the dependence of the energy reflection coefficient on the angle of incidence was determined by computer simulation for uranium in the 100 eV range. Here for $\theta \gtrsim 50°$ the major contribution of the reflected energy comes from back-scattered atoms [5.168]. The results presented in the following are based, however, on (5.18).

For calculating the backward sputtering yield further isotropic elastic scattering is assumed, corresponding to a differential cross-section of

$$d\sigma(E, T)=\frac{\sigma(E)}{T_{\mathrm{m}}}dT \tag{5.19}$$

with $\sigma(E)$ is the total cross-section. Inserting (5.17–19) into (5.10) and integrating over T we get for the *backward sputtering yield* by neutrons at normal incidence ($\vartheta=0$) [5.8]:

$$Y_{\mathrm{b}}(E,0)=\delta N\Lambda\sigma(E)\frac{T_{\mathrm{m}}}{2}, \tag{5.20}$$

where

$$\delta = \tfrac{14}{15}\gamma(0, T) + \tfrac{1}{30}, \tag{5.21}$$

giving values of $\delta = 1/16$ to $1/19$. Because we had assumed isotropic hard-spheres collisions, we may also write $T_m/2 = \langle T \rangle$ and for recoil energies well below $\simeq 1$ keV further, $\sigma(T)\langle T \rangle = D(E)$.

At oblique angles of incidence, $Y_b(E, \vartheta)$ increases according to [5.8], proportional to $(\cos\vartheta)^{-4.76}$ for small ϑ. For $\vartheta \lesssim \pi/2$, the backward sputtering yield should approach the forward sputtering yield ($\vartheta \gtrsim \pi/2$) which is given by (5.16).

The ratio between the backward and forward sputtering yields for neutrons at energies where elastic collisions dominate [5.8, 79–83] is given from (5.16, 20) with (5.21) by

$$\frac{Y_b}{Y_f} = \delta \frac{\sigma(E)\langle T \rangle}{\langle \sigma(E, T)\nu(T) \rangle} \tag{5.22}$$

which is close to δ for neutron energies below one MeV where $\nu(T) \simeq T$. For higher neutron energies, $\nu(T)$ becomes smaller than T. Further, $\gamma(0)$ and thus also δ decreases with increasing T. Finally, inelastic collisions become more important, causing enhanced emission of primary knockon atoms in the forward direction relative to elastic scattering. These effects may partly cancel, but Y_b/Y_f is expected to decrease with increasing neutron energy [5.79].

The greater forward than backward sputtering yield for energetic neutrons, as obtained with (5.22), is caused by the isotropic scattering which strongly favours energy transfer in the forward direction. For sputtering by ions, the interaction is via a Coulomb potential which, at high energy, favours transfer in the direction normal to the ion trajectory. Thus for ions the difference between backward and forward or transmission sputtering is much less pronounced [5.7, 8].

b) Transverse Sputtering

For grazing incidence, where ϑ approaches $\pi/2$, forward and backward sputtering yields become very large, and the exact determination of ϑ is critical. The maximum yield which can be obtained occurs at such small angles that each neutron collides within the depth R_p, i.e. $\cos\vartheta \lesssim N\sigma R_p$. Here each neutron contributes to sputtering and the yield lies in the range of 10–100 atoms per energetic neutron.

In an experiment the neutron flux J_n from a source is given. By tilting the sample surface to grazing incidence the fluence at the surface decreases proportional to $\cos\vartheta$, but the number of atoms removed per unit time, which is given by $J_n(\cos\vartheta)\, Y \simeq J_n N \Lambda D(E)/2$, stays constant, independent of ϑ. This corresponds to half the forward sputtering yield for the case that the neutrons would penetrate the same surface at normal incident.

For a cube we may define a *transverse sputtering yield* for one side-surface with respect to the neutron fluence passing the front and back surfaces nearly perpendicular by

$$Y_t(E) = N\Lambda \frac{D(E)}{2}. \tag{5.23}$$

For a cube this equation also holds exactly for the case that the neutrons do not penetrate, but just move parallel to the side surface inside the cube.

c) Sputtering by an Isotropic Neutron Flux

If a sample of volume V and surface area s is subjected to an isotropic flux of neutrons (Fig. 5.1b), as occurs near the core of a nuclear reactor, the total number of primary knockon atoms with energy T produced per incident neutron is given by

$$N^{\mathrm{p}}_{\mathrm{tot}}(E,T)dT = Nl\, d\sigma(E,T), \tag{5.24}$$

where l is the average distance a neutron travels in the solid. This has been investigated in neutron physics, and for a convex solid we have [5.169, 170]

$$l = 4V/s \tag{5.25}$$

which can easily be verified for a sphere. For targets with linear dimensions much larger than the range of the collision cascades, the mean number of primary knockon atoms per unit depth is given by dividing $N^{\mathrm{p}}_{\mathrm{tot}}$ through the volume V and multiplying with the surface area s:

$$N^{\mathrm{p}}(E)dT = 4N\, d\sigma(E,T). \tag{5.27}$$

For the isotropic sputtering yield we now obtain from (5.10)

$$Y_{\mathrm{isotr.}}(E) = 4N\Lambda \int_{U_0}^{T_m} \frac{d\sigma(E,T)}{dT} dT \int_0^{\infty} F_{\mathrm{D}}(T,\theta,x)dx, \tag{5.28}$$

where x is the distance from the surface. Due to the isotropic neutron flux, the directions of the primary knockon atoms are also isotropic and the integral over $F_{\mathrm{D}}(T,\theta,x)$ gives $\nu(T)/2$, see (5.6, 15). We finally get for the *isotropic sputtering yield*

$$Y_{\mathrm{isotr.}}(E) = 2N\Lambda D(E). \tag{5.29}$$

For an isotropic neutron flux all surfaces contribute equally to sputtering. Sputtering on just one side of a sample will also be given by (5.29), reduced by the ratio of the selected to the total surface area.

From (5.16, 20, 23) we find that the total neutron sputtering yield of a cube subjected to a directed neutron flux at normal incidence on one side is a factor of 1.5 higher than the isotropic neutron sputtering yield. This is due to the relatively large contribution of transverse sputtering from the side surfaces.

Equation (5.28) can, in principle, also be applied if primaries are produced with thermal neutrons by (n, γ) processes or in fission processes. However, in the first case, due to the low primary knockon energies we are no longer dealing with the linear cascade regime so that (5.29) may overestimate the yields. For fission, $\nu(T)$ has to be replaced by $\nu_1(T_1)+\nu_2(T_2)$ where $\nu_1(T_1)$ and $\nu_2(T_2)$ are the mean energies deposited at the target nuclei by the two fission products. Generally the starting energies of fission products are in the MeV-energy range and the linear-cascade theory may underestimate the sputtering yield [Ref. 5.7, Chap. 2].

d) Comparison with Computer Simulation and Experimental Results

Sputtering yields by a directed beam of 14.1 MeV neutrons at normal incidence have also been investigated in some detail for niobium with the computer simulation program TRIM [5.80–83, 171]. The differential cross-section for the interaction of the neutrons with the Nb atoms, which was used, is shown in Fig. 5.16 [5.80–82, 172, 173]. The hard-sphere differential cross-section taken in (5.19) is also shown for comparison. The probability that primary knockon atoms created uniformly over the depth x, with energy T and angle θ will lead to the emission of an atom at the surface, was multiplied by the atomic density N and named specific sputtering yield. Its value was calculated with TRIM for each T and the corresponding θ by following the trajectories of up to 50,000 primary knockon atoms created uniformly in the volume of the material. A planar surface potential of $U_0=7.5$ eV was assumed at the surface. Figure 5.17 shows the results obtained for backward as well as for forward sputtering at normal neutron incidence [5.80–83]. These curves can also be calculated with the analytical theory from (5.11) with and (5.14, 17–19) giving $N\Lambda\nu(T)$ for forward and $N\Lambda T\gamma(\theta, T)$ with $\cos\theta=\sqrt{T/T_m}$ for backward sputtering. The values obtained with $\nu(T)$ from [5.147] are also introduced in Fig. 5.17. There is good general agreement, but the analytical results are typically higher by a factor of 2 to 3 than the TRIM results. In order to get the sputtering yields, these numbers have to be multiplied by $d\sigma(E, T)$ and integrated over T.

Such detailed data, as presented in Fig. 5.17, should well be described with the computer simulation program if the correct input data are used. For primary knock atoms with lower energy ($E \lesssim 10$ keV) the yields calculated by TRIM for forward and backward sputtering become equal. This is expected as these primary knockon atoms all start nearly perpendicular to the neutron beam. In the analytical calculations the forward yields are always larger than the backward yield. For the low-energy primary knockon atoms this may be caused by an overestimation of the damage energy which should be $\nu(T)/2$ instead of $\nu(T)$, see (5.14, 15). Also for the larger angles ($\theta \gtrsim 50°$) the energy

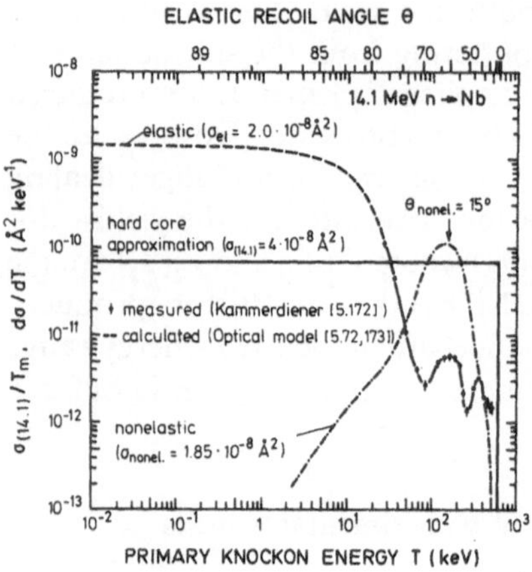

Fig. 5.16

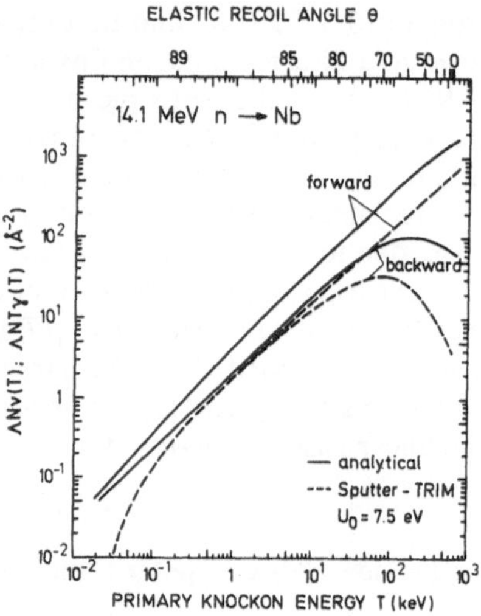

Fig. 5.17

Fig. 5.16. Differential cross section for 14.1 MeV neutrons bombarding Nb as a function of the transferred energy T or the recoil angle θ. The elastic and nonelastic cross section are shown together with the hard core approximation

Fig. 5.17. Calculated values for the probability that primary knockon atoms produced uniformly in the solid with energy T and angle θ will lead to the emission of a surface atom, multiplied by the atomic density N ([Ref. 5.7, Chap. 2] and [5.8, 80–82])

reflection coefficient $\gamma(\theta, T)$ is still uncertain. The total yields are, however, dominated by high-energy primary knockon atoms. Here a fraction of the primaries and the cascade atoms may be emitted with high energy, which leads to a lower deposited energy density in the near surface layers, than applied in the analytical theory without surface correction.

Calculated forward sputtering yields at normal neutron incidence ($\vartheta = 0$) for Au and Nb from (5.16) using values for the damage energy cross section $D(E)$ published by different authors [5.147, 153, 154, 161, 163], and the yield obtained with the aid of the computer simulation program TRIM [5.80, 83] are shown in Figs. 5.18 and 19. In these figures the values of the sputtering yields measured with fission reactor neutrons [5.64, 65] and with about 14 MeV neutrons (Figs. 5.6 and 7) are also introduced.

For Au the isotropic yield measured with fission neutrons is about a factor of 2 higher than the forward yield calculated from (5.16). This is in agreement with the theoretical estimate given by (5.28). The yield measured with 14 MeV neutrons from a Be(d, n) source is about a factor of 3 lower.

For Nb the yield value taken from measurements with 14-MeV neutrons, which have large uncertainties, is slightly lower than the values calculated from (5.16). The lowest yield values for Nb are obtained with the TRIM-simulation program. As

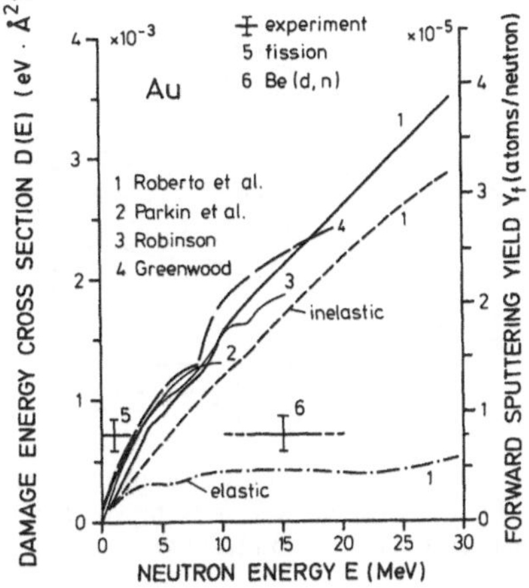

Fig. 5.18

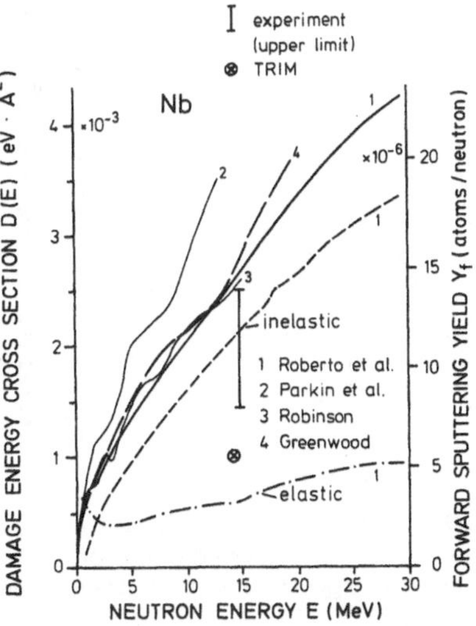

Fig. 5.19

Fig. 5.18. Damage energy cross section and forward sputtering yield for Au bombarded by neutrons at different energies. Mean values of the measured yields are included from Fig. 5.6

Fig. 5.19. Damage energy cross section and forward sputtering yield for Nb bombarded by neutrons at different energies [5.52, 153, 154, 161, 163]. The mean value of the measured yields according to Fig. 5.7 are included, together with one value calculated with the TRIM program [5.80–82]

already expected from the values of $N\Lambda\nu(T)$ in Fig. 5.17, they are about a factor of 3 below those from (5.16).

The reason for the differences between the analytical calculations and the TRIM results, both for forward and backward sputtering, are not yet quite understood; but one reason may be the neglection of surface corrections [Ref. 5.7, Chap. 2] in the analytical treatment. The TRIM program has also been used to calculate the forward and backward sputtering yields by 14 MeV protons incident on Nb. These numbers were in good agreement with the results of experiments performed with protons to simulate 14 MeV neutron sputtering [5.80–83].

Forward neutron sputtering yields calculated for normal incidence from (5.16) for several elements if interest are shown in Figs. 5.20 and 21. They all lie between 5×10^{-6} and 5×10^{-5} atoms neutron.

The ratio of backward to forward sputtering yields obtained from the experiments is of the order of 1/3 to 1/4 and thus much larger than the analytical results which agree with those from the TRIM program, giving a value of about 1/15 to 1/20. The higher experimental ratios are presumably due to sputtering by energetic recoils leaving the backward collector and hitting the target together with the neutrons [5.80–82].

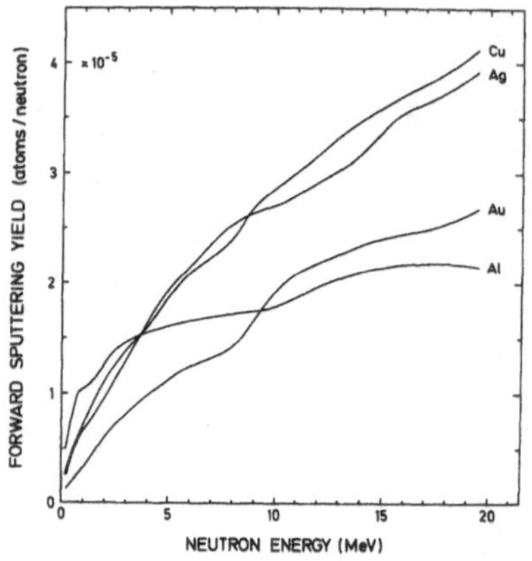

Fig. 5.20. Calculated forward sputtering yields for Cu, Ag, Au, Al, using values for the damage energy cross section from [5.52, 163]

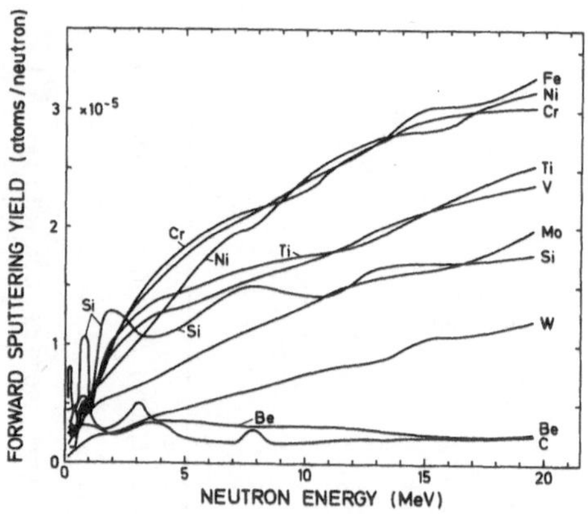

Fig. 5.21. Calculated forward sputtering yields for C, Be, W, Si, Mo, V, Ti, Fe, Ni, and Cr, using values for the damage energy cross section from [5.52, 163]

5.3.3 Emission of Radioactive Primary Knockon Atoms

In sputtering experiments with neutrons, the primary knockon atoms can be producted in a nuclear reaction such as (n, γ), $(n, 2n)$, (n, p), (n, α) or a fission event. For calculating the transferred energy, this reaction can mostly be approximated by the formation of a compound nucleus which moves in the same direction with the momentum of the neutron and has an energy of [5.153–155]

$$T = E(1+A)^{-1}. \tag{5.29}$$

For $A \gg 1$ this is approximately equal to E/A or $T_m/4$. The particle emission in the subsequent decay of the compound nucleus will introduce a straggling in direction and energy, but this can be small if the reaction energies Q are small compared to the kinetic energy of the compound nucleus [5.160].

The radioactive primary knockon atoms starting with an energy T in the direction ϑ of the neutron may have a range distribution $F_R(T, \vartheta, x)dx$ in the direction x normal to the surface. If they start uniformly distributed in the solid of thickness c at distances x_0 from the surface, the probability for an atom to reach the surface and be emitted is given by

$$\int_0^c dx_0 \int_0^\infty F_R(T, \vartheta, x + x_0)dx.$$

If the thickness c is much larger than the mean range this can be transformed to

$$\int_0^\infty x F_R(T, \vartheta, x)dx.$$

Here the surface potential is neglected, assuming that the majority of the atoms arrive at the surface with much higher energies. In calculating the radioactive primary knockon emission yields Y^{p}, the geometry has to be taken into account.

a) Directed Neutron Flux

The primary knockon emission yields in the forward $Y_{\mathrm{f}}^{\mathrm{p}}$ (and backward $Y_{\mathrm{b}}^{\mathrm{p}}$) directions are obtained similarly to sputtering yields for amorphous media, (5.12), and are given by (Fig. 5.1)

$$Y^{\mathrm{p}}_{\mathrm{f,(b)}}(E,\vartheta)=\frac{N}{\cos\vartheta}\sigma_{\mathrm{react.}}(E)\int\limits_{0,(-\infty)}^{\infty,(0)} xF_{\mathrm{R}}(T,\vartheta,x)dx\,, \tag{5.30}$$

with T from (5.9). We introduce the mean projected range of a primary knockon atom starting with T in a solid in a direction ϑ relative to the surface normal, given by

$$\langle R_{\perp}(T,\vartheta)\rangle\frac{\int\limits_0^{\infty} xF_{\mathrm{R}}(T,\vartheta,x)dx}{\int\limits_0^{\infty} F_{\mathrm{R}}(F,\vartheta,x)dx}\,. \tag{5.31}$$

For nearly perpendicular neutron incidence ($\vartheta=0$) the denominator is close to 1 and $\langle R_{\perp}\rangle\simeq\langle R_{\mathrm{p}}\rangle$. For single-energy neutrons we get for the *forward emission yield* of radioactive primary knockon atoms

$$Y_{\mathrm{f}}^{\mathrm{p}}(E)=N\sigma_{\mathrm{react.}}(E)\left\langle R_{\mathrm{p}}\left(\frac{E}{1+A}\right)\right\rangle. \tag{5.32}$$

At bombardment with neutrons having an energy distribution, Eq. (5.32) with (5.31 and 3) lead to (5.1). For the backward emission we regard the mean range $\langle R_{\mathrm{b}}\rangle$ in the backward direction which is given for normal neutron incidence ($\vartheta=0$) by

$$\langle R_{\mathrm{b}}(T)\rangle=\frac{\int\limits_{-\infty}^{0} xF_{\mathrm{R}}(T,0,x)dx}{\int\limits_{-\infty}^{0} F_{\mathrm{R}}(T,0,x)dx}\,. \tag{5.33}$$

Here the denominator is equal to the particle reflection coefficient R_{N} for ions bombarding a surface with energy T, being of the order of a few % [5.158, 174, 175]. For the *backward emission yield* of radioactive primary knockon atoms at single-energy neutron bombardment we obtain

$$Y_{\mathrm{b}}^{\mathrm{p}}(E)=N\sigma_{\mathrm{react.}}(E)R_{\mathrm{N}}\left\langle R_{\mathrm{b}}\left(\frac{E}{1+A}\right)\right\rangle. \tag{5.34}$$

For oblique angles of incidence ($\vartheta > 0$), the number of radioactive primary knockon atoms per unit depth and per incident neutron increases with $1/\cos\vartheta$, whereas the emission probability of these primaries decreases with respect to forward emission, due to the shorter range in the direction to the surface.

For the emission yield of radioactive primary knockon atoms at nearly grazing neutron incidence we may define, similarly to the sputtering yield (5.23), a *transverse emission yield* Y_t^p *for a cube* (Fig. 5.1), given by

$$Y_t^p(E) = N\sigma_{\text{react.}}(E)\frac{\left\langle R_t\left(\frac{E}{1+A}\right)\right\rangle}{\pi}. \tag{5.35}$$

Here $\langle R_t(T)\rangle$ is the mean transverse range (Fig. 5.14), defined by

$$\langle R_t(T)\rangle = \int_0^\infty x F_{R_t}(T, 0, x)dx, \tag{5.36}$$

where F_{R_t} is the range distribution of the primaries in the direction perpendicular to the starting direction. The factor π in the denominator enters because the transverse ranges are distributed on a cylinder around the starting direction, while only the components of the direction normal to the surface enter into the emission yield.

The relation of (5.35) has been used to measure the transverse range of Au atoms with a mean energy of 91.4 keV created by an $(n, 2n)$ reaction in gold (Sect. 5.2.5b) [5.131].

b) Isotropic Neutron Flux

An isotropic angular distribution of primary knockon atoms is obtained for an isotropic neutron flux or if the primary knockon atoms are produced by fission or by an (n, γ) reaction with thermal neutrons. In this case the mean range $\langle R_\perp\rangle$ for atoms in the outward direction, normal to a surface is given by [5.6]

$$\langle R_\perp\rangle = \frac{\langle R_p\rangle}{4}. \tag{5.37}$$

Using the same quantities as introduced for sputtering by an isotropic neutron flux see (5.23), we obtain the *isotropic radioactive primary knockon emission yield* from a convex probe

$$Y_{\text{isotr.}}^p(E) = N\sigma_{\text{react.}}(E)\left\langle R_p\left(\frac{E}{1+A}\right)\right\rangle. \tag{5.38}$$

Similar to sputtering, the emission yield of radioactive primary knockon atoms is uniform at all surfaces. The emission yield of a given surface area is reduced compared to (5.38) by the corresponding ratios of the given surface to the total surface area.

c) Primary Knockon Emission and Sputtering Yield

For many purposes it is useful to calculate the primary knockon emission probability per primary knockon atom produced in a solid, $Y^{\mathrm{p}}/N^{\mathrm{p}}$. This is given for one energy by (see also Fig. 5.1)

$$\text{forward emission} \quad \frac{Y_{\mathrm{f}}^{\mathrm{p}}}{N_{\mathrm{tot}}^{\mathrm{p}}} = \frac{\langle R_{\mathrm{p}} \rangle}{c}, \tag{5.39}$$

$$\text{backward emission} \quad \frac{Y_{\mathrm{b}}^{\mathrm{p}}}{N_{\mathrm{tot}}^{\mathrm{p}}} = \frac{\langle R_{\mathrm{b}} \rangle R_{\mathrm{N}}}{c}, \tag{5.40}$$

$$\text{transverse emission (for a cube)} \quad \frac{Y_{\mathrm{t}}^{\mathrm{p}}}{N_{\mathrm{tot}}^{\mathrm{p}}} = \frac{\langle R_{\mathrm{t}} \rangle}{\pi a}, \tag{5.41}$$

$$\text{isotropic emission} \quad \frac{Y_{\mathrm{isotr.}}^{\mathrm{p}}}{N_{\mathrm{tot}}^{\mathrm{p}}} = \frac{s \langle R_{\mathrm{p}} \rangle}{4V}. \tag{5.42}$$

At nuclear reactions the left-hand side is the ratio of the number of radioactive atoms emitted at the surface to the number of radioactive atoms produced in the bulk, see Eq. (5.2) and [5.6]).

Further, the sputtering yield per emitted radioactive primary knockon atom, Y/Y^{p} can be calculated [5.8]. This is meaningfull if radioactive primary knockon atoms make the major contribution to sputtering. In this case $D(E) = \sigma_{\mathrm{react.}}(E)\nu(E/(1+A))$. For the linear cascade regime [Ref. 5.7, Chap. 2] this is given in a first approximation by:

$$\text{forward sputtering} \quad \frac{Y_{\mathrm{f}}(E)}{Y_{\mathrm{f}}^{\mathrm{p}}(E)} = \Lambda \frac{\nu\left(\frac{E}{1+A}\right)}{\left\langle R_{\mathrm{p}}\left(\frac{E}{1+A}\right)\right\rangle}, \tag{5.43}$$

$$\text{backward sputtering} \quad \frac{Y_{\mathrm{b}}(E)}{Y_{\mathrm{b}}^{\mathrm{p}}(E)} = \Lambda \frac{\delta \cdot \frac{E}{1+A}}{R_{\mathrm{N}}\left\langle R_{\mathrm{b}}\left(\frac{E}{1+A}\right)\right\rangle}, \tag{5.44}$$

$$\text{transverse sputtering (for a cube)} \quad \frac{Y_{\mathrm{t}}(E)}{Y_{\mathrm{t}}^{\mathrm{p}}(E)} = \Lambda \frac{\pi}{2} \frac{\nu\left(\frac{E}{1+A}\right)}{\left\langle R_{\mathrm{t}}\left(\frac{E}{1+A}\right)\right\rangle}, \tag{5.45}$$

$$\text{isotropic sputtering} \quad \frac{Y_{\mathrm{isotr.}}(E)}{Y_{\mathrm{isotr.}}^{\mathrm{p}}(E)} = \Lambda \frac{2\nu\left(\frac{E}{1+A}\right)}{\left\langle R\left(\frac{E}{1+A}\right)\right\rangle}. \tag{5.46}$$

Generally, all quantities on the right hand side can be obtained from range and damage calculations and measurements of energetic ions in solids [5.52, 147, 151, 152–154, 158, 161–163, 174, 175]. Thus, the sputtering yield caused by the radioactive primary knockon atoms can be obtained from (5.43–46) by measuring their emission yield Y^{p}.

The relation (5.46) is especially used in neutron sputtering of fissile material where the total number of sputtered atoms per fission fragment emitted at the surface is measured [5.78, 84]. On the rhs, the contribution of both fission fragments has to be taken into account.

5.3.4 Mechanisms for Emitting Microparticles

While the release of single atoms in physical sputtering of metals via a collision cascade is reasonably well understood [5.7, 8], the mechanisms by which microparticles are emitted are still open. The basic questions, especially for larger particles, are how the energy, the energy density and the momentum needed to create the new surface area of such a microparticle can be provided by the neutrons and how the particle will be knocked off [5.34, 176–178]. Why does a microparticle break loose at certain planes and what determines its size? Several possible mechanisms can be regarded for the emission of small atomic clusters and microparticles.

In sputtering of solids with keV ion beams, the emission of *large molecules* consisting of up to 20 atoms, or in the case of organic materials up to 100 atoms, has been observed in SIMS investigations [5.100, 101, 179–182]. The observation of molecules consisting of only a few atoms is explained by a simultaneous emission of several single atoms which may recombine to form the molecule outside the solid. This picture is confirmed in the sputtering of single crystals by finding more molecules released in focusing directions [5.180]. However, larger clusters containing more than a few atoms must be sputtered together, presumably by breaking an area of weak bonding.

In bombarding solids consisting of atoms with large atomic number Z, with ions of large atomic number such as Au^{+}, or with molecular ions at energies in the range of 30–100 keV, where the nuclear stopping power has a maximum, the cascades which are created can be very dense, i.e., a *collision spike* may develop. For such spikes the slowing down of the cascade atoms is slower compared to the linear cascade regime because cooling takes place only at the outer part of the spike region ([Ref. 5.7, Chap. 2] and [5.183, 186]). The possible existence of such spikes has been used to explain the measured energy spectra of sputtered atoms [5.183, 187], the larger sputtering yield found for high energy heavy ions compared to the yield predicted from linear cascade theory ([Ref. 5.7, Chap. 2] and [5.8, 188–193]) and the extra large damaged areas in solids at bombardment with molecular ions [5.194, 195]. The effect of such spikes at the surface of a solid has recently been investigated in detail by *Merkle* and *Jäger* [5.72, 196, 197]. Figure 5.22 shows a TEM-picture of an Au foil after bombardment with

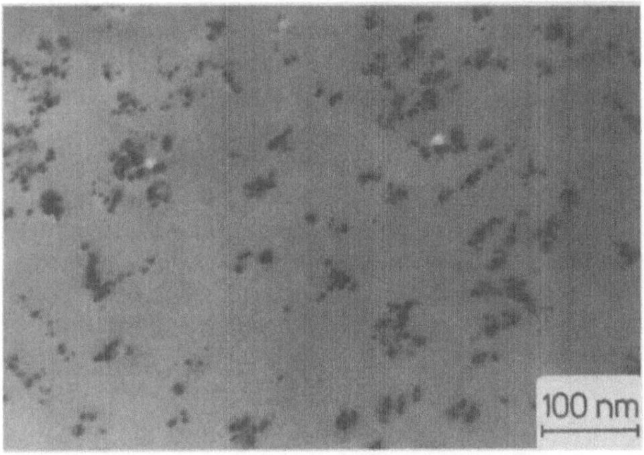

Fig. 5.22. Electron micrograph showing damage structures from individual cascades introduced by 500 keV Bi ions in a thin gold foil. The black spot-type damage represents mostly dislocation loop contrast from collapsed depleted zones. The different groupings of black spots within each bunch of subcascades show the great variation that exists from one cascade to another in the spatial arrangement of damage. Fluctuations in energy deposition are also thought to be responsible for the two clearly visible surface craters which appear in this micrograph in the form of two bright areas [5.72]

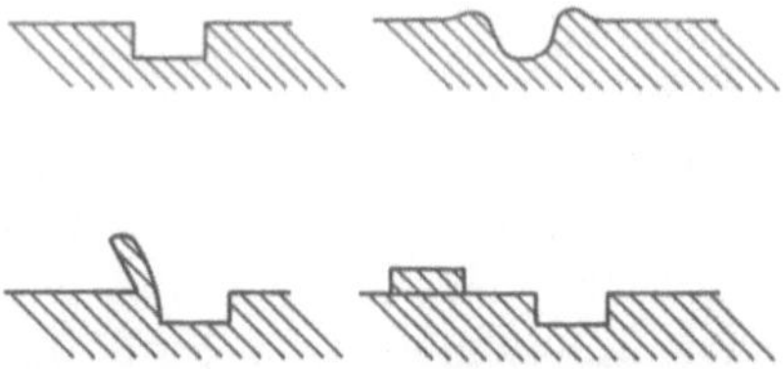

Fig. 5.23. Schematic cross-sectional views of the crater structures that have been observed by transmission electron microscopy. Most frequently, "regular" craters are observed whose contrast behavior merely indicates missing material at the surface. In this case the crater edges may be facetted or tapered (not shown). Additional structures are observed in high energy density spikes produced by molecular ion bombardment. In a few instances, craters seem to have a rim of extra material: evidence of plastic deformation in the form of a lid sticking out of the surface or a microparticle that has been deposited nearby point to the existence of explosive sub-surface spikes [5.72]

500 keV Bi^+ ions. The black spots represent mostly dislocation loops from collapsed depleted zones. Each group of spots corresponds to the impact of one Bi^+-ion. On the average, about every 10^{-2} ions produce a crater of about 5 nm diameter on the surface. These craters are seen as bright areas in the TEM picture. Several different shapes of such craters have been deduced from stereo TEM pictures and are shown in Fig. 5.23. From such a visible crater about 2000 atoms are removed as single atoms or as a small particle. The observed dependence of the crater yields on bombarding ion energy and energy density in the cascades, indicate that they may be formed in extra high density surface spikes. Such dense spikes can be explained by fluctuations in the energy

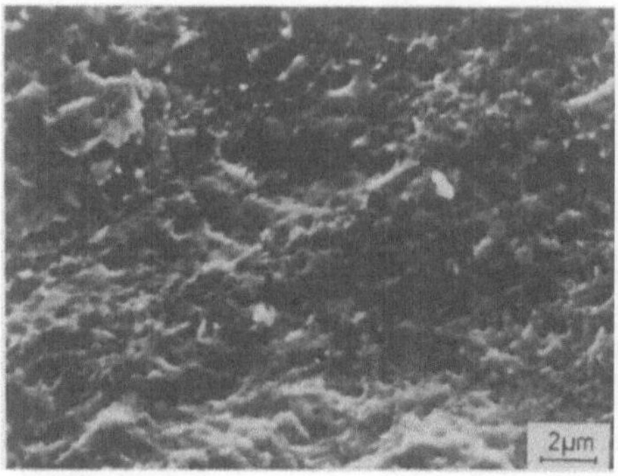

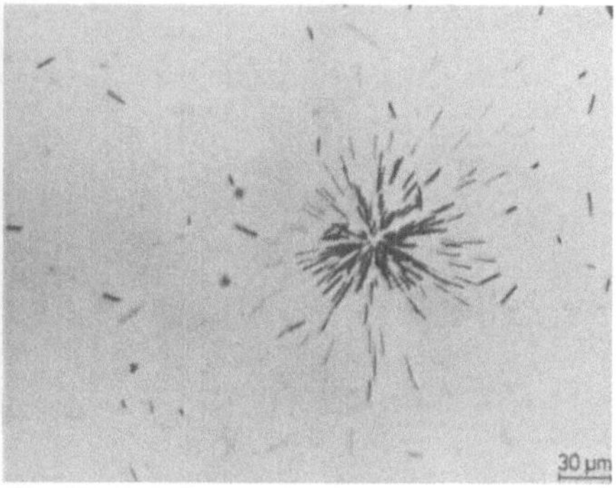

Fig. 5.24 **Fig. 5.25**

Fig. 5.24. SEM picture of a U(UO_2) surface after bombardment with 3×10^{17} 80 keV Ar^+/cm^2 [5.96]

Fig. 5.25. Fission tracks on muscovit mica from *U* which was sputtered by 40 keV protons collected on an Al foil and subsequently placed against the mica and irradiated with neutrons [5.96]

deposition in the cascades. As the primary knockon atoms produced in neutron sputtering may have energies of a few 100 keV, such spikes may also occur. There will be more spikes in Au than in Nb due to the higher atomic number of Au. However, such spikes are not able to explain the microparticle emission reported for 14 MeV neutron bombardment [5.66, 67, 73, 176].

In *sputtering of fissile material* such as uranium dioxide UO_2 or some glasses containing uranium by H, He, and Ar ions at energies of up to 120 keV [5.96] and by Fe^+-ions of 100 keV [5.97, 98], the removed material was deposited on a collector and detected in the same way as in neutron sputtering, i.e., by fission tracks in mica. For heavily damaged surfaces which were produced by prebombardment with 3×10^{17} 80 keV Ar cm^{-2}, shown in Fig. 5.24, the very same star patterns were observed on the mica foils as in neutron sputtering of heavily damaged surfaces. This can be seen by comparing Fig. 5.25, which shows the results obtained for 40 keV proton bombardment [5.96], with Fig. 5.8, obtained for neutron bombardment [5.78]. Such star patterns were also found in investigations by a second group. They reported, in addition, that these star patterns disappear if the target surface is flooded with electrons for compensation of the charging by the ion beam [5.97, 98].

The measurements show that the emission of the microparticles from uranium, uranium oxide or some glasses is not a special feature of neutron bombardment. Especially for hydrogen bombardment at 40 keV, the collision cascades are different from those produced in neutron bombardment. As microparticles are found only for rough surfaces, they may have been preformed on the surface of the brittle material during the previous bombardment. The disappearance of the microparticle emission at simultaneous electron bombardment leads to the conclusion that a surface charging of the insulating oxide particles may be one mechanism contributing to the microparticle emission. Such a charging may also occur in neutron bombardment due to secondary electrons from the intense γ-radiation which is always present in neutron sources.

It can be suggested that similar mechanisms as for UO_2 are also responsible for the *emission of the larger microparticles or chunks* reported to be found at 14 MeV neutron bombardment for specially prepared Nb, V, and Si surfaces. The scanning electron microscopy pictures taken from such samples show that microparticles have been pre-formed on the surface during the mechanical treatment. They may then become partly loose by the light chemical etch and be bound to the surface mainly by adhesion forces. Such microparticles can easily be released by Couloumb repulsion due to charging up by secondary electron emission. A pre-forming of the microparticles is necessary because the energy which can be transferred to target atoms by 14 MeV neutrons is orders of magnitude lower than the energy needed to create the new surface of micron-size chunks released from an initially plane surface [5.177, 178].

It had also been suggested that elastically stored energy in the surface layer, originating from the mechanical treatment, may add to the emission of chunks. Collision spikes created in the near-surface layer could trigger the release of the stored energy [5.177, 178]. However, it is not clear how this energy is concentrated to finally break off microparticles.

5.4 Conclusion

The release of atoms by sputtering from the surface of a solid due to neutron bombardment, being part of the radiation damage effects in the near surface layers, has been investigated both experimentally and theoretically with some success within the last few years. The theoretical treatment of neutron sputtering for metals is based upon collision cascade theory. First-order analytical solutions from Boltzmann transport equations in the linear cascade regime show that neutron sputtering yields are proportional to the damage energy cross section, $D(E) = \langle \sigma(E, T)\nu(T) \rangle$. As this has been investigated for metals in some detail with respect to bulk radiation damage, sputtering yields in different directions with respect to the incident neutron flux can be calculated with some confidence. For a directed neutron beam the highest yields are obtained in the forward direction, while transverse sputtering yields are about a factor of 2 and backward yields a factor of 10–20 smaller than forward yields. For an isotropic neutron flux, the calculated yields are a factor of two higher than the forward yields. In one case, i.e., 14.1 MeV neutron bombarding Nb, the computer simulation program TRIM was used to calculate forward and backward sputtering yields. The values obtained are about a factor of 3 smaller than those from the analytical theory. This discrepancy may partly be due to the neglection of surface corrections in the analytical calculations and could be resolved by a more detailed comparison of both calculations.

In the sputtering measurements performed up to now, the neutron fluences have been low and, generally, less than the equivalent of 1 monolayer was removed from the target surface and deposited on the collectors. For sputtered nonradiative atoms, the detection sensitivity is limited by contaminants on the collector, resulting in large uncertainties in the measured sputtering yields. The collector technique, which was used in all experiments, has the additional disadvantage that the neutrons passing the collector also release collector atoms by sputtering. Among these, in particular primary knockon atoms emitted in a forward direction, may have a large energy and cause more backward sputtering at the target surface than the neutrons. This is probably the reason that the experimentally determined ratios of forward to backward sputtering yields are smaller than predicted by both the analytical theory and the computer simulation.

In general, the best values of the measured forward sputtering yields for fission neutrons and for 14 MeV neutrons on Au and Nb agree with calculated yields within the experimental errors. In the future, more accurate measurements of the sputtering yields using neutron fluences of $\gtrsim 10^{19}$ neutrons cm^{-2} and well-defined surfaces are desirable. For measurements of backward sputtering it will be necessary to use collectors with a hole for the neutron beam to pass through and good neutron shielding on all other surfaces.

Radioactive-primary knockon or radioactive-recoil emission yields in forward, backward and transverse directions, have mostly been measured with $\simeq 14$ MeV neutrons for several solids with high sensitivity. More detailed investigations at different directions of neutron incidence and different neutron energies could give information on the ranges of these atoms in different directions in the solid.

Some investigators have reported the emission of microparticles for surfaces such as U, UO_2 and Nb, V and Si. For U, UO_2 such microparticle emission was also found in light and heavy ion sputtering, but for neutron sputtering only at very high fluences, resulting in highly damaged surfaces. One possible explanation is electrostatic charging of the microparticles pre-formed by cracks at the brittle and insulating UO_2 surfaces. For Nb, V, and Si the emitted microparticles were observed on the collector at low neutron fluences. Their occurrance can be understood if they are pre-formed on the target surface by the mechanical treatment and are also released by electrostatic charging due to secondary electrons. Microparticle emission does not seem to be a problem for 14 MeV neutrons bombardment at the highest fluence used in today's experiments.

Acknowledgements. The author would like to thank especially P. Sigmund, M. T. Robinson, J. P. Biersack, and O. K. Harling for many helpful comments and suggestions in finishing this review. A. Riccato, T. Thomas, W. Styrus, K. L. Merkle, W. Jäger, H. H. Andersen, K. Böning, W. Möller, and L. R. Greenwood contributed with several discussions and provided me with information prior to publication and are kindly acknowledged. Mrs. G. Daube and E. Krauss patiently typed several versions of the manuscript.

References

5.1 D.S.Billington, J.H.Crawford, Jr.: *Radiation Damage in Solids* University Press, (Princeton NJ 1961)
5.2 R.R.Coltman, C.E.Klabunde, J.K.Redman: J. Appl. Phys. **33**, 3509 (1962)
5.3 E.Segre: *Nuclei and Particles* (Benjamin, New York 1965)
5.4 N.Bohr: K. Dan. Vidensk. Selsk, Mat. Fys. Medd. **18**, No. 8 (1948)
5.5 O.Hahn: Naturwiss. **17**, 296 (1929)
5.6 S.Flügge, K.E.Zimen: Z. Phys. Chem. (Leipzig) B**42**, 179 (1939)
5.7 R.Behrisch (ed.): *Sputtering by Particle Bombardment* I, Topics Appl. Phys., Vol. 47 (Springer, Berlin, Heidelberg, New York 1981)
5.8 P.Sigmund: Phys. Rev. **184**, 383 (1969); **187**, 768 (1969)
5.9 S.I.Taimuty: Nucl. Sci. Eng. **10**, 403 (1961)
5.10 R.A.Schmitt, R.A.Sharp: Phys. Rev. Lett. **1**, 445 (1958)
5.11 J.Biersack, K.E.Zimen: Z. Naturforsch. **16a**, 849 (1961)
5.12 O.Selig, R.Sizmann: Nucleonik **8**, 303 (1966)
5.13 J.Biersack: Z. Phys. **211**, 495 (1968)
5.14 J.C.Ward: Oak Ridge National Laboratory ORNL-3152 (1961)
5.15 H.Mitsui: J. Nucl. Sci. Technol. **1**, 203 (1963)
5.16 O.K.Harling, M.T.Thomas, R.L.Brodzinski, L.A.Rancitello: J. Appl. Phys. **48**, 4328 (1977)
5.17 J.A.Davies, W.M.Campbell: AECL Chalk River Nuclear Labs. Internal Report 244 (1959); J.A.Davies, J.Friesen, J.D.McIntre: Can. J. Chem. **38**, 1526 (1960)
5.18 K.Keller: Plasma Phys. **10**, 195 (1968)
5.19 M.Kaminsky (Coordinator): Proc. Int. Working Session on Fusion Reactor Technology, Oak Ridge, Tenn. (1971) and IEEE Trans. NS-**18**, 208 (1971)
5.20 R.Behrisch: Nucl. Fusion **12**, 695 (1972)
5.21 R.Behrisch, B.B.Kadomtsev: Proc. Int. Conf. on Plasmaphysics and Controlled Nuclear Fusion, Tokyo (1974), IAEA, Vienna **2**, 229 (1975)
5.22 R.Behrisch, O.K.Harling, M.T.Thomas, R.L.Brodzinsky, L.H.Jenkins, G.J.Smith, J.F.Wendelken, M.J.Saltmarsh, M.Kaminsky, S.K.Das, C.M.Logan, R.G.Meisenheimer, J.E.Robinson, M.Schimotomai, D.A.Thompson: J. Appl. Phys. **48**, 3914 (1977); BNWL 2194 (1976)
5.23 D.J.Rose: Nucl. Fusion **9**, 183 (1969); ORNL-TM 2204 (1968)
5.24 D.Steiner: Nucl. Fusion **11**, 305 (1971)
5.25 D.Steiner: Nucl. Fusion **14**, 33 (1974)
5.26 R.G.Mills (ed.): "A Fusion Power Plant", Princeton Plasma Physics Laboratory, MATT-1050 (1974)
5.27 B.Badger et al.: "UWMAK-I", A Wisconsin Toroidal Fusion Reactor Design (1973)
5.28 B.Badger et al.: "UWMAK-II", A Conceptual Tokamak Power Reactor Design (1975)
5.29 B.Badger et al.: "UWMAK-III", A Noncircular Tokamak Power Reactor Design (1976)
5.30 R.W.Conn: J. Nucl. Mater. **76/77**, 103 (1978)
5.31 J.L.Cecchi: J. Nucl. Mater. **93/94**, 28 (1980)
5.32 H.Vernickel, B.M.U.Scherzer, J.Bohdansky, R.Behrisch: *Proc. Intern. Symp. on Plasma-Wall Interaction*, Jülich (1976) (Pergamon, London 1977) p. 209
5.33 D.Dusza, S.K.Das, M.Kaminsky: In *Radiation Effects on Solid Surfaces*, ed. by M.Kaminsky, Advances in Chemistry Series, No. 158 (The American Chemical Society 1976)
5.34 R.Behrisch: Nucl. Instrum. Methods **132**, 293 (1976)
5.35 O.K.Harling, M.T.Thomas, R.Brodzinsky, L.Rancitelly: J. Nucl. Mater. **63**, 422 (1976)
5.36 R.Behrisch: J. Nucl. Mater. **108/109**, 43 (1982)
5.37 R.Behrisch: Erg. Exakt. Naturwissensch. **35**, 295 (1964)
5.38 D.W.Norcross, B.P.Fairand, J.N.Anno: J. Appl. Phys. **37**, 621 (1966)
5.39 K.Verghese: Trans. Am. Nuclear Soc. **12**, 544 (1969)

5.40 C.E.Klabunde, B.C.Kelley: ORNL-P-2160 (1966)
5.41 P.Persiani (ed.): Proc. Intern. Conf. on Radiation Test Facilities for CTR Surface and Materials Program, Argonne National Lab., July 1975, Report ANL-CTR-75-4
5.42 H.Ullmaier (ed.): "High energy and high intensity neutron sources", Nucl. Instrum. Methods **145** (1977) special issue
5.43 M.A.Kirk (ed.): Proc. Intern. Conf. on Neutron Irradiation Effects in Solids, Argonne Nat. Lab. (1981), J. Nucl. Mater. **108/109** (1982)
5.44 P.Grand, A.N.Goland: Nucl. Instrum. Methods **145**, 49 (1977)
5.45 M.J.Saltmarsh, C.A.Ludeman, C.B.Fulmer, R.C.Styles: Nucl. Instrum. Methods **145**, 81 (1977)
5.46 M.Nakagawa, K.Böning, P.Rosner, G.Vogl: Phys. Rev. B**16**, 5285 (1977)
5.47 L.R.Greenwood, R.R.Heinrich, R.J.Kennerley, R.Medrzychowski: Nucl. Technol. **41**, 109 (1978)
5.48 L.R.Greenwood, R.R.Heinrich, M.J.Saltmarsh, C.B.Fulmer: Nucl. Sci. Eng. **72**, 175 (1979)
5.49 J.B.Roberto, C.E.Klabunde, J.M.Williams, R.R.Coltman, Jr., M.J.Saltmarsh, C.B.Fulmer: Appl. Phys. Lett. **30**, 509 (1977)
5.50 M.A.Kirk, L.R.Greenwood: J. Nucl. Mater. **80**, 159 (1979)
5.51 M.A.Kirk, R.C.Birtcher, T.H.Blewitt, L.R.Greenwood, R.J.Popek, R.R.Heinrich: J. Nucl. Mater. **96**, 37 (1981)
5.52 L.R.Greenwood: Proc. Intern. Conf. Neutron Irrad. Effects, Argonne Nat. Lab. (1981), J. Nucl. Mater. **108/109**, 21 (1982)
5.53 F.T.Binford, T.E.Cole, E.N.Cramer: "The high-flux Isotope Reactor", Oak Ridge Nat. Lab., ORNL-3572 (1968)
5.54 "Neutron research facilities at the high flux reactor of the ILL", Institut Max von Laue Paul Langevin, Grenoble (Jan. 1975)
5.55 R.Booth, J.C.Davis, C.L.Hanson, J.L.Held, C.M.Logan, J.E.Osher, R.A.Nickerson, B.A.Pohl, B.J.Schumacher: Nucl. Instrum. Methods **145**, 25 (1977)
5.56 J.B.Hourst, M.Roche, J.Morin: Nucl. Instrum. Methods **145**, 19 (1977)
5.57 R.Gähler: Diplomathesis, Technische Universität München (1974)
5.58 R.Behrisch, R.Gähler, J.Kalus: J. Nucl. Mater. **53**, 183 (1974)
5.59 A.C.Holmholz, E.M.McMillan, D.C.Sewell: Phys. Rev. **72**, 1003 (1947)
5.60 R.Serber: Phys. Rev. **72**, 1008 (1947)
5.61 M.T.Thomas, D.Styrus, O.K.Harling: Battelle Specific North West Laboratory (private communication); and J. Appl. Phys. (to be published)
5.62 J.M.Carpenter: Nucl. Instrum. Methods **145**, 91 (1977)
5.63 T.S.Baer, J.N.Anno: J. Appl. Phys. **43**, 2453 (1972)
5.64 M.A.Kirk, T.H.Blewitt, A.C.Klank, T.L.Scott, R.Malewicki: J. Nucl. Mater. **53**, 179 (1974)
5.65 M.A.Kirk, R.A.Conner, D.G.Wozniak, L.R.Greenwood, R.L.Malewicki, R.R.Heinrich: Phys. Rev. B**19**, 87 (1979)
5.66 M.Kaminsky, J.H.Peavey, S.K.Das: Phys. Rev. Lett. **32**, 599 (1974)
5.67 M.Kaminsky, S.K.Das: J. Nucl. Mater. **53**, 162 (1974)
5.68 O.K.Harling, M.T.Thomas, R.L.Brodzinsky, L.A.Rancitelli: Phys. Rev. Lett. **34**, 1340 (1975)
5.69 O.K.Harling, M.T.Thomas, R.L.Brodzinsky, L.A.Rancitelli: J. Appl. Phys. **48**, 4315 (1977)
5.70 M.D.Rogers: J. Nucl. Mater. **12**, 332 (1964)
5.71 R.Behrisch, J. Roth, J.Bohdansky, A.P.Martinelli, B.Schweer, D.Rusbüldt, E.Hintz: J. Nucl. Mater. **93/94**, 645 (1980)
5.72 K.L.Merkle, W.Jäger: Philos. Mag. **44**, 741 (1981) and private communication
5.73 M.Kaminsky, S.K.Das: J. Nucl. Mater. **60**, 111 (1976)
5.74 M.Koedam: Physica **25**, 747 (1959)
5.75 R.S.Nelson, M.W.Thompson: Proc. R. Soc. London **259**, 458 (1961)
5.76 W.O.Hofer: Radiat. Eff. **21**, 141 (1974)
5.77 G.Nilsson: J. Nucl. Mater. **20**, 215 and 231 (1966)
5.78 J.Biersack, D.Fink, P.Mertens: J. Nucl. Mater. **53**, 194 (1974)
5.79 H.H.Andersen: J. Nucl. Mater. **76/77**, 190 (1978)
5.80 J.P.Biersack, E.Santner, R.Neubert, J.Ney: J. Nucl. Mater. **63**, 443 (1976)

5.81 J.P. Biersack, W. Kaczerowski, J. Ney, B.K.H. Rahim, A. Riccato, G.R. Thracker, H. Uecker: J. Nucl. Mater. **76/77**, 640 (1978)
5.82 H. Uecker, A. Riccato, G.R. Thacker, J. Ney, J.P. Biersack: J. Nucl. Mater. **93/94**, 670 (1980) J. Biersack: Private communication
5.83 A. Riccato: Technische Universität Berlin, Habilitationsschrift (1983) private communication
5.84 J.B. Niday: Phys. Rev. **121**, 1471 (1961)
5.85 M.A. Kirk, R.A. Conner: Proc. Intern. Conf. on Fundam. Aspects of Radiation Damage, ed. by M.T. Robinson, F.W. Young, Jr., USERDA Oak Ridge (1976)
5.86 S. Yosim, T.H. Davies: J. Phys. Chem. **56**, 599 (1952)
5.87 R.R. Coltman, C.E. Klabunde, J.K. Redman: Phys. Rev. **156**, 715 (1967)
5.88 J. Roth: Diplomathesis, Technische Universität München (1970)
5.89 R.R. Coltman, C.E. Klabunde, J.K. Redman, A.L. Southern: Radiat. Eff. **16**, 25 (1972)
5.90 R.I. Garber, G.P. Dolya, V.M. Kolyada, A.A. Modlin, A.I. Fedorenko: JETP Lett. **7**, 296 (1968)
5.92 L.H. Jenkins, T.S. Noggle, R.E. Reed, M.J. Saltmarsh, G.J. Smith: Appl. Phys. Lett. **26**, 426 (1975)
5.93 L.H. Jenkins, G.J. Smith, J.F. Wendelken, M.J. Saltmarsh: J. Nucl. Mater. **63**, 438 (1976)
5.94 F.S. Laptev, B.V. Ershler: Sov. At. Energy **1**, 513 (1956)
5.95 M.D. Rogers, J. Adam: J. Nucl. Instrum. **6**, 182 (1962)
5.96 R. Gregg, T.A. Tombrello: Radiat. Eff. **35**, 243 (1978)
5.97 K. Thiel, H. Külzer, W. Herr: Nuclear Tracks **4**, 19 (1980)
5.98 K. Thiel, U. Saßmannshausen, H. Külzer, W. Herr: Radiat. Eff. **64**, 83 (1982)
5.99 W.-K. Chu, J.W. Mayer, M.A. Nicolet: *Backscattering Spectrometry* (Academic Press, New York 1978)
5.100 *Secondary Ion Mass Spectrometry*, SIMS II, eds. by A. Benninghoven, C.A. Evans, Jr., R.A. Powell, R. Shimizu, H.A. Storms (Springer, Berlin, Heidelberg, New York 1979)
5.101 *Secondary Ion Mass Spectrometry*, SIMS III, eds. by A. Benninghoven, J. Giber, J. Làszló, M. Riedel, H.W. Werner (Springer, Berlin, Heidelberg, New York 1982)
5.102 A.W. Czanderna (ed.): *Methods of Surface Analysis*, Vol. 1 (Elsevier, Amsterdam 1975)
5.103 L. Reimer, G. Pfefferkorn: *Rasterelektronenmikroskopie* (Springer, Berlin, Heidelberg, New York 1977)
5.104 Ph. Staib, G. Staudenmaier: Proc. 7th Intern. Vac. Congr. and 3rd Intern. Conf. Solid Surfaces, eds. by R. Dobrozemsky, F. Rüdenauer, F.P. Viehböck, A. Breth (Vienna 1977) 2355
5.105 R.G. Meisenheimer: J. Nucl. Mater. **63**, 429 (1976)
5.106 K. Keller, R.V. Lee, Jr.: J. Appl. Phys. **37**, 1890 (1966)
5.107 M. Kaminsky: Plasmaphys. and Controlled Nuclear Fusion Research, Proc. Intern. Conf. IAEA, Wien **2**, 287 (1975)
5.108 R.I. Garber, V.S. Karasev, V.M. Kohjada, A.I. Fedorenko: Sov. At. Energy **28**, 510 (1970)
5.109 M. Kaminsky, S. Das: J. Nucl. Mater. **66**, 333 (1977)
5.110 T.S. Noggle, J.O. Stiegler: J. Appl. Phys. **33**, 1726 (1962)
5.111 P.B. Price, R.M. Walker: J. Appl. Phys. **33**, 3400, 3407 (1962)
5.112 A.D. Whapham, M.J. Makin: Philos. Mag. **7**, 1441 (1962)
5.113 R.L. Fleischer, P.B. Price, R.M. Walker: *Nuclear Tracks in Solids* (University of California Press, Los Angeles 1975)
5.114 M.D. Rogers: J. Nucl. Mater. **15**, 65 (1965)
5.115 M.D. Rogers: J. Nucl. Mater. **16**, 298 (1965)
5.116 T.H. Blewitt, M.A. Kirk, D.E. Busch, A.C. Klank, T.L. Scott: J. Nucl. Mater. **53**, 189 (1974)
5.117 S.F. Mughabghab, D.I. Garter: "Neutron Cross Sections", Vols. I and II, BNL 325 3rd ed., National Neutron Cross Section Center, Brookhaven, National Laboratory Associated Universities (1973)
5.118 *Sputtering by Particle Bombardment* III, ed. by R. Behrisch, Topics Appl. Phys. (Springer, Berlin, Heidelberg, New York) (to be published)
5.119 J.P. Biersack, D. Fink: In *Atomic Collisions in Solids*, Vol. 2, eds. by S. Datz, B.R. Appleton, C.D. Moak (Plenum Press, New York 1974) p. 737; *Application of Ion Beams to Metals*, eds. by S.T. Picraux, E.P. EerNisse, F.L. Vook (Plenum Press, New York 1974) p. 307

5.120 J.H.Kinney: J. Nucl. Mater. **103/104**, 1331 (1981)
5.121 M.W.Guinan, J.H.Kinney, Lawrence Livermore National Labs.: J. Nucl. Mater. **108/109**, 95 (1982) private communication (1981)
5.122 K.Nishizawa, Y.Morita, Y.Sensui, T.Matsuura: J. Nucl. Sci. Technol. **16**, 684 (1979)
5.123 T.Matsuura, Y.Sensui, K.Nishizawa, Y.Morita, M.Shinagawa: *Proc. Symp. on Sputtering* (SOS), eds. by P.Varga, G.Betz, F.P.Viehböck (Techn. University, Vienna 1980) p. 876
5.124 J.L.Thompson, W.W.Miller: J. Chem. Phys. **38**, 2477 (1963)
5.125 R.Gähler, J.Kalus, R.Behrisch: Nucl. Instrum. Methods **130**, 203 (1975)
5.126 D.L.Lesson, M.T.Thomas, O.K.Harling: J. Appl. Phys. **48**, 4337 (1977)
5.127 K.E.Zimen, D.Ertel: Nucleonik **4**, 17 (1962)
5.128 D.Ertel, K.E.Zimen: Nucleonik **5**, 256 (1963)
5.129 W.Wiechmann, D.Ertel, K.E.Zimen: Nucleonik **6**, 235 (1964)
5.130 R.Oberhauser, W.Wiechmann: Nucleonik **8**, 59 (1966)
5.131 J.B.Roberto, M.T.Robinson: J. Appl. Phys. **51**, 4589 (1980)
5.132 J.Csikai, P.Bornemisza, I.Hunyadi: Nucl. Instrum. Methods **24**, 227 (1963)
5.133 G.H.Kinchin, R.S.Pease: Rep. Progr. Phys. **18**, 1 (1955)
5.134 F.Seitz, J.S.Koehler: Phys. Status. Solidi **2**, 307 (1956)
5.135 R.S.Pease: Nuovo Cimento Suppl. XIII (1960)
5.136 M.W.Thompson, R.S.Nelson: Philos. Mag. **7**, 2015 (1962)
5.137 J.Lindhard, V.Nielsen, M.Scharff, P.V.Thomsen: K. Dan. Vidensk. Selsk, Mat. Fys. Medd. **33**, No. 10 (1963)
5.138 J.Lindhard, M.Scharff, H.E.Schiøtt: K. Dan. Vidensk. Selsk, Mat. Fys. Medd. **33**, No. 14 (1963)
5.139 G.Leibfried: *Bestrahlungseffekte in Festkörpern* (Teubner, Stuttgart 1965)
5.140 M.T.Robinson: Philos. Mag. **12**, 145, 741 (1965)
5.141 W.Brandt, R.Laubert: Nucl. Instrum. Methods **47**, 201 (1967)
5.142 R.S.Nelson: *The Observation of Atomic Collisions in Crystalline Solids* (North-Holland, Amsterdam 1968)
5.143 M.W.Thompson: *Defects and Radiation Damage in Metals* (University Press, Cambridge 1969)
5.144 P.Sigmund: Rev. Roum. Phys. **17**, 823, 969, 1079 (1972)
5.145 J.Leteurtre, Y.Quéré: *Irradiation Effects in Fissile Materials* (North-Holland, Amsterdam 1972)
5.146 Chr.Lehmann: *Interaction of Radiation with Solids and Elementary Defect Production* (North-Holland, Amsterdam 1977)
5.147 M.T.Robinson: In *Nucl. Fusion Reactors* (Brit. Nucl. Energy Society, London 1970) p. 364
5.148 K.B.Winterbon, P.Sigmund, J.B.Sanders: K. Dan. Vidensk. Selsk, Mat. Fys. Medd. **37**, No. 14 (1970)
5.149 M.T.Robinson: *Radiation-induced Voids in Metals*, AEC, Proc. of the 1971 Intern. Conf. at Albany, New York, AEC Symp. Ser. **26**, 397 (1972)
5.150 M.M.R.Williams: In *Progress in Nuclear Energy*, Vol. 3, eds. by M.M.R.Willians, R.Sher (Pergamon Press, London 1979) p. 1
5.151 ENDF/B, Evaluated Nuclear Data File, National Neutron Cross Section Center, Brookhaven National Laboratory, Upton, New York 11973. These files have been improved with time and several versions have been used
5.152 D.I.Garber, L.G.Strömberg, M.D.Goldberg, D.E.Cullen, V.M.May: *Angular Distributions in Neutron Induced Reactions*, Vols. I and II, National Neutron Cross Section Center, Brookhaven National Laboratory, USAEC (1970)
5.153 J.B.Roberto, M.T.Robinson: J. Nucl. Mater. **61**, 149 (1976)
5.154 J.B.Roberto, M.T.Robinson, C.Y.Fu: J. Nucl. Mater. **63**, 460 (1976)
5.155 C.Y.Fu, F.G.Perey: J. Nucl. Mater. **61**, 149 (1976)
5.156 L.B.Groshev, A.M.Demidov, V.N.Lutsenko, V.I.Pelekhov: *Atlas: Gamma-Ray Spectra from the Capture of Thermal Neutrons* (The Chief Bureau of the Use of Atomic Energy, Council of Ministers USSR 1959)

5.157 V.J.Orphan, N.C.Rasmussen, T.L.Harper: *Line and Continuum Gamma-Ray Yields from Thermal Neutron Capture in 75 Elements*, Gulf General Atomic GA-10248 (1970)
5.158 K.B.Winterbon: *Ion Implantation, Range and Energy Deposition Distributions*, Vol. 2 (IFI Plenum, New York 1975)
5.159 D.Jackson: in *Proc. Symp. on Sputtering (SOS)*, ed. by P.Varga, F.P.Viehböck (Techn. University, Vienna 1980) p. 2
5.160 J.B.Marion, F.C.Young: *Nuclear Reaction Analysis, Graphes and Tables* (North-Holland, Amsterdam 1968) p. 141
5.161 D.M.Parkin, A.N.Goland: Brookhaven National Laboratory BNL 50434 (1974)
5.162 C.E.Klabunde, R.R.Coltman, Jr.: J. Nucl. Mater. **108 109**, 21 (1982)
5.163 L.R.Greenwood: Argonne National Laboratory, Argonne Il, USA (1981) private communication
5.164 M.T.Robinson, O.S.Oen: J. Nucl. Mater. **110**, 147 (1982) [In this reference $v(T)$ is the average number of displaced atoms and for the damage energy $E_1(T)$ is used]
5.165 P.Sigmund: Can. J. Phys. **46**, 731 (1968)
5.166 H.H.Andersen: Appl. Phys. Lett. **13**, 85 (1968)
5.167 H.H.Andersen: Radiat. Eff. **7**, 179 (1971)
5.168 M.T.Robinson: J. Appl. Phys. **54**, 2650 (1983)
5.169 A.M.Weinberg, E.P.Wigner: *The Physical Theory of Neutron Chain Reactions* (The University of Chicago Press 1958) pp. 715, 716, and ref. to A.Cauchy (1884)
5.170 M.V.Meghreblian, D.K.Holmes: *Reactor Analysis* (McGraw-Hill, New York 1960) pp. 375, 376, and ref. to P.A.M.Dirac (1943)
5.171 J.P.Biersack, L.G.Haggmark: Nucl. Instrum. Methods **174**, 257 (1980)
5.172 J.L.Kammerdiener: Lawrence Livermore Laboratory Report UCRL-51232 (1972)
5.173 G.M.Perey, F.G.Perey: Phys. Rev. **132**, 755 (1963)
5.174 J.Bøttiger, J.A.Davies, P.Sigmund, K.B.Winterbon: Radiat. Eff. **11**, 69 (1971)
5.175 J.Bøttiger, K.B.Winterbon: Radiat. Eff. **20**, 65 (1973)
5.176 M.T.Robinson: J. Nucl. Mater. **53**, 201 (1974)
5.177 M.Guinan: J. Nucl. Mater. **53**, 171 (1974)
5.178 J.E.Robinson, B.S.Yarlagadda, R.A.Sacks: J. Nucl. Mater. **63**, 432 (1976)
5.178a Y.Hayashiuchi, Y.Kitazoe, T.Sekiya, Y.Yamamura: J. Nucl. Mater. **71**, 181 (1977)
5.179 R.F.Herzog, W.P.Poschenrieder, F.G.Satkiewicz: Radiat. Eff. **18**, 199 (1973)
5.180 G.Staudenmaier: Radiat. Eff. **13**, 87 (1972); **18**, 181 (1973)
5.181 J.Schou, W.O.Hofer: Appl. Surf. Sci. **10**, 383 (1982)
5.182 J.F.K.Huber (ed.): *Proc. 9th Int. Mass Spectrometry Conference*, Intern. J. Mass Spectrom. Ion Phys. (1982)
5.183 M.W.Thompson: *Le Bombardement Ionique*, ed. by J.J.Trillard (Centre Nat. Tech. Sci., Belevue 1961)
5.184 P.Sigmund: Appl. Phys. Lett. **25**, 169 (1974); **27**, 52 (1975)
5.185 R.Kelly: Radiat. Eff. **32**, 91 (1977)
5.186 P.Sigmund, C.Claussen: J. Appl. Phys. **52**, 990 (1981)
5.187 M.Szymonski, A.E.de Vries: Phys. Lett. **63**A, 359 (1977)
5.188 K.L.Merkle, P.P.Pronko: J. Nucl. Mater. **53**, 231 (1974)
5.189 H.L.Bay, H.H.Andersen, W.O.Hofer, O.Nielsen: Nucl. Instrum. Methods **133**, 301 (1976)
5.190 H.H.Andersen, H.L.Bay: Radiat. Eff. **19**, 139 (1974)
5.191 H.H.Andersen, H.L.Bay: J. Appl. Phys. **45**, 953 (1974); **46**, 2416 (1975)
5.192 D.A.Thompson, S.S.Johar: Appl. Phys. Lett. **34**, 342 (1979)
5.193 D.A.Thompson, S.S.Johar: Nucl. Instrum. Methods **170**, 281 (1980)
5.194 J.B.Mitchell, J.A.Davies, L.M.Howie, R.S.Walker, K.Winterbon, G.Foti: In *Ion Implantation in Semiconductors*, ed. by S.Namba (Plenum Press, New York 1975)
5.195 D.A.Thompson, R.S.Walker: Radiat. Eff. **36**, 91 (1978)
5.196 K.L.Merkle: In 35th Annual Proc. *Electron Microscopy Society of America*, ed. by G.W.Vailey (Claiters Publ. Div., Baton Rouge, La 1977) p. 36
5.197 W.Jaeger, K.L.Merkle: *Proc. of the 9th Intern. Conf. on Electr. Microscopy*, Toronto; the Electron Microscopy Society of Canada **1**, 378 (1978)

6. Heavy Ion Sputtering Induced Surface Topography Development

George Carter, Boris Navinšek, and James L. Whitton

With 20 Figures

The generation of surface features during heavy ion sputtering is outlined. In the first part of this chapter a brief historical review and definition of the forms of feature development is given. The recently appreciated well-defined experimental conditions necessary to obtain reproducible and definitive data are then summarised. The associated surface topography development with defect production at the surface and in the bulk are then described and it is shown how initially small features can become enlarged and elaborated by increasing ion bombardment. Effects occurring between grains and within grains of polycrystalline substrates are discussed with particular reference to depressed, protuberant and repetitive structures. Finally, a first-order theory of erosion is outlined and compared with available experimental data.

6.1 Historical Review

In preceding chapters detailed consideration has been given to the theory of sputtering and to experimental studies of sputtering phenomena. The great majority of theoretical estimates assume a well-defined and atomically unperturbed surface whilst the most reliable experimental data for comparison with theory have been obtained with relatively perturbed surfaces. Since the initial observation of a sputtering process by *Grove* [6.1] in 1852, it has become increasingly recognised that the phenomenon causes both atomic scale and larger scale morphological changes to surfaces. Such morphological changes are not only of fundamental physical interest but can be of profound technical importance and can create severe problems if not fully understood or controlled. A specific example where undesirable morphological changes can create interpretational difficulties is the application of sputtering to effect layer removal in solids and impurity profiling, a topic discussed in a forthcoming Topics volume. A clear understanding of morphology development is also necessary to fully exploit the technique of ion beam milling or micromachining and in predicting first wall behaviour in fusion reactors.

Surface topographic changes result in *all* ion solid interaction situations, the degree to which they are *observable*, however, depends upon the detailed parameters of the solid and its environment and the incident ion flux. The macroscopic effect of sputtering is one of target erosion and, were the target amorphous and uniform and the ion flux uniform, one could expect a resulting uniform erosion of the material with an initially plane surface remaining macroscopically plane during the ion bombardment.

However, since targets are very seldom amorphous or uniform but frequently crystalline, an initially flat surface does not remain flat during bombardment but develops many irregular features due, in part, to differences in sputtering yield from grains of different orientations, in part to grain boundaries and to other existing defects and to bombardment induced defects in the solid.

Such topographic effects may be regarded as "*intrinsic*" since they arise from interactions of the ion-solid system alone. There are, however, "*extrinsic*" effects which arise from perturbations of the ion solid interactions by the presence of impurities, contaminants and inclusions both on the solid surface and within the bulk. It will be a major purpose of this chapter to attempt a clear separation of these intrinsic and extrinsic effects and to review, in detail, those investigations where intrinsic processes dominate and from which clearer understanding of the ion-solid interaction may emerge.

These intrinsic effects may be of "int*er*granular" form and arise from differences in sputtering yield from grains of different orientation and grain boundaries or of "int*ra*granular" origin arising from existing defects or bombardment induced defects in single grains of the solid.

In this chapter, emphasis will be put on surface topography caused by heavy ion bombardment. In this case, the ranges of the ions in the solid are relatively small while damage production and sputtering is relatively large so that the surface topography which develops is dominated by erosion. For light ion bombardment, the ion ranges in the solid are larger while sputtering is much smaller and high concentrations of the implanted gas can accumulate in the near surface layer. The surface topography developed under such conditions is in the form of blistering, flaking and spongy structures and will be dealt with in Chap. 7 by *Scherzer.*

The earliest observation of well-defined features on solid surfaces due to ion bombardment was made in 1942 by *Güntherschulze* and *Tollmien* [6.2]. They concluded from the angular dependence of the optical reflectivity that conical protuberances had formed on many metals. Since then there have been many and various such observations and the confusion existing in this field of study is mainly due to many investigators making such observations of effects created on material of unknown quality, by impure ion beams, in ill-defined atmospheres. An example of the complex topography developed on such an initially poorly defined surface is shown in Fig. 6.1 which illustrates features revealed after $5 \times 10^{19}\,\mathrm{cm}^{-2}$ 40 keV Ar^+ ion bombardment of a fine-grained polycrystalline Cu surface. It is quite clear that a detailed understanding of fundamental processes is unlikely to result from observations of this nature.

The initial techniques for studying surface topographic changes employed direct visual observations, since sputtering often induces changes in optical reflection (Rayleigh scattering). These observations were later supplemented by

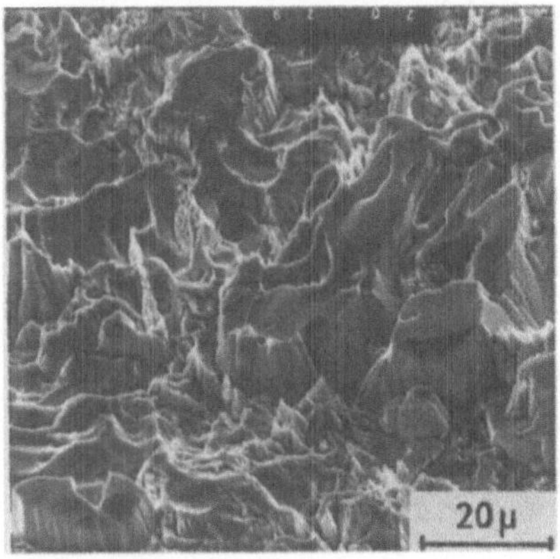

Fig. 6.1. Structures developed on a mechanically polished fine grained polycrystalline Cu substrate by $5 \times 10^{19}\,\mathrm{cm}^{-2}$ 40 keV Ar^+ ion bombardment

optical microscopy but still many of the features were beyond the resolution of these microscopes. The use of replica techniques in transmission electron microscopy (TEM) brought orders of magnitude improvement in resolution but the many artefacts associated with one or two-stage replica preparation allowed the forms of surface irregularities to be misinterpreted. The advent of the scanning electron microscope with its much larger depth of field, although having poorer lateral resolution than the transmission electron microscope, marked the beginning of a new series of investigations into bombardment induced surface structures and a new appreciation of the complexities. One of the most significant advances was made by *Stewart* [6.3] who first bombarded targets directly in the chamber of a scanning electron microscope (SEM).

Studies employing such high resolution techniques have helped establish that the important criteria dictating the nature and form of surface features are concerned with both the ion beam and the solid and their interaction and environment. Thus, ion species, mass, energy, dose and dose rate and incident angle relative to the target surface play their separate and combined roles. Additionally, target material, structure, orientation and temperature, the presence of impurities on the surface or as inclusions in the material and point and extended defects, either native to the solid or induced by irradiation, may all be important in determining the nature and extent of topographical features developed by energetic heavy ion bombardments. Such investigations have revealed the fascinating variety of surface features which can be developed on ion bombarded surfaces, including etch pits, furrows, ridges, facets, hillocks, plateaux, cones and pyramids.

This topic has been included in several books and review articles, e.g., by *Wehner* [6.4], *Behrisch* [6.5], *Kaminsky* [6.6], *Carter* and *Colligon* [6.7], *Navinšek* [6.8], *Wilson* [6.9], *McCracken* [6.10], *Townsend* et al. [6.11], and *Yurasova* [6.12] and *Trillat* [6.13]. Topographic studies have also been mentioned in some review articles on surface damage [6.14, 15], ion implantation [6.16, 17] and sputtering of thin films [6.18].

In all these works examples of surface features were collected and displayed and only little attempt was made to systematise the different observed features. A first attempt at such a classification was recently made by *Carter* et al. [6.19] whilst most of the other recent systematic work was concerned with specific artefacts such as cone formation [6.20, 21], the theory of development of surface morphology on amorphous solids [6.22] and dependence of sputter-etch pit form on crystal orientation [6.23]. Systematic investigations of polycrystalline and single crystal substrates by *Whitton* et al. [6.24] and *Hauffe* [6.25, 26] have further emphasised the complexity of the problem. It is hoped in this chapter to sort the chaff from the wheat and examine what, to our knowledge, are the most reasonable and well-substantiated conclusions that can be drawn from these spot observations and from the very few systematic studies made on this subject.

6.2 Experimental Conditions

Much of the confusion in the results reported on this subject is due to experiments being conducted with targets of unknown quality, purity, orientation, etc., unknown quality of the target environment and the ion beam. It is thus appropriate to give an account of the philosophy of the mechanics of surface preparation, and experimental practice.

6.2.1 Target Preparation

Most of the polycrystalline metals used in studies of surface topography are obtained in the form of as-rolled material. It thus follows that during the rolling/deformation process, considerably damage in the form of, e.g., dislocations is introduced, as are impurities from the rollers and in almost every case, the foils have a preferred orientation with the long axis of grains parallel to the rolling direction. The density of this damage is higher at and near the surface. Although total removal of the bulk damage may be difficult or impossible to achieve, removal of the near surface damage and the surface or nearsurface impurities is essential if systematic studies of ion bombardment induced effects are to be attempted.

The most straightforward and well-proven method for this is that of standard metallographic practice. In this, successive layers are removed by mechanical abrasion, the first layer removed being thicker than the roller-induced damage, the second layer thicker than the damage introduced during removal of the first layer and so on. The thickness of each layer may be decided by the size of polishing medium and time of polishing. The number of layers to be removed depends on the material but a procedure adaptable to most common materials is as follows: begin with 220 SiC grit, followed by 400 and then 600 and for the next stage use diamond paste (successively 6, 3, 1, 0.25 μm) before chemical or electropolishing. This sequence applies equally well to polycrystalline rolled foils, polycrystals cut from bulk metal and single crystals, whether cut by fine saw or by spark machining. Alternatively, a full sequence of careful chemical polishing may be undertaken [6.27] which may be more appropriate for single crystal targets than for polycrystals. Information concerning the depth of damage introduced into various materials may be found in the book by *Samuels* [6.28] and a comparison and assessment of the various methods of cutting, abrading, polishing and examining the end product has been made by *Whitton* [6.29]. The standard reference for chemical and electropolishing procedures is that of *Tegart* [6.30].

The above procedure, when done carefully, ensures a reasonably flat, specular surface, free of the roller induced damage and having only a normal air-formed thermal oxide. Generally, these polishing procedures are not performed carefully enough, when at all, so the specimen should be examined by a variety of techniques to deduce the nature and extent of remnant damage.

Thermal annealing treatment can help in removing residual damage but this should be done only in good vacuum conditions, otherwise adsorption of different gases can take place. Other problems which may result from annealing are thermal facetting of the surface and, in the case of nonelemental solids, a preferential loss of one component resulting in a nonstoichiometric surface layer. It is possible, of course, to go to greater lengths to ensure damage and contaminant free surfaces, by for example, chemical vapour corrosion or by using low energy heavy ion bombardment to remove the surface oxide (in ultrahigh vacuum) followed by thermal annealing to remove the damage caused by the ion bombardment. However, since (almost) all of the results to be reported here have been done in poor to high (but not ultrahigh) vacuum, we consider here real surfaces, i.e., the surface one expects on following the treatments outlined earlier on specimens finally exposed to air before the sputtering/topographical studies begin.

In order to remove surface contaminants, simple procedures can be used such as washing first with distilled water then with Analar grade ethyl alcohol which should dry by evaporation. Ultrasonic cleaning should *never* be used as this results in the introduction of high concentrations of vacancies into the metal which in turn cause high local stresses and even gross deformation.

6.2.2 Target Environment

The mounting of specimens, in a suitable target chamber capable of maintaining a pressure of $\lesssim 10^{-6}$ mm Hg, should be made in a way that does not introduce stresses into the specimen as the sputtering yield from stressed regions may be much higher than from nonstressed zones. The target chamber must have a cold trap otherwise hydrocarbons from pump oil can be recoil-implanted into the specimen by the bombarding ions as can other residual

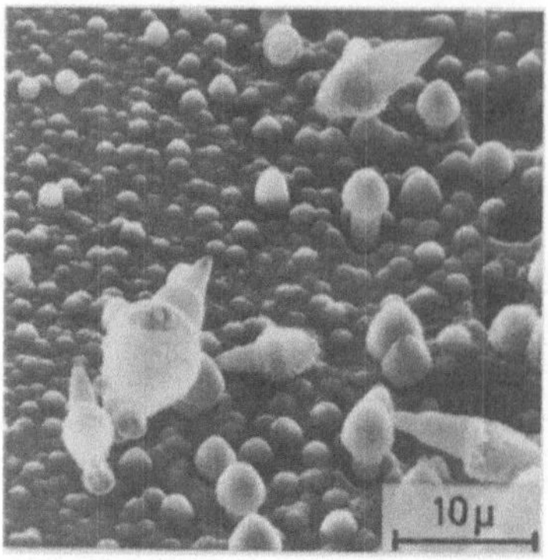

Fig. 6.2. Structures developed on hydrocarbon contaminated stainless steel by 10^{19} cm^{-2} 40 keV Ar^+ ion bombardment

contaminants. This is generally true only when oil diffusion pumps are used and problems may be diminished if turbomolecular, getter and sputter ion pumps are employed, provided that any mechanical pumps in the system are themselves well trapped. A possible result of this process is shown in Fig. 6.2 where a most unusual ejection of discrete volumes of material was observed on an argon ion bombarded stainless steel surface. This results from polymerised hydrocarbons incorporated in the steel, being ductile, swelling as a result of argon gas pressure build up, and in some cases being ejected. This phenomenon disappeared when a cold trap was used in the system but with identical bombardment conditions. Control of target temperature is important since this parameter may be critical in determining the forms of morphology development. Heated or refrigerated target stages may be employed with careful temperature measurement. It is inadvisable to employ high beam current densities to maintain "clean" surface conditions since this may lead to undesirable heating of the target.

6.2.3 Ion Beam Characteristics

The final consideration is for the ion beam. The most important condition, whenever systematic or comparative experiments are being made, is that the beam must be mass analysed. Different components in the ion beam possess different sputtering yields and it is unthinkable that any sensible interpretation of results can be attempted if mass analysis is not employed. Similarly, the beam should be stabilised and monoenergetic, since sputtering yield is known to change with energy. A beam, with local spatial and temporal variations in beam density, results in different rates of sputtering and surface decontamination, hence surface discontinuities, across the bombarded area. The beam current must be maintained constant during the course of an experiment since dose rate may be a sensitive parameter in the development of surface topography. Thus, under conditions to be elaborated on later, the surface of a copper crystal can become covered with pyramids when the beam current is $>100\,\mu A\,cm^{-2}$ but, with the same total dose with a beam current of $<100\,\mu A\,cm^{-2}$, only occasional pyramids are produced.

Finally, since sputtering yield changes with angle of incidence to the surface, the ion beam should be well collimated but, at the same time, collimators should be some distance from the target to reduce, or avoid completely, the possibility of the aperture material being sputtered on the target.

While it is true that all polycrystalline and single crystal materials will undergo some form of change in surface topography as a result of ion bombardment, the effect can be minimised by judicious choice of bombardment environment. When reactive gases, such as nitrogen and oxygen, are bled in to the normal vacuum environment, a marked reduction in surface roughness of the ion bombarded specimen is observed. *Bernheim* and *Slodzian* [6.31] have demonstrated this by bombarding aluminium with argon ions in a background oxygen pressure of 5×10^{-5} Torr. A marked reduction in surface roughness was observed compared to the effect caused by a similar bombardment before the admission of oxygen. The reduction in roughness is attributed to the formation of an amorphous oxide layer on the aluminium surface with the subsequent reduction in rate of nonuniform sputtering because of lack of orientation effects. It should, however, be noted that changes in sputtering yield occur due to the presence of oxide layers.

6.3 Defect Production and the Stages of Development of Surface Topography

A topography will almost *inevitably* develop upon heavy ion irradiated surfaces, even under the more carefully controlled conditions specified in the preceeding section. The process of ion bombardment induced sputtering creates at least atomic scale discontinuities at the surface, and such effects may be observed for individual ions and for fluences up to the order of 10^{16} ions cm^{-2}, depending upon ion and substrate parameters. As ion fluence increases above this level and the density of atomic scale discontinuities increases, the features become microscopically observable and of size in the range 100–10,000 Å. At fluences of 10^{17} ions cm^{-2} when, presumably, sufficient bombardment induced defects are built up within the crystal, local variations in sputtering yield occur resulting in major changes in surface topography, as evidenced by the development of etch pits, cones or pyramids and ripples. Further, some grains sputter at a faster rate than others resulting in a striking 3-dimensional mosaic [6.25]. At very large ion fluences, i.e., above the order of 10^{19}–10^{20} ions cm^{-2}, features assume macroscopically observable dimensions of order of submillimetric size.

Except for very low incident energies, ions will penetrate into the solid and may dissolve individually in the solid or form precipitates and clusters. In the case of heavy ions, penetration ranges are small and sputtering yields relatively large, thus inhibiting the buildup of high impurity concentrations. However, defect generation rates are expected to be large and it is anticipated that effects associated with defect generation will predominate over those associated with implant accumulation.

The defects will be generated at the surface and in the bulk of the solid by displacement processes. If temperatures are sufficiently high, such defects may migrate and anneal during the time scale of an experiment but even under such circumstances there will be a continuous dynamic competition between generation and annihilation of atomic scale topography. The defect migration may also result in annihilation at fixed sinks such as surfaces, grain boundaries and dislocations and recombination with other defect types (e.g., vacancy – interstitial recombination), but may also be able to form clusters such as dislocations and voids. Under conditions of very high energy deposition density which would result from bombardment of heavy targets with heavy, relatively low energy ions, the displacement process would be highly concentrated over localised volumes (i.e., energy spikes would occur) and lead to dynamic amorphisation or phase change of such volumes. For certain solids at low enough temperatures, notably semiconductors at room temperature, such amorphised volumes are known to be stable against recrystallisation. Thus, if the ion dose is increased, the near-surface region of the solid may be gradually transformed from the crystalline to the amorphous stage [6.32–34]. Ion impact may also cause the reverse process to occur and create crystalline regions near the surface of initially amorphous solids. The process of ion bombardment is thus perceived to be able to create, at least dynamically, a quite imperfect parasurface region in the irradiated solid. If such imperfections result in changes in either or both projectile stopping power and atom binding energies to the surface, then, as described in [Ref. 6.35, Chap. 2] local or general variations in sputtering yield will occur, thus providing the necessary initial conditions for differential sputtering processes and topography development.

Defect and disorder production processes operate from the lowest possible ion doses and are generally observable at much lower doses than those at which surface morphological changes are observed by the majority of experimental techniques. Detailed discussion of the generation of defects and their agglomerates is outside the scope of the present chapter but all of the characteristic disorder parameters outlined above have been studied directly by transmission electron microscopy and electron diffraction or inferred from measurements of electron spin resonance, light ion Rutherford backscattering, ion trapping, post bombardment gas evolution spectrometry and physical property changes of solids (e.g., electrical resistance, work function, optical transmittance and reflectance, etc.). Summaries of such studies may be found in the books of *Carter* and *Colligon* [6.7], *Thompson* [6.33], *Nelson* [6.14], *Dearnaley* et al. [6.17], *Mayer* et al. [6.34] and *Carter* and *Grant* [6.36] and an example of some defect structures as observed by TEM [6.37, 56, 123] are shown in Fig. 6.3.

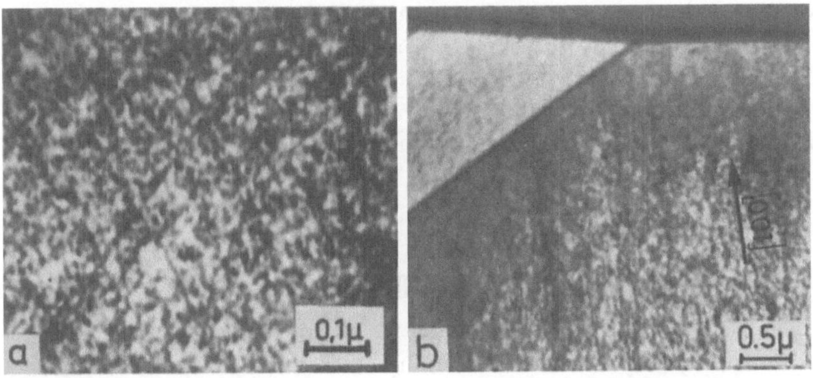

Fig. 6.3a, b. Dislocation structures generated by ion bombardment and observed by TEM. (**a**) 4 keV Ar^+ on W, 5×10^{17} ions cm^{-2} [6.123], (**b**) 80 keV Xe^+ on Cu, $>10^{16}$ ions cm^{-2} [6.37, 56] (note the dislocation alignment)

The defect production and annihilation process would generally attain a quasidynamic equilibrium between production, conversion and disappearance mechanisms in the absence of surface erosion at ion fluence in the range 10^{13}–10^{17} ions cm^{-2} for heavy ions. The majority of surface features are generally too small to be amenable to direct observation until doses above these levels. However, the act of surface atomic removal will ensure that surface and subsurface damaging processes continue to occur throughout continuing irradiation since the macroscopic surface is being constantly removed. Thus, a dynamic correlation of both disordering and surface morphological events may very well occur and has indeed been beautifully displayed in on-line transmission electron microscopic studies of both bulk and surface disordering processes using television display and recording techniques by *Nelson* and *Mazey* [6.37]. Such dynamic techniques are undoubtedly the most likely to eventually succeed in producing a clearer understanding of the fundamental processes responsible for the detailed development of surface features. Several laboratories have set up or are planning similar dynamic on-line studies largely using SEM instruments but few studies of this nature have been published. The study by *Nelson* and *Mazey* [6.37] only highlights and discusses some of the more clearly revealed features and potential mechanisms. In this section, mainly off-line bombardment condition data will be summarised in the order of phenomena observed as ion dose is increased.

In the following section, a brief discussion of the near atomic (< 100 Å) scale topographic studies will be given. A more detailed presentation of microscopic (100–10,000 Å) scale topographic studies will be given in Sect. 6.5 followed in Sect. 6.6 by a discussion of more macroscopic effects (> 10,000 Å). This latter section will also summarise some of the theoretical approaches to the prediction of surface morphological changes which arise from, for example, variations of sputtering yield $Y(\theta)$ with incidence angle θ of the ion flux to the surface normal as discussed in previous chapters, or variations in the spatial intensity of the ion current density J. Although these variations are also important to microscopic processes, they are most properly applicable to an uncontaminated surface only when the solid has achieved some degree of atomic randomisation. This will generally occur at higher ion doses or where macroscopic variations of the target surface parameters (orientation) dictate further macroscopic development.

6.4 Atomic Scale Topography ($\lesssim 100$ Å)

There are currently very few "direct" techniques available for the observation of atomic scale surface topography. Field Electron Emission Microscopic studies of ion bombarded W by *Vernickel* [6.38] have revealed the general atomic scale roughening of the surface but this technique does not allow individual atom identification. In fact, only the technique of Field Ion Microscopy can be truly said to resolve surface structure on the atomic scale and even with this method, only relatively few studies by *Brandon* and *Bowden* [6.39], *Müller* and *Sinha* [6.40, 41], *Hudson* and *Ralph* [6.42], *Gregov* and *Lawson* [6.43], *Walls* et al. [6.44], and *Current* and *Seidman* [6.45] involving refractory metal targets have been reported. A problem with this technique is that many surface crystal planes each of small area are automatically and simultaneously exposed to any form of ion "beam" irradiation. Thus, although studies of differential effects on different planes are potentially possible, few atoms occupy each plane so that sputter ejection may occur from several surface planar orientations simultaneously whilst the incident ion beam impacts upon different planes under different incidence conditions. The transition from undamaged to completely damaged surface planes occurs rapidly with low ion doses [6.44] but, if care is exercised, individual surface damage events may be followed. The major qualitative conclusion of such studies is that ion irradiation causes the creation of surface and bulk vacancies, near-surface interstitials and also leads to the generation of dislocations which can intersect the surface. Some detailed row by row and plane by plane examinations of sputter erosion of atoms have already been followed with interesting results. Thus, *Hudson* and *Ralph* [6.42] observed that about 1 in 10^3 sputtering events produced by 100 keV Xe^+ irradiation of Ir created surface craters of dimensions equivalent to more than 10^3 atomic volumes, whilst *Current* and *Seidman* [6.45] noted that the vacancy density and spatial distribution in near-surface displacement cascades generated by 30 keV Cu^+ bombardment of W were very similar to cascades generated in the bulk of the specimen. It is suggested that such studies could be very valuable in understanding the early stages of preferential erosion.

Other techniques have also been employed to indicate the generation of atomic scale morphology such as ion scattering spectrometry [6.46] and low energy electron diffraction [6.47–49] which generally responds to the average topography over a large area ($\sim 1\ \text{mm}^2$, for example). In the former type of study it has been shown that heavy inert gas ion irradiation of Cu and Ni single crystals leads to the generation of surface steps of atomic dimensions although the precise height and spatial extent of such steps is unable to be determined. It has been suggested that both ledges or island plateaux raised above the "*mean*" surface and depressed pits below the mean surface can be produced. LEED studies of semiconductor (Ge) [6.47] targets have revealed considerable, and indeed, almost total apparent randomisation of the surface as a result of low energy ($\lesssim 1$ keV) inert gas ion irradiation to comparatively small doses (10^{15} cm^{-2}) and apparent changes in the surface orientations of Ta [6.48] and W [6.49] single crystals where facetting was believed to occur. In both ion scattering and LEED studies, however, it was found that irradiation at elevated target temperatures could completely inhibit the generation of surface disorder and morphology change, presumably because of enhanced surface atomic mobility and defect annihilation. Even less direct measurements, which indicate that surface morphology can be perturbed on the atomic scale by heavy ion irradiation, include observations of changes in reactive gas chemisorption and compound formation [6.50] where bonding configurations are geometrically perturbed by surface defect structures.

Other techniques such as very high resolution transmission and scanning electron microscopy are potentially capable of revealing changes in surface topography on the tens of atomic unit scale but, thus far, have not been employed to investigate sputter etching phenomena although, as will be noted in the following section, have been successfully exploited on a larger scale.

6.5 Microscopic Scale Topography (100–10,000 Å)

The most direct and useful technique for observing ion bombardment induced changes in surface topography on a microscopic scale, i.e., from around 100 Å to micron size, is that of scanning electron microscopy (SEM). In particular, the large depth of field allows the simultaneous observation of many micron sized features, so an overall impression of even quite rough surfaces can be obtained very quickly. Modern scanning electron microscopes, even table top models, are capable of resolving features of 70 Å and the more specialised super SEM has a resolution of about 20 Å but no reports of the detailed application of these instruments for the surface topography are yet available.

In scanning electron microscopy pictures, particular care must be exercised in selecting the specimen tilt angle with respect to the beam and the detector. Figure 6.4 shows the very striking change in appearance of a single pyramid on the surface of argon ion bombarded copper when viewed (a) at 40° to the electron beam, and (b) normal to the electron beam. Such apparent differences can easily lead to misinterpretation of bombardment induced features. Thus, it is very possible that interpretative errors of visible images may have accounted

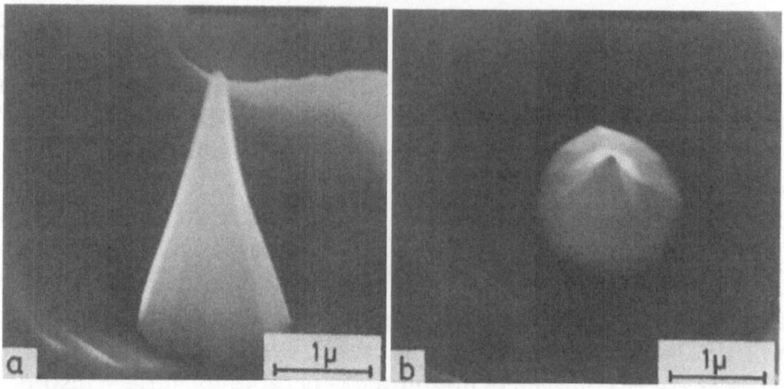

Fig. 6.4a, b. A single pyramid developed upon Cu by 40 keV Ar^+ ion bombardment as viewed by scanning electron microscopy with (**a**) viewing at 40° to the macroscopic Cu surface, (**b**) viewing normally to the macroscopic Cu surface

in part for the postulation during the past decade of different models [6.12, 51, 52] for the formation of cones or pyramids on heavy ion bombarded surfaces.

There are two interesting and opposite effects observed when examining ion bombarded crystals as a function of fluence. Between fluences of 10^{16}–10^{17} ions cm^{-2}, initially somewhat rough surfaces become smooth; it is only later at $>10^{17}$ ions cm^{-2} when, presumably, sufficient bombardment induced defects are built up within the crystal, that local variations in sputtering yield occur resulting in major changes in surface topography. It is well known that the stress fields around defects can lead to lower surface binding energies, hence higher sputtering yield. On the other hand, when the initial surface is relatively poorly prepared, initial roughness may be amplified to give rise to ridges, cones, trenches and pits originating from scratch marks and other major perturbations. These features are not unlike those developed with better clarity and resolution on an initially well-polished sample.

On well defined surfaces, as fluence is further increased, so more and more features become visible. First there is a marked effect of grain orientation, some grains sputtering at a faster rate than others, resulting in a striking 3-dimensional mosaic [6.25].

A simultaneous effect, to that of the intergranular height elaboration, is the development of intragranular features such as sputter etch pits which are very similar in appearance, if not of identical habit, to pits observed on chemical etching of polycrystalline material. These pits, originating presumably from intrinsic defects, have very well-defined shapes which are identical within each grain but differ in shape from grain to grain. On clean crystals it is often observed that these pits are precursors of some cones or pyramids. An increase in fluence of bombarding ions usually results in the appearance of more small pits and the already existing pits grow in size eventually overlapping with others but, until that happens, always retaining their characteristic shape.

At around the same fluence as pits, so too are finer scale ripples, often of well-defined periodicity and orientation. Again there is a strong orientation dependence, ripples being observed on some grains but not on others. Ripples are also observed within etch pits with orientations bearing apparently simple geometrical relationships to the orientations of the best developed pit edges.

A further increase in fluence to $>10^{19}$ argon ions cm^{-2} leads to the development of new small pits, enlargement of existing pits and the formation of cones and pyramids at the pit edges and corners which are then observed to stand independently as the pit enlarges. Continued bombardment results in an accumulation of pits and pyramids until the entire bombarded area may be covered with these features.

6.5.1 Feature Classification

The surface structures developed by ion bombardment can generally arise from both intrinsic and extrinsic processes.

Extrinsic processes result from perturbations of the sputtering process by local impurities, inclusions and inhomogeneities in the solid.

Intrinsic processes may be defined as those which result from the variations of sputtering yield with crystal orientation and from grain boundaries and dislocations present in the solid before irradiation as well as dislocations induced by irradiation and precipitates of the irradiating ion species.

In the following section, attention will be devoted entirely to a review of the development of those features which, it is believed, result from processes intrinsic to the ion-solid interaction. Two levels of classification will be employed. The first distinguishes between int*er*granular effects which

(i) result from the presence of different crystallites;
(ii) occur on a single crystal or on one single crystallite in polycrystalline substrates.

The second distinguishes different types of int*ra*granular features such as

(i) etch pits,
(ii) pyramids, and
(iii) ripples.

In addition to these intrinsic phenomena, however, there is quite conclusive evidence that extrinsic processes can also occur, often simultaneously with the intrinsic effects. Thus, cone and plateau development arising from the initial shadowing of the substrate by a contaminant which first allows the formation of a pillar then transforms to a cone, which itself may then be eroded away, is a very well documented process [6.3, 9, 53].

Most of the studies made by *Wehner* and co-workers [6.54] have led to the suggestion that the presence of an impurity of lower sputtering yield than the matrix is the only process which allows cone forms to develop by forming surface discontinuities, which then develop into conical habit as a result of the variation of sputtering yield with ion flux incidence angle.

Yurasova and co-workers [6.12, 20] have also suggested that surface impurity is essential for the initiation of all surface cones and that cone growth is due to surface atom migration. When a balance is achieved between growth by surface migration and erosion due to sputtering it is claimed that cones or pyramids stable against sputtering are observed. When the surface migration becomes greater than the sputtering rate, it is suggested that the sometimes observed [6.20] formation of extended or twisted bent cones can become possible.

Hauffe [6.26] also invokes the presence of impurities in the ion beam or in the residual gas in the target chamber as providing suitable conditions for the initiation of cone formation. He then proposed, however, that cones further develop by a redeposition of sputtering material and the effects of local differences in ion reflection. Unfortunately, this proposed mechanism appears to be supported by only isolated features but the careful work by *Hauffe* [6.26] does indicate the need for an extended investigation of his proposed mechanism.

It is interesting to note that while both *Yurasova* and co-workers [6.20] and *Hauffe* [6.26] invoke the need for an impurity of one sort or another for the initiation of cones, the form of the features they observed is, nonetheless, influenced by the crystallography of the single crystals they use. As an example, *Yurasova* observed pyramids having 8-fold symmetry, quite unlike the cones of circular cross section reported by *Wehner* and co-workers [6.51, 54], on a small grained polycrystalline material. It is, therefore, quite possible that some pyramid formation studies which have been ascribed to extrinsic processes are indeed the result of intrinsic differential erosion processes occurring in the radiation damaged solid. The fact that cone covered surfaces can undoubtedly be formed by contaminant protection irrespective of grain orientation, unlike the clean system discussed earlier, shows that the covering of surface by an impurity (often introduced by "seeding" from nearby apertures) can completely obliterate the effects of crystal orientation. Thus, while providing a mechanism for producing cone covered surfaces, it leaves little or no opportunity for studying the physics of intrinsic cone or pyramid production processes. It should be noted, however, that this type of impurity-induced cone is easily distinguished from the more regularly shaped pyramids produced in "clean" systems.

The cones observed by *Wehner* et al. [6.51, 54] bear a close similarity to those observed on semiconductors, e.g., on Si and on Ge, InSb, and GaAs [6.9]. In these cases, the ion bombardment used to produce a surface feature is more than sufficient to amorphise the surface, and near surface, to a depth of a few tens of nanometers; therefore, crystallographic effects might be expected to be absent.

There will, of course, be cases where, for example, the injected species accumulate to form precipitates of lower yield then the surrounding matrix with the ensuing possibility of impurity protection cone generation processes [6.54]. However, it will be assumed, although there is little direct proof of this (excepting studies by *Mazey* et al. [6.56] and *Nelson* and *Mazey* [6.37] of the substantial similarity of features generated by Xe^+, Pb^+, and Cu^+ ion irradiation of Cu), that implant incorporation processes are of secondary importance only. It will also be assumed, generally with justification, that effects due to, or enhanced by, interaction of the ambient environment are also absent in the studies reviewed. Finally, major attention will be focussed upon elemental substrates, although some relevant investigations with compounds will be mentioned, since species segregation effects undoubtedly often play an important role in feature development.

6.5.2 Int*er*granular Effects

Intergranular effects describe those processes which occur as a result of differential macroscopic sputtering yields between neighbouring grains because of the dependence of sputtering yield upon crystal orientation and give rise to the creation of grains of different elevation across a surface. The relative elevation difference between neighbouring grains after sputtering depends in detail upon ion incidence direction, since the $Y(\theta)$ function will also depend upon crystal orientation. Grain boundaries, being essentially low-angle dislocation networks, sputter preferentially and at different rates from the grains and this effect may be enhanced by segregation of impurities to grain boundaries. This differential orientation effect on erosion rates quickly establishes grains at different levels and the cliff-like features may form ridge-like protuberances when they intersect deepening etch pits and these ridges between different grains can provide sites amenable to the cutting out and later isolation of conical features. It is often found that the boundaries between grains adopt well-defined orientations relative to the initial plane surfaces of the grains and the incident ion flux direction (e.g., Fig. 6.5c). In some, but not all cases, well-defined trenches or depressions occur in the region of sputtered grain boundaries. *Hauffe* [6.25, 26] has conducted the most detailed studies of these intergranular effects but as yet no definitive conclusions as to the detailed elaboration and motion of boundaries can be made. Specific crystalline orientations have not been measured or controlled, nor have grain boundary elevation angles or planar areas been followed dynamically as a function of ion dose. This area provides definite possibilities for future experimental study since the boundaries are usually well defined and represent the type of edge structures discussed in Sect. 6.6.1.

6.5.3 Int*ra*granular Effects

In intragranular effects, the concern is with processes which occur within specific crystalline grains and are, therefore, identical to processes which occur on single crystal specimens. In this level of classification, two further categories may be defined. Firstly, features may be recognised as either depressed or protuberant relative to their mean surroundings and secondly, features may be isolated or repetitive, the latter frequently giving the appearance of periodic structures. In the following discussions, depressed structures will be considered first, since the sputtering process is one of erosion and a depression may be regarded as resulting from a locally enhanced erosion process. A protuberance generally results from the effects of enhanced erosion around the protuberance. Although repetitive structures may frequently result from high densities of overlapped isolated structures (as will be noted subsequently), they may also be independent of such features while they certainly contain both apparent elevated and depressed elements. Thus, such repetitive structures will be discussed separately although their relationship to isolated structures will be illustrated.

a) Depressed Structures (Etch Pits)

The most clearly defined depressed structures elaborated by sputtering are etch pits. Such structures were, in fact, reported in early literature on cathodic sputtering and were employed for the delineation of dislocations at surfaces [6.4, 57] as an alternative method to chemical etching. Such studies did not materially aid the understanding of the process of such etch pit development nor were they exhaustive enough to provide a general description of pit morphologies on different crystal surfaces under various irradiation conditions. Generally speaking, however, the morphological habit of pits was shown to be significantly different from pits developed by chemical dissolution of identically prepared surfaces and for this reason it was suggested, without real proof, that ionic erosion selectively elaborates emergent screw dislocations at surfaces to produce etch pits. Despite considerable recent studies, the fundamental processes of pit development are little better understood, although the evidence still suggests that pits do develop at local regions in the surfaces associated with some form of dislocation and at which atomic binding energies are reduced, thereby enhancing sputtering. Whether such dislocations are initially present in the solid before irradiation or are formed by radiation induced defect production and agglomeration is still open to debate. As illustrated in Fig. 6.5a, etch pits developed by 40 keV Ar^+ ions on a long grain of a Cu polycrystal [6.19, 21, 58] form in a collinear array parallel to the direction of original mechanical drawing, which is known to produce a similar alignment of dislocations. On the other hand, *Tanovic* et al. [6.58] have also reported that increased pit densities occur with higher ion dose rates which would result in higher defect generation rates and probably increased dislocation generation rates. There is no doubt that pit morphology is intimately correlated with crystal type and orientation and irradiation conditions. Thus, Fig. 6.5b illustrates the differing pit morphologies developed by 4 keV Ar^+ bombardment on two neighbouring grains of a W polycrystal [6.8]. In both cases, the pit

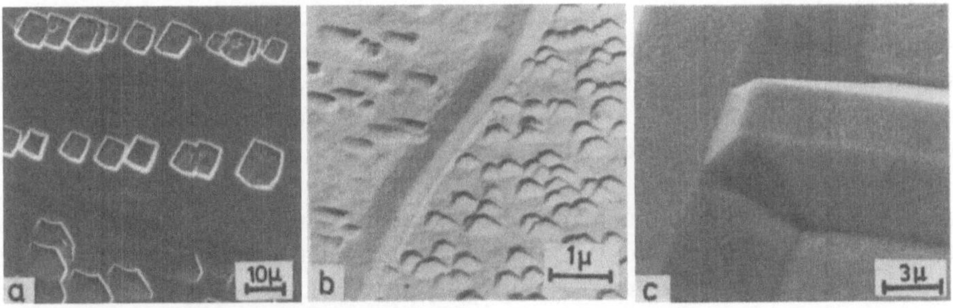

Fig. 6.5a–c. Etch pit structures developed by ion bombardment: (**a**) 40 keV Ar^+ on Cu [6.21], (**b**) 4 keV Ar^+ on W [6.8], (**c**) 10 keV Ar^+ on Cu [6.25]

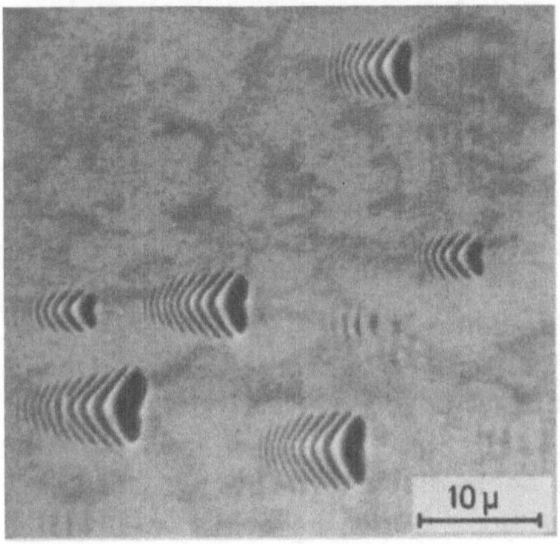

Fig. 6.6. The structure of etch pit trains developed by 40 keV Ar^+ ion bombardment of Si. 10^{20} ions cm^{-2} at 45° to surface normal [6.60]

morphology is clearly grain orientation dependent and this is generally true of most irradiated solids although an exception has been reported in the case of Si. In this case, *Nelson* and *Mazey* [6.37] and *Carter* et al. [6.19, 60] indicated that both Xe^+ and Ar^+ bombardment of different Si crystal surfaces with the ion beam incident at various angles between 30° and 60° to the surface normal, and at doses of the order 10^{19}–10^{20} cm^{-2}, led to identical etch pit patterns independent of the crystal orientation. An example of the rather interesting pit structure obtained is shown in Fig. 6.6 which shows that each feature is composed of a train of etch pits (which have the appearrance of a set of circular ripples). Further studies by *Carter* et al. [6.19] did reveal, however, that single etch pits could also be formed under apparently identical irradiation conditions, and that again the morphology of such pits was apparently crystal orientation independent. The reasons for these differing observations is not yet fully understood although incidence angle relative to the surface normal is certainly important [6.61]. The independence of crystal orientation upon pit morphology on Si may be associated with the pits being formed in an essentially amorphous near-surface region which is produced by irradiation. Thus, similar results to Si have been reported [6.61, 62] with initially amorphous solids such as Vitreous carbon and soda glass.

Very few other studies of pit morphology have been undertaken excepting some by *Tanovic* and *Perovic* [6.23] and *Whitton* et al. [6.52, 58], both with Cu substrates. In the former study [6.23], single crystals were employed and the pit morphology studied by optical microscopy and the orientation of the planar faces of the pit inferred from studies of the angular distribution of sputtered atoms. It was shown that on (110), (111), and (100) Cu surfaces, pits of two, three and four-fold symmetry, respectively, were formed and possessed shapes of rectangular or trapezoidal, trigonal and hexagonal inverted pyramids and square or hexagonal inverted pyramids, respectively. It was also concluded that the bounding faces of all these pits were probably parallel to (100) or (110) planes. In *Whitton* et al. [6.52, 58] pit studies on both single crystal and polycrystalline Cu samples, many complex shapes were observed. In general, however, for a given surface orientation, pits were of symmetric shape, the order of symmetry depending upon the surface orientation. The number of distinctive faces in each pit was generally comparatively low, however, ($\leqq 8$) and some faces were clearly mirror reflections of others. The pit symmetry and low face numbers suggest that the pit shapes are controlled by crystal symmetry requirements and that, as indicated by *Tanovic* and *Perovic* [6.23], faces correspond to low index planes. It is, therefore, probable that following pit initiation at a dislocation, the geometry of the pit is dictated by requirements for all faces to sputter at equivalent rates and thus equivalent or near-equivalent planes become selectively exposed. Since dislocations will probably align on quite specific crystal planes and will be different relative to each surface orientation, this will dictate to what extent local sputtering yields are perturbed and thus the probability of pit initiation and their growth rate. The studies of *Whitton* et al. [6.52, 58, 63] clearly revealed that pit size and density increased with increasing ion dose, presumably as either further native dislocations approached the receding macroscopic surface during erosion or other dislocations were generated by defect agglomeration. Pit densities were often found to be larger near grain boundaries than towards grain centres, again either due to higher initial or irradiation induced dislocation densities near the boundaries. The more macroscopic effect of such higher pit densities near boundaries, however, is to produce pit overlap and the formation of terraced structures on some grains near a boundary. This frequently leads to an overall grain curvature from the boundary into a grain and may well be responsible for the trenches often perceived, e.g., *Hauffe's* studies [6.25, 26] at some grain boundaries. There also appear to be correlations between pit form and density and the macroscopic erosion rate of different grains, as illustrated by Fig. 6.7a, which indicate different elevations of neighbouring grains together with different pit habit, size and density. The macroscopic erosion rate thus appears, quite reasonably, to be associated with the presence in, or ability of, a grain to sustain dislocations which enhance the sputtering yield.

Finally, in addition to growth in pit dimensions with increasing ion fluence as shown in the sequence [6.58] of Fig. 6.8a, b it is notable that pits themselves are frequently decorated with more internal microscopic features. These include

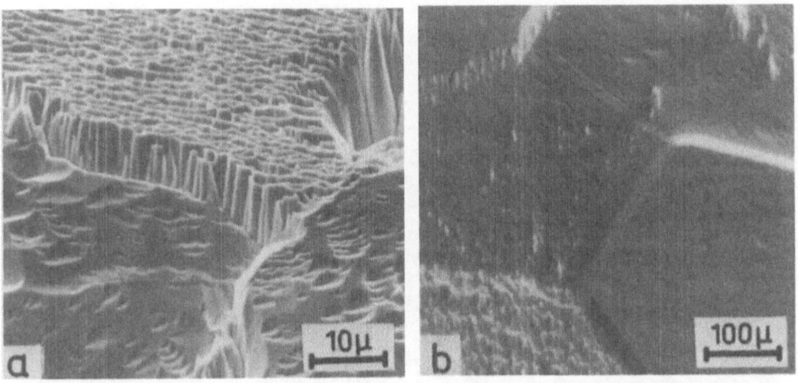

Fig. 6.7

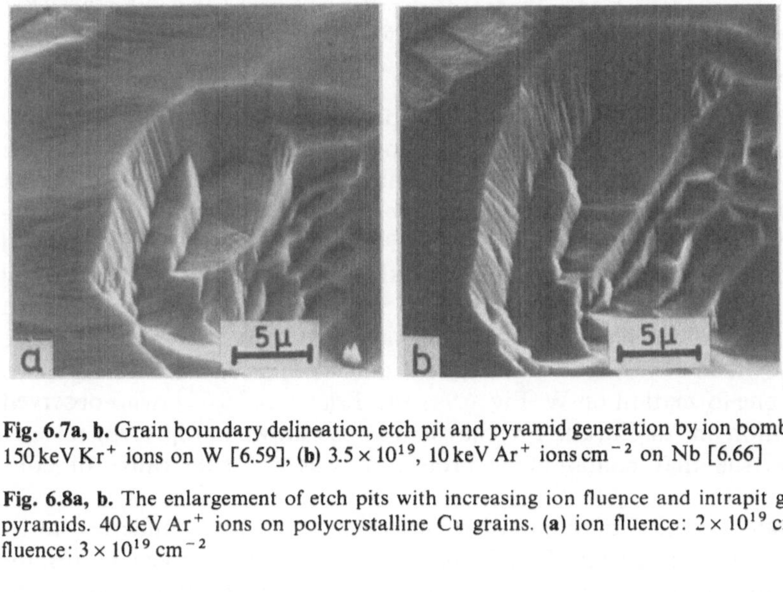

Fig. 6.8

Fig. 6.7a, b. Grain boundary delineation, etch pit and pyramid generation by ion bombardment. (**a**) 150 keV Kr^+ ions on W [6.59], (**b**) 3.5×10^{19}, 10 keV Ar^+ ions cm^{-2} on Nb [6.66]

Fig. 6.8a, b. The enlargement of etch pits with increasing ion fluence and intrapit generation of pyramids. 40 keV Ar^+ ions on polycrystalline Cu grains. (**a**) ion fluence: 2×10^{19} cm^{-2}, (**b**) ion fluence: 3×10^{19} cm^{-2}

striations on the boundaries both intersecting the macroscopic surface and upon the lower surfaces. Such striations appear as ripples on these lower surfaces and may be composed of well-defined facet planes, as also they may upon pit-pit bounding surfaces. This micro-facetting may, therefore, represent a further attempt for the detailed surface morphology to adapt to uniform and optimum sputtering erosion conditions.

b) Protuberant Structures (Pyramids)

In addition to intergranular effects which, due to differential macroscopic erosion rates, give rise to differently elevated grains and thus frequently the appearance of isolated plateau or ridges, intragranular protuberances may also be generated. Such features are generally of conical form although more

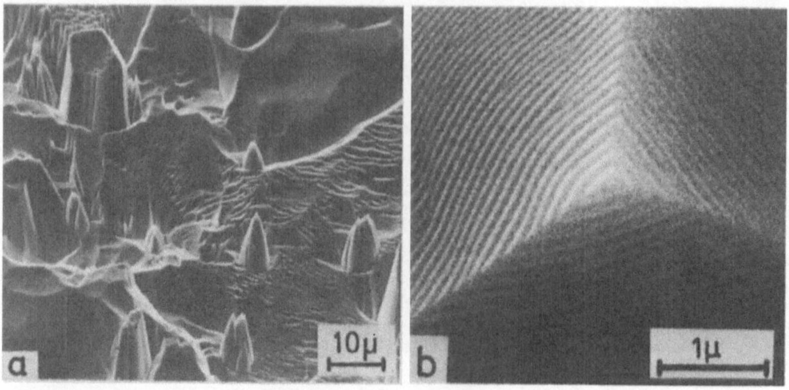

Fig. 6.9a, b. Ion bombardment induced cone and pyramid development. (**a**) 4×10^{18}, 150 keV Kr^+ ions cm^{-2} on W [6.59], (**b**) 40 keV Ar^+ ions on Pb [6.65]

detailed studies show that they frequently possess crystallographic symmetry and are thus more properly classified as pyramids, as seen already in Fig. 6.4.

Many observations of cones or pyramids have been made and earlier studies attributed their presence entirely to contaminant protection mechanisms. Some more recent studies [6.12, 20, 25, 26, 59, 64–67] with both metal and semiconductor substrates still make this association, although careful studies by *Whitton* and his colleagues [6.24, 52, 58, 68] cast doubt upon such a unique interpretation. Thus, in some of the above studies, together with those already shown in Fig. 6.1 and in Fig. 6.9 due to *Staudenmaier* [6.59] (who observed cone formation on W, Fig. 6.9a) and *Erlenwein* [6.65] (who observed cone formation on Au, Zn, and Pb; the case of Pb is shown in Fig. 6.9b), it is not necessarily true that contaminant protection was the sole cause of cone formation.

The major studies of alternative processes have been undertaken by *Whitton* and colleagues [6.23, 24, 52, 58, 63] with the following general results. Concentrating their attention upon Cu substrates, these authors observed pyramid formation at three major sources (i) from ridges formed at grain boundary-etch pit intersections, (ii) from ridges formed by neighbouring overlapping etch pits, and (iii) at corners of jogs in the boundaries of etch pits and from ripple or facet structures at the bases of etch pits [6.69]. In all cases the pyramids form from regions elevated above their immediate surroundings such as the base of a grain or an etch pit, but were never higher than the upper boundary from which they had clearly originated, i.e., the surface of a grain or a terrace within an etch pit. Such observations led these authors to conclude that pyramids resulted from an enhanced erosion in their immediate neighbourhood rather than an initially retarded erosion above them as would occur in contaminant protection.

Moreover, the form of such pyramids was always observed to possess crystallographic symmetry rather than the right circular or convoluted profile

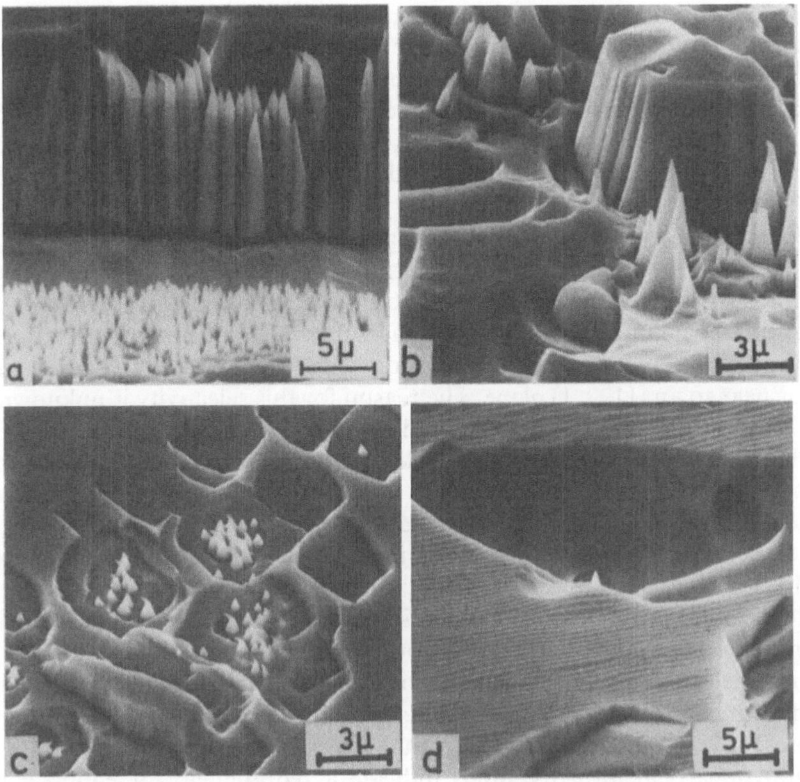

Fig. 6.10a–d. Sources of pyramid generation by 40 keV Ar^+ ion bombardment of polycrystalline Cu. (**a**) At grain boundaries (ion fluence: $2 \times 10^{19}\,cm^{-2}$, (**b**)–(**d**) within and between etch pits (ion fluence: $10^{19}\,cm^{-2}$)

usually associated with contaminant protection (cf. the earlier discussion in Sect. 6.5.1). Examples of these sources of pyramids are shown in Fig. 6.10. Thus, Fig. 6.10a indicates the emergence of pyramids from a ridge formed between a grain boundary and deepened etch pits near the boundary in a Cu grain [6.21]. Figure 6.10b indicates the development of pyramids from the boundary of an elevated grain in a fine grained Cu sample and residual pyramids surrounding this grain. Figure 6.7a illustrates the same source of pyramids in such a region as observed on W by *Staudenmaier* [6.59], while Figure 6.7b shows similar sources in polycrystalline Nb [6.66]. Figures 6.10c and 6.7b illustrate sources at interpit ridges and within etch pits, whilst further indications of pyramids within pits may be perceived in Figs. 6.5a and 6.8a.

That such pyramids are facetted and possess crystallographic structure is well illustrated in Fig. 6.4, which shows two different views of a pyramid observed on Cu by *Whitton* et al. [6.58, 63], and upon Pb by *Erlenwein* [6.65], as shown in Fig. 6.9b. This latter figure clearly indicates the finer terracing structure which can also decorate the planar facets of a pyramid.

The studies of *Whitton* and his colleagues [6.21, 24, 52, 58, 63] have all indicated that pyramid formation can occur when elevated discontinuities arise in surface perfection (e.g., ridges between boundaries and pits, pit-pit boundaries and at pit boundaries) but have not yet been able to specify in detail the necessary geometric requirement of such discontinuities. Thus, the earlier work of *Whitton* et al. [6.52, 63] clearly illustrated that whilst etch pits formed on all crystal faces of a bombarded Cu polycrystal, only pits on specific faces contained pyramids (an example is shown in Fig. 6.5a of neighbouring grains). Subsequent, more detailed studies by *Whitton* et al. [6.19, 21, 58] employing large-grained polycrystal and single crystal Cu targets showed quite definitively that pyramids formed only rarely upon surfaces for which the initial orientation was low index but readily and in profusion when the surface was oriented parallel or near to an (11, 3, 1) plane. The reason for this selectivity is unknown but observations do indicate that pyramid symmetry is usually similar to the pits in which they form, again suggesting that pyramids form in such a manner that their facets possess equivalent erosion rates. It thus appears that pyramids form when appropriate defect structures can be generated by irradiation (e.g., dislocation networks) and that these result in enhanced local erosion leading to both ridge (in boundary-pit and pit-pit intersections) and jog or corner morphologies (within pits) and that such structures are further elaborated to form pyramids. It is possible that striations observed on both grain and pit boundaries reflect a more regular dislocation network intersecting such boundaries which provide the further necessary enhanced erosion paths, or that they represent a more detailed facet structure of congruent erosion rates. Although symmetrical form pyramids indicate a requirement for equivalent facet erosion rates, it is not yet known which, if any, crystal planes are represented by such facets, nor is it known to what extent the elevation angles correspond to the requirements of erosion slowness theory predictions which will be outlined in Sect. 6.6. Preliminary studies by *Tanovic* et al. [6.58] of elevation angle for pyramids developed upon an (11, 3, 1) Cu surface by Ar^+ bombardment suggest that this is much more weakly dependent upon ion energy in the range 10–80 keV than may be anticipated from theory. Again therefore, there is a fruitful area for further study.

Three final features of pyramids are noteworthy. Firstly, when they do form, their axes are invariably aligned with the ion beam direction relative to the surface, independent of the particular crystal orientation or the symmetry of the pyramids. This again suggests that erosion slowness theory is of some importance in determining initiation and form. Secondly, experiments clearly suggest [6.64, 70, 71] that pyramids are unstable and that they are continuously generated, assume regular form and are then eroded away, being replaced by further generations of pyramids. Studies by *Witcomb* [6.72] of 18-8 stainless steel indicate that a quasi-equilibrium at least of pyramid density may be achieved during continued sputtering.

Finally, many pyramids are often observed to bend and twist and adopt convoluted rather than regular appearance, as shown in Fig. 6.11, although

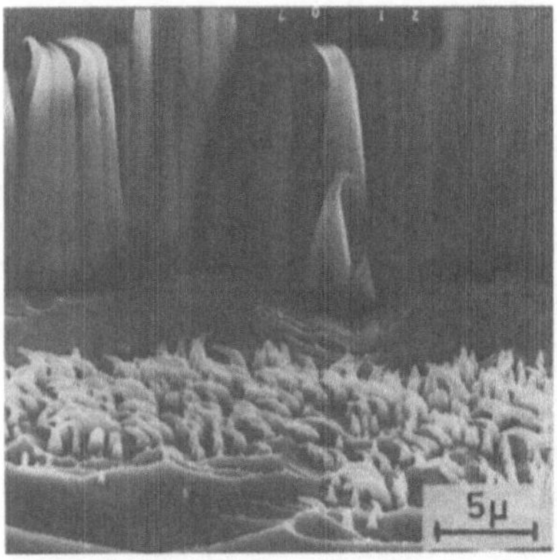

Fig. 6.11. Bent and twisted cone (pyramid) formations developed upon 40 keV Ar^+ ion bombardment polycrystalline Cu substrate. This is a different area of the same substrate shown in Fig. 6.10

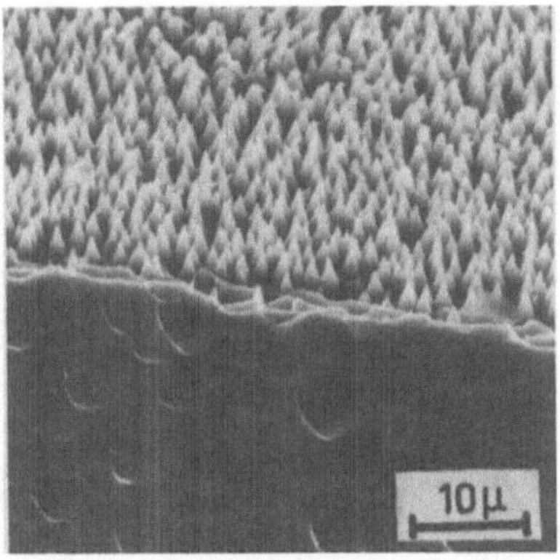

Fig. 6.12. Dense pyramid structure developed by $2 \times 10^{19}\,cm^{-2}$ 40 keV Ar^+ bombardment of (11, 3, 1) oriented grain of a copper polycrystal [6.24]

they still remain facetted. Studies by *Tanovic* et al. [6.21], *Carter* et al. [6.19], *Gvosdover* et al. [6.20] and *Yurasova* [6.12] on Cu all show these peculiar forms although there is no currently viable explanation for their occurrence.

c) Repetitive Structures (Ripples and Facetting)

Repetitive structures possess three forms. Firstly, collections of depressions, secondly, collections of protuberances and a third category where depressions and protuberances are difficult to distinguish. Since protuberance collections are quickly summarised, they are considered first.

When pyramids form in high density, they possess the appearance of an extensive family or forest. *Gvosdover* et al. [6.20], *Yurasova* [6.12], *Whitton* et al. [6.58, 63], *Tanovic* et al. [6.21], and *Carter* et al. [6.19] have reported such dense collections on Cu surfaces irradiated with Ar^+ ions, and *Whitton* et al. [6.24] showed quite conclusively that these structures form with highest probability if the irradiated surface is close to (11, 3, 1). Dense arrays of this type are illustrated in Fig. 6.12. Close inspection of such arrays indicates, generally, that the pyramids form in isolated or parallel linear sequences, an understandable form since they generally emerge from linear boundaries (as at grain boundaries), sequentially at jogs or corners of pits as these enlarge and from pit base surface facet structures, or at linear ridges between pits. If present in high density, such arrays on Cu are optically interesting since they appear red under white light illumination, presumably due to multiple reflection and selective absorption. These arrays may also be potentially useful in catalysis, corrosion, and surface tribological applications and for the absorption of radiation and charged particles.

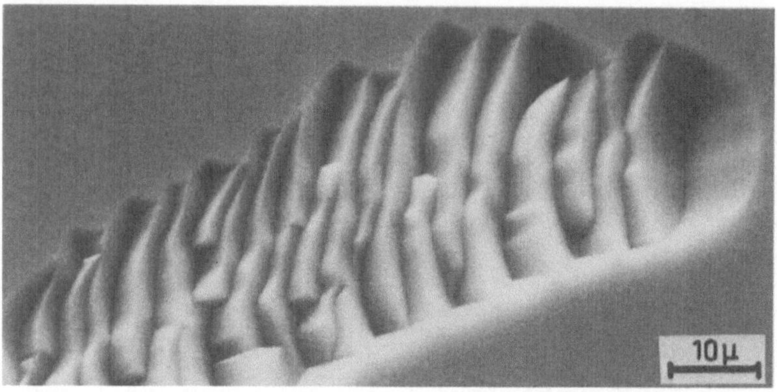

Fig. 6.13. Overlapped etch pit structure developed by $10^{19}\,cm^{-2}$ 40 keVr$^+$ Ar^+ ion bombardment of Si at 45° to the surface normal [6.19]

In the case of high densities of depressed structure, individual etch pits increase in size and number and in so doing may eliminate common boundaries to finally present a surface with only the major faces of the individual pits remaining. This overlapped pit structure then assumes the appearance of a terraced structure on a plane surface which may be corrugated, as in the example of 40 keV Ar^+ bombarded Si shown in Fig. 6.13 and a terraced grain boundary as shown in Fig. 6.7. When such an overlapped pit structure entirely covers a grain, the appearance then assumes a completely facetted structure, the third category of repetitive features. An example of this type is observable in Fig. 6.7b for Nb [6.66]. Facet structures developed completely over grains have been observed on many solids and further examples for a wide range of materials are shown in Fig. 6.14a–d. However, it is not certain whether any or all of the facetted structures displayed in these figures or observed by other investigators [6.8, 13, 73–78] are a consequence of extensive pit overlapping or originate from another process which forms regular rippled or terraced structures. Thus, it was already noted in the discussion of etch pits that these are generally decorated at a finer scale with striations or ripples whilst Fig. 6.10d clearly demonstrates the existence of a finer scale ripple morphology on the macroscopic surface of a grain outside the region of an etch pit. A number of authors [6.19, 21, 37, 52, 56, 63, 77] have reported the existence of such ripple structures, their frequent coincidence with, but on a smaller scale than, etch pits, and their differing appearance and alignment on different crystal orientations of the same solid and on different solids.

Although there is substantial data on this topic, little is comparative and only generalised conclusions may be reached. Thus, *Mazey* et al. [6.37, 56] concluded that on (110) and (310) Cu surfaces, the ripples were aligned parallel to ⟨001⟩ directions, that a crossed rippled system or cellular morphology with components parallel to $\langle 0\bar{1}1 \rangle$ and ⟨111⟩ directions formed on a (211) surface

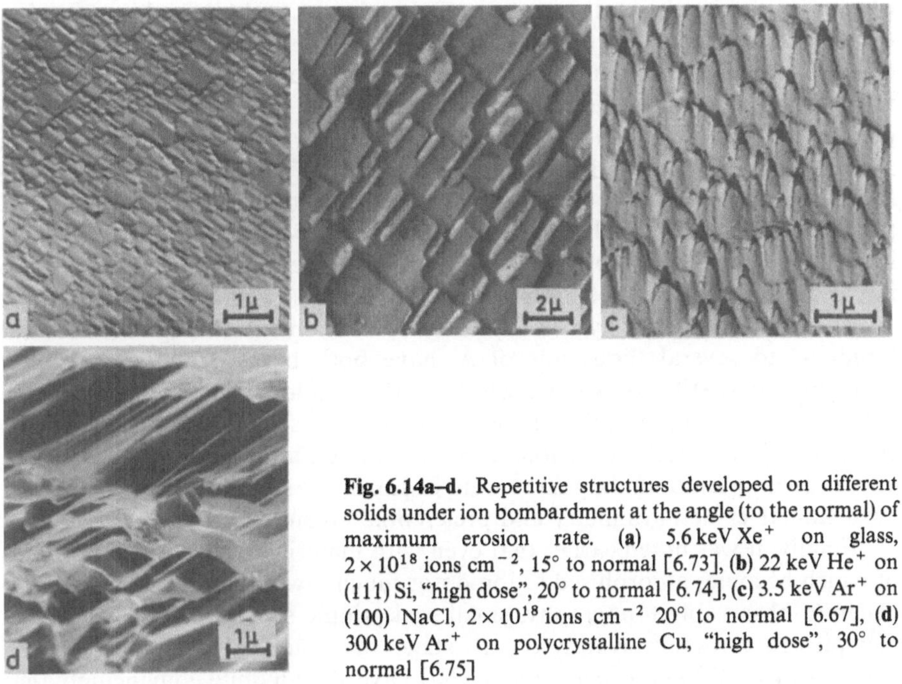

Fig. 6.14a–d. Repetitive structures developed on different solids under ion bombardment at the angle (to the normal) of maximum erosion rate. (**a**) 5.6 keV Xe^+ on glass, 2×10^{18} ions cm^{-2}, 15° to normal [6.73], (**b**) 22 keV He^+ on (111) Si, "high dose", 20° to normal [6.74], (**c**) 3.5 keV Ar^+ on (100) NaCl, 2×10^{18} ions cm^{-2} 20° to normal [6.67], (**d**) 300 keV Ar^+ on polycrystalline Cu, "high dose", 30° to normal [6.75]

and that no ripples formed on a (100) surface. Detailed transmission electron microscopic studies as indicated in Fig. 6.3b, led these authors to suggest that the ripple alignment was parallel to underlying aligned dislocation arrays in the solid formed by defect production and agglomeration, and that the structure of the rippled surface was such as to expose (100) facet planes to the irradiation flux. Similar ripple or facet alignments parallel to ⟨001⟩ directions (and ⟨110⟩ directions) were observed in studies of Ar^+ ion irradiation of several different Ag surface orientations by *Haymann* and *Waldburger* [6.76], while *Hermanne* and *Art* [6.77] observed what may be either overlapped square section pyramids, etch pits, or ripple-like facet structures aligned parallel to ⟨110⟩ directions. In studies of Ar^+ bombarded (100) oriented Cu at different temperature and ion incidence angles, *Elich* et al. (79) observed both isolated pit structures and well-developed facet structures depending upon irradiation conditions, but concluded that major facet planes which developed were parallel to (110) orientations in agreement with earlier studies by *Fluit* and *Datz* [6.80]. However, on Au surfaces, *Cunningham* and *Ng-Yelim* [6.81] suggested that (100) surfaces become facetted parallel to (114) planes, that (111) surfaces facetted to (114) and (110) and that (110) and (114) surfaces facetted parallel to the initial surface. Thus, even for the same fcc structural type of material, there are somewhat conflicting suggestions as to the final surface morphology – arising undoubtedly from different conditions under which the studies were conducted. Nevertheless, it is not unreasonable to conclude at this stage that the fine scale ripple morphology which, when observed at high resolution exhibits a facetted structure, may initiate from stable dislocation networks which locally perturb the surface to induce the necessary conditions for preferential sputtering. This association is very clearly revealed in the work of *Mazey* et al. [6.37, 56] and in a recent study by *Art* [6.82] who observed a one to one correspondence between ripples and dislocations. Which specific facet planes develop preferentially, and why, is still open to debate, however. It is also interesting to note that both the larger etch pit structures (which may develop into relatively long wavelength periodic structures) and the facetted ripple structures, which generally develop with wavelengths in the region of several hundreds to several thousands of Å, have both been attributed to defect structures. It is still not known what are the fundamental differences in the defect structures responsible for these two types of structure. It would not be surprising, however, if it were found that facet boundaries of pits and of ripples and, indeed, grain boundary faces were of similar habit since they must reflect both similar crystal symmetry and preferential erosion conditions.

Finally, it was noted earlier that even with materials such as Si, the surface of which becomes amorphised at ion doses much lower than those required to generate observable topography, small scale ripple structure and overlapped etch pit structures form readily at early stages of erosion. At later stages, *Nelson* and *Mazey* [6.37] and *Carter* et al. [6.60] have shown quite convincingly that the individual structures further overlap to produce quasi-continuous wave structures (sometimes corrugated as shown in Fig. 6.13). Quasi-periodic structures have also been observed on other semiconductors such as Ge [6.64] and amorphous solids such as glass [6.61, 62, 83], Vitreous carbon [6.61] and glass-ceramics [6.84] (as shown in Fig. 6.15a, b) and Araldite [6.61, 85]. For near-normal ion incidence, the final structure of such solids is generally one of random bumps and depressions. At larger incidence angles, the structure becomes of the transverse wave type and at nearer grazing incidence, the waves become aligned parallel with the surface projection of the ion flux. There is currently no clear understanding of these effects.

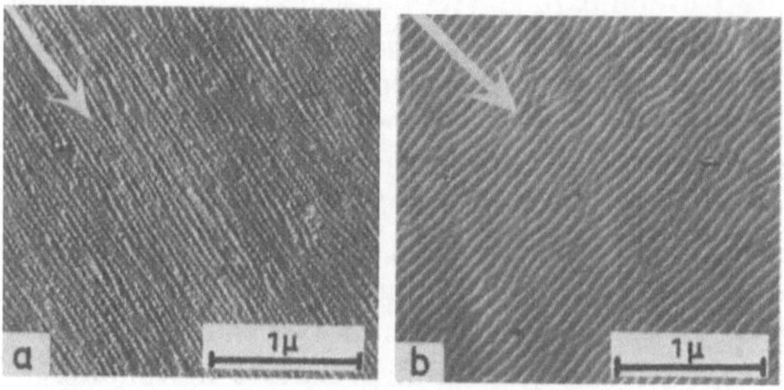

Fig. 6.15a, b. Repetitive surface structure developed by Ar^+ ion bombardment of glass [6.83] (**a**) at 80° to normal, (**b**) at 30° to normal

6.6 Macroscopic Scale Topography (>10,000 Å) and Theoretical Considerations

In the present discussion the emphasis will be upon macroscopic features in which the sputter etch effects may induce differential dimensional changes in the order of 10^{-4}–10^{-2} m. At such levels of dimensional change, theoretical models of morphology change become more meaningful and a brief review of a first-order theory of topography elaboration will therefore be given and compared with experimental results. These experimental investigations reveal the necessity of including secondary effects due to ion flux variation, sputtered atom redeposition and migration and these will also be discussed. Clearly, in this regime, surface morphology changes are readily determined by quite standard precision metrology techniques such as optical microscopy, optical interferometry and surface profilometry employing sensitive stylus displacement methods. This latter technique can nowadays also be used for assessing surface morphology changes normal to the surface with a resolution of the order of 10^{-8} m for hard materials, as illustrated in Fig. 6.16, which shows a profile measured on a crater sputtered in Si.

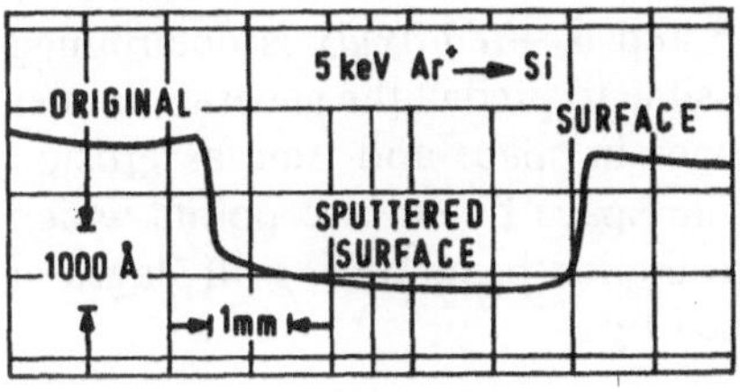

Fig. 6.16. Talystep trace of a crater produced in Si by 5 keV Ar^+ ion bombardment

6.6.1 First-Order Erosion Theory

If, for simplification, uniform irradiation with an ion flux density J in the $-0y$ direction of a three-dimensional surface of a random isotropic solid, of which a surface section normal to the $x0y$ plane is considered as shown in Fig. 6.17, then, ignoring secondary effects such as local flux variations due to scattering from neighbouring surface elements, atomic redeposition, surface atomic transport, etc., it is readily shown that the rate of erosion or recession $(-\partial m/\partial t)$ of a surface element normal to the surface is given by

$$-\frac{\partial m}{\partial t}=\frac{JY(\theta)\cos\theta}{N}, \tag{6.1}$$

where $Y(\theta)$ is the sputtering yield (in atoms per ion) of the solid for an angle of incidence θ between the ion flux and the normal to the surface at the point considered, and N is the solid density.

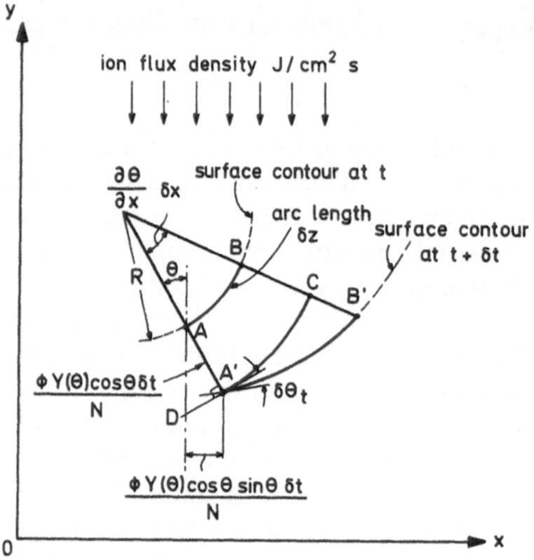

Fig. 6.17. The erosion of a surface section in the $x0y$ plane subjected to bombardment with a uniform ion flux J cm^{-2}s^{-1} in the $-0y$ direction. The motion of two neighbouring points with normals included at angles θ and $\theta+\delta\theta$ along their normals in a time δt is shown [6.86]

If the differences in rate of recession along the surface normal directions for neighbouring points of initial orientation θ and $\theta+(\partial\theta/\partial x)dx$ is determined from the relationship such as in (6.1), it is possible to predict the general manner in which the complete surface profile changes in space and time as erosion proceeds. In particular, the paths followed in space by surface points which maintain a constant orientation θ relative to the ion beam during all stages of erosion have been shown [6.86] to be defined by

$$\left|\frac{\partial\theta}{\partial t}\right|_x \Big/ \left|\frac{\partial\theta}{\partial x}\right|_t = \left|\frac{\partial x}{\partial t}\right|_\theta = v_x = \frac{J}{N}\frac{dY}{d\theta}\cos^2\theta \tag{6.2}$$

and

$$\left|\frac{\partial\theta}{\partial t}\right|_y \Big/ \left|\frac{\partial\theta}{\partial y}\right|_t = \left|\frac{\partial y}{\partial t}\right| = v_y = \frac{J}{N}\left(\frac{dY}{d\theta}\sin\theta\cos\theta - Y\right). \tag{6.3}$$

Carter et al. [6.86] recognised that these equations indicated that the surface morphology developed in a wave-like manner but *Barber* et al. [6.87] pointed out that such waves were of kinematic type and therefore amenable to detailed analysis introduced by *Lighthill* and *Whitham* [6.88, 89] and successfully applied by *Frank* [6.90, 91] to studies of the chemical dissolution of solids. Thus, (6.2, 3) define completely the "characteristics" of points on a real surface which, during erosion, maintain a given, constant orientation θ. The velocity of such points is given by

$$v_\theta = (v_x^2 + v_y^2)^{1/2} \tag{6.4}$$

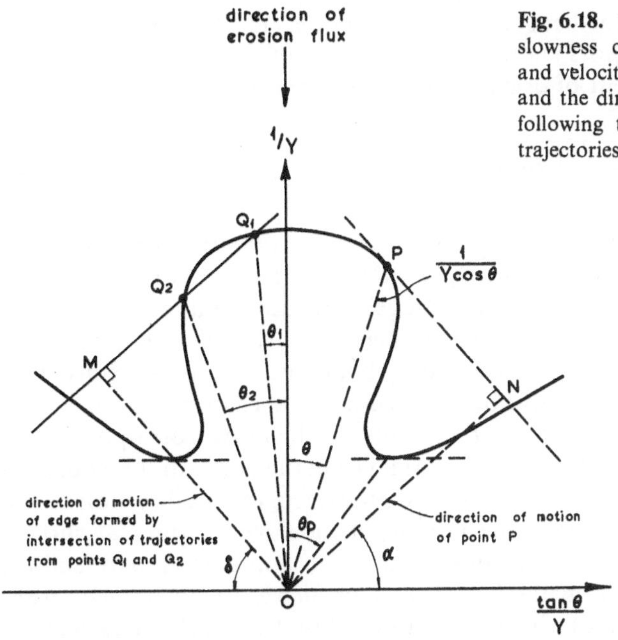

Fig. 6.18. The non-normalised erosion slowness curve depicting the direction and velocity of a point P at orientation θ and the direction and motion of an edge following the intersection of two point trajectories [6.93, 94]

and the direction of motion by

$$(\tan\alpha)_\theta = \frac{v_y}{v_x} = \frac{[dY(\theta)/d\theta]\sin\theta\cos\theta - Y(\theta)}{[dY(\theta)/d\theta]\cos^2\theta}. \tag{6.5}$$

Barber et al. [6.87] demonstrated that if for given sputtering parameters the $Y(\theta)$ function was known, then a normalised erosion slowness curve or the polar plot of $[Y(\theta)\cos\theta]^{-1}$ as a function of θ could be constructed, as shown in Fig. 6.18, which is derived from a typically observed form of $Y(\theta)$. As discussed in earlier chapters, $Y(\theta)$ generally increases from a minimum value of $Y(0)$ at $\theta=0$, initially approximately as $Y(0)\sec\theta$, to a maximum at $\theta=\theta_p$ and then declines towards zero at $\theta=\pi/2$. The importance of this erosion slowness curve is that, for a given value of θ, if a tangent is drawn to the corresponding point on the curve and the normal from the origin to this tangent constructed, then the reciprocal of the length of this normal is just Nv_θ/J and the angle α between the normal and abscissa of the erosion slowness plot is as defined by (6.5). Thus, the erosion slowness curve can be employed directly to follow, by geometrical construction, the time sequential development of a surface during sputter erosion. This is achieved for all points each with different orientations θ initially present on the real surface by deducing the distance moved $v_\theta\delta t$ in a time step δt (for given values of J and N) along the direction of the normal in Fig. 6.18, and after this time step, by reconstructing the surface and then iterating the process for further time steps. *Barber* et al. [6.87] also showed that since characteristics can cross, certain orientations can be eliminated from the surface as erosion progresses and this was further studied by *Ducommun* et al. [6.92] and *Carter* et al. [6.22, 93]. These authors showed that the time rate of change of radius of curvature $\mathcal{R}$ of an element of orientation θ was given by

$$\left|\frac{\partial \mathcal{R}}{\partial t}\right|_{\theta} = \frac{J}{N}\left(\frac{d^2 Y}{d\theta^2}\cos\theta - \frac{2dY}{d\theta}\sin\theta\right). \tag{6.6}$$

Equation (6.6) can then be used to show that for a given orientation θ and an initial radius $\mathcal{R}(\theta, 0)$, the radius can potentially be reduced to zero in real, positive time. This process is known as "edge" formation and indicates a local discontinuity in the developing surface. *Barber* et al. [6.87] showed that if such an edge formed with orientations θ_1 and θ_2 bounding the discontinuity, then the motion of the edge could also be deduced from an erosion slowness curve by constructing the chord between the points on the curve corresponding to orientations θ_1 and θ_2 and erecting the normal from the origin to this chord. As in the case of the continuous curve, the reciprocal length and direction of this normal then completely specify the motion of the edge on the real surface. *Nobes* [6.94] compared the various theoretical approaches and showed that the erosion slowness method of *Barber* et al. [6.87], their own [6.86] differential surface analysis [which led to (6.2, 3)] and the earlier treatment of surface morphological changes by *Stewart* and *Thompson* [6.53] could, in fact, be rendered identical. Indeed, it was shown that the *Stewart* and *Thompson* [6.53] approach which had considered the motion of intersecting surface planes was essentially a treatment of edges and their motion. As a result of this early approach, *Stewart* and *Thompson* [6.53] had predicted that certain orientations given by $\theta = \pm\theta_p$ would form preferentially upon sputtered surfaces and this was later confirmed and extended to show that $\theta = 0$ orientations would also result in the differential and erosion slowness treatments of *Carter* et al. [6.86] and *Barber* et al. [6.87], respectively. These results are particularly important since they indicate that two-dimensional triangular forms of elevation angle $\pm\theta_p$ (semivertical angle of $\pi/2 - \theta_p$) or their three-dimensional analogue of conical forms on the one hand, or plane surface normal to the sputtering flux on the other hand, should result from erosion.

Careful geometric construction studies using the erosion slowness curve by *Barber* et al. [6.87], computational studies by *Catana* et al. [6.95], *Ishitani* et al. [6.96], and *Ducommun* et al. [6.97] and more recent analytical studies by *Carter* et al. [6.98] have revealed the validity of these predictions. But they have also shown that the conical forms possess only transient stability and erode with other protuberant and depressed features until the stable end form of the surface is planar. Thus, Fig. 6.19 illustrates the predicted sputtering of a sphere using the *Barber* et al. [6.87] erosion slowness geometric construction method and Fig. 6.20 shows [6.92] the predicted time development of initially sinusoidal surfaces of several amplitude : wavelength ratios. In both cases, triangular forms appear, at least transiently, and the elevation angles of such forms are generally found to correspond quite closely to the value of θ_p of the associated $Y(\theta)$ function. In the case

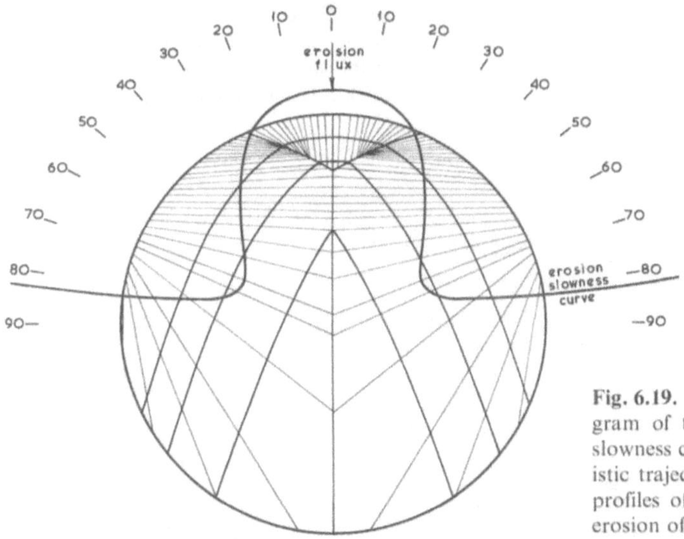

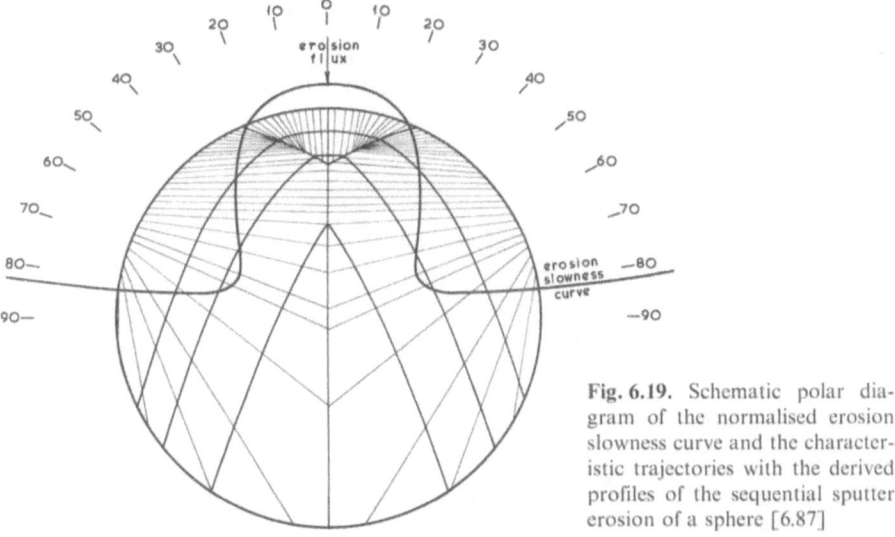

Fig. 6.19. Schematic polar diagram of the normalised erosion slowness curve and the characteristic trajectories with the derived profiles of the sequential sputter erosion of a sphere [6.87]

of a sphere, *Witcomb* [6.72] has recently noted that θ_p values are unlikely to be completely achieved because of the finite boundaries whereas, as indicated earlier by *Nobes* et al. [6.99], a hemispherical cap to an infinitely long cylinder would form such a conical end with an elevation angle of θ_p. A further feature of Fig. 6.20 is the clearly defined edge geometry which develops but finally vanishes into the plane form.

6.6.2 Comparison of Theory with Experimental Studies

The erosion theory predicts the formation of conical forms, edges and flat planes and each of these has been observed on the macroscopic sputtering scale and indeed at the microscopic level. In the following discussion, a brief comparison of experimental studies with theory will be given and some necessary developments of the simply erosion theory outlined. Comparison will be made of both macroscopic studies where theory is most applicable, but some examples of microscopic studies which are qualitatively explained by theory will also be given.

It should be noted first that erosion theory predicts cone or pyramid development only when the angle of maximum sputtering yield is present (or potentially present) within initial protuberances above the mean surface. The initial smoothing of a surface of low roughness is compatible with such a prediction whilst the subsequent generation of cone and pyramid forms at larger ion fluences when high angle surfaces form from either contaminant protection or from facet, etch pit and grain boundaries, are also predictable. If the initial surface is itself rather rough with poor polishing, then ridge and pyramidal structures may develop *ab initio* from scratch marks, etc.

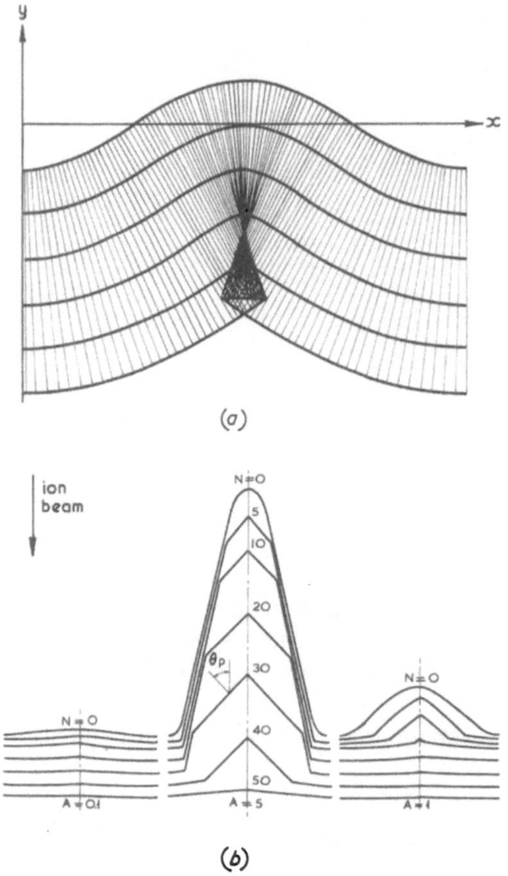

Fig. 6.20a, b. The computed erosion of initially sinusoidal functions where (**a**) $y=0.1 \sin x$ and (**b**) $y=A \sin x$ (where $A=0.1$, 5.0, and 1.0) [6.92]

Furthermore, although experimental evidence tends to confirm theoretical predictions qualitatively, there are few experimental studies at either macroscopic or microscopic scale where direct comparisons of, for example, measured cone (or pyramid) elevation angles have been made with measured angles of maximum sputtering yield. The development of the impurity protected cone generation described in Sect. 6.5.3 was first observed [6.2] and predicted by *Wehner* [6.4] on a qualitative basis and on a quantitative basis by *Stewart* and *Thompson* [6.53]. The temporal development of microscopic cones has also been studied by *Wilson* [6.9, 64], *Witcomb* [6.100], and *Hauffe* [6.26]. At the macroscopic level, conical forms were observed by *Wehner* [6.4] to result from the sputtering of small metal spheres while triangular cross sections have been produced in the sputtering of initially cylindrical silica fibres by *Meckel* et al. [6.101]. The production of edges and their subsequent motion has been studied experimentally by *Ducommun* et al. [6.92, 97] during sputtering of initially contoured Si surfaces and shown to agree closely with the predictions of erosion slowness theory, whilst there are numerous examples in the current literature [6.102–104] of ion beam machining which exhibit edge forms which have been explained in terms of erosion slowness.

The final planar polishing of surfaces has been fully described by *Barber* et al. [6.87] and is a quite conventional method of removing asperities and depressions from macroscopic surfaces. It is important to note that theory predicts this rapid polishing to a flat surface of all structures, initially depressed relative to plane surroundings. The well-documented evidence of etch pit generation and expansion can thus only be rationalised if local zones of higher sputtering yield can exist. These regions must be considered to be either, or both, native and bombardment induced defect structures.

In the preceding theoretical discussion it was assumed that no local (or general) variations in either J or $Y(\theta)$ occurred. Several authors have allowed for fluctuations in these parameters as follows:

If, for example, a cone-like feature is produced during sputtering, then ions may forward scatter from the sides of the cone and effectively enhance the local ion flux around the pedal region of the cone. This may give rise, as suggested by *Bayly* [6.105] and *Wilson* [6.9], to the depressed regions surrounding cones which are frequently observed, particularly with the impurity protection mechanisms (see, for example, Fig. 6.9a). Such a flux enhancement process may be somewhat offset by the condensation of forward sputtered material from the sides of the cone as discussed by *Bayly* [6.105] and *Wilson* [6.9, 64]. Both processes may also operate on a more generally contoured surface (e.g., the facetted surfaces at the microscopic level discussed in the preceding section) and *Hauffe* [6.26] and *Littmark* and *Hofer* [6.106] have considered theoretically the effect of such processes on the development and stability of facets, giving detailed generalised equations similar to (6.1).

Carter et al. [6.107] and *Smith* and *Walls* [6.108] have also considered the effects of a general spatial variation of the ion flux J which may result from the use of nonuniform ion beams, where J must be replaced by a variable $J(x, y)$. If this is effected, then geometric construction (or computer simulation based upon this construction) can be readily employed but with appropriate values of $J(x, y)$ used at each time step for each surface point. The general results of such a treatment are that, if initially, the ion flux impinges everywhere almost normally to a plane surface, then the sputter etch profile formed is the mirror image of the spatial ion flux variation. If, however, large angles are initially present or form from the production of a deep pit by sputtering, then the variations of $Y(\theta)$ can assume equal or greater importance than the spatial variations of $J(x, y)$, and the surface contour development no longer mirrors the flux profile but is also conditioned by erosion slowness considerations. The detailed shape of surface contours can only be determined by full knowledge of both $Y(\theta)$ and $J(x, y)$ functions and must be evaluated for each erosion case separately since *no* stable end forms ever result. Clearly this approach becomes very necessary at the macroscopic erosion level and could be important in micromachining applications. It may be noted that the contaminant induced cone generation process may be regarded as a special case of nonuniform flux conditions. In the substrate surface zone protected by the contaminant, the incident flux is effectively zero so that only the surrounding substrate sputters until the contaminant is itself removed. If, before this decontamination, the boundary between the sputtered and unsputtered zones can develop to a sufficiently high inclination angle, then following plateaux formation and inward regression of the boundary to exclude the top surface of the plateaux, cones of semivertical angle θ_p will result. If decontamination occurs before sufficiently large angled plateaux boundaries develop, the plateaux will be excluded without cone formation.

The above modifications to first-order erosion theory are equally valid at both macroscopic and microscopic scales. In this latter regime and at the atomic scale, other processes may also assume importance.

Sigmund [6.109] pointed out a difficulty of the simple erosion theory in that it assumes that sputtering is a macroscopic rather than an atomic and local process. Thus, when an ion enters a solid and creates a collision cascade, sputtering of individual atoms may occur remote from the physical point of ion impact. On average, however, there will be a reasonably well-defined spatial distribution of partial or differential sputtering yields relative to ion impact points, but the distribution will depend upon the angle of ion incidence to the surface. Effectively, sputtering yields vary locally on a surface and if, for example, the dimensions of a surface feature are similar to those of the displacement cascade within the solid, then this cascade may be unable to develop and the local sputtering yield will be perturbed. Following considerations of this type, *Sigmund* [6.109] predicted that the conical forms discussed earlier should develop more sharply than envisaged by simple $Y(\theta)$ theory, i.e., θ (elevation of the tip) $<\theta_p$ and that the pedal regions of cones should be depressed relative to the surroundings. It is difficult, from experiments with microscopic cones, to determine whether the *Sigmund* [6.109] mechanism or the reflected flux enhancement mechanism is most responsible for pedal depressions, but *Wilson* [6.64], *Yurasova* [6.12], *Lewis* et al. [6.70], and *Auciello* et al. [6.71] have presented some experimental evidence which supports the cone tip sharpening mechanism.

Other secondary processes which may limit the validity of simple erosion theory include the reorganisation of the surface by either thermal or radiation enhanced diffusion and experimental studies of such processes have been conducted at the microscopic level by *Teodorescu* et al. [6.110, 111] and *Elich* et al. [6.79]. Theoretical studies of effects associated with concurrent surface and volume atomic migration during sputtering, which are also applicable at the macroscopic level, have been reported by *Carter* et al. [6.112, 113] and suggest that such effects tend to assist more rapid surface flattening during erosion in agreement with experiment.

Defect migration and agglomeration processes have also analysed by *Hermanne* [6.114] who evaluated the conditions necessary to allow the generation of a regular dislocation network during simultaneous erosion of the surface and the nascent dislocation array. *Hermanne* [6.114] and *Nelson* and

Mazey [6.37] as indicated earlier, postulated that such dislocation networks were responsible for surface facetting processes and noted that ion energy and type, flux density and fluences and substrate orientation and temperature could all influence the development of such features. No stringent tests of these models have yet been performed, however.

In the preceding discussion, including theoretical analysis of topography evolution, it has been assumed that it is sufficient to treat only two-dimensional surface sections. More recently three-dimensional analysis, treating the surface as a moving wave, as suggested by *Carter* et al. [6.22, 94], have been undertaken. Spatially variable ion flux incident upon amorphous [6.115] or crystalline [6.116] substrates have been considered and further generalisation to include time-dependent ion flux, and species- and spatially-dependent substrate properties have been developed [6.117–119]. This latter behaviour is representative of polycrystalline substrates, dislocations in single crystal substrates and contaminated (surface or bulk) solids. The general analysis [6.117–119] also implicitly includes processes additional to erosion, such as atomic redeposition and atomic diffusion. Such analysis is shown to reduce to the two-dimensional approach outlined in the present discussion. Further details are found in [6.115–119] for the features observed experimentally on real three-dimensional solids.

It should be noted, finally, that although no suitable theory yet exists which satisfactorily predicts experimentally measured $Y(\theta)$ variations, which are of course basic to the erosion slowness modelling of surface morphological change, several authors [6.115–117] have discussed the relationship between the angle (θ_p) at the maximum yield and surface channelling or reflection processes. One of these theories with good credibility appears to be that of *Chadderton* [6.122] which distinguishes the difference between measured θ_p angles, which are conditioned both by the changing spatial development of the recoil cascade within the solid and the variation of ion reflection from the surface, and predicted planar channelling angles ψ_{min} which were associated only with ion reflection from a continuum potential.

6.7 Conclusions

This review has attempted to summarise both the major techniques employed and the most clearly definable results of studies of surface topography generated by ion bombardment on the atomic, microscopic and macroscopic scales which are generally associated with low, intermediate and high ion doses. At the atomic level, features occur by single atom removal and migration to probably produce small scale ledges, islands and troughs at surfaces. At intermediate doses, the great majority of results indicate that extended defects, either native to or radiation induced, interact with the surface to give rise to local perturbations in sputtering yield and the production of etch pits and pyramids and a denser collection of such individual features. It appears that both overlapped etch pits and repetitive microfacetting associated with dislocation arrays can produce the commonly observed facetting of crystal surfaces. Differences in etch pitting and facetting may be partially responsible for differences in macroscopic erosion yields of different crystal surfaces and thus, for the elevation relief observed between grains or polycrystalline targets. Although defect structures are probably of major importance in initiating intrinsic features, variation of sputtering yield with incidence angle on individual crystal planes are probably of similar importance in determining intrinsic feature development with heavy ion bombardment, although some evidence of blister formation is becoming available. Extrinsic effects at the microscopic level are generally associated with impurity or contaminant protection and lead to cone formation, whilst component segregation in sputtered compounds may lead to similar effects.

At high doses, variations of sputtering yield with ion incidence angle and spatial variations of ion flux density are the major parameters influencing macroscopic surface morphology and erosion slowness theory is relatively successful in predicting such surface contour developments. There are clearly major areas of uncertainty and a lack of detailed understanding at all stages of topography development and, in view of both fundamental physical interest and substantial technological importance, it is believed that further well-defined research and endeavour is worthwhile.

References

6.1 W.R.Grove: Trans. R. Soc. London **142**, 87 (1852)

6.2 A.Güntherschulze, W.V.Tollmien: Z. Phys. **119**, 685 (1942)

6.3 A.D.G.Stewart: PhD Thesis. University of Cambridge (1962)

6.4 G.K.Wehner: Adv. Electron. Electron. Phys. **7**, 239 (1955)

6.5 R.Behrisch: Erg. Exakten Naturwiss. **35**, 295 (1964)

6.6 M.Kaminsky: *Atomic and Ionic Impact Phenomena on Metal Surfaces* (Springer, Berlin, Heidelberg, New York 1965) Chap. 10, p. 142

6.7 G.Carter, J.S.Colligon: *Ion Bombardment of Solids* (Elsevier, New York 1968) Chap. 7, p. 310

6.8 B.Navinšek: In *Physics of Ionized Gases*, ed. by M.Kurepa (Institute of Physics, Belgrade 1972) p. 221
B.Navinšek: Prog. Surf. Sci. **7**, 49 (1976)

6.9 I.H.Wilson: In *Physics of Ionized Gases*, ed. by V.Vujnovic (Institute of Physics, Zagreb 1974) p. 589

6.10 G.M.McCracken: Repts. Prog. Phys. **38**, 241 (1975)

6.11 P.D.Townsend, J.C.Kelly, N.E.W.Hartley: *Ion Implantation, Sputtering, and Their Applications* (Academic Press, London 1976) Chap. 6, p. 111

6.12 V.E.Yurasova: In *Physics of Ionized Gases*, ed. by B.Navinšek, University of Ljubljana (1976) p. 544

6.13 Le Bombardement Ionique, Proc. Conf. held at Bellevue, France 1961 (ed. by J.J.Trillat, CNRS France, 1962)

6.14 R.S.Nelson: *The Observation of Atomic Collisions in Crystalline Solids* (North-Holland, Amsterdam 1968) Chap. 6, p. 206

6.15 G.Carter, J.S.Colligon, W.A.Grant: Radiat. Res. Rev. **3**, 49 (1971)
6.16 G.Dearnaley: Rpts. Prog. Phys. **32**, 405 (1969)
6.17 G.Dearnaley, J.H.Freeman, R.S.Nelson, J.Stephen (eds.): *Ion Implantation* (North-Holland, Amsterdam 1973) p. 207
6.18 E.Kay: Adv. Electron. Electron. Phys. **17**, 245 (1962)
6.19 G.Carter, M.J.Nobes, J.L.Whitton, L.Tanovic, J.S.Williams: VII Intern. Conf. on Atomic Collisions in Solids, Moscow, Sep. 1977 (Moscow State University Publishing House, 1980) Vol. 2, p. 69
6.20 R.S.Gvosdover, V.M.Efremenkova, L.B.Shelyakin, V.E.Yurasova: Radiat. Eff. **27**, 237 (1976)
6.21 L.Tanovic, J.L.Whitton, G.Carter, M.J.Nobes, J.S.Williams: VII Intern. Conf. on Atomic Collisions in Solids, Moscow, Sept. 1977 (Moscow State University Publishing House, 1980) Vol. 2, p. 83
6.22 G.Carter, J.S.Colligon, M.J.Nobes: Radiat. Eff. **31**, 65 (1977)
6.23 L.Tanovic, B.Perovic: Nucl. Instrum. Methods **132**, 393 (1976)
6.24 J.L.Whitton, L.Tanovic, J.S.Williams: Appl. Surf. Sci. **1**, 408 (1978)
6.25 W.Hauffe: Phys. Status Solidi A: **4**, 111 (1971)
6.26 W.Hauffe: Proc. of the IBMM Conf. Budapest, Hungary (1979)
6.27 F.W.Young, T.R.Wilson: Rev. Sci. Instrum. **32**, 559 (1961)
6.28 L.E.Samuels: *Metallographic Polishing by Mechanical Methods* (*Pitman, London* 1968)
6.29 *J.L.Whitton: Proc. R. Soc. Edinburgh Sect. A:* **311**, 63 (1969)
6.30 W.J.McG.Tegart: *Electrolytic and Chemical Polishing of Metals*, 2nd ed. (Pergamon Press, London 1959)
6.31 M.Bernheim, G.Slodzian: Int. J. Mass. Spect. Ion. Phys. **12**, 93 (1973)
6.32 J.R.Parsons: Philos. Mag. **12**, 1159 (1965)
6.33 M.W.Thompson: *Defects and Radiation Damage in Metals* (Cambridge University Press, Cambridge 1969)
6.34 J.W.Mayer, L.Eriksson, J.A.Davies: *Ion Implantation in Semiconductors* (Academic Press, New York 1970)
6.35 R.Behrisch (ed.): *Sputtering by Particle Bombardment* I, Topics Appl. Phys., Vol. 47 (Springer, Berlin, Heidelberg, New York 1981)
6.36 G.Carter, W.A.Grant: *Ion Implantation of Semiconductors* (Edward Arnold, London 1976)
6.37 R.S.Nelson, D.J.Mazey: *Ion Surface Interactions*, ed. by R.Behrisch, W.Heiland, W.Poschenrieder, P.Staib, H.Verbeek (Gordon and Breach, London 1973) p. 199
6.38 H.Vernickel: Z. Naturforsch. **21**, 1308 (1966)
6.39 D.G.Brandon, P.B.Bowden: Discuss. Faraday Soc. **31**, 70 (1961)
6.40 M.K.Sinha, E.W.Müller: J. Appl. Phys. **35**, 1256 (1964)
6.41 E.W.Müller: Conf. J. Phys. Soc. Jpn. **18**, Suppl. II, 1 (1963)
6.42 J.A.Hudson, B.Ralph: In *Atomic Collisions Phenomena in Solids*, ed. by D.W.Palmer, M.W.Thompson, P.D.Townsend (North-Holland, Amsterdam 1970) p. 85
6.43 B.Gregov, R.P.W.Lawson: Can. J. Phys. **50**, 791 (1972)
6.44 J.M.Walls, R.M.Boothby, H.N.Southwork: Surf. Sci. **61**, 419 (1976)
6.45 M.I.Current, D.N.Seidman: Nucl. Instrum. Methods **170**, 377 (1980)
6.46 For a review of this technique see D.G.Armour, J.A.Van den Berg, L.K.Verheij: J. Radioanalytic. Chem. **48**, 359 (1979)
6.47 R.L.Jacobson, G.K.Wehner: J. Appl. Phys. **36**, 2674 (1965)
6.48 J.E.Boggio, H.E.Farnsworth: Surf. Sci. **1**, 399 (1964)
6.49 C.E.Haque, H.E.Farnsworth: Private communication (1965)
6.50 V.Krasevec, B.Navinšek: Surf. Sci. **45**, 39 (1974)
6.51 G.K.Wehner, D.J.Hajicek: J. Appl. Phys. **42**, 1145 (1971)
6.52 J.L.Whitton, G.Carter, M.J.Nobes, J.S.Williams: Radiat. Eff. **32**, 129 (1977)
6.53 A.D.G.Stewart, M.W.Thompson: J. Mater. Sci. **4**, 56 (1969)
6.54 G.K.Wehner, P.Yurista: In *Proc. Symp. Sputtering*, ed. by P.Varga, G.Betz, F.P.Viehböck (Inst. Allgemeine Phys., Tech. Univ. Wien 1980) p. 573
6.55 H.H.Andersen: Symposium on the Physics of Ionized Gases, Rovinj, (ed. by V.Vujnovic 1974) p. 361

6.56 D.J. Mazey, R.S. Nelson, P.A. Thackery: J. Mater. Sci. **3**, 26 (1968)
6.57 B.B. Meckel, R.A. Swalin: J. Appl. Phys. **30**, 89 (1959)
6.58 L. Tanovic, J.L. Whitton, H. Kofod: Symposium on the Physics of Ionized Gases, Dubrovnik, ed. by B. Navinšek (1978)
6.59 G. Staudenmaier: Radiat. Eff. **13**, 87 (1972)
6.60 G. Carter, M.J. Nobes, F. Paton, J.S. Williams, J.L. Whitton: Radiat. Eff. **33**, 65 (1977)
6.61 G.W. Lewis, M.J. Nobes, G. Carter, J.L. Whitton: Nucl. Instrum. Methods **170**, 363 (1980); **194**, 509 (1982)
6.62 M. Navez, C. Sella, D. Chaperot: In *Ionic Bombardment Theory and Applications*, ed. by J.J. Trillat (Gordon and Breach, New York 1964) p. 339
6.63 J.L. Whitton, G. Carter, M.J. Nobes, J.S. Williams: In *Physics of Ionized Cases*, ed. by B. Navinšek (J. Stefan Institute, Ljubljana 1976) p. 246
6.64 I.H. Wilson: Radiat. Eff. **18**, 95 (1973)
6.65 H.P. Erlenwein: Private communication, BSci. Thesis, University of Berlin (1977)
6.66 C.J. Altstetter, P.E. Tortorelli: J. Nucl. Mater. **63**, 235 (1976)
6.67 V. Marinkovic, B. Navinšek: In *Proc. 3rd Eur. Reg. Conf. on El. Micro.* (Publ. House, Czech. Acad. of Sci., Prague 1964) p. 311
6.68 L. Tanovic: Private communication
6.69 R. Shimizu: Jpn. J. Appl. Phys. **13**, 228 (1974)
6.70 G.W. Lewis, J.S. Colligon, F. Paton, M.J. Nobes, G. Carter, J.L. Whitton: Radiat. Eff. Lett. **43**, 49 (1979)
6.71 O. Auciello, R. Kelly, R. Iricibar: Radiat. Eff. Lett. **43**, 37 (1979)
6.72 M.J. Witcomb: J. Mater. Sci. **11**, 859 (1976)
6.73 H. Bach: Private communication
6.74 J. Punzel: Phys. Status Solidi **24**, K1 (1967)
6.75 K. Rodelsperger, A. Scharmann: Nucl. Instrum. Methods **132**, 355 (1976)
6.76 P. Haymann: PhD Thesis, Université de Paris, 1962
P. Haymann, C. Waldburger: In *Le Bombardment Ionique*, Proc. of Conf. held at Bellevue, France, 1961 (ed. by J.J. Trillat, CNRS France, 1962)
6.77 N. Hermanne, A. Art: Physica (Zagreb) **2**, Suppl. 1, 72 (1970)
6.78 K. Mihama: Metaux No. 412, 219 (1960)
6.79 J.J.Ph. Elich, H.E. Roosendaal, H.H. Kersten, D. Onderdelinden, J. Kistemaker, J. Elen: Radiat. Eff. **8**, 1 (1971)
6.80 J.M. Fluit, S. Datz: Physica **30**, 345 (1964)
6.81 R.L. Cunningham, J. Ng-Yelim: J. Appl. Phys. **40**, 2904 (1969)
6.82 A. Art: Private communication
6.83 M. Navez, C. Sella, D. Chaperot: Comptes Rendus **254**, 240 (1962)
6.84 H. Bach: J. Non-Cryst. Solids **3**, 1 (1970)
6.85 R.S. Dhariwal, R.K. Fitch: J. Mater. Sci. **12**, 1225 (1977)
6.86 G. Carter, J.S. Colligon, M.J. Nobes: J. Mater. Sci. **4**, 730 (1969); **8**, 1473 (1973)
6.87 D.J. Barber, F.C. Frank, M. Moss, J.W. Steeds, I.S.T. Tsong: J. Mater. Sci. **8**, 1030 (1973)
6.88 M.J. Lighthill, G.W. Whitham: Proc. Soc. Edinburgh Sect. A: **229**, 281 (1955)
6.89 M.J. Lighthill, G.W. Whitham: Proc. R. Soc. Edinburgh Sect. A: **229**, 317 (1955)
6.90 F.C. Frank: *Growth and Perfection of Crstals* (Wiley, New York 1958) p. 411
6.91 F.C. Frank: Z. Phys. Chem. (Neue Folge) **77**, 84 (1972)
6.92 J.P. Ducommun, M. Cantagrel, M. Moulin: J. Mater. Sci. **10**, 52 (1975)
6.93 G. Carter, J.S. Colligon, M.J. Nobes: Radiat. Eff. **31**, 65 (1977)
6.94 M.J. Nobes: PhD dissertation, Univ. of Salford (1976)
6.95 G. Catana, J.S. Colligon, G. Carter: J. Mater. Sci. **7**, 467 (1972)
6.96 T. Ishitani, M. Kato, R. Shimizu: J. Mater. Sci. **9**, 1227 (1974)
6.97 I.P. Ducommun, M. Cantagrel, M. Marchal: J. Mater. Sci. **9**, 725 (1974)
6.98 G. Carter, M.J. Nobes, J.L. Whitton: J. Mater. Sci. **13**, 2725 (1978)
6.99 M.J. Nobes, J.S. Colligon, G. Carter: J. Mater. Sci. **4**, 730 (1969)
6.100 M.J. Witcomb: J. Nucl. Sci. Technol. **9**, 551 (1974)
6.101 B.B. Meckel, T. Nenadovic, B. Perovic, A. Vlahov: J. Mater. Sci. **10**, 1188 (1975)

6.102 D.T.Hawkins: J. Vac. Sci. Technol. **12**, 1389 (1976)
6.103 E.G.Spencer, P.H.Schmidt: J. Vac. Sci. Technol. **8**, 552 (1971)
6.104 P.Norgate, V.J.Hammond: Phys. Technology **5**, 186 (1974)
6.105 A.R.Bayly: J. Mater. Sci. **7**, 404 (1972)
6.106 U.Littmark, W.O.Hofer: J. Mater. Sci. **13**, 2577 (1978)
6.107 G.Carter, M.J.Nobes, K.I.Arshak, R.P.Webb, D.Evanson, B.D.L.Eghawary, J.Williamson: J. Mater. Sci. **14**, 728 (1979)
6.108 R.Smith, J.M.Walls: Surf. Sci. **80**, 557 (1979)
6.109 P.Sigmund: J. Mater. Sci. **8**, 1545 (1973)
6.110 I.A.Teodorescu, F.Vasiliu: Radiat. Eff. **15**, 101 (1972)
6.111 F.Vasiliu, I.A.Teodorescu, F.Glodeanu: J. Mater. Sci. **10**, 399 (1975)
6.112 G.Carter, J.S.Collingon, M.J.Nobes: Symposium on the Physics of Ionized Gases, Dubrovnik, ed. by B.Navinšek (1976)
6.113 G.Carter: J. Mater. Sci. **11**, 1091 (1976)
6.114 N.Hermanne: Radiat. Eff. **19**, 161 (1973)
6.115 R.Smith, J.M.Walls: Philos. Mag. A**42**, 235 (1980)
6.116 R.Smith, T.P.Volkering, J.M.Walls: Philos. Mag. A**44**, 879 (1981)
6.117 G.Carter, M.J.Nobes: In *Ion Beam Modification of Surfaces*, ed. by O.Auciello, R.Kelly (Elsevier, Amsterdam 1983)
6.118 G.Carter, M.J.Nobes, G.W.Lewis, J.L.Whitton: Vacuum (1983) to be published
6.119 G.Carter, M.J.Nobes: Vacuum (1983) to be published
6.120 I.H.Wilson, M.W.Kidd: J. Mater. Sci. **6**, 1362 (1971)
6.121 M.J.Witcomb: J. Mater. Sci. **9**, 1227 (1974)
6.122 L.T.Chadderton: Radiat. Eff. **33**, 129 (1977)
6.123 B.Navinšek, V.Marinkovič, M.Osredkar, G.Carter: Radiat. Eff. **3**, 115 (1970)

7. Development of Surface Topography Due to Gas Ion Implantation

By Bernhard M. U. Scherzer

With 37 Figures

The surface structures which develop due to gas ion implantation into solids are predominantly caused by the action of stresses and internal gas pressure in the implanted layer while erosion by sputtering contributes only a minor part. The development of these structures is closely related to the range and the trapping mechanism of implanted gas ions and to the properties of the gas-solid system.

The phenomena observed at the surface and in the subsurface layer can be divided into four regions of ion fluence which are characterized by the following typical phenomena:

(i) At very low fluences the interaction of single gas atoms with trapping sites predominates.
(ii) At intermediate fluences, clustering of trapping sites and/or trapped gas atoms occurs leading to gas bubble formation and material swelling.
(iii) At high fluences the amount of the implanted gas trapped in the solid saturates. At the same time the surface layer may be deformed into blisters and flakes due to internal gas pressure and lateral compressive stress.
(iv) At very high fluences which are sufficient to sputter erode a surface layer which is approximately equivalent to the ion range, the disappearance of blisters and transition to a spongy surface is generally found. Under special conditions, repetitive flaking is observed.

This review is predominantly oriented toward helium and hydrogen implantation in metals. Effects in nonmetals are only marginally considered and chemical interactions are not included.

7.1 General Outline

Sputtering of solids by energetic ions and atoms is always accompanied by the implantation of the incident particle species into the target material. The concentration profile of the implanted atoms which builds up in the solid is determined by a number of parameters, the more important of which are:

— The speed of surface regression due to sputtering, by which the implant is set free;

— the range distribution of the incident ions;
— the mobility of the implanted atoms in the target material;
— the flux density of the incident ions;
— the density of trapping sites for the incident particles in the solid lattice;
— the chemical interaction between the implanted species and the solid;
— the target temperature.

If the diffusion of the implanted atoms in the solid is negligible because either the temperature is low enough or sufficient stable traps are available, the concentration of implanted atoms which can be achieved in the target depends predominantly on the sputtering yield Y. For sputtering yields smaller than one atom per ion, concentrations larger than one implanted atom per target atom can in principle be obtained. But if the implant is a gas which is not chemically bound by the target material, the concentration of implanted atoms saturates in most materials at about 0.3–1 gas atom per target atom. At these saturation concentrations a modification of the structure in the implanted layer occurs. Cracks or channels can be formed by which gas from deeper layers may be released to the surface. Also, the density of occupied trapping sites may become sufficiently high to allow the transport of gas atoms along chains of interacting sites toward the surface by percolation.

The structural modifications in the implanted layer are accompanied by stress formation and its relief by creep and swelling in the implanted layer. If the implanted ions have ranges less than about 10 nm (e.g., H^+ or He^+ with primary energy $E_0 < 1$ keV), this will only produce a rearrangement of the first atomic layers. For ranges larger than about 10 nm, characteristic surface structures develop caused by a deformation of the implanted layer and in many cases its partial separation from the bulk.

The development of such surface structures which we will designate as "evolutional" are typical for gas ion implantation in low diffusivity materials with very small sputtering yields ($Y \ll 1$). They are different in appearance and origin from the "erosional" surface structures which are produced by the non-uniform erosion and redeposition of sputtered material at larger sputtering yields ($Y \gtrsim 1$) described in the previous chapter by Carter, Navinšek, and Whitton.

Evolutional phenomena on surfaces generally show many special features which are determined by parameters referring to the target material (e.g., composition, temperature, structure, pretreatment), to the ion beam (e.g., energy, fluence, flux, angle of incidence) and to the properties of the gas-solid system (e.g., solubility, diffusivity, chemical interaction). The development of surface structure by gas ion bombardment can very roughly be subdivided into four stages determined by the ranges of ion fluence:

i) Very Low Fluence Range ($\lesssim 10^{18}$ ions m^{-2}). The implanted ions are primarily trapped as single gas atoms at vacancies, interstitial sites or dislocations. Concentrations of implanted gas and damage sites produced by the implanted ions are low and interaction between implanted gas atoms or between damage sites can be neglected. There are no visible surface effects at these fluences.

ii) Intermediate Fluence Range (10^{18}–10^{21} ions m^{-2}). Point defects (interstitials and vacancies) produced by the incident ions agglomerate in dislocation loops followed by the formation of tangles of dislocations as shown in Fig. 7.1a.

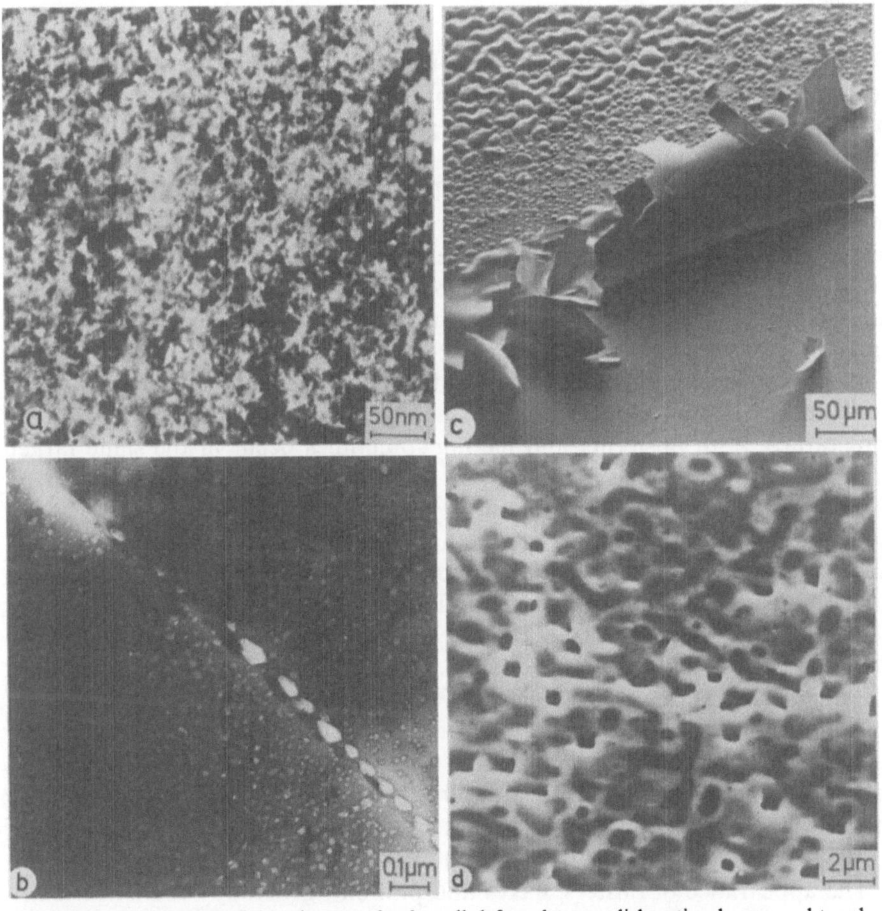

Fig. 7.1. (**a**) TEM bright field micrograph of small defect clusters, dislocation loops, and tangles of dislocations in Ni after bombardment with 3.1×10^{19} He-ions m^{-2} of 0.25 keV, 4.4×10^{19} He-ions m^{-2} of 0.5 keV, 7.5×10^{19} He-ions m^{-2} of 1 keV, 1.5×10^{19} He-ions m^{-2} of 2 keV, 3.3×10^{19} He-ions m^{-2} of 4 keV, and 10^{20} He-ions m^{-2} of 8 keV at room temperature [7.1]
(**b**) TEM-micrograph showing helium bubbles due to bombardment with 2×10^{21} He-ions m^{-2} of 20 keV on Mo at 1300 K. The bubbles have a faceted shape, and enhanced growth is observed at a grain boundary [7.2]
(**c**) Flakes formed by 100 keV $^3He^+$ bombardment of Nb at room temperature (SEM micrograph by Dr. H. Klingele, München) [7.3]
(**d**) Pinholes in a Nb(100) surface after bombardment with 3×10^{21} He-ions m^{-2} at 100 keV, $T = 1100$ K [7.4]

Clustering of occupied trapping sites may lead to the formation of gas filled *bubbles* (Fig. 7.1b) in the solid within the ion range. Depending on temperature and surface mobility of the target atoms, the bubbles may be cavities of spherical or of facetted shape with diameters between 1 and 100 nm. They preferentially grow along grain boundaries. At sufficiently low temperatures they sometimes arrange in a lattice. The formation of gas bubbles is accompanied by bulk swelling, stress formation and an outward movement of the surface.

iii) High Fluence Range (10^{21}–10^{24} ions m^{-2}). If the gas concentration inside the lattice reaches a critical value of about 0.3–1 gas atoms per target atom, the implanted surface layer may separate partly or completely from the bulk causing a surface modification known as *blistering* and *flaking*.

Blisters (Fig. 7.1c) are plastic dome-shaped bulgings of the surface layer of mostly circular, but sometimes irregular circumference. A lenticular cavity is included between the blister cap and the bulk material.

Flakes are rather uniform surface films that peel off the bombarded surface and are sometimes rolled up like a carpet, as shown in the lower part of Fig. 7.1c.

Blister covers and flakes are of rather homogeneous thickness named "deckeldicke", which depends on ion energy and range and on the target temperature. Both blistering and flaking sometimes occur in the same target spot (Fig. 7.1c) probably because of inhomogeneous ion beam current distribution.

In the course of blistering or flaking, the surface layer can be removed partially or completely leaving the underlying surface uncovered. The new surface may again undergo blistering or flaking. This process of *exfoliation* may occur in several successive layers causing an erosion which can be considerably larger than the erosion by sputtering.

After implantation of high fluences of gas ions at target temperatures, which are sufficiently high for surface migration or bulk diffusion of the target material to occur, pinholes are observed on the surface (Fig. 7.1d). Their edges are either rounded or aligned with some principal crystal directions. These pinholes are assumed to be formed due to gas bubble migration to the surface or due to the intersection of growing bubbles with the surface.

The appearance of blisters, flakes and pinholes generally occurs at a well-defined critical fluence and is accompanied by the onset of reemission of the implanted gas.

iv) Very High Fluence Range ($> 10^{23}$ ions m^{-2}). By bombardment to fluences much larger than the critical fluence for first blistering, blisters and flakes are eroded by sputtering (Fig. 7.2a, b). Thus, depth regions come to the surface where the implanted gas concentration is equal to saturation or to $1/Y$, whichever is lower if diffusion is assumed to be negligible. These surfaces show a *spongy structure* (Fig. 7.2c) containing a high density of small holes which

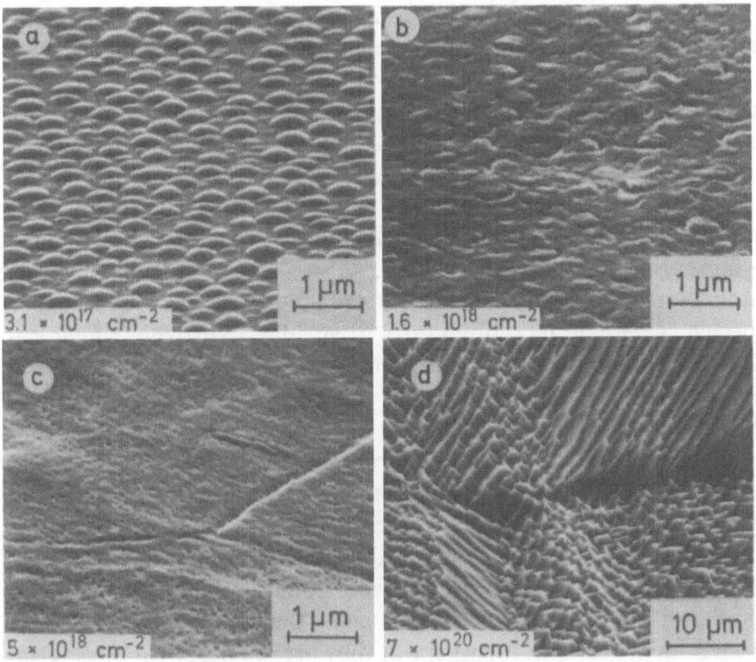

Fig. 7.2. Surface features on polycrystalline Nb after 9 keV He^+ bombardment with different fluences at room temperature and normal incidence (SEM micrograph by Dr. H. Klingele, München) [7.5]

may be remnants of gas bubbles exposed by the receeding surface. Although somewhat similar in appearance, spongy structures differ from high temperature pinholes and are caused by different mechanisms. The development of spongy structures may be delayed by exfoliation of successive layers.

With a further increase of fluence the surface structure becomes dominated by nonuniform sputter erosion and deposition (Fig. 7.2d). Steps, pyramids, ridges and grooves evolve, similar to the erosional structures found for heavy ion bombardment at much lower fluences as described in the previous chapter.

Evolutional surface structure development has been most extensively investigated for helium ion bombardment of metals in an energy range between 1 keV and several MeV. Very similar results have been found for hydrogen ion bombardment of materials having low solubility for hydrogen at temperatures where diffusion is sufficiently suppressed. Evolutional surface phenomena have also been observed for heavy gas ion bombardment (N^+, Ne^+, Ar^+), but generally only at energies above 100 keV.

Gas bubbles in solids and blisters on surfaces may also arise from reasons unrelated to ion implantation. They have been observed in the process of cooling of cast or hot rolled metals when the concentration of dissolved gas, especially hydrogen, exceeds the solubility limit and the surplus gas precipitates into bubbles [7.6–9]. Precipitation also occurs due to the chemical transformation of a soluble gas like hydrogen into an insoluble compound (e.g., CH_4 or H_2O) in the solid [7.10, 11].

Surfaces of sintered molybdenum and stainless steel (SS 321) exposed to large flux densities (10^{14}–10^{17} particles $cm^{-2}\,s^{-1}$) of hydrogen ions and atoms with energies between 0.2 eV and 25 keV were investigated in the scanning electron microscope (SEM) and have been reported as showing craters with a depth and width of about 1 μm [7.11a–c] or, in Mo, blisters with a "deckeldicke" 2 orders of magnitude larger than the ion range [7.11d, f]. This effect has been explained by the diffusion of atomic hydrogen into grain boundaries or other defects below the surface, where it recombines to molecular hydrogen which can build up sufficiently high pressures to deform the overlying surface layer or even to eject whole grains [7.11a–e].

Plastic deformations similar to blisters are observed on thin films deposited on substrates having different coefficients of thermal expansion under the action of compressive stress parallel to the film surface. On free standing films the whole layer may be moved perpendicularly to the surface, a phenomenon called buckling which has also been treated theoretically [7.12–14].

Extensive experimental information on evolutional surface structures due to gas ion implantation has been acquired over the last 10 years. Several theoretical models have been put forward but they are still not able to describe all the observed phenomena correctly. Nevertheless, many details of the processes involved are well established. In this chapter, the field of gas ion implantation produced surface phenomena is reviewed with special emphasis on helium and hydrogen implantation in low solubility materials. Surface modifications due to chemical interactions such as hydride or oxide formation will only briefly be mentioned. The chapter is subdivided into four parts according to the four fluence ranges described above. Since each stage of surface structure develops from the previous one this division provides a gradual introduction from atomistic interactions with single damage sites to collective interactions, and elastic and plastic deformations of larger areas. Since these phenomena are closely related to the state and the concentration profile of the implanted gas in the solid, we start with a short survey on ion ranges and on gas solubility and diffusivity. Because of the spectacular appearance and the ease of observation, a majority of experimental effort has been put into the observation of blistering and flaking at high fluence ranges. Accordingly, these phenomena will be treated more extensively. A critical evaluation of the different models for blistering and flaking, and an assessment of the contribution of exfoliation to erosion will be given.

To our knowledge no review of the complete field of gas ion induced surface topography has been made up to now. Special reviews on blistering and flaking have been published by *Roth* [7.15], *Das* and *Kaminsky* [7.16a], *Erents* [7.16b], *Kaletta* [7.17], *St.-Jacques* et al. [7.18], *Das* [7.19a], *Guseva* et al. [7.19b], and *Wilson* [7.19c]. A bibliography on blistering and ion erosion by light ions was collected by *Navinšek* and *Peternel* [7.20].

7.2 Historical Overview

The trapping of gaseous ions in solids was observed as early as 1858 by *Plücker* [7.21]. He observed that in dc gas discharges the colour of the discharge changes with time and sometimes the discharge extinguishes completely, an effect which he explained by the loss of gas from the discharge due to binding to the electrodes. For a long time the main interest in this field concentrated on the continuous loss of gas in gas discharges and x-ray tubes which was caused by the implantation of positive ions into regions of the tube with negative potential with respect to the plasma. This effect was called "gas cleanup" or "hardening" because of the higher discharge voltage needed in x-ray tubes at lower pressures. This early work on gas trapping was of considerable technological importance. The investigations were mostly phenomenological and the basic physical effects were obscured by poor surface conditions and ill-defined target preparation and implantation conditions. Since 1950 with the development of ultrahigh-vacuum techniques, an increasing amount of work was done under well-defined experimental conditions. The best reproduceable results were obtained for the implantation of noble gases. Comprehensive reviews were given by *Pietsch* [7.22], *Alpert* [7.23], *Strotzer* [7.24, 25], and *Carter* et al. [7.26–28].

The formation of gas bubbles in metals (Cu, Al, and Be) due to the implantation of helium ions and subsequent annealing was first described in 1958 by *Barnes* et al. [7.29–32]. They observed that concentrations of helium far beyond the solubility limit could be formed within the range of the ions. The bubble growth was ascribed to the collection of thermally produced vacancies. It was therefore assumed that bubble formation did not occur below some critical value of the target temperature at which mobile vacancies are produced at a sufficient rate. But *Sass* and *Eyre* [7.33], and *Mazey* et al. [7.34] showed that gas bubbles are also formed during helium implantation of molybdenum at room temperature where no thermal vacancies are produced. They found that the low temperature bubbles tended to form a lattice with identical structure and orientation as the atomic lattice of the crystal. The mechanism of low temperature bubble initiation obtained substantial clarification by the work of *Kornelsen* and *Edwards* [7.35, 36] and *Van Veen* and *Caspers* [7.37]. They showed that implanted He in W, Mo, and Ni is trapped at damage sites as vacancies either present or produced by the incident ions. Without the presence of damage the helium is not trapped at room temperature, it is quite mobile and may quickly leave the target. If a vacancy is filled with several helium atoms, additional helium atoms may cause the displacement of a neighbouring lattice atom to form a divacancy thus starting the growth of a bubble.

The first observations of surface deformations due to hydrogen ion bombardment with "canal" rays [1] of more than 10 keV were reported by *Stark* and *Wendt* [7.38]. On a number of minerals like rock salt, sylvite, mica, calcite

1 Ion beam obtained through a hole in the cathode of a dc discharge.

and fluorite as well as on glass they found surface roughening because of the formation of surface elevations with blister shape or surface pits due to the breaking of the blister cover. The exfoliation of thin layers of macroscopic extensions was also observed.

Stark and *Wendt* used the german word "Bläschen" for the gas filled surface elevations. The notation "blistering" has originally been used for surface deformations of metals due to hydrogen accumulation at laminar discontinuities, e.g., [7.39]. *Primak* [7.40] introduced the term "radiation blistering" for gas ion beam produced surface deformations.

Exfoliation at the surface of aluminium due to helium ion implantation was observed in preparing He-loaded targets for nuclear cross section measurements in accelerator laboratories [7.41]. The first more detailed investigation was done by *Primak* et al. [7.40] who found that targets of Si, corundum, spinel, rutile and peridot are blistered and pitted after bombardment with protons and helium ions of 100–600 keV. By infrared absorption measurements they observed the formation of a reflecting layer which they attributed to the formation of a gas layer at the end of the range of the ions. Blistering of highly polished Al-coatings by implantation of hydrogen and helium ions has also been studied because of its influence on the reflectance of spacecraft coatings [7.42b].

At a very early stage the release of gas from a Pt sample during heating in vacuo was observed to be accompanied by the emission of metal particles which formed a film on a collector [7.42a]. This was explained as a mechanical pulverisation of the hot metal due to the formation of gas filled foam at the surface. *Kaminsky* [7.43] found that the changes in surface morphology were accompanied by gas release. During the bombardment of a Cu single crystal by 125 keV D^+ gas bursts of 10^9 to 10^{10} particles each were observed which were attributed to the migration of bubbles to the target surface where they "explode" and leave etch pits.

Since about 1972, a great number of investigations have been made of the phenomena of blistering and flaking under hydrogen, helium and to a lesser extent nitrogen, neon and argon ion bombardment of many materials. One major reason for these investigations was to assess the possible contribution of these phenomena to first wall erosion in the thermonuclear fusion plasma devices [7.16, 19, 44–48]. Only very little of this work extends to ion fluences which are sufficiently high to show the influence of sputter erosion on blistered surfaces. *Martel* et al. [7.49] first found the disappearance of blisters due to sputter erosion of a surface layer of about the same thickness as the blistered layer. Consequently, a very rough and pockmarked surface structure develops. *Behrisch* et al. [7.5] and *Gusev* et al. [7.50] confirmed the effect of blister disappearance for He bombardment of Nb in the energy range from 9 to 100 keV. They showed that the pockmarks are actually holes in a spongy surface structure.

A qualitative model of blister formation was proposed by *Stark* and *Wendt* [7.38] who assumed that implanted ions are thermalized in the target and accumulate at some positions after having been neutralized. They exert a pressure on the surface layer which then either forms a vault-like chamber or will break away from the bulk. The blistering model of *Primak* and *Luthra* [7.51] is also based on the action of gas pressure in sub-surface cavities by which a surface layer is plastically deformed. The influence of lateral stresses due to swelling or compaction of the implanted layer was already indicated by these authors. In later work, blistering and flaking was explained by the action of pressure alone [7.52–60]. *Behrisch* et al. [7.61] assumed lateral compressive stress as contributing essentially to blister formation in order to explain why the "deckeldicke", i.e., the thickness of the blister cover, was larger than the projected ion range. The formation of such stress and its relief during blistering was measured by *EerNisse* and *Picraux* [7.62]. In this work, and independently by *Risch* et al. [7.63], the final shape of blisters was calculated by continuum mechanical methods. Large swelling of He implanted metal layers was observed by *Blewer* and *Maurin* [7.64] and by *St.-Jacques* et al. [7.65, 66] and was taken as evidence in favor of the pressure model, but a final decision between pressure and stress models has not yet been obtained.

7.3 Implantation of Gaseous Ions in Solids

7.3.1 Ranges of Light Ions in Solids

The interaction of energetic ions with a solid can to a first approximation be separated into the interaction with the electrons and the collisions with the nuclei of the solid [7.67–69]. While the interaction with electrons contributes most of the energy loss, the deflections at the target atoms determine the trajectory of the particle. The total path length R of an ion with primary energy E_0 in a solid is given by

$$R=\int_{E_0}^{0}\frac{dE}{S(E)}. \tag{7.1}$$

The differential energy loss $S(E)$ contains two parts, one due to collisions with the target nuclei and another due to collisions with electrons:

$$S(E)=\left(\frac{dE}{dx}\right)_{\mathrm{n}}+\left(\frac{dE}{dx}\right)_{\mathrm{e}}. \tag{7.2}$$

For light ions at not too low energies, $(dE/dx)_{\mathrm{e}} \gg (dE/dx)_{\mathrm{n}}$. In this case the total path length is determined mainly by the energy loss to electrons. Energy loss data have been compiled for H, D [7.70] and for He [7.71] in all elements based on semiempirical values. For energies above $\sim$500 keV/nucleon, these data are generally uncertain within $\pm 5\%$ while for low energies the uncertainty may be much larger. Below 20 keV there are only very few experimental data.

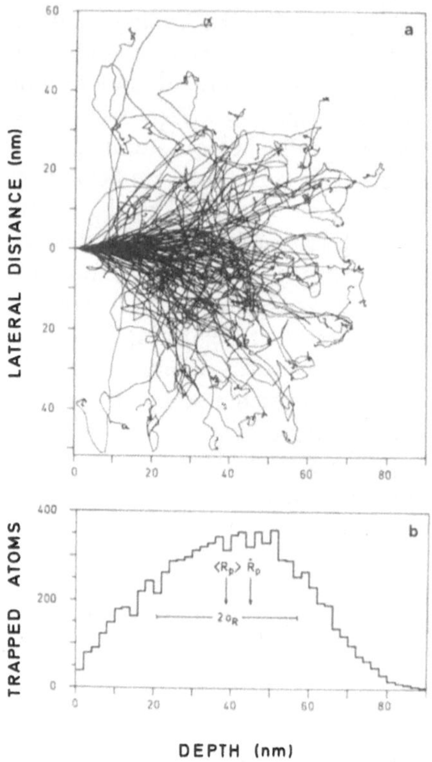

Fig. 7.3. (**a**) Particle paths of 10 keV ^{4}He-ions in Nb for normal incidence. Projection of particle paths on a plane parallel to the direction of incidence. TRIM-calculation [7.85] including electronic energy loss from *Ziegler* [7.71]. A total of 120 particle paths have been calculated, 102 particles come to rest in the target. The trajectories of the reflected particles are not shown. (**b**) Projected range distribution for equal conditions as given in (**a**) as calculated by TRIM. Out of 10^4 primary particles, 8874 come to rest in the solid while the rest is backscattered [7.86]

Two different approaches have been made to calculate projected ranges R_p; one is the approximate solution of the transport equation in an infinite homogeneous amorphous medium [7.69, 72–78], the other is a Monte Carlo simulation for single crystalline and polycrystalline material [7.79–82], as well as for amorphous material [7.82–85].

The trajectories of 10 keV He-ions in amorphous Nb projected on a plane normal to the surface as calculated with the Monte-Carlo program TRIM [7.85] are shown in Fig. 7.3a [7.86]. The trajectories of those particles which eventually leave the target before being thermalized, i.e., the kinetically reflected particles, have been omitted in the figure. They amount to $(11.3 \pm 0.4\%)$ of all incident particles. Figure 7.3b shows the corresponding projected range distribution. The mean projected range $\langle R_p \rangle$, the most probable projected range $\hat{R}_p$ and the mean range straggling σ_R are also shown in the figure.

Data for mean values of total path length $\langle R \rangle$, projected range $\langle R_p \rangle$ and range straggling σ_R of H, D, and He in all elements are published in the compilations of *Ziegler* and *Anderson* [7.70, 71].

For light ions in heavy target materials, $\langle R_p \rangle \lesssim \hat{R}_p$ and $\langle R_p \rangle \ll \langle R \rangle$ due to large angle deflections at the target nuclei. Only at very high energies ($E_0 \gtrsim 1$ MeV/nucleon) do $\langle R \rangle$, $\langle R_p \rangle$, and $\hat{R}_p$ agree within a few percent because the particle trajectories are almost straight lines for the greater part of the path.

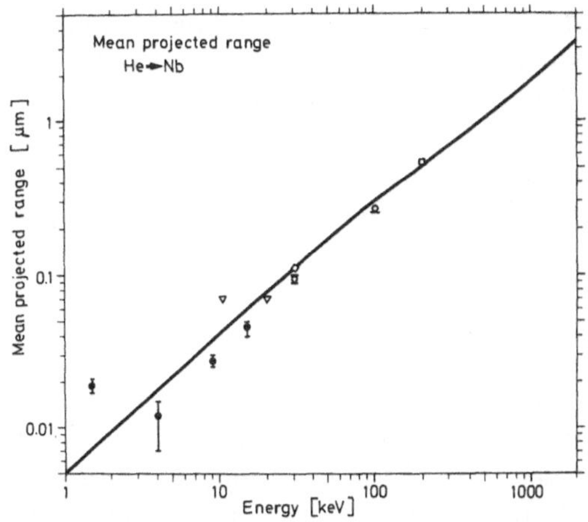

Fig. 7.4. Comparison of mean projected ranges as a function of particle energy for He in Nb. Semiempirical values: (——) *Ziegler* [7.71]; nuclear reaction: (●) *Behrisch* et al. [7.61]; (○) *Risch* et al. [7.105]; (◇) *Scherzer* et al. [7.106]; (□) *Biersack* et al. [7.98]; ERD: (▽) *Terreault* et al. [7.107]; (□) *St.-Jacques* et al. [7.66]

Range measurements for light ions have been made by sputtering combined with secondary ion mass spectroscopy (SIMS) [7.87–90], by proton elastic scattering (PES) [7.19, 92], Rutherford backscattering (RBS) [7.92a, b], nuclear reaction techniques [7.61, 92b–99], elastic recoil detection (ERD) [7.101, 102], interferometric determination of the depth of the lattice discontinuity produced by the implanted gas [7.40] and by equating the most probable projected range $\hat{R}_p$ to the "deckeldicke" of blisters and flakes [7.103, 104]. A review of depth profiling methods by using ion beams was prepared by *Bøttiger* [7.100]. The "deckeldicke" method is only applicable for hydrogen and helium ions with energies $E_0 > 100$ keV because at lower energies considerable differences between $\hat{R}_p$ and the "deckeldicke" have been observed as will be discussed later. The agreement with calculated values is generally better at high energies ($E_0 > 20$ keV) than at low energies ($E_0 < 20$ keV) where only very few experimental data on stopping power exist and the relative experimental error in range determination becomes large. A comparison of measured mean projected ranges as a function of energy with *Ziegler's* [7.71] tabulations is shown in Fig. 7.4 for He in Nb as an example. For 20–60 keV He in Al, Si, V, Ni, and Zr measured values of the most probable ranges $\hat{R}_p$ were found up to 30% smaller and range straggling σ_R in some cases smaller and in some cases larger than values calculated by transport theory [7.108]. The reason for this discrepancy between transport theory and experiment is not understood. Inadequate surface correction, inaccurate electronic stopping powers and scattering cross sections used in the calculations may contribute.

7.3.2 Solubility and Diffusivity

The behaviour of gas atoms after being slowed down in a solid to a level where they come into thermal equilibrium with the surrounding lattice is to a first approximation determined by the solubility S and the diffusivity D of the gas in the material. It is further influenced by the density and binding energies of trapping sites as lattice defects and radiation damage. In the present context we are mainly interested in the properties of noble gases (He, Ne, Ar) and of hydrogen in metals.

If the gas forms a true solution in the solid the equilibrium gas concentration c in the solution at temperature T and an external gas pressure p is given by

$$c=p^{\beta}S_0\,\mathrm{e}^{-Q_s/kT} \quad \text{(Sievert's law)}, \tag{7.3}$$

where S_0 is the solubility constant and Q_s the heat of solution which is positive if the solution is an endothermic process. β is a stoichiometric factor: $\beta=1$ for an atomic solution of a monatomic gas, $\beta=1/2$ for an atomic solution of a diatomic gas. Equation (7.3) is derived from the law of mass action, e.g. [7.109], assuming an ideal gas and an ideal solution.

For rare gases in undamaged metals the solubility is generally below the experimental limit of detection [7.110, 111a–c] because the heat of solution is very large (several eV for He in Metals [7.112a]).

For hydrogen the solubility depends strongly on the material. In some metals and semiconductors (e.g., Be, Zn, Ga, Ge, Cd, In, Sn, Te, Tl, and Pb) it is extremely low or even zero [7.109, 110]. In another group of metals (e.g., Mg, Al, Si, Cr, Mn, Fe, Co, Cu, Mo, Ag, W, Pt, and Au) hydrogen forms a simple solution or α-phase [7.109, 110, 113]. Equation (7.3) is valid for concentrations below about 1 at.%. In these metals no hydride formation is observed for the gas-solid system. The solution is endothermic ($Q_s>0$) and supersaturated systems produced, e.g., by quenching of samples loaded with hydrogen at high temperatures tend to form hydrogen precipitates in the form of gas bubbles [7.9, 114].

In a third group of metals (e.g., the subgroups IA: alkali; IIA: alkaline earths; IIIB: La, Ce, etc.; IVB: Ti, Zr; VB: V, Nb, Ta of the periodic system, the actinides and Ni and Pd), one or more hydride phases are formed when the simple solution or α-phase is saturated [7.109, 110, 113, 115]. In most of these metals the solution of hydrogen is strongly exothermic ($Q_s<0$). High relative saturation concentrations with $N_H/N_{Met}\gtrsim 1$ characterise the metals of this group. N_H and N_{Met} are the atom densities of hydrogen and target metal, respectively, but Sievert's law, (7.3), applies only for the α-phase, i.e., at low concentrations. At higher concentrations one or more hydride phases may coexist with the solution phase and the pressure-concentration relation shows steps which can be assigned to the development of the hydride phases [7.109, 113, 116, 117]. By hydride formation mechanical, chemical, electrical, and structural properties of the material are drastically changed [7.113, 115].

The diffusivity D of gas atoms in a solid has generally also an exponential dependence on temperature T and is given by

$$D=D_0\,\mathrm{e}^{-Q_D/kT}, \tag{7.4}$$

where Q_D is the activation energy for diffusion and D_0 is a constant. D_0 and Q_D change with the type of sites by which the atoms move through the solid. It is, therefore, influenced by phase transitions occuring, for example, in the interaction of the hydride forming metals with hydrogen [7.115]. Since $Q_D > 0$, the logarithm of the diffusivity $\ln D$ decreases linearly with increasing reciprocal temperature $1/T$ (Arrhenius plot), as long as no phase changes take place. This means that diffusion can be suppressed in any material at sufficiently low temperature.

Hydrogen data on solubility and diffusivity in most metals are published in a number of reviews and monographs, e.g., [7.113, 115–120]. For rare gases there is little information on the diffusivity in metals. For He in Ni, *Philipps* et al. [7.121] recently determined $D_0 = (2.7 \pm 1) \times 10^{-3}\,\mathrm{cm^2/s}$ and $Q_D = (0.81 \pm 0.05)\,\mathrm{eV}$ for $1173\,\mathrm{K} \leqq T \leqq 1523\,\mathrm{K}$ where thermal vacancies are the dominant He traps. The diffusivities are influenced by lattice defects and radiation damage and this influence is more effective for rare gases [7.35, 36] than for hydrogen [7.122], as will be discussed later.

7.3.3 Gas-Solid Systems Formed by Implantation

If the gas is introduced into the solid by implantation, strong nonequilibrium states of the gas metal system may be obtained. These depend firstly on the rate at which the gas is implanted, given by the implantation current density and, secondly on the rate at which it is removed by diffusion from the depth where it comes to rest. The implanted ions, in addition, create radiation damage such as Frenkel defects, point defect clusters and dislocation networks which have a strong influence on the binding energy and on the diffusion of the gas atoms in the lattice. The properties of gas-solid systems produced by ion implantation are therefore generally very different from those where the gas is introduced into the undamaged material by diffusion from the surrounding gas.

Three typical situations may be distinguished:

i) *The diffusion of the implanted gas is negligible* either because the temperature is sufficiently low or because the implanted gas is trapped in damage sites with high binding energies. In this case, the amount of gas particles collected per bombarded surface area as a function of incident fluence will behave as shown in Fig. 7.5 for the case of He-implantation into Nb at room temperature [7.123]. At low fluences all gas particles are collected with the exception of those which are kinetically reflected. In this fluence range the amount of collected gas atoms increases linearly. At higher fluences the amount of collected atoms saturates gradually and at high fluences the increase becomes very small or even zero. Similar behaviour is also found for hydrogen implanted in metals at low temperatures [7.124–127].

A number of models have been developed to explain the observed trapping and release behaviour of gas ions in such gas-solid systems where surface

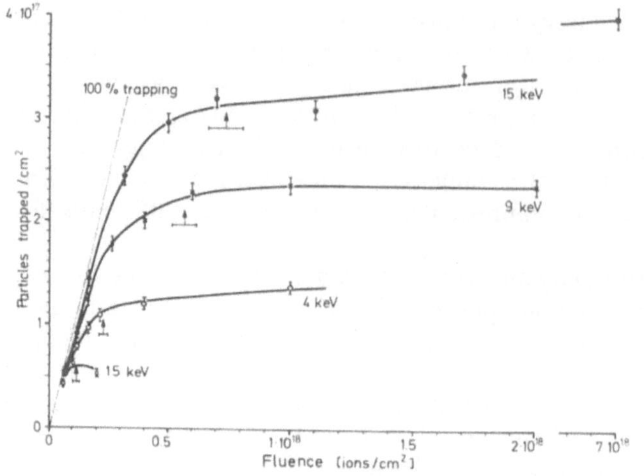

Fig. 7.5. Trapped ^{3}He in Nb as a function of fluence [7.123]

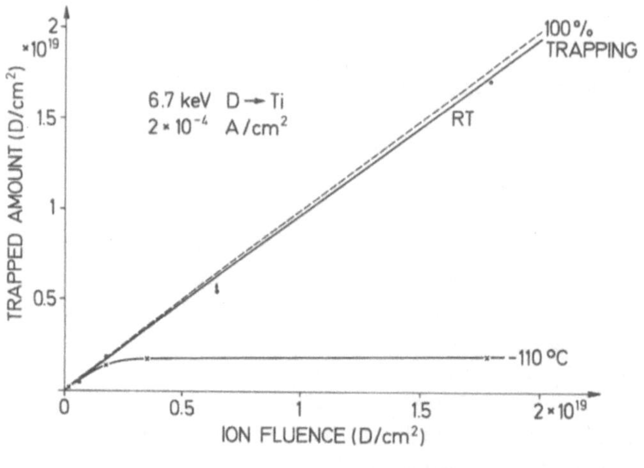

Fig. 7.6. Dependence of the trapped amount on ion fluence for D in Ti at different temperatures [7.135]

sputtering and thermal diffusion can be neglected [7.128–133]. In these models the influence of blisters or flakes is generally neglected. It is assumed that the implanted gas is bound to trapping sites where it is immobile. During ion bombardment, previously trapped gas atoms may be activated to leave their trapping site and diffuse to the next free trapping site or to the surface. The experimentally observed shape of trapping curves (Fig. 7.5) can be satisfactorily explained by these models.

ii) *The diffusion of the implanted gas is high but the solubility is low.* In this case the implanted gas will diffuse to the surface and to a lesser extent into the bulk. The amount of collected gas depends on the implantation current density and it will decrease due to losses through the surface after stopping the bombardment. This behaviour is found, for example, for hydrogen implantation in metals that do not form hydrides, e.g., in stainless steel at sufficiently high temperatures to activate diffusion [7.134].

iii) *Diffusion and solubility are high.* The implanted gas is collected with high efficiency up to very high fluences because the gas is distributed over the whole bulk of the material as soon as the implanted range distribution becomes saturated. This behaviour is observed for hydrogen implantation of hydride-forming metals at temperatures where the diffusion is activated. A typical example is shown in Fig. 7.6 for deuterium implantation of Ti at room temperature [7.135] where the trapped amount increases linearly with fluence while at 163 K, saturation of the trapped amount is found for $f \gtrsim 3 \times 10^{22}$ D m^{-2} similar to He in Nb shown in Fig. 7.5.

An important parameter characterizing the damage produced in materials is the relationship between the gas concentration obtained, for example, by implantation and the radiation damage produced in a lattice [7.136]. This relationship is expressed by the ratio between the number of implanted gas atoms per target atom (GPA) and the number of displacements per atom (DPA) in the implanted region. This parameter can be very different for different irradiation conditions of materials. It is as low as 10^{-6} for materials irradiated in a fission reactor [7.17] while it becomes 10^{-5} for irradiation by 14 MeV neutrons because of the strong increase of the cross sections for gas production reactions such as (n, α) and (n, p) at neutron energies above ~6 MeV [7.17]. For gas implantation as described here and which occurs at the first wall facing a hot thermonuclear plasma, the parameter is extremely high and may reach values of 0.1 to 1.

7.4 The Trapping of Gas at Very Low Fluences ($\lesssim 10^{17}$ ions m^{-2})

In the following the development of internal structures and of evolutional surface phenomena will be described for materials in which the chemical interaction between gas atoms and target atoms can be neglected. Consequently, most results will refer to helium implants but also some results on hydrogen isotopes and some singular results with other gases will be described.

The interaction of single gas atoms with damage sites can be most effectively studied if the gas concentration in the lattice is low enough to prevent too many gas atoms being trapped at one site and to prevent interaction between trapping sites. At fluences $< 10^{17}$ ions m^{-2}, the local gas concentration stays well below 1 at.% for all primary energies and the above condition is generally fulfilled. A review of defect trapping of gas atoms in metals was published by *Picraux* [7.137].

The damage in a crystal providing trapping centers for gas atoms may be produced either by the implanted gas ions themselves or by previous bombardment with heavy ions or other projectiles or by other damage producing processes. For He a trapping mechanism has been observed which is independent of radiation damage.

The state of He in metals at very low fluences has been investigated by thermal helium desorption spectroscopy (THDS) [7.138a, 139], a technique developed from the early thermal desorption work of *Kornelsen* et al. [7.35, 36, 138a]. In these measurements He was implanted at energies below the threshold energy for damage production into samples which have been predamaged by some high energy He or heavy ion bombardment.

In the fluence range below 10^{17} ions m^{-2}, the interaction of helium with point defects and small defect clusters has been studied in a number of metals (W, Mo, Ni) by thermal desorption spectrometry [7.35, 36, 138a, c, 139, 141–143]. These measurements give a number of well-defined desorption peaks which can be allocated to the dissociation of He from different defect configurations. Figure 7.7 shows how a desorption spectrum of He from (100) W can be fitted by 6 theoretical first-order desorption transients. The discrete binding states are given in the figure caption and in Table 7.1a. In Table 7.1a–c the experimentally determined activation energies for some of these reactions are given. The experimental results receive strong support from atomistic model calculations [7.37, 144–146]. These calculations which have been reviewed by *Reed* [7.147] give similar energies of activation for the proposed reactions as were determined experimentally. The theoretical values are given in Table 7.1a–c for comparison.

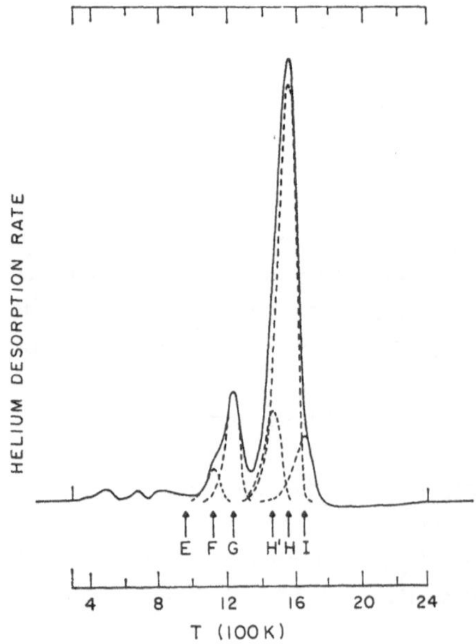

Fig. 7.7. Thermal desorption spectrum of He for a predamaging bombardment of 8×10^{11} cm^{-2} Kr-ions of 5 keV and injection of 8×10^{12} cm^{-2} He-ions of 250 eV. (---) indicate theoretical desorption transients:

peak *H* HeV → He + V
peak *G* He_2V → He + HeV
peak *I* $He_2V_2^{II}$ → 2He + 2V
peak *F* He_3V → He + He_2V
peak *E* He_4V → He + He_3V
peak *H'* $HeKrV_2^{II}$ → He + KrV_2^{I}

Description of notations is given in Table 7.1 [7.35]

Table 1a–d. Activation energies for thermal desorption measured by thermal desorption spectrometry compared to theoretical atomistic model calculations. (**a**) He–W, (**b**) He–Mo, (**c**) He–Ni, (**d**) comparison of calculated binding energies of a helium atom to a vacancy and activation energy of interstitial motion

Notations:

He: one interstitial helium atom
He_n: *n* interstitial helium atoms
V: one host lattice vacancies
V_2^I: first neighbour divacancy
V_2^{II}: second neighbour divacancy
HeV (KrV): one He (Kr) atom in a vacancy, substitutional occupancy
$HeKrV_2^I$: mixed substitutional occupancy of a divacancy

a) **He–W**

Reaction	Calculated binding energy [eV] [7.145]	Calculated activation energy [eV] [7.146]	Experimental activation energy [eV] [7.35]
$HeV \rightarrow He+V$	4.39	5.07	4.05
{ $He_2V_2^{II} \rightarrow He+\ldots$			4.37
{ $He_2V_2^I \rightarrow He+HeV_2^{II}$	4.42		
$He_2V_2^I \rightarrow He+\ldots$			4.8
{ $HeKrV_2^{II} \rightarrow He+KrV_2^I$ [a]			3.84
{ $HeKrV_2^I \rightarrow He+KrV_2^I$	4.40		
$He_2V \rightarrow He+HeV$	2.89	3.43	3.14
$He_3V \rightarrow He+He_2V$	2.52	3.02	2.88
$He_4V \rightarrow He+He_3V$	2.50	2.94	2.41
$He_5V \rightarrow He+He_4V$	2.31		
$He_6V \rightarrow He+He_5V$	2.52		
$He_7V \rightarrow He+He_6V$	1.03		

b) **He–Mo**

Reaction	Calculated activation energy [eV] [7.37]	Experimental[b] activation energy [eV] [7.139, 143]
$HeV \rightarrow He+V$	4.2	3.05
$He_2V \rightarrow He+HeV$	2.82	2.5
$He_3V \rightarrow He+He_2V$	2.50	2.3
$He_{4,5,6}V \rightarrow He+He_{3,4,5}V$	2.37	2.05
$He_{1,2,3}NeV \rightarrow He+He_{0,1,2}NeV$		2.1
$He_nV_m \rightarrow He+\ldots$ clusters		3.2–5

c) **He–Ni**

Reaction	Experimental activation energy [eV] [7.36]
$HeV \rightarrow He+V$	2.10
$He_2V \rightarrow He+HeV$	1.70
$HeNeV_2^{\text{!!}} \rightarrow He+\ldots$	2.35
$He_2V_2 \rightarrow He+\ldots$	2.70

d) **HeV binding energy and activation energy for interstitial motion in different metals**

Material	Binding energy HeV [eV] [7.112, 149 h]	Activation energy for interstitial motion of He [eV]
Ni	3.16	0.08
Cu	1.84	0.57
Pd	3.16	1.74
Ag	1.53	0.86
V	2.96	0.13
Fe	3.75	0.17
Mo	3.87	0.23
Ta	3.30	0.0
W	4.42	0.24

[a] Different reactions are assumed to explain one experimentally observed desorption peak.
[b] The dissociation energy (first order desorption; $\nu=5\times10^{13}\,s^{-1}$) is approximated by $E^D \approx T_m/390$, (T_m) desorption peak temperature [K], E^D in [eV].

As a result, it was shown that helium atoms are mobile in Ni, Cu, V, Fe, Mo, Ta, and W in interstitial positions with an activation energy of only a few tenth of an eV, in agreement with the early experimental result for He in W of *Erents* et al. [7.148a] and the observation of *Wagner* and *Seidman* [7.148b] that subthreshold He in W is mobile at 96 K. Also in Ni thermal desorption measurements have shown that interstitial He is mobile below 80 K [7.148c, d]. Strong binding (several eV) occurs between helium atoms and vacancies (Table 7.1d). A single vacancy was shown to bind 4–6 helium atoms with energies of more than 2 eV, the binding of the nth He atom being generally weaker than that of the $(n-1)^{th}$. Still higher binding energies than for single vacancies are found for helium in divacancies and vacancy clusters.

It has been shown by THDS that such clusters are formed by mutation of vacancies at a certain filling degree with He. In W this mutation from He_nV to He_nV_2 occurs at $n \geqq 4$ [7.35, 138c], in Mo it occurs at $5 \leqq n \leqq 8$ [7.139, 149b]. During further He implantation the helium vacancy clusters accumulate further He. Multiple occupation of He-vacancy clusters up to $n \approx 100$ was observed in

Mo [7.139, 149b] and W [7.138c] by the THDS technique. With increasing filling degree the lattice atoms around the cluster rearrange to provide room for the highly insoluble He atoms. In Mo the He vacancy clusters develop initially in the form of two-dimensional platelets instead of three-dimensional spheres [7.149b–d]. The growth of the He-vacancy clusters is considered as the nucleation stage of gas bubbles.

The formation of the He-vacancy complex can also take place in the absence of any lattice damage. *Thomas* et al. [7.149e–g], have observed trapping of He in vacancy clusters formed by spontaneous precipitation when the He is introduced into the metal either by the decay of dissolved tritium (the so-called tritium trick) or by subthreshold implantation into well annealed material. Atómistic model calculations [7.149h] show that in Ni interstitial He is bound to a cluster of interstitial He atoms with energies rising with the number n of He atoms in the cluster. At a critical number of 5 clustered interstitials a near Frenkel pair is formed and He arranges in the vacancy while the self-interstitial atom stays close to the cluster. At $n=8$ a second vacancy is formed and so on. The self-interstitial atoms themselves cluster close to the "bubble" to reduce the strain energy.

Trapping of He in W at substitutional heavy rare gas atoms has been measured by *Kornelsen* et al. [7.138c]. It is shown that the binding energy to the traps decreases with the mass of the rare gas atom from 3.1 eV at a HeV trap to 1.2 eV at Xe. Contrary to the behaviour of He in vacancies the binding energy increases with the number of He atoms in the trap i for Ar, Xe, and Kr up to $i=7$.

Thermal desorption spectrometry of helium has also been performed among others on Nb [7.150], Re and Pt [7.151a], Pd, Cu, Ni [7.151b, c], and stainless steel [7.152, 153, 151b, c]. But only when extremely low incident ion fluences and high quality single-crystal targets are employed can desorption peaks be related to specific trapping sites [7.149i]. Large numbers of additional binding states due to dislocations [7.154] grain boundaries, impurities and radiation damage complicate the interpretation of thermal desorption spectra. High binding energies to trapping sites have also been found for the heavy rare gases (Ne, Ar, Kr, Xe) implanted in W and Ni [7.35, 36, 155]. Again, clustering of several gas atoms is observed at higher fluences.

Hydrogen interacts less strongly with damage sites than the rare gases. Atomistic model calculations by *Baskes* and *Melius* [7.158a] and *Bisson* and *Wilson* [7.158b] show that the binding energy of H to a vacancy is 0.03 eV in fcc and 0.31 eV in bcc lattices. It is thus smaller or equal to the activation energy of interstitial mobility. There are only few measurements on activation energies for thermal desorption for hydrogen in metals implanted in the fluence range at or below 10^{17} hydrogen ions m^{-2}. At higher fluences, hydrogen trapping at damage sites has been observed in Ni [7.158c] and Al [7.158d]. Trapping is believed to be associated with vacancies in Mo and W [7.159], in Fe [7.160], and in stainless steel (SS 304 LN) [7.161, 162]. Damage trapping of hydrogen by preimplanted He, O, and Ne was found by *Picraux* et al. [7.163],

Besenbacher et al. [7.164], and by *Schulz* [7.133]. The nature of these trapping sites is not yet known. According to the calculations of *Bisson* and *Wilson* [7.158b], the presence of helium in a vacancy tends to repel hydrogen.

7.5 Trapping and Structural Modifications at Intermediate Fluences (10^{18}–10^{21} ions m^{-2})

With increasing gas atom concentration in the implanted layer of a solid, the available trapping sites become multiply occupied. In thermal desorption spectrometry the appearance of high temperature desorption stages is observed which can be explained by the formation of helium vacancy clusters with high binding energy (Fig. 7.8, Table 7.1). It is still not possible to describe the trapping and thermal release behaviour in this fluence range by an atomistic model because the large number of desorption steps overlap into a broad peak.

The fluence range of 10^{18}–10^{21} ions m^{-2} is characterised by complete trapping of the implanted gas within the ion range for helium in metals at or below room temperature and for hydrogen at low enough temperatures to suppress thermal diffusion. At these fluences sufficient damage is present in the lattice so that all atoms coming to rest in the lattice are trapped. This is illustrated by the linear increase of the trapped amount of 1.5–15 keV He in Nb at room temperature, shown in Fig. 7.5 corresponding to 100% trapping efficiency if corrected for kinetic reflection. It is this fluence range which is referred to in this paragraph. The upper end of this fluence range is approxi-

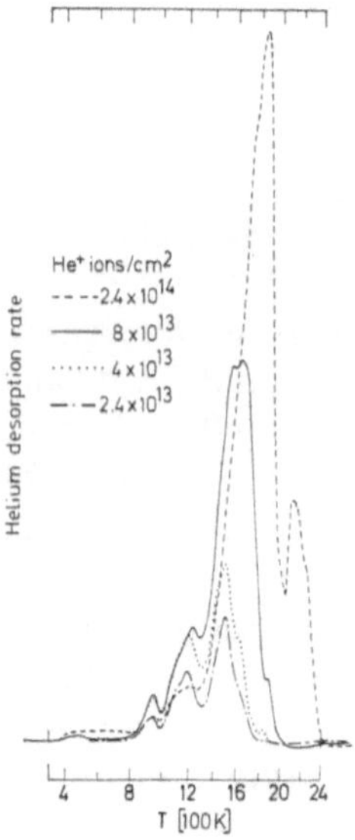

Fig. 7.8. Thermal desorption spectra for intermediate helium doses at 250 eV. Predamage as in Fig. 7.7 [7.35]

mately determined by the starting deviation from 100% trapping and the attainment of almost saturation concentration at the peak of the implantation profile. Saturation concentrations range from 0.3 He-atoms/metal-atom in room temperature fcc metals [7.165] to 0.6 He-atoms/metal-atom in bcc metals [7.61], and about 1 hydrogen-atom/metal-atom in metals like Mo or 316SS [7.124, 125] at 150 K. These gas concentrations in a gas-solid solution would correspond to extremely high pressures. In an undamaged lattice, precipitation will occur under these conditions. But, as shown in the preceeding chapter, the binding energy of helium to vacancies is much larger than to interstitial positions. Thus, in a lattice with a large vacancy concentration, much more He can be retained than in a well-annealed lattice.

If the gas concentration in the lattice exceeds several percent, the vacancies filled with He tend to form larger clusters. As the number of He atoms in a cluster becomes $i \gtrsim 10$ the binding energy to the trap increases [7.138c]. The thermal desorption spectrum for He in W becomes quite independent of the original trap type.

Hydrogen behaves similarly to helium in some materials of very low solubility like Al and Cu [7.166, 167]. No bubbles could be detected after proton bombardment of stainless steel even at very low temperatures [7.168].

Gas bubbles have, however, been observed in metals which dissolve hydrogen endothermically like Cu, if a supersaturated solution of hydrogen is produced by quenching. The hydrogen solution tends to return to thermodynamic equilibrium by precipitation into bubbles within the metal [7.114].

Like void formation, bubbles necessarily cause swelling of the material which leads to an expansion normal to the implanted surface and the formation of compressive stress parallel to it.

7.5.1 Experimental Observation of Bubble Formation

Gas bubble formation was first observed by TEM in Cu implanted with 30 MeV α-particles to a concentration of about 1 at.% and subsequent heating to 900 K [7.29]. The bubbles were found to form first along the edges of the implanted helium range before they form in the center where the helium concentration has its maximum. This can be understood as the trapping of vacancies arriving from all sources outside the range at the borders of the helium rich area. Because of this observation it was first assumed that the thermal production and mobility of vacancies is a necessary condition of bubble growth [7.32].

Gas bubbles have since been found by TEM after helium implantation in many metals as in Cu [7.29], Al [7.30, 169], V [7.170–172], Ni [7.173], stainless steel [7.174], Nb [7.17], and Au [7.176]. Heavy rare gas bombardment has also been found to produce gas bubbles in Pt [7.177]. Bubble formation due to proton bombardment was observed in Al [7.166] and in Cu [7.114, 167, 178a].

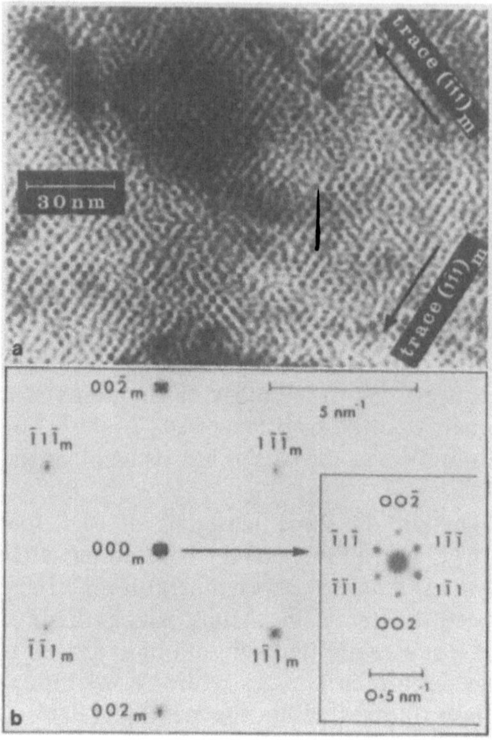

Fig. 7.9. (**a**) TEM micrograph of He gas bubbles in Cu following He ion irradiation to a fluence of 4×10^{17} He-ions cm^{-2} of 30 keV at 300 K. The electron beam is along ⟨110⟩ in the matrix and is normal to the plane of the bubble layer. Directions of the traces of the (111) matrix planes in the (110) plane are indicated [7.183].
(**b**) Selected area electron diffraction pattern from the specimen of (a). The electron energy is 100 keV. The (110) matrix pattern is shown on the left and the supterlattice diffraction pattern centered on the (000)-transmitted beam is shown enlarged on the inset. In the latter slight changes in specimen tilt and translation have been made to obtain more uniform excitation [7.183]

Van Swijgenhoven et al. [7.178b] observed bubbles in *Metglas* [7.178c] due to 5 keV Ar bombardment.

Most observations of bubbles have been made by TEM, but some alternative methods have been applied. The presence of gaseous H in Cu and of He in Al has been demonstrated by specific heat measurements below 3 K. The λ-anomaly found for H in Cu suggests the presence of molecular hydrogen in bubbles [7.179]. For He in Al, *Katyal* et al. [7.180] found a characteristic λ-peak at 2.15 K in the presence of sufficiently large gas bubbles. Following a model of *Ohtaka* and *Lucas* [7.181], the density of He gas in Al films was measured by observation of the shift and broadening of the $1\,^1S_0\rightarrow1\,^1P_1$ transition resonance in absorption (584 Å). He densities of more than 10^{29} He m^{-3} are obtained from samples containing less than 4 at.% average He concentration, indicating an agglomeration in bubbles.

The distribution of gas bubbles is found to be inhomogeneous at low concentrations (10^{-5} He-atoms/metal-atom) with preferential nucleation at grain boundaries and dislocations (Fig. 7.1b). At concentrations of 10^{-3} He-atoms/metal atom, homogeneous nucleation in the whole metal lattice is observed [7.182]. In Mo bombarded with 30 keV He-ions, preferential trapping and the development of larger bubbles at grain boundaries become more apparent with increasing temperature [7.34].

Bubble formation also proceeds at temperatures well below the temperature limit for thermal vacancy mobility. Mo, Cu, Ni, and stainless steel show bubbles after helium bombardment at room temperature. The bubbles arrange in a superlattice structure parallel to the lattice of the solid (Fig. 7.9a, b) [7.1, 2, 33, 60, 183, 184a]. The ordered superlattice structure of the bubbles is, however, restricted to small regions of only a few adjacent bubble planes [7.184d]. At room temperature these bubbles have diameters of about 2 nm and lattice parameters of 6–9 nm and are smaller and more difficult to observe [7.185] than the high temperature bubbles. Bubble densities of $C_b \simeq 10^{25}$ m^{-3} are thus obtained.

The basic processes involved in bubble lattice formation are not yet clear. However, it seems probable that there is a common mechanism responsible for the ordering in a lattice similar to the case of the void superlattice [7.186, 187]. In the latter case, the following sequence has been proposed for the development of the superlattice [7.188]: (i) the initial formation of many small, randomly distributed voids; (ii) the growth of large voids at the expense of small ones; (iii) the appearance of local ordered regions; (iv) the spread of order to adjacent regions. In the case of the bubble superlattice, a fifth stage is invoked involving the interconnection of some bubbles to form elongated pipe or disk-like channels especially close to surfaces [7.1, 184a] (Fig. 7.10) and a sixth stage in which a porous layer of interconnected channels has evolved [7.1], (Fig. 7.11) has been found. For He in room temperature Cu a somewhat different development has been found [7.184c]. In this case the bubble lattice breaks up when a bubble radius $r_b = 1.5$ nm is reached and bubbles grow by coalescence.

The target temperature has no observable influence on the bubble lattice formation up to 470 K in Ni [7.1] and up to 970 K in Mo [7.2]. Thermally activated vacancy mobility starts at 390 K in Ni [7.189] and at about the same temperature in Mo [7.190]. Therefore, thermally activated vacancy migration seems to have no influence on the formation of bubble superlattices. Above 570 K in Ni and above 970 K in Mo, the bubbles are found to arrange in a random structure followed by coalescence of bubbles [7.1, 2].

At high irradiation temperatures a size distribution of gas bubbles with depth is observed. At 850 K, *Ehrlich* and *Kaletta* [7.172] measured an increase by more than a factor of 2 of the mean bubble diameter with depth in V irradiated with 240 keV He. The largest bubbles occur at a depth of 1.2 µm which is ~50% larger than the mean projected ion range.

A detailed study of bubble density and size as a function of fluence and depth for 500 keV He^+ on Ni at 773 K [7.191, 192] shows quite complicated behaviour. At fluences $\lesssim 5 \times 10^{19}$ ions m^{-2}, a surface region of ~0.2 µm is free of bubbles. Beyond this zone the bubble size generally decreases with depth while the bubble density has a maximum of about 10% beyond the calculated mean projected range. The maximum of the volume swelling almost coincides

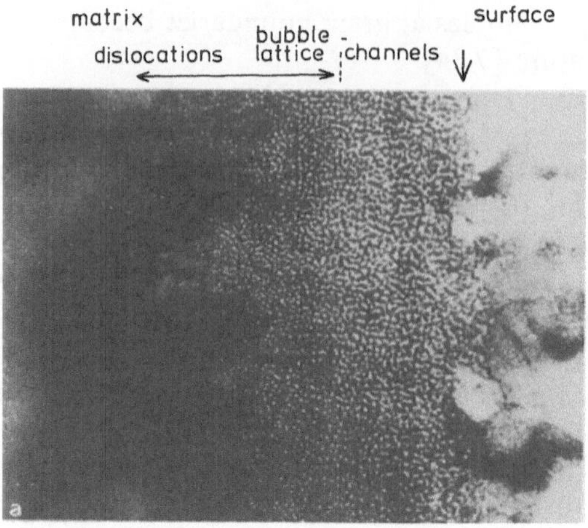

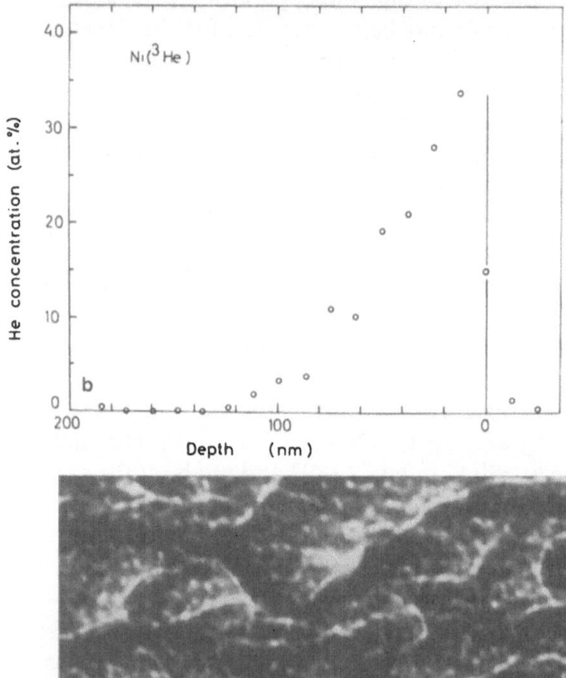

Fig. 7.10. (**a**) TEM micrograph of transverse section of a bombarded Ni surface after implantation of 3×10^{21} He-ions m^{-2} of 8 keV at 300 K. The same preimplantation as in Fig. 7.1a was applied.
(**b**) Depth profile of ^3He in Ni after implantation of 10^{21} He-ions m^{-2} of 8 keV at 300 K as obtained by $^3He(D, \alpha)$ H nuclear reaction [7.1]. The same preimplantation as in Fig. 7.1a was applied

Fig. 7.11. TEM defocus contrasts of interconnected channels and He bubbles in Ni after bombardment with 10^{22} He-ions m^{-2} of 8 keV at 300 K [7.1]. The same preimplantation as in Fig. 7.1a was applied

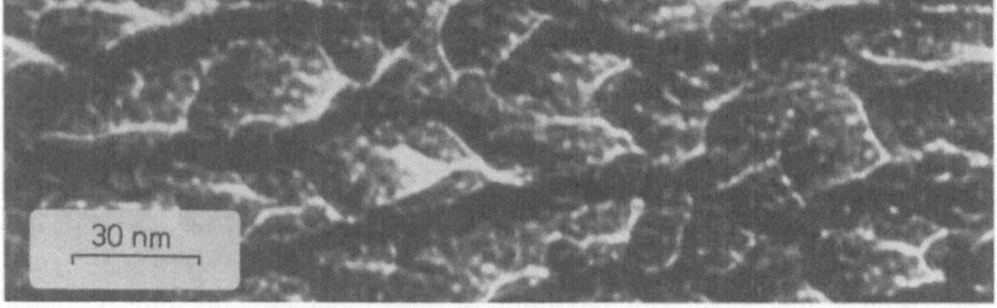

with that of the bubble density. At higher fluences bubble coalescence occurs near the end of the ion range and the bubble density is strongly reduced there. The maximum swelling then coincides with a maximum of the bubbles size.

7.5.2 Swelling and Stress

The formation of damage and of gas bubbles due to ion implantation causes swelling of the material and stresses may build up in the implanted lattice. The relative volume expansion $\Delta V/V$ can be measured by two different methods. One is to assess the total bubble volume per unit volume of the implanted material as obtained from TEM micrographs. This may lead to an underestimation of swelling because it does not include bubbles (and vacancy clusters) which probably exist but are too small to be detected in the TEM. Local swelling ratios obtained are of the order of 5–10% for helium implantation into Mo ($300\,\mathrm{K} \leqq T \leqq 973\,\mathrm{K}$), Cu, Ni and stainless steel at room temperature [7.1, 183, 184, 193] as well as into 973 K V [7.194a]. Peak values of 15% were reported for 20 keV He bombardment of 773 K Ni [7.173].

Another method measures the expansion of the implanted area normal to the surface. A linear expansion of 0.74% per atomic per cent of He is found up to a concentration of 10 at.% He or a fluence of $3 \times 10^{21}\,\mathrm{He\,m^{-2}}$ in Er bombarded with 160 keV He; at higher fluences the expansion becomes nonlinear [7.64]. More recently, *St.-Jacques* et al. [7.65, 66] measured large surface expansion of room temperature Nb under 5–25 keV He bombardment by Tolansky interference microscopy (Fig. 7.12). These measurements have recently been extended to higher temperature (773 K) and to Ti alloy [7.194b]. The surface displacement was found to increase linearly with fluence, independent of the primary energy (Fig. 7.13). Neglecting particle reflection, an average swelling of $1.2 \times 10^{-2}\,\mathrm{nm}^3$ per implanted He-atom is obtained, equivalent to $0.67\,\Omega$ (Ω: volume of a Nb atom), up to a volume expansion of

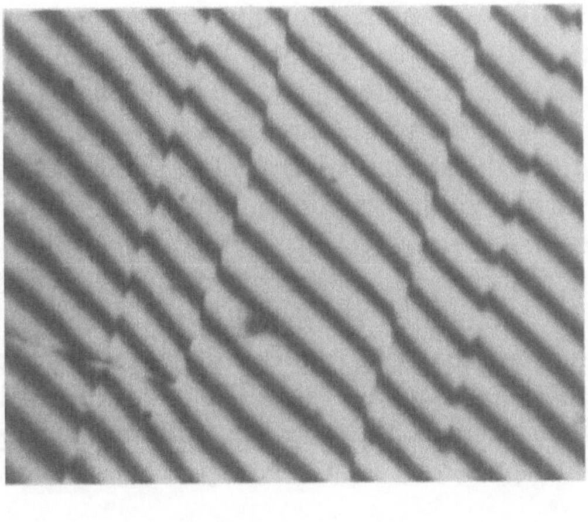

Fig. 7.12. Interference micrograph of electropolished Nb implanted with 20 keV He^+ to a dose of $3.1 \times 10^{18}\,\mathrm{He\,cm^{-2}}$. The surface was partially masked with a TEM grid during implantation (175 ×) (courtesy of Dr. *St.-Jacques*)

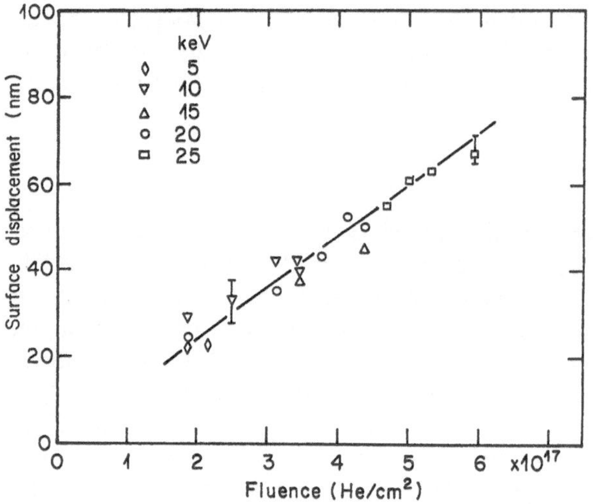

Fig. 7.13. Surface displacement of Nb vs incident He-fluence at energies from 5 to 25 keV. Only the initial linear portion of the curves are plotted [7.66]

$\Delta V/V \simeq 37\%$. An average swelling of $0.02\,\mathrm{nm}^3$/He atom equivalent to $1.1\,\Omega$ estimated by *St.-Jacques* et al. [7.66] implies a particle reflection coefficient $R_N = 0.41$ which is a factor of 2–6 higher than those observed in this energy range [7.195a]. The actual swelling per helium atom will be between these two extreme values. No TEM measurements of Nb-swelling by He-bombardment have been made. But assuming similar dimensions for the bubble lattice as were found in other metals ($r_b \simeq 1.5\,\mathrm{nm}$, $C_b \simeq 10^{25}\,\mathrm{m}^{-3}$), about 70% of the total surface expansion observed by *St.-Jacques* et al. must be due to vacancy clusters which are below the resolution limit of the TEM. *Veilleux* et al. [7.194b] have compared the relative surface expansions of different metals. They conclude that the relative surface expansion increases with the temperature and with the softness of the metal.

For an implanted surface, swelling can only take place in the direction of the surface normal. Lateral forces are compensated by the counter forces from the surrounding unaffected lattice except for self-supporting films in which large area swelling due to He implantation is observed [7.195b]. It depends on the behaviour of the interstitials to what extent lateral stress will form in the implanted layer [7.2]. If the interstitials diffuse to the surface, only swelling will result without stress. The same is true if the interstitials form loops on inclined planes which glide to the surface driven by image forces. Only if interstitial loops are retained in the specimen can they result in the formation of considerable stress in the implanted surface as measured by *EerNisse* and *Picraux* [7.62] for He in Al, Nb, and Mo and by *Hartley* [7.196a] for N_2^+ on stainless steel and by *Primak* and *Monahan* [7.196b] for ceramics. The formation of interstitial loops has also been found for He in Mo by *Mazey* et al. [7.2]. The maximum values of lateral stress that can be built up during irradiation are limited by stress relief processes such as creep or plastic flow

[7.2] or by mechanical failure preferentially near defects if the local stress surpasses the yield value [7.62]. It has been discussed by *Mazey* et al. [7.2] and by *Wolfer* [7.206] that irradiation creep is too small to be significant for stress relief in He-bombarded layers. The former [7.2] assume that plastic flow is more important.

7.5.3 Gas Pressure in Bubbles – Mechanisms of Bubble Growth

The nucleation of gas bubbles in solids can be understood from the work of *Kornelsen* [7.35, 36] and *Caspers* et al. [7.143, 146] as the transmutation of initially single vacancies filled by several gas atoms into gas filled vacancy clusters. For the further growth of bubbles, a number of models have been put forward [7.197–201].

a) High Temperature (Thermal Equilibrium) Model

Bubble growth occurs by collection of vacancies. This may proceed in different ways. At very high temperatures, thermally produced vacancies will arrive at, or leave, a bubble according to thermodynamic equilibrium. The equilibrium condition is that the work done by the expanding gas in the growing bubble is compensated by the increase in surface free energy. Similar to bubbles in liquids, the equilibrium pressure p is given by, e.g. [7.202, 203].

$$p=\frac{2\gamma_s}{r_b}, \tag{7.5}$$

where γ_s is the surface free energy and r_b the radius of the bubbles. The existence of equilibrium state in bubbles at high temperatures has been derived from a number of experimental observations: *Barnes* and *Mazey* [7.202] found that in He implanted Cu at 1073 K, the radius r_b of a bubble obtained by coalescence of n bubbles with radii $r_{b,n}$ is

$$r_b=(\sum r_{b,n}^2)^{1/2}. \tag{7.6}$$

This means that the bubble surface area is conserved during coalescence rather than the volume, in agreement with the equilibrium condition of an ideal gas with the surface free energy where the equilibrium number of gas atoms m relates to the radius r_b as [7.202]

$$m=\frac{8}{3}\frac{\pi\gamma r_b^2}{kT}. \tag{7.7}$$

Smidt and *Pieper* [7.171] find reasonable consistency between the helium amount implanted in V at 1473 and 1673 K and the helium content in bubbles, assuming equilibrium at pressures determined by ideal gas law. Further, the fact that no strain field around gas bubbles can be seen in TEM in He implanted V at 973 K [7.194a] and Nb at 1473 K [7.204] was also taken as evidence that there is no excess pressure in the bubbles.

In the nonequilibrium case, the virtual work argument leading to (7.5) can not be applied. Instead, the sum of surface and bulk stresses counteract the pressure p. Therefore, γ_s is then the surface stress rather than the surface energy, a distinction which is conceptually important. Numerically, both are of the same order of magnitude [7.205, 206].

b) Low Temperature (Overpressurized Bubble) Model

At temperatures where thermal vacancies are not available, two mechanisms have been proposed by which new vacancies can be acquired by a bubble to increase its volume. One is the capture of vacancies produced by the incident beam or other damaging radiation [7.201, 207, 208]; the other is the punching of prismatic interstitial loops out of one side of the bubble by the internal gas pressure [7.197, 198, 208, 209a], an effect similar to the punching of dislocation loops at a Frank-Read source under sufficient shear stress (Fig. 7.14). This is energetically favourable if

$$\left(p - \frac{2\gamma_s}{r_b}\right)\pi r_b^2 b > \mu_s b^2 \frac{r_b}{2} \ln(r_b/r_d), \tag{7.8}$$

where μ_s is the shear modulus, b the burger's vector and $r_d \simeq b$. In Fig. 7.15 the minimum pressure at which loop punching can occur

$$p_{LP} = \frac{\mu_s b}{2\pi r_b} \ln(r_b/r_d) + \frac{2\gamma_s}{r_b} \tag{7.9}$$

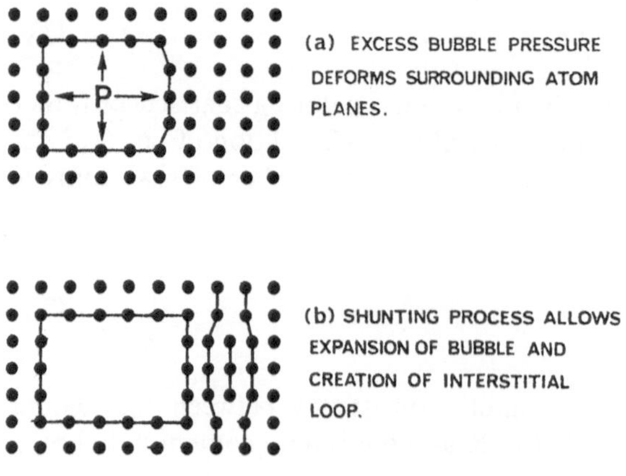

P
(a) EXCESS BUBBLE PRESSURE DEFORMS SURROUNDING ATOM PLANES.
(b) SHUNTING PROCESS ALLOWS EXPANSION OF BUBBLE AND CREATION OF INTERSTITIAL LOOP.

Fig. 7.14a, b. Model for punching of interstitial loops from a gas filled cavity [7.198]

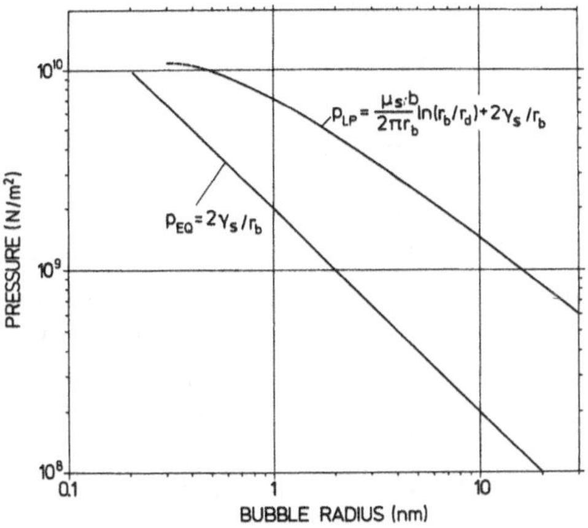

Fig. 7.15. Limiting pressure for loop punching P_{LP}; (7.9) compared to equilibrium pressure P_{eq}; (7.5) as a function of bubble radius. At large bubble radii $P_{LP} \gg P_{eq}$, while at small radii $P_{LP} \simeq 2P_{eq}$

is compared to the equilibrium pressure p_{eq}, (7.5), as a function of bubble radius for a set of γ, μ, and b typical for metals [7.118]: $\gamma_s = 1\,\mathrm{j\,m^{-2}}$, $\mu_s = 10^{11}\,\mathrm{j\,m^{-3}}$, $b = 0.2$ nm. For bubble diameters of 1–2 nm, p_{LP} is in the upper $10^9\,\mathrm{N\,m^{-2}}$ range.

Interstitial loop punching has been observed for hydrogen bubbles obtained by quenching a saturated hydrogen solution in Cu. Thereby, large oversaturation was obtained which resulted in the precipitation of excess gas as bubbles [7.114]. Further evidence of loop punching from pressurized He bubbles was obtained from neutron irradiated AlLi alloy [7.209b].

c) Percolation Model

Another mechanism of bubble growth without vacancy production has been suggested by *Baskes* and *Wilson* [7.199, 148e]. They propose spontaneous rearrangement of random defects in close proximity to each other leading to the formation of bubbles. It is assumed that when two or more "vacancies" which are filled with up to 6 helium atoms per vacancy occur within an interaction distance r, they spontaneously rearrange into a common bubble with a decrease in free energy. Statistical calculations of vacancy production in a helium bombarded lattice showed that at a critical fluence some of these bubbles grow catastrophically. This was related to the onset of blistering.

At present none of these models is sufficiently supported by experimental evidence to allow a critical evaluation.

d) Equation of State for Gas in Bubbles

In most of the work on gas bubble growth it has been assumed that the equation of state of the gas can be approximated by either the ideal gas law or by van der Waal's equation. For He, these equations of state are no longer valid for atom densities $n_{He} > 10^{27}$ He-atoms m^{-3}.

The actual He density in gas bubbles has recently been measured by different methods: *Donnelly* et al. [7.210a, b] have measured the blue shift and broadening of the $^1S_0 - ^2P_1$ transition of He due to high gas density in bubbles by optical absorption spectroscopy. In Al foils implanted with 1.45 at.% He they find a He density of $1.18 \times 10^{29}\,\mathrm{m}^{-3}$. The blue shift of the same transition was measured by electron energy loss spectroscopy (EELS) in Al and Ni [7.210c–g]. He densities up to $1.4 \times 10^{29}\,\mathrm{m}^{-3}$ (Al) and $2 \times 10^{29}\,\mathrm{m}^{-3}$ (Ni) were found. These measurements are supported by small angle x-ray scattering spectroscopy (SAXS) on single crystal nickel foils implanted uniformly with He. Bubbles with uniform radius are found which increases with the He fluence while the bubble distance remains unchanged. A He density of $1.8 \times 10^{29}\,\mathrm{m}^{-3}$ is found and a corresponding pressure of $3 \times 10^{10}\,\mathrm{N\,m}^{-2}$ [7.184d].

For He several pressure-density relations have been proposed. The empirical relation of *Rowlinson* [7.211]

$$p = 4.83 \times 10^7 \exp(5.15 \times 10^{-29}\, n_{\mathrm{He}})\,[\mathrm{N\,m}^{-2}], \tag{7.10}$$

where n_{He} is in atoms m^{-3}, is frequently used. *Wolfer* [7.212] has derived an equation of state from the theory of the liquid state and the known interatomic potential for helium. The implicit equations which can only be solved numerically are in good agreement with measurements up to a pressure of $10^8\,\mathrm{N\,m}^{-2}$ [7.213] but pressure increases too strongly with density above $10^8\,\mathrm{N\,m}^{-2}$ compared to recent experimental data of *Mills* et al. [7.214], while (7.10) is about a factor of 2 too high in the whole region of interest. Recently *Trinkaus* [7.210e] has developed a high density equation of state for He which is in good agreement with the high pressure measurements of *Mills* et al. (Fig. 7.16).

It is still an open question whether all implanted He is eventually trapped in microscopic bubbles or whether submicroscopic cavities may take up an essential fraction. Assuming that all trapped gas is completely contained in the bubble space visible in the TEM which corresponds to a swelling of 5–10%, gas densities of $3\text{–}6 \times 10^{29}\,\mathrm{m}^{-3}$ would result. The corresponding pressure is $> 10^{11}\,\mathrm{N\,m}^{-2}$ and therefore about an order of magnitude larger than P_{LP}. As a consequence, the larger fraction of the implanted gas must be assumed to be trapped in submicroscopic bubbles or vacancy clusters. This conclusion was first drawn by *Kaletta* [7.194a] for He in 973 K Vanadium, it was supported by *Terreault* et al. [7.216a] and *Johnson* et al. [7.215] for Cu, by *Fenske* et al. [7.192], *Jäger* and *Roth* [7.1], and *Johnson* and *Mazey* [7.184a, 215] for Ni and stainless steel. On the other hand the measurements of *Manzke* et al. [7.210e, g] and of *Haubold* and *Lin* [7.184d] seem to exclude the presence of submicroscopic bubbles which could account for the excess gas.

Van Swijgenhoven et al. [7.216b] have obtained a good fit between their measured bubble volumes, He-concentration profile and values calculated using the loop punching model (7.9) in Ni, if they assume a fraction of He in visible bubbles increasing from 0.3 to 0.55 with increasing fluence.

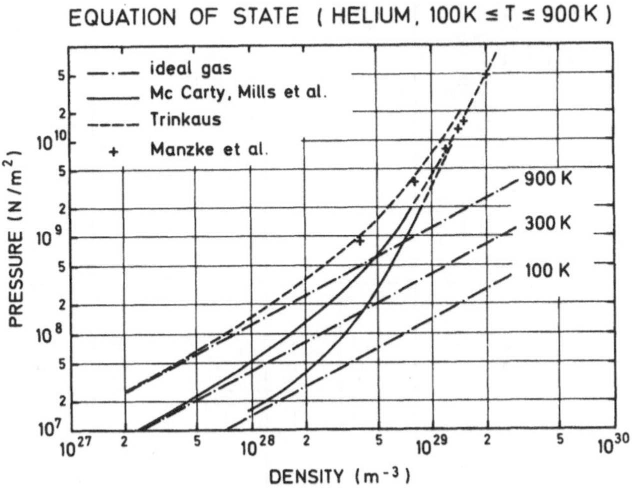

Fig. 7.16. Equation of state for He (100 K ≦ T ≦ 900 K). Ideal gas (–·–), experimental values by *McCarty* [7.213], *Mills* et al. [7.214] (—), *Manzke* et al. [7.210e, g] (+), and high-density equation of state by *Manzke* et al. [7.210e] (---)

7.6 High Fluences (10^{21}–10^{23} ions m^{-2}); Blistering and Flaking

When the implanted fluences reach a level of ~10^{21} to 10^{23} ions m^{-2}, a saturation of trapped gas in a certain part of the implanted range is attained and a sudden transition to a new surface state has been observed in almost all metals for helium bombardment at energies above 1 keV. This is characterised by the appearance of blisters with circular or irregular shape [Fig. 7.1c (upper part) and Fig. 7.2a], large irregular flakes (Fig. 7.1c, lower part) or surface pinholes (Fig. 7.1d), depending among other parameters on the ion species, target material and temperature. The formation of these surface structures can be correlated with either burst-like [7.43, 217] or gradual emission of implanted gas. For hydrogen bombardment of metals, surface modifications occur at temperatures where diffusivity is low. Also in nonmetals, surface effects depend largely on diffusivity.

In Table 7.2, ion/target combinations are listed in which blistering and flaking has been investigated. Due to its possible contribution to first wall erosion in fusion reactors, the materials graphite, Al, V, Nb, Mo, stainless steel, Ni, and nickel base alloys have been most extensively studied. Ion energies range from ~1 keV to many MeV and a wide range of temperatures are investigated.

In the following, an attempt has been made to give a survey of the more general trends observed. Special features of single materials (Table 7.2) are not exhaustively treated.

Table 7.2. Experimental observation of blistering and flaking

Type, structure preparation	Temp. [K]	Energy [keV]	Special features	Ref.
a) Single element Targets				
H^+–Be				
Sintered	R.T.	33	Cracks	[7.218, 219]
D^+–Be				
	473	6	Circular blisters	[7.220]
Sintered, mech. polished	R.T.	15	Circular blisters, cracked	[7.221]
He^+–Be				
Sintered, mech. polished	R.T.	15	Circular blisters	[7.221]
Hot rolled, annealed, unannealed	300; 550	20; 25	Blistering at 300 K Exfoliation at 550 K	[7.222]
Polycryst. foil (a) hot rolled	R.T.	10–100	$d \propto t^{1.25}$	[7.223]
(b) plasma sprayed (c) sintered	R.T.	100	Reduction of blistering in low density material	[7.224, 225]
Plasma sprayed	475; 675	20; 300		[7.226]
H^+–C				
Pyrolytic	523	$\sim 10^{-3}$		[7.228]
Pyrolytic	320–420	0.5–7.5	No blistering	[7.227]
Cloth		20		[7.229]
Glassy	873	10, 20, 80	Cracks	[7.230]
Pyrolytic	R.T.	100	Blistering	[7.231, 232]
D^+–C				
Pyrolytic	R.T.–793	6.7; 15	No blistering	[7.233]
Coating	R.T.	10, 20	Growth features	[7.234]
Coating		40–120		[7.223, 235]
ATJ, pyrolytic Cloth	R.T., 873	100	Small flakes	[7.236]
He^+–C				
Cloth	R.T.–920	0.5–1.1	Blistering (?)	[7.237]
Coating	R.T.	10, 20	Growth features	[7.234]
Glassy	873	10–80	Cracks	[7.230]
Cloth		20		[7.229]
Coating, ATJ, pyrolytic	Ambient	40–120		[7.238]
Pyrolytic	R.T.	100	Cones	[7.231, 232]
Single crystal, pyrolytic	293–793	70–100		[7.239, 240]
Pyrolytic	$\sim$520	2000, 4000		[7.228, 241, 242]
Pyrolytic		1900	Grooves, flakes	[7.243, 244]
Carbosital USB-15	370	10–100	No blisters or cracks	[7.418]
Pyrolytic	300–1000	70–100	Blisters only on thin films	[7.419]
N^+–C				
Pyrolytic		1100		[7.243, 244]
Ar^+–C				
Pyrolytic		450		[7.243, 244]
D^+–Si				
Single crystal	100, 300	200		[7.245]
Single crystal		1–1000		[7.246]
Ar^+–Si				
Single crystal		$\geqq 100$		[7.247]
Ar^+–Ge		450		[7.248]

Table 7.2 (continued)

Type, structure preparation	Temp. [K]	Energy [keV]	Special features	Ref.
He^+–Al				
Single Crystal	Beam heated	25–4000	Blistering	[7.51]
Polycrystalline	77–473	7–450	Blistering, pitting	[7.54, 249]
Polycrystalline	288	100		[7.250]
		20–100		[7.251]
		120		[7.218,252]
D^+–Al				
Polycrystalline		1–1000	Blistering (?)	[7.246]
He^+–Al				
Foil	Beam heated	100–800	Exfoliation	[7.41]
Single crystal	Beam heated	25–4000	Blistering	[7.51]
Annealed, SAP	R.T.	100	Blistering, reduced in SAP	[7.253]
Polycryst. foil, SAP	R.T.–600	100–500	Exfoliation	[7.225,254]
Polycrystalline	80–775	20	Blistering	[7.255]
			Exfoliation	[7.180]
			Multilayer exfoliation	[7.256,257]
Coating foil, SAP	150–773	300		
SAP		100	Reduction of erosion	[7.258]
		20–100		[7.251,218]
Electrodeposited	30–850	10–80	Blistering, flaking	[7.259]
		40	Stress relief by blistering	[7.62]
Evap. film	70–R.T.	5		[7.260]
			Microstructural eff.	[7.410]
Single crystal	R.T.	150	Mult. layer spalling	[7.420]
Polycrystalline	390	175	Alt. tensile and compressive loading	[7.338b]
Foils		5		[7.421]
Ne^+–Al	R.T.	9	Blistering after heating	[7.261]
Xe^+–Al				
Vapour depos.	R.T.	≦160	No blisters	[7.262]
H^+–Ti				
Polycrystalline	320–420	20	Blisters	[7.263]
	78–500		No blistering at R.T.	[7.264]
Single crystal		20–35	Blistering	[7.265]
D^+–Ti				
Polycrystalline	140–500	6.6	Blistering at 160 k	[7.135]
He^+–Ti				
Annealed, cold worked	350–675	300	Blistering, flaking	[7.266]
Porous, dense		10–40	More flaking on dense mat.	[7.267]
		20–100		[7.268]
H^+–V				
MARZ grade, polycrystalline	183–390	150		[7.269]
MARZ grade, polycrystalline	170–390	150	No blistering	[7.270]
Cold rolled, vacuum annealed	R.T.	150	Blistering (?)	[7.271]
		20–100		[7.251]
Single crystal		10–40		[7.267]
Foil, polycryst.		Thermal	Bubbles, blisters, grain ejection by rapid heating	[7.422]
D^+–V		60	No blistering	[7.272,273]

Table 7.2 (continued)

Type, structure preparation	Temp. [K]	Energy [keV]	Special features	Ref.
He^+–V				
MARZ grade, polycrystalline	673–1473	300		[7.269]
MARZ grade, polycrystalline	673–1473	300	Blistering, flaking, holes	[7.270]
MARZ grade, polycrystalline	673–1473	300	Flaking, holes	[7.204]
MARZ grade, polycrystalline	673–1473	300	Flaking, swelling	[7.274]
Cold rolled foil, bulk material	373–1073	80	Blistering, flaking	[7.275]
Polycrystalline	723, 973	200, 2000	Bubbles, blistering, flaking	[7.194a, 254]
Polycrystalline	R.T.	100–1000	"Deckeldicke"	[7.276]
	R.T.	100–1500	"Deckeldicke"	[7.277]
		20–100		[7.251, 268]
	R.T.	500		[7.272, 278 279, 320]
		20		[7.280]
Single crystal		10–40		[7.267]
	370–1070	80		[7.281]
Polycrystalline	350–500	40	Flaking, pinholes	[7.282]
Polycrystalline		20	Multienergy, multi-angle bomb.	[7.423]
Ar^+–V				
Polycrystalline	<570	300, 400	Multiple energy	[7.283]
Ar^+–Fe				
Single crystal	R.T.	10	Blistering (?) on (110) face	[7.284]
H^+–Ni				
Polycrystalline		40–140	No blistering	[7.51]
		20–100		[7.251]

Type, structure preparation	Temp. [K]	Energy [keV]	Special features	Ref.
D^+–Ni				
		150–400		[7.285, 286]
Polycrystalline	773	50	Simultaneous bomb. with He^+	[7.287]
Polycrystalline	120	150–400		[7.288]
MARZ grade, polycrystalline	33	10		[7.127]
He^+–Ni				
Polycrystalline	773	100	Simultaneous bomb. with D^+	[7.287]
Single crystal	R.T.–1230	500	Flaking, blistering, pinholes	[7.289]
Polycryst. foil	773	20,500	Bubble distrib.	[7.173]
Polycryst. foil	R.T.	20, 500	$d \propto t_B^{1.15}$	[7.223]
Polycrystalline		40–140	Exfoliation	[7.51]
		20–100		[7.251]
Single crystal				[7.165]
	670–920	20	Swelling, fracture	[7.290]
Polycrystalline	R.T.	36	Blistering	[7.52]
	673–973	20–80	Blistering, swelling	[7.291]
Polycrystalline, foil	R.T.–1173	20, 500	Bubble distribution	[7.191]
Single + polycryst.	R.T.	40	Reemission + surf.str.	[7.388b]
Single + polycryst.	R.T.–1200	20–150	Reemission + surf.str.	[7.257b]
Polycrystalline foil	200–600	5	Development of bubble distrib.	[7.216b]
Polycrystalline		50	Incr. of hardness	[7.424]
Predamaged with C^{++}	770	40	Increased flaking	[7.425]
N^+–Ni				
		1500		[7.292]
Ar^+–Ni				
	900	1000		[7.292]

Table 7.2 (continued)

Type, structure preparation	Temp. [K]	Energy [keV]	Special features	Ref.
H^+–Cu				
Polycrystalline		100	Blistering	[7.44]
Polycrystalline		70–140	Exfoliation	[7.51]
Polycrystalline	R.T.	250	Blisters, holes, by beam heating	[7.293b]
Polycrystalline	R.T.	325	Blistering, grain boundary movement	[7.293a]
D^+–Cu				
Single crystal		125	Surface pitting	[7.43]
Polycrystalline	620	200		[7.193, 294]
		1–1000	Plasma focus	[7.246]
Polycrystalline		60	Blistering	[7.273]
	120	200	Crit. fluence	[7.295]
Polycrystalline	100–400	200	3 blister-regions	[7.426]
Polycrystalline	120	200	Blister ring at periphery of impl. spot	[7.427]
Foil	120	200	Blisters + flakes ($t_B > t_F$)	[7.184c]
He^+–Cu				
Evaporated film	R.T.	25	Double peak	[7.296]
Foil	R.T.	1–20	Double peak	[7.216a, 297]
OFHC, thin foil evaporated film	300, 500	20	Double peak	[7.222]
Ar^+–Cu				
Polycrystalline	R.T.	10	Blistering	[7.284]
He^+–Zr				
		2000		[7.298, 299]
H^+–Nb				
Polycrystalline	973–1273	150	Crow foot pattern	[7.300, 301]
Cold rolled, vacuum annealed	R.T.	150	Blistering	[7.271]

Type, structure preparation	Temp. [K]	Energy [keV]	Special features	Ref.
D^+–Nb				
Single crystal				[7.272]
Polycrystalline		500	Crow foot pattern	[7.302]
	520–1000	250–300	Blistering	[7.303]
He^+–Nb				
Single crystal	150–1300	0.5–9		[7.304]
Polycrystalline	R.T.	1–15		[7.49]
Polycrystalline	R.T.	500	Flaking	[7.56]
Single crystal	1173	500	Crowfoot pattern	[7.305]
Single crystal	R.T.–1173	500–1500	Crowfoot pattern	[7.306]
Polycrystalline	1073–1673	6		[7.307]
	R.T.	50–350		[7.98]
Single crystal	163–1273	0.5–9		[7.308, 309]
Polycrystalline	673–1473	300		[7.204, 310]
Foils	R.T.	10		[7.107]
Polycrystalline	293–1273	9, 15		[7.311]
Cold rolled	R.T.	3–500	Multienergy implant.	[7.312]
Cold rolled	R.T.–973	5–15	Multienergy implant.	[7.313]
Polycrystalline	R.T.	10–1500		[7.314]
		100		[7.254]
MARZ grade, cold rolled	R.T.–973	1–15		[7.315]
Polycrystalline	R.T.	100–1500	t_B related to $\langle R_p \rangle$	[7.103, 277]
Cold rolled foil		5–25	Swelling	[7.65]
	1000–1100	20	Blister bursting	[7.316]
	1073	20		[7.317]
		20–100		[7.251]
Single crystal		30–100		[7.165]
		20–1500	$d = f(t_B)$ replacement	[7.318]
Polycrystalline	R.T.	30, 50		[7.319]
	R.T.–1170	500		[7.278]
Single crystal	R.T.	1.5–15		[7.61, 123]
Polycrystalline	R.T.	9–100	Very high fluence	[7.5, 320]

Table 7.2 (continued)

Type, structure preparation	Temp. [K]	Energy [keV]	Special features	Ref.
He$^+$–Nb				
Single crystal				[7.272, 279, 231]
			Multienergy	[7.322]
			Multienergy	[7.323]
Single crystal				[7.324]
Polycrystalline		500–1500	Crowfoot pattern	[7.302]
		15		[7.325]
	100–770	450		[7.326]
Annealed, cold worked	R.T.		Multienergy	[7.327]
		1500		[7.328]
		60	Stress relief	[7.62]
Single crystal		2–4	Blistering	[7.393]
Polycrystalline	570–973	20–500	$d \propto t_B^{1.19}$	[7.329, 392]
Polycrystalline	R.T.	5–25		[7.66]
Single crystal	573	1000		[7.330]
		0.5–3500	Multiple energy impl.	[7.331]
MARZ grade	300	20	Double peak	[7.222]
Foil	773	10	Large bubble development at $f>f_c$	[7.416]
Polycrystalline	570	100	Alt. tensile and compressive load	[7.338b]
Ne$^+$–Nb				
Single crystal	140	850		[7.332]
H$^+$–Mo				
Polycrystalline	R.T.	100–200	Rough surface	[7.333]
Polycrystalline	100–1473	150, 300		[7.269]
Polycrystalline	100–1473	150		[7.270]
Cold rolled, annealed	R.T.	150	Blistering	[7.271]
		20–100		[7.251, 334]
	R.T.	100	Scratched surface	[7.232]

Type, structure preparation	Temp. [K]	Energy [keV]	Special features	Ref.
H$^+$–Mo				
Polycryst., rolled	<420	0.02–20	Deckeldicke = $100 \times \langle R_p \rangle$	[7.11d]
Sintered	R.T.–770	25	RT-370 K: blisters, sputt. at high fluence 420–520 K: "ravaged" surf., craters 570–770 K: no blisters or craters	[7.11a]
Zone refined	R.T.	5	Observation of gas bubbles	[7.428]
D$^+$–Mo				
Polycrystalline	R.T.	15		[7.221]
		60		[7.273]
Polycryst. foils		40–120		[7.335]
MARZ grade, polycrystalline	33	10	Blistering	[7.127]
Zone refined	R.T.	0.02, 0.3	Blistering caused by diff.	[7.11f]
He$^+$–Mo				
Polycrystalline	R.T.	15		[7.221]
Polycrystalline	293–1273	18–60	Bubble lattice	[7.34]
Polycrystalline	R.T.	100		[7.333]
Polycrystalline	R.T.	75–350		[7.336]
Polycrystalline	673–1473	150, 300		[7.204, 269, 270]
Rolled foil, zone refined		25–100	Bubble lattice	[7.337a]
	R.T.	18		[7.57]
		36		[7.52]
		20–100		[7.251]

Table 7.2 (continued)

Type, structure preparation	Temp. [K]	Energy [keV]	Special features	Ref.
He^+–Mo				
Rolled foil, zone refined	R.T.	100–400	Scratched surface	[7.232, 338–340]
Thin film	<1223	200		[7.341]
		10, 100	Stress relief	[7.62]
Polycrystalline	R.T.	60–200	Sequential energy implantation	[7.342]
Polycrystalline	920–970	20	Rough surface	[7.394]
Single crystal {111}	R.T.	200	Spalling and blist.	[7.429, 430]
Polycrystalline	R.T.	distr. <40	Irreg. shaped blist.	[7.431]
Single crystal	R.T.	200	Prestressing	[7.432]
CVD-coatings	670–870	40	Blisters only on polished coatings	[7.433]
Chem. dep.	R.T.	40	Blisters only on polished surface	[7.434]
Zone refined	R.T.	5	Bubbles $24 \times R_p$ (also "Deckeldicke")	[7.428]
Polycrystalline		20	Multienergy, multi-angle bomb.	[7.423]
Ar^+–Mo				
Single crystal		300	Blistering	[7.343]
		450		[7.344]
H^+–Pd				
	78–500		No blistering at R.T.	[7.264]
D^+–Pd				
MARZ grade	33	10	Blistering	[7.127]
He^+–Pd				
Polycrystalline	93–470	300		[7.345, 346]
Polycrystalline	R.T.	4800–5700	Irregular domes	[7.347]
Polycrystalline	R.T.	5000	Microtomes	[7.348]
Ar^+–Ag				
Single crystal	R.T.	10	Blistering	[7.284]
H^+–Sn				
Single crystal	R.T.–500	1500–4000		[7.349]
He^+–Sn				
Single crystal	R.T.–500	1500–4000		[7.349]
He^+–Er				
Evaporated film	R.T.	160	Swelling	[7.64]
			Blistering	[7.350]
H^+–Ta				
Single crystal	78–500	20–35		[7.264, 265]
D^+–Ta				
MARZ grade	33	10		[7.127]
H^+–W				
Single crystal	<500			[7.264, 265]
He^+–W				
Single crystal	78	0.5–2.5		[7.351]
Single crystal		0.2–3		[7.352]
Zone refined	R.T.	5		[7.428]
He^+–Re				
Polycrystalline	300–1200	21		[7.353]
He^+–Pt				
Polycrystalline	R.T.	36	Blistering	[7.52]
H^+–Au				
Polycrystalline		71		[7.51]
He^+–Au				
Polycrystalline		70		[7.51]
	R.T. (?)	3520	Structural investigat. in blister cover and bottom	[7.435]
Ar^+–U		15, 25		[7.354]

Table 7.2 (continued)

Type, structure preparation	Temp. [K]	Energy [keV]	Special features	Ref.
b) Vanadium, Titanium, Zirkonium, Niobium, Molybdenum alloys				
Deuterium				
TZM		60	Less blistering than in Mo	[7.273]
TZM	Uncontr.	40, 60	Little blistering	[7.335]
TZM	Uncontr.	0.6	Shock tube	[7.355]
T-111	Uncontr.	0.6	Shock tube	[7.355]
Helium				
V-3% Ti		200,2000		[7.194a]
V-20% Ti				
Nb-alloy	700–800	20		[7.316]
VN-2AE				
V–Zr–C	300–1400	20		[7.280]
Nb–Mo–Zr				
Nb–Zr–C				
TZM				[7.356]
TS–M–6	390	40	Eff. of heat treatment	[7.436]
V Ti	1100	200–2000	Microhardness changes	[7.437]
c) Steel and Ni-alloys				
Hydrogen				
SS 316	100–605	20		[7.263, 357]
SS 302				
SS		0.5–7.5		[7.227]
SS	300–1000	20, 25	Disintegrated blist.	[7.358]
SS 316	100–1000	150, 300	Blistering only below R.T.	[7.269, 270]
Hydrogen				
EP–838	520			[7.264]
SS 304 films	R.T.	1–20	Buckling	[7.359]
SS 316	320–930	0.3–6	High fluence	[7.89]
SS 321	320	Thermal H	Pits or bubbles, not with Pd coatg.	[7.11c]
Deuterium				
SS 304 films	R.T.	1–20		[7.359]
SS 316	150	7	Blistering	[7.124]
SS 316		60	No blistering	[7.273]
4301	R.T.	15		[7.221]
SS		5–20		[7.360]
PE 16	77	33		[7.361]
Inconel	77	33		[7.361]
SS 304				[7.362]
SS 316	R.T.	0.6	Shock tube	[7.365]
AISI 4130				
PE 16				
SS	470–770	20	Simult. bomb. with He	[7.438]
Helium				
EP-838	520	175	Complete exfoliat.	[7.363]
SS 304	R.T.–900	12	Equil. struct.	[7.364]
SS 316				
PE 16				
Inconel 600				[7.365]
Hastelloy B				
SS 316		300	Simultaneous with Ar^+-sputtering	[7.366]
SS 304 films	R.T.	1–20	Buckling	[7.359]
SS 316				[7.298]
Inconel 718		2000		[7.299]

Table 7.2 (continued)

Type, structure preparation	Temp. [K]	Energy [keV]	Special features	Ref.
Helium				
$Cr_{18}Ni_{40}Mo_5$	293–1073	40		[7.367]
		20–80	Angular dependence	[7.395]
SS 304		80	In situ TEM	[7.368]
SS 316				
Cr_{16}, Ni_{14}, Mo_3 Cr–Ni-alloy	373	40		[7.369]
SS 316	723	100	15 Layer exfoliat.	[7.370]
SS 316	575–775	3–20	Multienergy impl.	[7.371]
SS 316	100–970			[7.217]
SS				[7.279, 372]
SS	300–1000	20, 25		[7.358]
SS	<370	40, 100	Rep. blistering	[7.50]
SS	480–700	175	Blistering, flaking	[7.373]
SS		20–100		[7.251]
SS	700–800	15, 20		[7.316, 317]
SS 304	570	3, 10, 15	Recrystallization	[7.374]
SS	R.T.–920	6, 10		[7.364]
4301	R.T.	15		[7.221]
SS 316	100–1000	150, 300		[7.269, 270]
SS		5–20	Blister explosion	[7.360]
SS	R.T.	40	3 blister generations	[7.375]
Inconel 625	R.T.	100	39 layer exfoliation	[7.406]
SS 304, polycryst.	Water cooled	2000	Exfoliat., spher. struct. in exfoliated area	[7.439]
Inconel 625		20–100	Suppression of rep. blistering	[7.440]
SS 304, SS 316	R.T.	100	Exfoliation	[7.441]
SS 316	R.T.	2–10	Multienergy impl.blister disappear.	[7.442]
Ni 201-alloy		50	Incr. of hardn.	[7.424]
NIMONIC 90		50	Incr. of hardn.	[7.424]

Type, structure preparation	Temp. [K]	Energy [keV]	Special features	Ref.
Helium				
SS	370–1070	800–4000	Large exfoliat.	[7.443]
INCO-625	370–1070	800–4000	Large exfoliat.	[7.443]
SS	470–770	40	Simult. bomb. with D	[7.438]
SS predamaged by C^{++}	770	40	Increased flaking	[7.425]
SS		40	Blistering due to dissolved H	[7.444]
SS 304	300–770	100	Surf. roughness by blisters	[7.445]
SS 316				
Nitrogen, argon				
SS	900	1000		[7.292]
SS		100, 200		[7.196]

d) Metallic Glasses

Type, structure preparation	Temp. [K]	Energy [keV]	Special features	Ref.
Helium				
Metallic glass $Ni_{45}Fe_5Co_{20}Cr_{10}Mo_4B_{16}$	R.T.	50	$f_c = 4 \times 10^{17}$ He cm^{-2}	[7.446]
Metglas 2826A $Fe_{32}Ni_{36}Cr_{14}P_{12}B_6$	R.T.	40	2. gen. blisters in 1. gen. exfoliated spots	[7.447]
Metallic glass $Fe_{40}Ni_{38}Mo_4B_{18}$	300	5	Irregular shape	[7.216b]
Vitrovac $B_{20}Fe_{40}Ni_{40}$	130–430	2–20	Blistering for $E > 18$ keV $f_c = 8 \times 10^{17}$	[7.448]
Argon				
Metglas 2826 MB $Fe_{40}Ni_{38}Mo_4B_{18}$	R.T.	5	Bubbles at 6 nm depth, blisters at $f_c = 10^{16}$ cm^{-2}	[7.178b]

Table 7.2 (continued)

Type, structure preparation	Temp. [K]	Energy [keV]	Special features	Ref.
e) Platinum alloy				
Helium				
Pt-10% Rh			Whiskers	[7.376]
Pt-10% Rh			Elongated blisters (?)	[7.449]
Argon				
PtSi	R.T.	20–160	Repeated bubble formation	[7.377]
f) Compounds (Carbides, Oxides, Nitrides etc.)				
Hydrogen				
MgO		140		[7.51, 378]
SiC				
B_4C				
Si_3N_4				
TiB_2				
Al_2O_3				
ZrO_2				
$BaTiO_3$				
SiC		10		[7.379]
KCl	300–670	300		[7.380]
$LiNbO_3$	R.T.	100	Surf. Acoust. waves	[7.450]
Deuterium				
BeO		6		[7.220]
SiC		140		[7.378]
B_4C				
Si_3N_4				
TiB_2		140		[7.378]
Al_2O_3				
ZrO_2				
$BaTiO_3$				

Type, structure preparation	Temp. [K]	Energy [keV]	Special features	Ref.
Deuterium				
SiC		10–15		[7.381]
TiB_2	R.T.	5–100		[7.234]
TiN				
HfN	973	40		[7.382]
TiB_2				
TiB_2				[7.411]
Helium				
SiC		600–3500		[7.383]
Ceramic coatings				[7.276]
SiC	390–870	50–175	Exfoliation	[7.384]
Corundum				
Spinel				
Rutile, Peridot		100–140	Exfoliation	[7.40]
SiC		600		[7.385]
SiC	300–1000	70–100	Flaking at high T	[7.419]
SiC		140		[7.378]
B_4C				
Si_3N_4				
TiB_2				
Al_2O_3				
ZrO_2				
$BaTiO_3$				
Glass		150		[7.386]
KCl	300–670			[7.380]
TiB_2	R.T.	5–100		[7.234]
$LiNbO_3$	R.T.	100	Surf. acoust. waves	[7.450]
Mo_2C	R.T.	40	Blister enhancement by polish	[7.434]
Mo–TiC	R.T.	100–400		[7.451]
TiC				

Table 7.2 (continued)

Type, structure preparation	Temp. [K]	Energy [keV]	Special features	Ref.
Neon				
$LiNbO_3$	R.T.	100	Surf. acoust. waves	[7.450]
Argon				
KCl	300–670	300		[7.380]
$LiNbO_3$	R.T.	100	Surf. acoust. waves	[7.450]
Mo–TiC	R.T.	100–400		[7.451]
Xenon				
Orthoclase,	R.T.	50		[7.452]
Labradorite				
Augite, Bytownite				
Olivine, Anorthite				
Alleite				
Oligoclase				
Andesite				

Type, structure preparation	Temp. [K]	Energy [keV]	Special features	Ref.
Coatings				
Hydrogen				
SiC	770	1		[7.453]
TiC	320–770	100		[7.454]
Deuterium				
SiC on Mo (CVD)		7		[7.455]
TiC (CVD)	R.T.	20, 40, 60	Blistering, flaking	[7.456]
SiC on Mo				
SiCAlO on Mo		7	$f_c > 10^{18}$ cm^{-2}	[7.457]
Helium				
TiC (CVD)	R.T.	20, 40, 60	Blistering, flaking	[7.456]
TiC	320–770	200		[7.454]

7.6.1 Conditions for Blistering

Blistering, flaking and pinhole formation of bulk materials by ion bombardment have only been observed for gas ion bombardment. While it is almost always found for He bombardment of metals, special conditions (e.g., low temperature and high flux density) have to be met for hydrogen implantation in many materials in order to produce evolutional surface structures. For heavy gases (N, Ne, and Ar), the formation of blisters has been observed only at higher energies ($E_0 \gtrsim 100$ keV) [7.196a, 243, 244, 247, 248, 283, 292, 332, 343, 344, 380], while at $E_0 < 100$ keV only a few cases of blistering have been reported [7.178b, 284, 261, 354] (see also Table 7.2).

Two major conditions must be fulfilled for these phenomena to occur:

i) A critical concentration of implanted gas must be achieved in the solid with $0.3 \lesssim c_{\text{crit}} \lesssim 1$ gas atoms/target atom [7.15, 52, 61, 63, 107, 264, 366]. As the maximum concentration is also limited by particle reflection and surface regression due to sputtering, blistering and flaking is only observed if

$$\frac{1-R_N}{Y} \geqq c_{\text{crit}}\,. \tag{7.11}$$

Here, R_N is the reflection coefficient and Y the sputtering yield. $(1-R_N)/Y$ in the maximum implant concentration that can be obtained if saturation effects are neglected.

ii) The critical concentration must first be reached at a distance of at least several tens of nanometers from the surface in order to produce blisters [7.371]. Blistering and flaking is suppressed if c_{crit} is first obtained at the surface and continuously proceeds into the bulk.

Due to sputtering, the maximum of the implantation profile may have shifted to the surface before c_{crit} is reached. For a Gaussian range profile, negligible diffusion and Y independent of fluence this occurs [7.387] for

$$c_{\text{crit}} \geqq \frac{1}{Y}\operatorname{erf}[\hat{R}_p/(\sqrt{2}\sigma_R)]\,. \tag{7.12}$$

For high implantation energies ($\hat{R}_p/\sigma_R > 3$), both (7.11 and 12) are equivalent while at low energies, (7.12) is more stringent. For hydrogen and helium ions the right-hand side of (7.12) is larger than c_{crit} for all energies $E_0 > 500$ eV. For heavy gas ions (e.g., Ar) it is close to one for $E \lesssim 100$ keV and is larger than one at higher energies.

7.6.2 Parameters Influencing Surface Structure

The development of surface structure by gaseous ion bombardment is influenced by a number of interrelated parameters, the more important of which

Table 7.3. Parameters affecting gas-ion induced surface structure

Target related parameters	Target material and composition Temperature Method of production and pretreatment Bulk structure: crystal grain size damage concentration inclusions and precipitates Mechanical properties: elastic modulus yield modulus prestressing Target orientation Surface roughness
Projectile related parameters	Type of projectile Energy Flux density Fluence Angle of incidence
Gas-solid related parameters	Solubility Diffusivity Phase state

are listed in Table 7.3. They can be divided into 3 groups according to their relation to the target, the ions and the gas-solid system. Some of these parameters are closely interrelated, so that a separate discussion of all of them appears impractical. Since the more important influences of the gas-solid related parameters, i.e., solubility, diffusivity and phase have already been discussed in Sect. 7.3, they will not be treated here. The target material, its method of production and the pretreatment have their main influence on the bulk structure and the mechanical properties and are therefore treated together. From the wealth of experimental information available, only those effects which are most important for the type of structure observed have been discussed in more detail.

a) Temperature

One of the most important parameters for evolutional surface structure is the target temperature. It was first shown by *Erents* and *McCracken* [7.52] for 36 keV He bombardment of Mo that roughly four temperature ranges (Fig. 7.17) can be distinguished in which different developments of surface structure are observed:

i) At low temperatures (300 K for Mo), circular blisters are found which are partly exfoliated;
ii) at intermediate temperatures (800 K for Mo), strong flaking of irregular areas is observed;

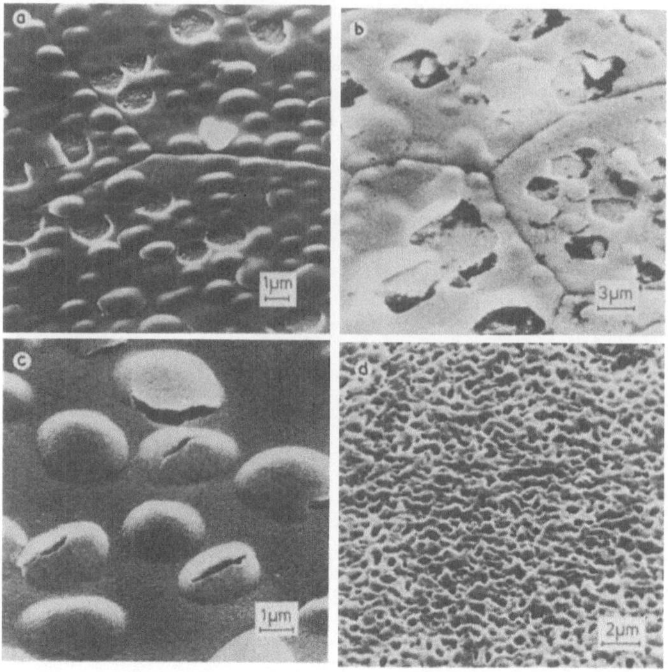

Fig. 7.17a–d. Blistering of molybdenum surfaces after bombardment with 36 keV He^+ ions at different temperatures (**a**) 300 K; (**b**) 800 K; (**c**) 1100 K; (**d**) 1600 K [7.52]

iii) at high temperatures (1100 K for Mo), again circular blisters occur which are somewhat larger than at low temperature and show little exfoliation; and

iv) at very high temperatures (1600 K for Mo), pinhole formation is found.

These four temperature ranges were also found by *Roth* [7.304] for 4 keV He bombardment of a Nb single crystal. At higher energies the transitions from blistering to flaking and back to blistering with increasing temperature and a maximum exfoliation at intermediate temperatures was observed for 80 keV He on V [7.275], for 500 keV He on V, Nb, V-20% Ti, 304 stainless steel [7.320, 372, 378, 379] and for 300 keV He on V, Nb, Pd, and 316 stainless steel [7.217, 270, 274, 345]. A very similar sequence of the four temperature ranges has recently been obtained for 500 keV He on Ni by *Sinha* et al. [7.289]. *Bauer* [7.48] has characterised these ranges in terms of the homologous temperature T/T_m (T_m: melting temperature) (Fig. 7.18) including He-implantation in 7 different materials between 20 and 300 keV. The scheme can only be taken as approximate, the limits of the different zones may be shifted by more than 0.1 T/T_m from one system to the other.

The temperature dependence of surface structures of He implanted metals is not yet well understood. *Erents* and *McCracken* [7.52] have shown that for He

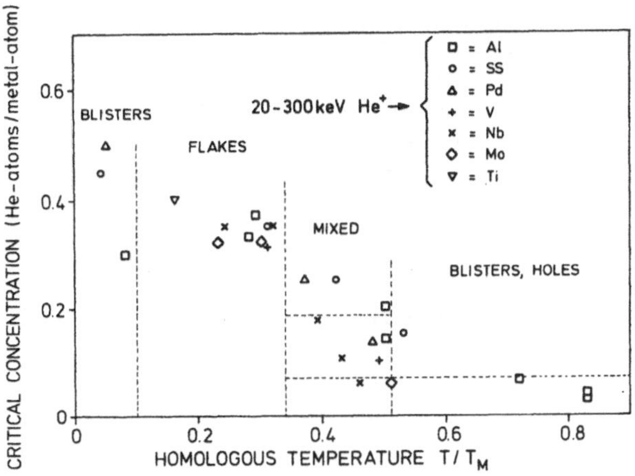

Fig. 7.18. Summary of He implantation data for seven different materials implanted at energies from 20–300 keV. The critical concentration is determined from the dose required to produce surface deformation, while the implantation temperatures are expressed as a function of the melting temperature [7.48]

the flaking range occurs at a temperature just *below* the first peak in the thermal desorption spectrum (1000 K), while pinhole formation starts close to the onset of self-diffusion of lattice atoms (0.59 T_m for Mo). For He in Pd, *Thomas* and *Bauer* [7.345] found the flaking zone *above* the temperature of the low temperature desorption peaks which are below room temperature and at 273 K, respectively, in this case.

Also, the helium re-emission *during* implantation shows characteristic differences in the different temperature ranges [7.217, 274]. The critical fluence for surface deformation which approximately coincides with the onset of gas emission [7.388] decreases with increasing temperature. In the flaking region the re-emission peak occurs at rather constant intervals when a new layer is exfoliated, as will be discussed in more detail later.

The characteristic features of gas re-emission during helium ion implantation in the different temperature ranges have been correlated to the mobility of vacancies and the growth of gas bubbles [7.204, 389]. The critical concentration for surface deformation decreases with increasing temperature as shown in Fig. 7.18. The same trend is observed for the critical fluence f_c [7.269, 308].

For hydrogen ion bombardment the influence of temperature on surface structures of metals is less well investigated. Generally, blistering similar to the low temperature stage of He bombarded metals is found in metals with very low or endothermic hydrogen solution at or below room temperature, i.e., Be [7.221], Al [7.252], Ni [7.285], Cu [7.43, 51, 294], Mo [7.221, 270], Ni [7.285], Au [7.390], and stainless steels [7.221, 270]. Exfoliation is much less severe than by He bombardment. At higher temperatures, the critical concentrations required to produce surface deformations can no longer be achieved because of the exponential increase of diffusivity. The maximum temperatures for hydrogen blistering of some metals are listed in Table 7.4. The temperature limit will be shifted to higher temperatures for larger current density as discussed later, but this shift is small.

Table 7.4. Maximum temperatures for blistering of metals with $Q_s > 0$

Material	T_{max} [K]	Ref.
Be	<473	[7.220]
Al	$\gtrsim 623$	[7.252]
Ni	>120	[7.285]
	<293	
Cu	>350	[7.167]
Mo	$\gtrsim 293$	[7.221, 251]
	<573	[7.335]
Au	$\gtrsim 293$	[7.221]
TZM	≈ 300	[7.335]
Stainless steel	≈ 300	[7.227]

On metals with $Q_s < 0$ (exothermic solution) and high enough temperatures corresponding to a hydrogen mobility which allows the implanted H to diffuse into the bulk fast enough (V, Nb, Ta), no surface deformation due to implanted hydrogen is generally found. Blistering reported for 20 and 150 keV H^+ bombardment of room temperature V [7.251, 271] and for 300 keV D^+ bombardment of Nb at $823 \leqq T \leqq 973$ K [7.303] could not be confirmed by other authors [7.270]. These surface features were probably produced by either impurities as oxygen or carbon in the ion beam or on the target which would strongly reduce the diffusivity of H in the implanted layer. Similarly, the star or crowfoot shaped patterns observed for hydrogen and helium bombardment of Nb at $900 \text{ K} \leqq T \leqq 1200$ K [7.302, 304] have been shown to be due to carbides [7.300, 301] and hydrides on the surface. No hydrogen bombardment measurements are known for V and Nb at temperatures low enough to suppress hydrogen diffusion, but precipitation at low hydrogen concentrations has been found in Nb at temperatures below 240 K with Rutherford backscattering and channeling [7.391]. In Ta the formation of very few blisters was observed during 10 keV D^+ bombardment at 35 K up to fluences of 4×10^{22} D m^{-2} [7.127]. In Fig. 7.19, blister formation is shown on a Ti surface bombarded with 6.6 keV D^+-ions at 160 K. At this temperature D is not mobile in the TiD_2 layer and a critical concentration of 1 D-atom/TiD_2 molecule is obtained.

b) Flux Density

The flux density of the implanted gas ion beam influences the development of surface topography in two ways:

i) the deposition of heat in the implanted layer increases with flux density. This may cause a temperature rise if the heat cannot be dissipated fast enough into the bulk of the target.
ii) the gas concentration in the implanted layer is determined by the flux density and the speed of diffusion to the bulk and to the surface.

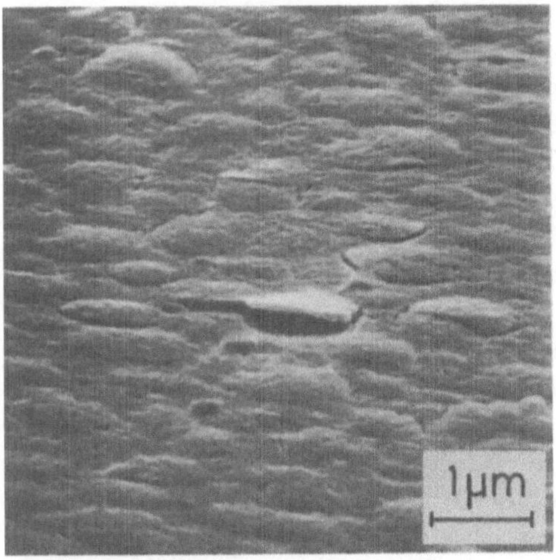

Fig. 7.19. Blistering of a Ti surface after bombardment with 6×10^{18} cm^{-2} D^+-ions of 6.6 keV at 160 K. Surface concentrations of 3 D-atoms/Ti-atom are obtained, indicating a saturation concentration of 1 D-atom/TiD_2-molecule [7.135]

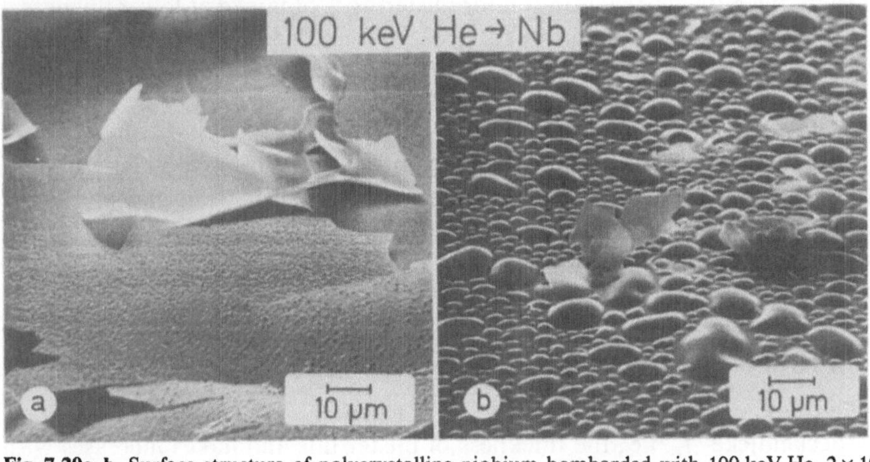

Fig. 7.20a, b. Surface structure of polycrystalline niobium bombarded with 100 keV He, 2×10^{22} He^+ m^{-2}, at 1 mA cm^{-2} (**a**), and 0.3 mA cm^{-2} (**b**) [7.5]

Both effects have been experimentally observed. *Das* and *Kaminsky* [7.272] observed an increase in density of large size blisters with increasing flux density for 500 keV He^+ ion bombardment of V and Nb at 1173 K while the small size blister density stays unaffected. It was shown by *Behrisch* et al. [7.5] that the effect of flux density on surface topography development is at least partly due to heating of the blister or flake covers by the incident ion beam. At 100 keV He^+ ion bombardment of polycrystalline niobium with 1 mA cm^{-2} (equivalent to 100 W cm^{-2}), a sudden rise of surface temperature above 1300 K in the bombarded spot is observed at the critical fluence when blisters or flakes appear. This temperature rise has been explained by the sudden interruption of thermal contact between the covers and the bulk material. Figure 7.20a, b show surface structures corresponding to 1 and 0.3 mA cm^{-2} bombardment, respectively. In the high flux density case, the flakes show pinhole formation similar to high temperature bombardment. Perforated blister covers due to pinhole formation were also observed for 250 keV H^+ bombardment of Cu [7.293b]. The effect shown in Fig. 7.1c that blistering and flaking occur in adjacent areas of the beam spot are probably due to a nonuniform current distribution in the bombarded area. *Fahlstrom* and *Sinha* [7.336] have reported ion beam heating of Mo by 250 keV He^+ bombardment between 343 and 453 K for 6.7×10^{13} and 1.2×10^{15} ions $cm^{-2} s^{-1}$, respectively, and they observed an increase of blister diameters with target temperature. Calculations of beam spot temperatures assuming a reduced thermal conductivity due to subsurface damage and microstructures like bubbles, dislocations and microcracks were made by *Yadava* [7.293c].

A decrease of critical fluence f_c for blistering with increasing flux density was found for 15 keV H^+ bombardment of Mo at room temperature [7.221]. Because of the loss of implanted gas through the surface and into the bulk by diffusion, the critical concentration for blistering is obtained at lower fluences if the flux density is high. *Behrisch* et al. [7.227] found no blistering on a stainless steel surface bombarded at room temperature with H^+ of low flux density while at high flux density (several mA cm^{-2}), blistering occured.

Möller et al. [7.285] reported the opposite effect for the bombardment of Ni at 120 K with 300–400 keV D^+. They assume that the observed increase of f_c with flux density is due to an increase in gas bubble density. The incident power density in this experiment is 100–600 W cm^{-2} and an increase in target temperature, though it was not observed, might explain the increase in f_c due to increasing diffusion loss at higher flux density.

c) Particle Energy ("Deckeldicke", Blister Diameter)

The particle energy determines the depth at which the ions are implanted. Thus, the "deckeldicke" t_B, i.e., the thickness of blister lids and flakes increases with increasing energy. This is shown for 15–250 keV He bombardment of Nb in Fig. 7.21. The "deckeldicke" t_B has been measured by scanning electron microscopy and by Rutherford backscattering, either with the channeling technique [7.105, 165, 304, 308] or by thickness measurements of removed flakes [7.259]. In Fig. 7.22 the "deckeldicke" t_B as a function of particle energy E_0 is shown for He bombardment of Nb at room temperature, where most measurements were made. Good agreement between the results of different groups is found. At low energies the "deckeldicke" appears considerably larger than the mean projected range $\langle R_p \rangle$ of the ions which is shown for comparison, while at high energies the agreement between t_B and $\langle R_p \rangle$ is good.

The low energy discrepancy between t_B and $\langle R_p \rangle$ has provoked a discussion on the blistering mechanism. One of the essential points of this discussion is the question of which technique gives the correct value for t_B. The "deckeldicke" measured by SEM is obtained in the dimension of a length and is therefore

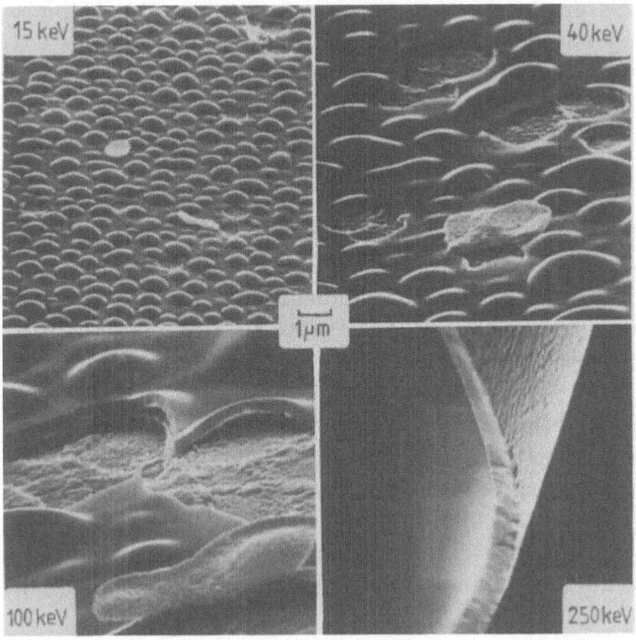

▲

Fig. 7.21. Surface structures after 15, 40, 100, and 250 keV He bombardment of Nb at room temperature [7.165]

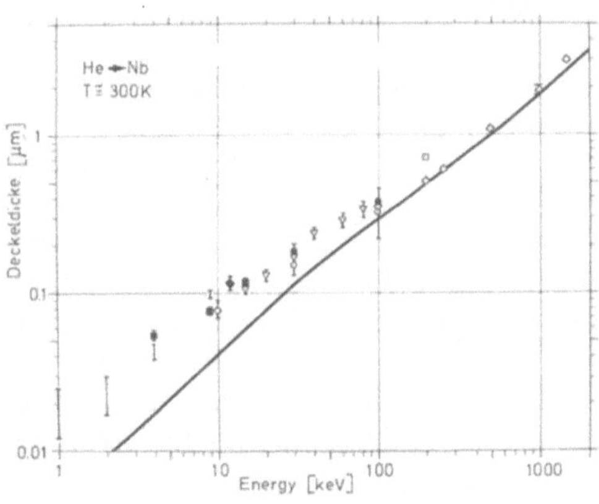

◄

Fig. 7.22. "Deckeldicke" as a function of energy for He bombardment of Nb at room temperature. (I) [7.308] (RBS, SEM); (■) [7.61]; (◆) [7.15] (RBS); (♀) [7.314] (SEM); (○) [7.165] (SEM); (●) [7.165] (RBS); (◇) [7.103, 392] (SEM); (□) [7.98] (SEM)

affected by swelling. At very low energies where no exfoliation takes place, SEM measurements are not possible and, if only very few blisters exfoliate, their thickness may be different from the average due to local inhomogeneities. The RBS-channeling technique determines the areal density (atoms m^{-2}) of a crystalline surface layer misaligned by the blistering process [7.304]. It may be influenced by damage in the subfracture zone.

Further measurements of $t_B(E)$ have been made for He on Be and Ni [7.223] and Al and V [7.254]. A low energy (20–60 keV) deviation of t_B from $\langle R_p \rangle$ is found for Ni while for Be, Al, and V no significant deviation is found.

For hydrogen bombardment only few measurements of the "deckeldicke" as a function of energy have been made. While *Primak* and *Luthra* [7.51] reported a "deckeldicke" equivalent to the maximum ion range for H^+ on MgO, *Johnson* and *Armstrong* [7.294] and *Möller* et al. [7.285] find it to be smaller than the most probable ion range in Cu and Ni, respectively. These observations are not yet fully understood but they are probably caused by the different trapping mechanism of hydrogen in the different bombarded solids.

The blister diameters show a more or less wide distribution. d depends on the energy of the implanted ions (Fig. 7.21) [7.16, 52, 165, 308, 336]. It was found to increase to a first order linearly with implantation energy. It is further proportional to t_B^m with $0.85 \leqq m \leqq 1.5$ [7.15, 63, 165, 223, 318, 392]. Some implications of the relations $d(E)$ and $d(t_B)$ will be discussed later in connection with blister models.

The critical fluence for blistering increases with energy as has been shown for a number of target materials at H and He bombardment [7.16, 52, 165, 285, 308]. This behaviour can be qualitatively understood because a higher fluence is needed to reach the critical gas concentration at higher energies because of the larger width of the range distribution of the implanted ions [7.198]. A proportionality of f_c with the square root of E_0 was found experimentally [7.165] in agreement with this model.

d) Target Orientation

The parameters of blister formation such as critical fluence, blister size, density and shape are dependent on the orientation of the target crystal. This effect was demonstrated on annealed polycrystalline Mo surfaces bombarded with 15 keV He-ions by *Verbeek* and *Eckstein* [7.221]. The blister size and density differs from one grain to the other. Some grains are almost unaffected while others are heavily blistered (Fig. 7.23). The differences in critical fluence between grains of different orientation was demonstrated conspicuously by in situ SEM observation of a Mo surface bombarded by 60–200 keV He-ions [7.342]. The orientation dependence of blistering has also been observed for other materials. *Milacek* and *Daniels* [7.249] have shown for hydrogen bombarded Al that blister density is lowest in grains with orientation near {111} and {100}. They argue that the blister density increases with the amount of retained gas which is assumed to be highest for implantation along open directions because of the greater range (e.g., ⟨110⟩ in a fcc lattice).

In Nb single crystals, *Roth* et al. [7.393] and *Risch* [7.165] showed that the blister size increases for He bombardment parallel to close-packed directions as compared to random directions. Blisters due to random and ⟨100⟩ incidence for 15 keV helium are shown in Fig. 7.24. At 4 keV the blister diameter after bombardment in a ⟨100⟩ direction is twice as large as in a random direction, while the "deckeldicke" determined by Rutherford backscattering is only 5% larger. At 9 and 15 keV the increase due to aligned bombardment is about 30% for the diameter and 20% for the "deckeldicke". The critical fluence is higher for

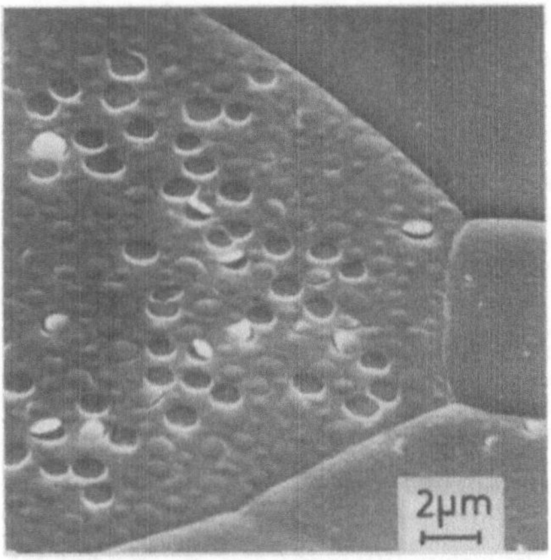

◀ **Fig. 7.23.** SEM micrograph of annealed polycrystalline molybdenum irradiated with 15 keV He^+, 2.5×10^{22} He^+ m^{-2}, $j = 0.12$ mA cm^{-2} [7.221]

▼ **Fig. 7.24.** Blistering of monocrystalline niobium due to 15 keV He^+ bombardment in a random and a ⟨100⟩ direction of incidence [7.165]

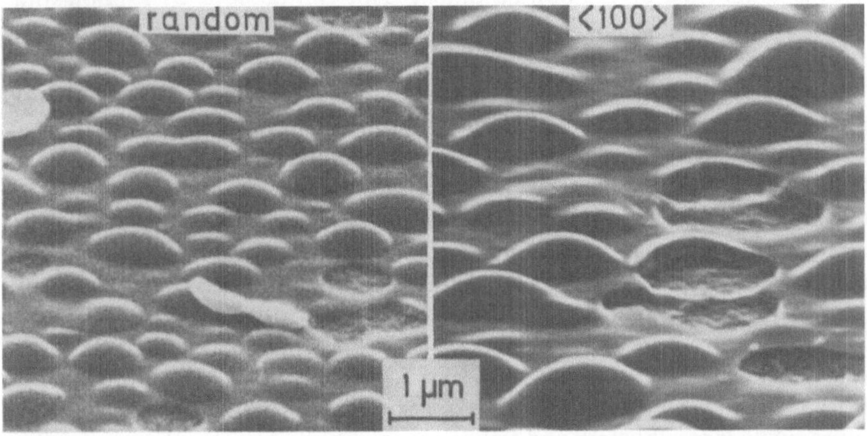

aligned bombardment but the blister density decreases. The influence of the direction of ion injection is certainly connected with the ion range but no consistent model is yet available.

e) Bulk Structure, Pretreatment, and Surface Roughness

Large differences of surface structure are found for materials which have been processed in different ways. Well-annealed Al shows strong repetitive flaking during bombardment with 100 keV He^+ at room temperature at a fluence of 1.0 C cm^{-2}, while sintered aluminium powder (SAP 895) with 10.5% Al_2O_3 develops small blisters of which only few are ruptured under equal bombardment conditions [7.255, 408]. At 670 K flaking of annealed Al was reduced to one layer but again SAP 895 shows only a few ruptured blisters at this temperature. The difference in flaking behaviour compared to annealed Al was much less pronounced in SAP 930 with only 7% Al_2O_3 content for 300 keV He^+-bombardment in the temperature range from 150 K to 473 K at fluences up to 6×10^{22} He-ions m^{-2} [7.225, 257a]. Also, in sintered Be bombarded with 100 keV He^+ at room temperature and at 600 °C, exfoliation of blisters was orders of magnitude lower than in vacuum cast material [7.224].

Generally, a trend is observed toward larger blister size, stronger flaking and exfoliation as the crystalline structure of the material becomes more perfect. This was first shown by *Das* and *Kaminsky* [7.56, 320] for He bombarded V and Nb, and was confirmed by other authors [7.257a, 281]. The reasons for this behaviour may be the influence of the different elastic and plastic properties of the material and the differences in gas trapping and bubble formation in differently treated samples of different structure and pretreatment.

Blistering, flaking and exfoliation is considerably reduced if the target surface is roughened before bombardment. This can be done by sandblasting, abrading, scratching, chemical etching or by deposition of a rough surface coating, e.g. by chemical vapor deposition [7.411]. *Bauer* and *Thomas* [7.310] have shown that a rather stable pinhole structure similar to the one shown in Fig. 7.1d reduced flaking considerably. Figure 7.25 shows the appearance of Mo surfaces which were electropolished (a) or roughened with emery paper of #1200 (b) and #400 (c) prior to bombardment by 100 keV He^+. On the electropolished surface, blistering and some exfoliation is observed. On surfaces abraded with emery paper, exfoliation (b) and even blistering (c) is completely suppressed [7.232]. It has been tentatively explained by the fact that no continuous flat gas layer can be built up under a rough surface. Therefore, the roughness structure must have a length scale which is comparable to the blister diameter. This is demonstrated for a multigrooved Mo-surface in Fig. 7.26 [7.338a]. The blister diameter always remains smaller than the groove distance while larger blisters and flakes are found on ungrooved material. The reduction of blistering by mechanical surface treatment may also be due to

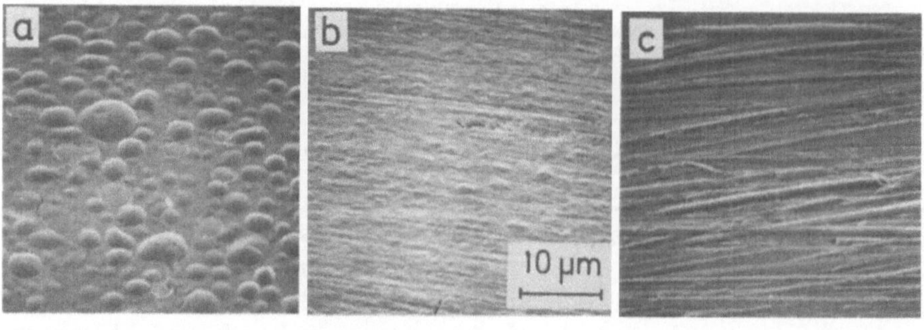

Fig. 7.25a–c. SEM micrographs of molybdenum surfaces bombarded with 100 keV helium ions to a fluence of 10^{23} He^+ m^{-2}. The surfaces were previously (**a**) electropolished, (**b**) roughened with emery paper #1200, (**c**) roughened with emery paper #400 [7.232]

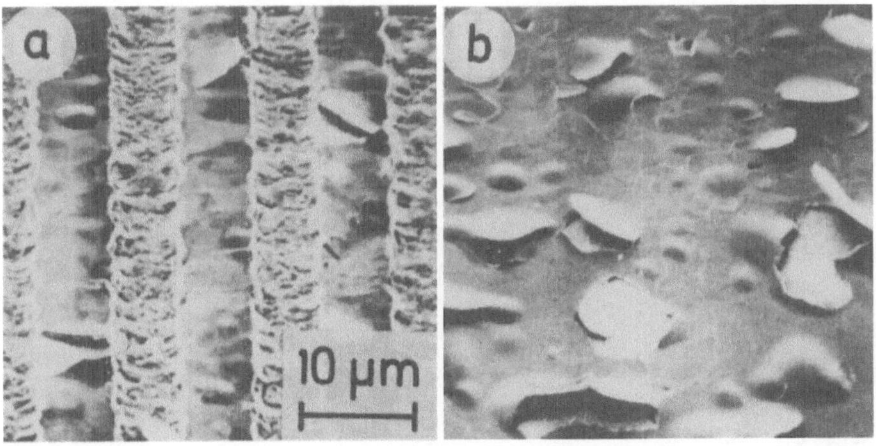

Fig. 7.2a, b. SEM micrographs of molybdenum surfaces after bombardment with 100 keV helium ions to a fluence of 1.4×10^{22} He^+ m^{-2} at 770 K. (**a**) Multigrooved surface, (**b**) ungrooved surface [7.338a]

structural modifications in a μm thick layer produced by this treatment. Evans has given a simple model showing that for a given "deckeldicke" t_B a blister needs a minimum diameter d, below which it cannot be formed [7.337b].

f) Prestressing

It was shown by *Ivanov* et al. [7.373] that prestressing of stainless steel with loads of 4×10^7 to 2×10^8 N m^{-2} in a direction normal to the implantation direction increases the critical fluence for blistering or flaking. At higher fluences the prestressed surfaces show flaking under conditions where unstressed material shows blistering (175 keV He^+, $T = 700$ K). Under the applied load it can be assumed that the bulk material yields plastically and the stress in the more brittle implanted layer may be much larger than the nominal average load. An increase of the critical fluence is also found in Al and Nb if the material is alternatively put under tensile and compressive stress during He bombardment [7.338b].

g) Angle of Incidence

The dependence of "deckeldicke" and blister diameter on the angle of ion incidence to the surface normal was recently measured by *Risch* et al. [7.105] for 30 and 100 keV He on Nb. Blister diameter d and "deckeldicke" t_B decrease strongly with increasing angle of incidence (Fig. 7.27). In a CrNi-alloy, *Guseva* et al. [7.395] showed a decrease of "deckeldicke" t_B with the angle of incidence α as $t_B \propto \cos\alpha$, while the blister diameter d decreases less rapidly than $\cos^{3/2}\alpha$. This result is explained in terms of increasing reflection of the incident He-ions with increasing angle of incidence α.

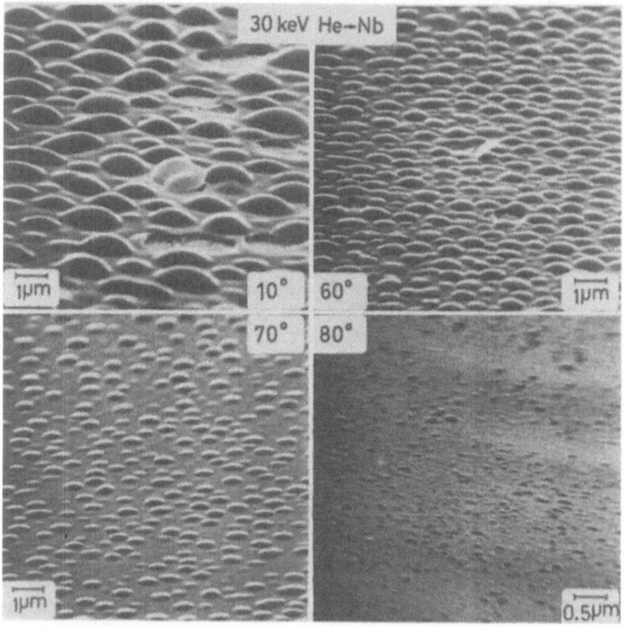

Fig. 7.27. SEM micrographs of niobium bombarded with ^{3}He-ions at different angles of incidence and energies of 30 keV [7.105] (SEM micrographs by Dr. H. Klingele, München)

7.6.3 Gas Emission During and After Blistering and Flaking

It was already mentioned in Sect. 7.3.3 that the trapping behaviour of ions implanted into materials with very low diffusivity (Fig. 7.5) has been satisfactorily described by theoretical models neglecting the influence of blistering and flaking. However, the existence of a strong correlation between gas re-emission and surface deformation was observed for He bombardment of metals [7.52, 204, 217, 257a, 269, 310, 326].

Recently, several attempts have been undertaken to check the correlation between the onset of gas re-emission and blister formation. *Behrisch* et al. [7.123] found that gas re-emission from Nb implanted by low energy helium (1.5–15 keV) actually begins before the first blisters are observed. Also, the critical fluence for blistering of Mo by 200 keV He bombardment was measured as being almost 50% larger than the fluence at which gas re-emission starts [7.342]. Simultaneous measurements of surface deformation by diffuse scattering of laser light and the amount of retained gas showed that in Ni, Pd, and Ta implanted with 10 keV D^+ at 35 K, the critical fluence for surface deformation coincided with the onset of deuterium re-emission while in Mo the gas re-emission started at a lower fluence [7.127]. *Ehrenberg* [7.257b, 388] has combined the laser scattering method with a direct measurement of gas re-emission. Within the experimental errors, the surface deformation and the gas re-emission

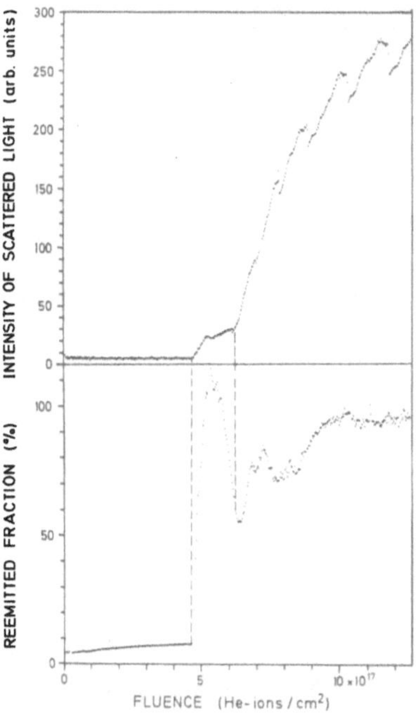

Fig. 7.28. Surface deformation as measured by diffuse scattering of light and gas re-emission as a function of fluence during 40 keV He^+ implantation in Ni at normal incidence and room temperature [7.388]

of Ni implanted with 40 keV He start simultaneously (Fig. 7.28). The differences found in some experiments between the onset of gas re-emission and blister formation are not understood. Inhomogeneous current density distribution or insufficient resolution to observe very small blisters may explain some of the results.

The gas re-emission behaviour depends on the type of surface deformation [7.257a]. In the low temperature blistering range ($T \lesssim 0.1\, T_m$, see Fig. 7.18) it may pass through one or more initial maxima (Fig. 7.28) before it slowly approaches the 100% steady-state. Simultaneous measurements of the total emitted gas and the blistered area have shown that probably already in the initial phase of re-emission the gas is released from the total implanted area and not only from the blisters alone [7.257b]. This may be caused by an interconnected network of channels forming when local saturation with gas is reached near the maximum of the ion range.

In the intermediate temperature range where flaking occurs ($\sim 0.1\, T_m$–$0.4\, T_m$), re-emission often starts with a peak. In metals like Al, V, Nb, Mo and stainless steel, the burst-like re-emission of He is repetitive with approximately constant fluence intervals [7.204, 217, 257a, 269, 310]. This effect is illustrated by the 670 K re-emission curve for 300 keV He^+ bombardment of V [7.204] in Fig. 7.29 and by the 264 K re-emission of 20 keV He^+ bombardment of Al [7.255] (Fig. 7.30). Burst-like re-emission is less well developed at lower energies. No more than about 5–20% of the gas retained in the sample is

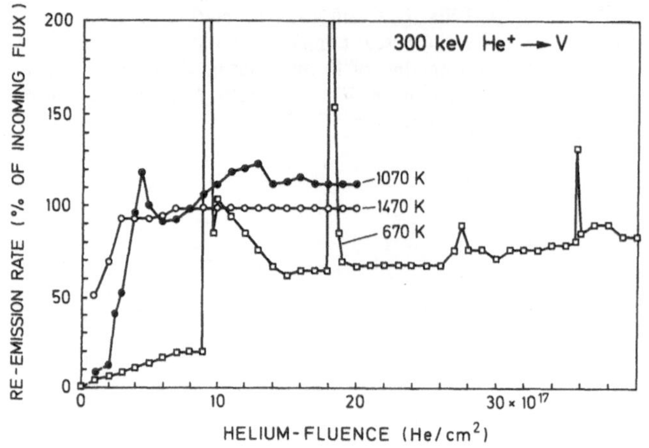

Fig. 7.29

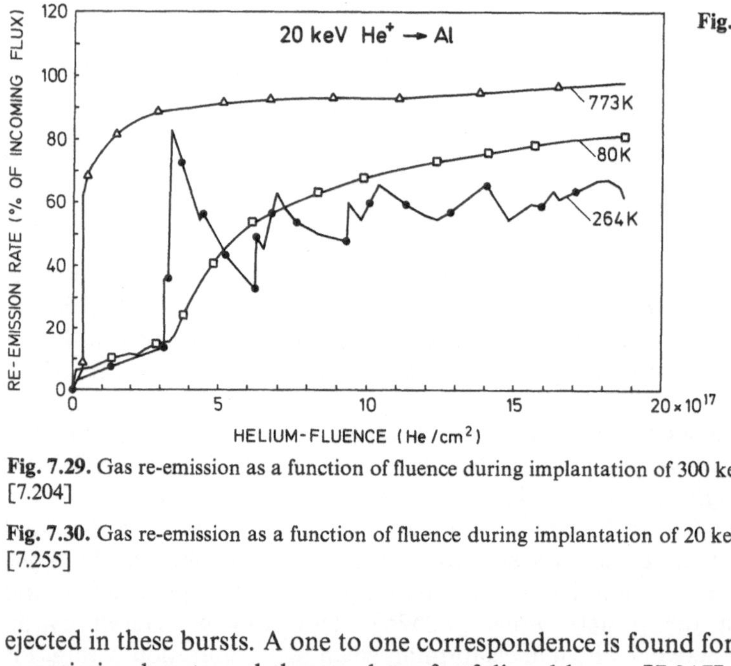

Fig. 7.30

Fig. 7.29. Gas re-emission as a function of fluence during implantation of 300 keV He in vanadium [7.204]

Fig. 7.30. Gas re-emission as a function of fluence during implantation of 20 keV He in aluminum [7.255]

ejected in these bursts. A one to one correspondence is found for the number of re-emission bursts and the number of exfoliated layers [7.217].

In the high temperature ranges ($T \gtrsim 0.4\, T_m$), the re-emission shows a steep initial increase with a single peak or a continuous transition into a stationary state with close to 100% re-emission (Figs. 7.29, 30).

7.6.4 Depth Profiles of Implanted Gas After Blistering

In Nb, Ni and stainless steel, the depth profiles of 1.5–30 keV He ions have been found to agree with the range distribution of the ions within the experimental error [7.61, 106, 165] up to the critical fluence f_c. At higher fluences the distribution develops a plateau with the saturation concentration extending over a depth range from the surface to the maximum projected range. At 40 keV He in Ni, *Ehrenberg* et al. [7.388b] observed a peak value of the He concentration at the critical fluence which is about 30% higher than the plateau value obtained at higher fluences.

Terreault et al. [7.216a, 222, 296, 297] have observed double peaking in the depth profile of 20 and 25 keV He in Cu after blistering has occurred. A minimum in the gas profile at about the mean projected range $\langle R_p \rangle$ develops at supercritical fluences. At the same time the total quantity of trapped gas decreases by up to 50%. This effect is explained by the release of gas from the blister cavities due to crack formation in the blister lids causing a depletion of gas around the depth at which the cap separates from the bulk. Double peaking of 80 keV He-ions in flaked V at 673 K was also observed by *Blewer* and *Langley* [7.275]. But in this measurement, no loss of gas during development of the double peak was observed. The peak positions corresponded to 0.4 $\langle R_p \rangle$ and close to $\langle R_p \rangle$, respectively. The phenomenon was in this case attributed to the migration of implanted helium into the maximum of the damage profile. Since double peak profiles have not been reported for other materials and are dependent on target pretreatment, they cannot unambiguously be attributed to gas emission by blisters.

7.7 Models for Blistering and Flaking

The phenomena of blistering and flaking are confined to gas ion implantation into materials of low diffusivity. They are observed after implantation of a surface layer to critical concentrations which are far above the solubility limit. Part of the implanted gas has been observed to precipitate in gas bubbles or other defect clusters. Consequently, the origin and development of blisters and flakes has initially been looked at as due to the action of high internal gas pressures in gas filled cavities alone [7.51, 52, 54, 56, 198, 208]. But the large swelling and the formation of lateral compressive stress accompanying gas ion implantation have led some investigators to consider this stress as well, presumably combined with the action of gas pressure in cavities, in order to explain the observed structures [7.15, 61–63].

There is experimental evidence for the assumption that the blistering and flaking process occurs in two stages:

i) the formation of an interface of reduced strength between the surface layer and the bulk;
ii) the deformation of the surface layer.

The interface of reduced strength can be seen directly on SEM micrographs taken from microtome sections normal to the implanted surface [7.64]. It has been shown by *Jaeger* et al. [7.1] that a network of interconnected channels is formed within the bubble arrangement as soon as the critical gas concentration is reached. This channel network interconnects large parts if not the whole implanted area laterally at some depth below the surface [7.257b, 388c]. As mentioned already in the introduction, surface features similar to blisters are observed on thin films evaporated on solid substrates. In such films the interface between the evaporated layer and the bulk provides an easy fracture and, upon ion implantation blisters separate at this interface although the ion range is either much larger [7.412] or much smaller [7.413a] than the evaporated layer. Finally, strong deformations of free standing thin films are observed due to ion implantation which does not appear to be the result of bubble swelling [7.413b].

Models for the reduced-strength interface should be able to explain the observed values of the critical fluence f_c and the "deckeldicke" t_B, as well as their dependence on particle energy, target temperature, ion flux density, etc., while models for the second stage should account for the surface features observed as blisters, flakes, pinholes, the blister diameter d, the blister shape, surface crack formation and exfoliation.

A number of models have been put forward to describe blistering and flaking some of which are quite sophisticated [7.51, 52, 54–60, 62, 63, 198, 199, 207, 208, 244, 396–400]. A final evaluation of these models is not yet possible. This is particularly due to the large uncertainties about the state, the mechanical properties of the highly damaged lattice and the behaviour of the high concentration of gas precipitates.

7.7.1 Initial Stage Models

Almost all of the older models on blister initiation start from the assumption of excessive growth of large gas bubbles either by the collection of vacancies and vacancy clusters from their environment or by coalescence of smaller bubbles [7.51, 52, 54–60]. Assuming that the total bubble volume does not change during the coalescence of two or more bubbles, the internal gas pressure will be constant while the equilibrium pressure, (7.5), decreases with increasing volume. This gives rise to an excess pressure which will be able to rupture the material and to deform or rupture the covering surface layer.

a) Critical Swelling Model

In an early model, *Evans* et al. [7.58–60] started from the experimental evidence that rupture and deformation of the surface layer occur at a rather well defined critical fluence f_c and that at $T \lesssim 0.3\, T_m$ the gas bubbles arrange in a lattice [7.1, 2]. They assumed that individual bubbles grow during implantation. Bubble coalescence takes place when neighboring bubbles touch each other. In an ideal bcc lattice of equally sized bubbles this occurs at a critical value of local swelling $(\Delta V/V)_c = 68\%$. Assuming that all implanted He is collected in bubbles and the gas pressure in the bubbles is in thermodynamic equilibrium with surface free energy γ_s, (7.5), the critical fluence can be calculated:

$$f_c = \varrho \sigma_R \sqrt{2\pi} \frac{(\Delta V/V)_c}{1-(\Delta V/V)_c}. \tag{7.13}$$

ϱ is the gas density in bubbles and σ_R the standard deviation of the projected range. It was found that agreement with experimental data of f_c is best for $(\Delta V/V)_c = 40\%$ instead of the ideal value of 68%. This discrepancy is explained by the imperfection of the bubble lattice. The maximum swelling is assumed to take place at the peak of the He-distribution. The rupture between bulk and surface layer will therefore take place in this plane.

This model is in contradiction with experimental evidence in two respects: (i) as pointed out above, the total amount of implanted He cannot be accommodated in the gas bubble volume, the local swelling by visible bubbles never exceeding $\Delta V/V \simeq 10\%$ except at high temperatures; (ii) the rupture plane does not always occur at the depth where the implanted gas concentration has its maximum.

b) Interbubble Fracture and Loop Punching Model

This model of blister formation, therefore, has been modified by *Evans* [7.208] to explain the experimental observation that the "deckeldicke" t_B is larger than the most probable projected range of the ions [7.61, 63, 165, 314].

Evans assumes that the gas pressure p in a bubble of radius r_b exceeds the equilibrium pressure (7.5). With regard to the bubble lattices observed in a number of metals, he considers a set of coplanar bubbles in a simple cubic array having equal radii and excess pressure (Fig. 7.31a). If the tensile stress normal to the plane due to the excess pressure is distributed homogeneously over the plane area not occupied by the bubbles, the condition for the pressure to overcome the fracture value $\sigma_{\perp F}$ of the tensile microstress normal to the surface is

$$p_F = \sigma_{\perp F}[(\pi r_b^2 C_b^{2/3})^{-1} - 1] + 2\gamma_s/r_b, \tag{7.14}$$

where C_b is the bubble density. If condition (7.14) is met, an internal crack occurs in the bubble plane (Fig. 7.31b). It is further assumed that individual bubbles of adjacent layers break into the crack (Fig. 7.31c) which will be widened by this process (Fig. 7.31d) forming a detached zone above a flat cavity which is the nucleus of a blister (Fig. 7.31e, f). The excess pressure in bubbles and hence the stress on the interconnecting material depends on bubble sizes and on the mechanism of bubble growth and the maximum stress does not necessarily occur at the most probable range $\hat{R}_p$ of the implanted ions. *Evans* [7.208] has considered two mechanisms to compete (i) the loop punching

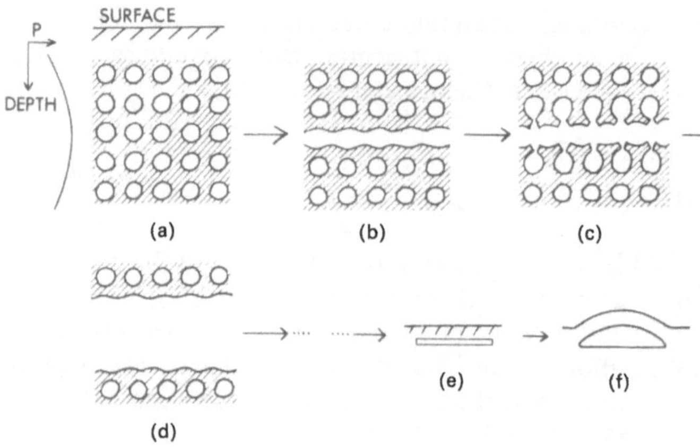

Fig. 7.31a–f. Interbubble fracture mechanism: (**a**) high density of overpressurised bubbles, (**b**) crack formation, (**c**) bubbles adjacent to original crack become involved to (**d**) widen the crack and increase the pressure, (**e**) penny-shaped crack which either extends to cause flaking or (**f**) develops into dome shaped blister [7.198]

mechanism and (ii) the acquisition of vacancies formed in the bubble surface (vacancies are assumed to be immobile). Since the second mechanism, which tends to reduce the internal pressure in the bubble, is most efficient at the maximum of the damage distribution, loop punching will become more dominant toward the deep end of the implanted distribution. Comparing the pressures $p_{\rm LP}$ for loop punching (7.9) and $p_{\rm F}$ for interbubble fracture (7.14) it can be shown that $p_{\rm LP} < p_{\rm F}$ for small bubble radii $r_{\rm b}$. In this case a pressure increase will cause bubble growth by loop punching. At a critical radius $r_{\rm b}^*$, $p_{\rm F}$ becomes smaller than $p_{\rm LP}$ and the bubbles can no longer grow by loop punching because interbubble fracture takes place. For Mo, where $C_{\rm b} = 1.5 \times 10^{25}$ bubbles m^{-3}, $\mu_{\rm s} = 125.6 \times 10^9$ N m^{-2}, $_{\rm F} = 7.2 \times 10^9$ N m^{-2} [7.118] and $b = 0.2725$ nm *Evans* [7.198] obtains $r_{\rm b}^* = 1.7$ nm and the associated pressure $p^* = 7.0 \times 10^9$ N m^{-2}. Assuming all He gas being trapped in bubbles the local swelling at the critical value is $(\Delta V/V)^* = 0.3$ and using the equation of state for He (Fig. 7.16) the atomic ratio $(n_{\rm He}/n_{\rm Mo})^* = 0.48$ which are in good agreement with measured values in Nb [7.66, 165].

In a later paper [7.198], in view of the large swelling obtained by *St.-Jacques* et al. [7.65], the inter-bubble fracture model was modified regarding loop punching as the only mechanism of bubble growth and giving a fracture depth equal to $\hat{R}_{\rm p}$.

c) Refined Interbubble Fracture Model

Wolfer [7.206] has refined the model of *Evans* [7.208] by superposition of an average tensile microstress $\bar{\sigma}_\perp$ surrounding the bubbles normal to the surface and a lateral compressive macrostress $\sigma_{||}$ across the bombardment layer parallel to the surface. $\sigma_{||}$ saturates prior to the onset of blistering but remains the driving force for blister dome formation as will be discussed later. On the other hand, when $\bar{\sigma}_{\perp}$ reaches a value of 0.003 μ for fcc metals [7.401] or 0.009 μ_s for bcc metals [7.402], interbubble fracture is expected to occur. These values are considerably smaller than the fracture stress $\sigma_F \simeq 0.06\mu_s$ assumed by *Evans.*

Wolfer started from a generalization of (7.14) given by *Evans* relating the gas pressure within the bubbles p to the tensile microstress

$$p = \bar{\sigma}_{\perp}(A^{-1} - 1) + 2\gamma_s/r_b\,, \tag{7.14a}$$

where the fractional area of intersected bubbles A is equal to $\pi r_b^2 C_b^{2/3}$ for a simple cubic array [7.208] or to $\pi r_b^3 C_b$ for a random bubble arrangement (*Underwood* [7.458]). Evaluating (7.14a), *Wolfer* assumed (i) the equation of state for He based on the interatomic potential and liquid state theory, (ii) a bubble arrangement of uniform bubble size and either cubic or random structure, (iii) trapping of all implanted helium in these bubbles, each helium atom going to the bubble closest to the point where it is stopped so that the helium concentration profile roughly reflects the ion range profile and (iv) a total swelling close to $\Delta V/V = N_{He}/N_{Met}$ [7.66]. With these assumptions and the fracture values of $\bar{\sigma}_{\perp}$ given above, he calculated critical helium concentrations for 500 and 750 K Ni which are in reasonable agreement with measured saturation concentrations. This agreement may appear somewhat fortuitous in view of the fact that the measured bubble swelling is reported to be much smaller than assumed here and therefore only a small fraction of the implanted gas can be trapped in the visible bubbles. Nevertheless, the interbubble fracture mechanism seems to be the most adequate in explaining the initial stage of blister and flake formation.

d) Calculation of Local Stress Due to Differences in the Range and Damage Distributions

The stress built up in a metal by the implantation of He has been calculated by *Hall* [7.414] assuming that the lattice strain is proportional to the local He concentration. As a result plastic deformation according to this model occurs first at the maximum of the He concentration. Agreement with experimental critical fluences is obtained if each He atom requires 0.5 atomic volumes in reasonable agreement with measurements of surface expansion [65, 66] and in excellent agreement with EELS and SAXS-measurements [7.184d, 210g].

Recently *Martynenko* [7.207] calculated the average stress around gas bubbles as a function of depth from the relative densities of vacancies and helium atoms. Because of the shift of the defect profile corresponding to the concentration of vacancies produced toward the surface compared to the range profile, the maximum of the local stress profile is at a larger depth than the mean projected range $\langle R_p \rangle$. This effect is more pronounced at low energies ($E_0 < 10$ keV) while at high energy $t_B \simeq \langle R_p \rangle$, which is in good agreement with experimental observations (Fig. 7.22).

e) Development of Pinholes

At about $T \simeq 0.5\,T_m$ the development of pinholes in the surface is observed (Fig. 7.1d). Two mechanisms may be responsible for their appearance:

i) Thermal vacancies are available at these temperatures. Therefore, the bubble growth can proceed in equilibrium with internal gas pressure (7.5) until the bubble intersects the surface.
ii) Surface mobility allows the migration of bubbles. This motion may, under the influence of lateral stress, be directed toward the surface.

No decision can be made between these models at the moment.

7.7.2 Models of Surface Deformation

As soon as the surface layer has started to separate from the bulk at a certain depth it will be deformed by the forces acting on this layer. These forces are due to the gas pressure inside the crack acting in a direction normal to the crack surface and to the compressive stress in the surface layer generally parallel to the surface of the crack. Early models of blistering assume only forces due to the pressure of the included gas [7.51, 52, 54–56]. According to the flat bottom of blisters observed (e.g., Fig. 7.21), the cavity between the surface layer and the bulk is generally assumed to be initially of a lenticular or penny-type shape. The assumption of *Milacek* et al. [7.54] and *Auciello* [7.55] of a spherical cavity due to gas bubble agglomeration appears unrealistic in view of experimental evidence, except perhaps at high temperatures where pinhole formation dominates.

a) Spherical Shell Model

Primak and *Luthra* [7.51] assume that the blister cap approximates the shape of a spherical shell (Fig. 7.32) of inner chord d, height h, and inner radius r_s. If a fraction η_f of the implanted fluence f is released into the shell volume, the pressure inside the shell is given by

$$p = 6\eta_f d^2 f[h(4h^2+3d^2)]^{-1}$$
$$\simeq 2\eta_f f/h \quad \text{for} \quad 4h^2 \ll 3d^2 . \tag{7.15}$$

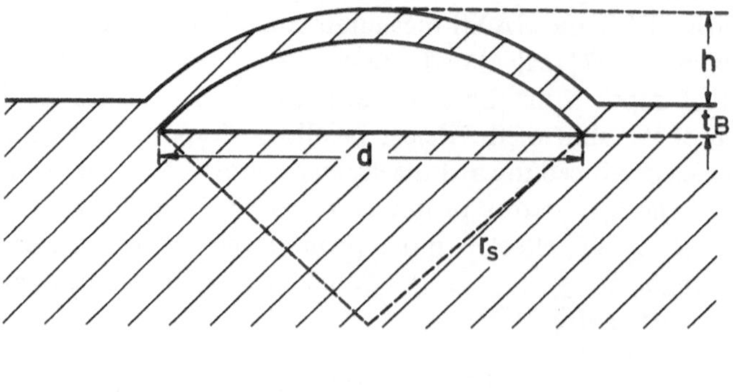

Fig. 7.32. Schematic diagram of a blister cap

The surface dilatation of the blister cap is so large that it can only be explained by plastic deformation. Neglecting the influence of surface tension, edge effects and shear, *Primak* and *Luthra* calculated the equilibrium shape of a blister cap by equating the work due to a virtual expansion of the volume below the blister shell $p\Delta V_s$ to the deformation work $\sigma_y t_B \Delta A_s$:

$$p\Delta V_s = \sigma_y t_B \Delta A_s \,, \tag{7.16}$$

where t_B is the "deckeldicke" and ΔA_s the increase in surface area. The quantity σ_y was not clearly defined by *Primak* and *Luthra*. They considered it to be either a yield modulus or, if the blister shatters, a rupture modulus. From (7.15, 16) the quantity σ_y can be obtained from experimentally determined values of p, h, d, and t_B:

$$\sigma_y = p\left[h^2 + \left(\frac{d}{2}\right)^2\right] / (4ht_B) \,. \tag{7.17}$$

For hydrogen and helium blistering on MgO, $\sigma_y \approx 10^9\,\mathrm{N\,m^{-2}}$ was obtained [7.51].

It has been pointed out by *Das* and *Kaminsky* [7.56] that by taking σ_y to be a yield stress, (7.17) is transformed into an equation derived earlier by *Hill* [7.403] for a circular diaphragm firmly clamped around its perimeter and loaded from one side with a pressure p. *Das* and *Kaminsky* calculated a pressure $p \simeq 24 \times 10^5\,\mathrm{N\,m^{-2}}$ for a typical blister in Nb of $d = 20\,\mu\mathrm{m}$, $h = 1\,\mu\mathrm{m}$, and $t_B = 1.2\,\mu\mathrm{m}$ (500 keV $^4He^+$) when $\sigma_y = 5 \times 10^7\,\mathrm{N\,m^{-2}}$ was assumed. The gas fraction η_f present in the blister volume is estimated to be about 14% in this example. In this model the pressure p increases strongly with decreasing blister diameter d.

b) Plastic Deformation Model

Erents and *McCracken* [7.52] assumed that all the implanted gas fluence is collected in a layer formed by a sudden coalescence of gas bubbles at a depth corresponding to the projected range of the ions. For a circular blister plastic deformation will occur first at a diameter d given by

$$\pi\left(\frac{d}{2}\right)^2 p = \pi d \langle R_p \rangle \sigma_y \,; \tag{7.18}$$

σ_y is the yield stress. With the further assumption that the average thickness of the coalescing bubble layer is equal to the width of the range distribution $2\sigma_R$, *Erents* and *McCracken* concluded that to a first approximation, p is independent of d and thus $d \propto \langle R_p \rangle$ contrary to (7.17) by *Primak* and *Luthra* [7.51].

c) Lateral Compressive Stress Model

Lateral compressive stress as an additional driving force of surface deformation has been introduced into blistering theory by *Risch* et al. [7.63] and *EerNisse* and *Picraux* [7.62]. They solved the differential equation for the elastic displacement $w(r)$ as a function of the distance r from the symmetry axis of a thin circular plate which is clamped at the circumference [7.13]:

$$M_r + r\frac{dM_r}{dr} - M_t + r\left(Q_B - S_i\frac{dw}{dr}\right) = 0, \tag{7.19}$$

where M_r and M_t are the radial and tangential bending moments, $Q_B = p \cdot r/2$ the shear force per length of circumference and S_i the compressive stress parallel to the surface integrated over the thickness t_B, as shown in Fig. 7.33. As a result, the displacement of the layer becomes infinite for

$$d = 7.66\sqrt{\frac{E_{el}}{12(1-\nu^2)S_i}}\, t_B^{3/2}. \tag{7.20}$$

E_{el} is Young's modulus and ν Poisson's ratio. At this value of d the plate is assumed to jump into a new equilibrium position. It has been shown experimentally [7.62] that the integrated compressive stress $S_{i,max}$ at the critical fluence f_c is nearly independent of the particle energy $S_{i,max} = 2.5 \times 10^{-6}\,[\mathrm{m}]\,\sigma_y$. Therefore, from (7.20) the relation

$$d \propto t_B^m \tag{7.21}$$

with $m = 1.5$ between the blister diameter and thickness is obtained. In the fluence range below f_c, S_i increases only very slowly [7.62]. This explains why

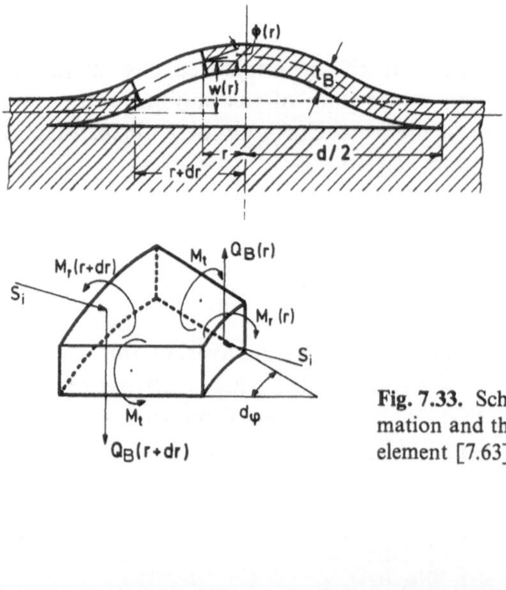

Fig. 7.33. Schematic drawing of a blister during formation and the forces and moments acting on a volume element [7.63]

Table 7.5. Experimental values for m in the relation (7.21)

Metal	m
Be	1.25
V	0.85
Ni	1.15
Nb	1.22

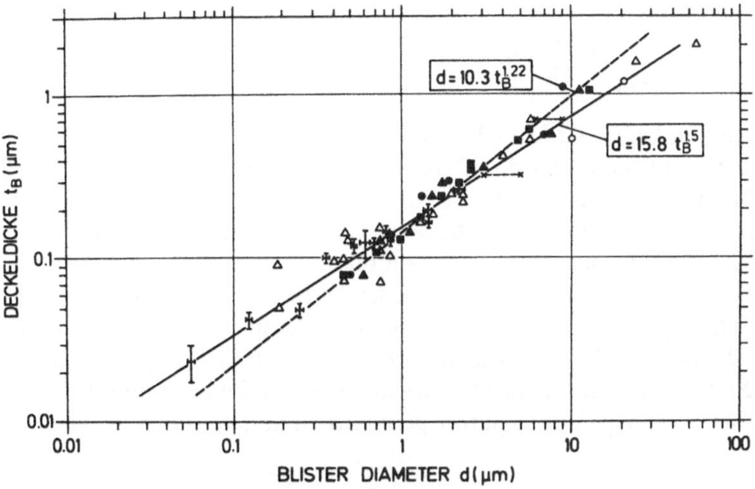

Fig. 7.34. The relationship between "deckeldicke" t_B and blister diameter d for He bombardment induced blistering of Nb at room temperature. (⊥) [7.308]; (△) [7.63]; (×) *Lutz* (private communication); (○) [7.302]; (■) polycrystalline, (●) (111), (▲) (100) [7.392]

$S_{i,max}$ is almost independent of energy while f_c increases roughly with $E_0^{1/2}$ as mentioned in Sect. 7.6.2c. It is, therefore, very likely that f_c is not determined by a critical value of S_i but rather by a process such as interbubble fracture described in Sects. 7.7.1c, d [7.206].

Relation (7.21) was found to be fulfilled approximately for He in Nb [7.15, 63]. More recent measurements by *Das* et al. [7.223, 318, 392] for He in several metals yielded the values for m given in Table 7.5. In Fig. 7.34 the experimental relation of d and t found by different groups for He bombardment of Nb at room temperature are plotted. Apparently a relation with $m=1.5$ is obtained if the low energy values by *Roth* et al. [7.308] and the high energy values by *Risch* et al. [7.63] and *Kaminsky* and *Das* [7.302] are considered. With the mechanical constants for unirradiated Nb at room temperature ($E_{el}=1.03\times10^{11}$ N m^{-2}, $\nu=0.38$, $\sigma_y=1.45\times10^8$ N m^{-2}) [7.392] we get from (7.20) $d=40.3\,t_B^{3/2}$ (d, t_B in μm). The proportionality factor is larger than that obtained from the measurement. This discrepancy is reduced if the calculations are done for circular plates which are not firmly clamped at the circumference [7.62].

It has been pointed out by *Wolfer* [7.206] that the inclusion of a stress gradient and of ligaments between the otherwise detached blister layer and the underlying material will give values for m which are somewhat different from $m=1.5$.
Watson and *Wolfer* [7.79] have shown that lateral compressive stress can produce an elastic surface buckling of a plate with characteristic buckling dimensions $a \propto t_B^m$ with $0.75 \leqq m \leqq 1.5$ depending on the attachment of the plate to the substrate metal by ligaments of varying strength. $m=0.75$ is the result for a very strong attachment, $m=1.5$ that for a weak one.

d) Crack Growth Model

Finally, *Kamada* and *Higashida* [7.399, 415a] have developed a sophisticated model for calculating the stress distribution around a two-dimensional crack parallel to the surface of a material based on the concept of continuously distributed edge dislocations. The stress field around the crack tip has a singularity, meaning that plastic deformation takes place in this region in a real material. As the crack spreads parallel to the free surface, the plastic zone at the tip increases. The growth of the crack is assumed to stop as soon as the plastic zone reaches the surface. It is shown that on the assumption of pure pressure loading of the cover over the crack the critical condition for stopping the crack growth contains two parts, one is purely geometrical and contains the relation

$$d \propto t_B , \tag{7.22}$$

while the other contains only $(p/\sigma_y)^2$. *Kamada* and *Higashida* argue that experimental deviations from (7.22) are due to a dependence of $(p/\sigma_y)^2$ on energy. They conclude that lateral compressive stress is not necessary to explain the deviations from (7.22) observed.

In this model the blister is represented by a lenticular bubble with the crack as its boundary. The pressures necessary for an extension of the crack in a homogeneous material were so high that they could never be achieved in actual blisters. Therefore the model was extended to the case where prior to crack formation an array of gas filled bubbles has been produced [7.415b, c]. In this case the ratio d/t_B increases strongly with

$$\kappa = \frac{X_0}{t_B} \frac{E_{el}}{\sigma_y} , \tag{7.22a}$$

where X_0 is the mean spacing of the gas bubbles.

e) Transition from Blistering to Flaking

The transition from blistering to flaking is not considered by most of the models, except for that of *Risch* et al. [7.63] and *Kamada* et al. [7.415]. The first authors showed that if the integrated stress $S_i = 0$, (7.19) has a simple solution for a plate clamped at its edges:

$$w(r) = \frac{p}{64N}\left[r^2 - \left(\frac{d}{2}\right)^2\right]^2 . \tag{7.23}$$

$N = E_{el} t_B^3 / 12(1-\nu^2)$ is the stiffness of the plate. The maximum radial bending moment which occurs at $r = d/2$ is

$$M_r(d) = -\frac{p}{32} d^2 . \tag{7.24}$$

Assuming that the number of gas atoms in the blister volume increases in proportion to the area $\pi(d/2)^2$, it can be shown that the pressure decreases with d^2 and M_r becomes independent of the diameter d. *Risch* et al. [7.63] conclude that in this case the crack propagation at the blister circumference proceeds until the surface layer will fail due to some material imperfection and as a result large surface flakes will appear. The formation of flakes by rapid crack propagation has also been suggested by *Thomas* and *Bauer* [7.266] from an in situ observation of a Ti surface during bombardment with 300 keV He in the scanning electron microscope.

In the model of *Kamada* et al. [7.415] the criterion for the transition from blistering to flaking is given by the parameter κ (7.22a). For $\kappa \lesssim 5$, d/t_B goes to

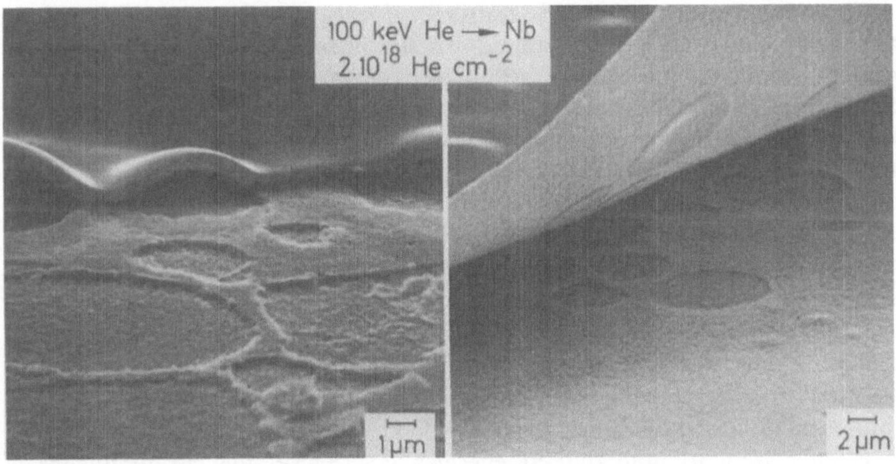

a b

Fig. 7.35a, b. Simultaneous production of blistering and flaking by changing the current density shortly before reaching the critical fluence during 100 keV He-bombardment of Nb at room temperature [7.63].

(**a**) Low after high current density,

$3.8\,\frac{\mu A}{mm^2} \rightarrow 0.2\,\frac{\mu A}{mm^2}$

(**b**) high after low current density

$0.2\,\frac{\mu A}{mm^2} \rightarrow 4.8\,\frac{\mu A}{mm^2}$.

infinity, which means flaking. Flaking should, according to this theory, occur for small values of X_0 and E and large σ_y. This result is difficult to understand since flaking usually occurs at larger temperature than blistering (Fig. 7.18) and usually X_0 increases with T and σ_y decreases with T.

Blistering and flaking can be produced in the same target spot by switching from high to low current density, or vice versa, shortly before reaching the critical fluence [7.63, 184c]. In Fig. 7.35 it is shown for the case of 100 keV He-bombardment of Nb that the "deckeldicke" t_B is by a factor of ~2 smaller in the flaked area than in the blisters. Similar differences in t_B for blisters and flakes can be seen in SEM micrographs of simultaneously blistered and flaked surfaces published by other authors, e.g., [7.184c, 361, 404]. These results to some extent support the stress assisted model of blister formation by *Risch* et al. [7.63, 165] and *EerNisse* et al. [7.62].

7.8 Very High Fluences ($>10^{23}$ ions m^{-2})

7.8.1 Disappearance of Blistering and Flaking

The development of surface structures of helium bombarded material at very high fluences was first investigated by *Martel* et al. [7.49] who implanted Nb with 1–15 keV He^+ up to fluences of 6×10^{23} He-ions m^{-2}. At a sputtering yield of $Y=7\times10^{-2}$ Nb-atoms/He-ion (15 keV) [7.405] and a mean projected range $\langle R_p\rangle=65$ nm or 3.6×10^{21} Nb-atoms m^{-2} [7.71], a layer with a thickness of almost 12 $\langle R_p\rangle$ has been sputtered. As the fluence is increased beyond the critical fluence for blistering f_c, the blisters are gradually reduced in size and finally disappear completely at a blister cut-off fluence sufficient to sputter a layer of 2–4 $\langle R_p\rangle$. At the same time, gas bubbles are observed to grow whereas the bubble density decreases [7.416].

Blister disappearance has also been observed for 9, 50, and 100 keV He on Nb (Figs. 7.2a–d and 7.36) [7.5, 323], 40 keV He on Cr–Ni alloy and stainless steel [7.50], and for 7.5 keV H on stainless steel and SiC [7.227]. In all cases studied, a cut-off fluence equivalent to sputtering of 1–2 t_B (t_B="deckeldicke") was sufficient for blister disappearance, even though more than one layer had exfoliated. The explanation is that blistering and flaking only take place during an initial fraction of the bombardment equal to several critical fluences so that sputtering of one or two times the deckeldicke removes any of these structures. It is interesting to note that at energies $E\leqq15$ keV, the cut-off fluence is determined by the "deckeldicke" t_B rather than the mean projected range $\langle R_p\rangle$ of the ions, the latter being smaller by about a factor of two. This gives some support to the assumption that low energy deckeldicke measurements by Rutherford backscattering are correct.

Under special conditions exfoliation may be repetitive for a great number of times. *Kaminsky* and *Das* [7.370] observed 17 successive exfoliations during

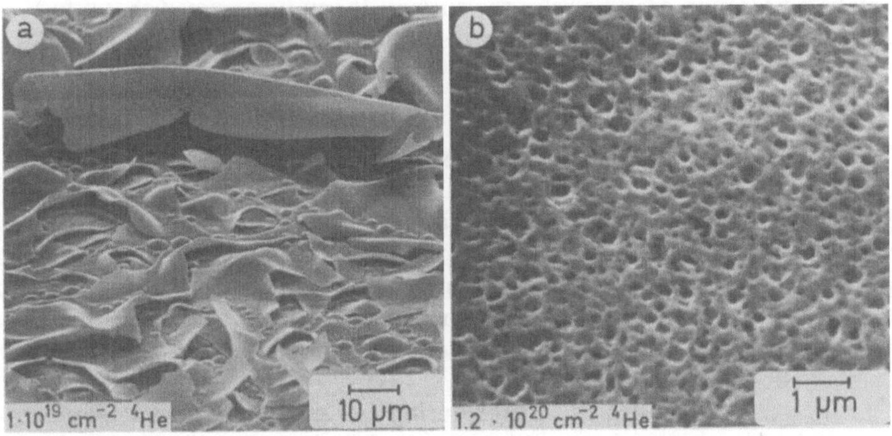

Fig. 7.36a, b. Disappearance of blisters and flakes produced by bombardment of room temperature Nb with 100 keV He-ions (**a**), after applying a very high fluence (**b**) [7.5]

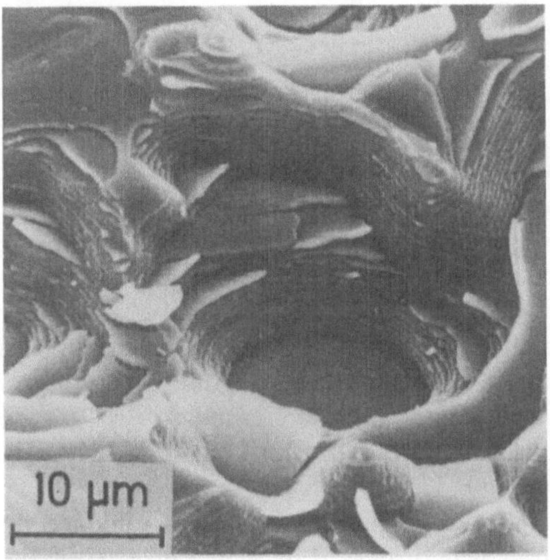

Fig. 7.37. Repetitive flaking of Inconel bombarded with 50 keV He-ions at room temperature at a current density of 640 μA cm^{-2}. Up to 39 flakes exfoliated [7.406]

bombardment of stainless steel at 720 K with 100 keV He and *Whitton* et al. [7.406] found even up to 39 repetitions in 50 keV He bombardment of Inconel at room temperature (Fig. 7.37). The fluences applied were sufficient to sputter one "deckeldicke" in both cases. The appearance of repetitive exfoliation seems to be enhanced by three circumstances: (i) a well-polished and annealed target surface, (ii) an intermediate target temperature in the range where flaking occurs (Fig. 7.18) and (iii) high particle energy and flux density. The areal fraction that exfoliates decreases rapidly with the number of repetitions. At the 17th repetition, *Kaminsky* and *Das* found a fraction of 10^{-2} to 10^{-3} that exfoliated forming deep holes in the surface. It can be assumed that a further increase of fluence would also finally result in a removal of the flaking structure by sputtering.

7.8.2 Contribution of Blistering and Flaking to Erosion

Exfoliation of blisters and flakes contributes to the erosion of the bombarded surface. Contrary to sputtering, this erosion is generally discontinuous, taking place in steps equal to the critical fluence f_c. In determining the erosion yield due to exfoliation it is, therefore, important to state at which fluence the yield was obtained. This requirement is not observed in many investigations on the exfoliation yield.

Since exfoliation occurs mostly during helium bombardment while hydrogen bombardment shows little or no exfoliation, we will refer mostly to helium in the following. Assuming that a surface fraction ξ exfoliates after the critical fluence f_c has been implanted, the yield Y_{ex} neglecting the effect of sputtering is given by

$$Y_{ex} = \frac{t_B \xi}{f_c} N_v \,, \qquad (7.25)$$

where t_B is the "deckeldicke" and N_v the atom number density of the material. Typical values $10^{21} \leqq f_c \leqq 10^{22}$ ions m^{-2}, $50 \leqq t_B \leqq 500$ nm, $5 \times 10^{22} \leqq N_v \leqq 10^{23}$ atoms cm^{-3} give $Y_{ex} \simeq 1$ for $\xi = 1$, a value which exceeds the sputtering yield by factors of 10 to 100. But complete exfoliation ($\xi = 1$) rarely occurs in more than the first critical fluence intervals. As mentioned above, *Kaminsky* and *Das* [7.370] reported exfoliation of type 316 stainless steel bombarded with 100 keV He^+ ions at 723 K. While the first two layers exfoliate completely ($\xi = 1$) at the 17th layer, the exfoliated fraction has decreased to $\xi = 10^{-2}$ to 10^{-3}. Similar results are obtained for Cr–Ni alloys and steel below 373 K at He energies between 40 and 100 keV [7.367].

The surface and bulk structure of the target has a strong influence on exfoliation and therefore on erosion as was discussed in Sect. 7.6.2e. Thus the erosion yield of Al is much larger than that of SAP [7.253, 410], similarly vacuum cast Be exfoliates much more strongly than sintered Be [7.224]. The erosion yield of coatings which have a rough surface structure due to the production process (e.g., CVD) can be strongly enhanced by mechanically polishing the surface as was shown for TiB_2-coatings [7.411].

Temperature has a strong effect on exfoliation [7.407]. It was discussed earlier that exfoliation occurs predominantly in the homologous temperature range $0.1 \leqq T/T_m \leqq 0.3$ [7.48]. It was shown by *Kalin* et al. [7.280] that maximum erosion by exfoliation occurs at 600 K ($T/T_m = 0.28$) in V and at 1000 K ($T/T_m \simeq 0.36$) in niobium rich alloys for 20 keV He-bombardment. Y_{ex} decreases strongly toward higher and lower temperatures.

Generally, normally incident monoenergetic ions were used to determine Y_{ex}. No specific investigations on the influence of the angle of incidence and the energy distribution of the incident ions on exfoliation are known. At very high fluences ($f > 10^{23}$ ions m^{-2}) where sputter erosion is no longer negligible, it was shown by *Behrisch* et al. [7.5] that exfoliation is reduced due to multi-angle incidence for 100 keV He on Nb. *Wilson* et al. [7.371] showed that exfoliation by 20 keV He ions of 316 SS at 575 K can be completely suppressed by pre-implantation of subcritical fluences of He at 3, 10, and 15 keV.

The competition between blister formation and sputter erosion at multiple energy implantation was investigated by *Bauer* et al. [7.366, 408] for the special energy distribution which produces the implantation profile expected in a *D–T* burning tokamak fusion reactor. The sputter erosion was simulated by a 5 keV Ar^+ beam. It was shown that Be will not blister as long as the mean particle energy in the plasma edge is above 60 eV; for 316 SS it must be above 200 eV and for Nb above 1000 eV to suppress blistering. The contribution of α-particle blistering to first wall erosion in fusion devices was also assessed in model calculations by *Fenske* et al. [7.409].

7.8.3 Postblister (Equilibrium) Structures

With the disappearance of blisters and flakes due to sputtering, new structures develop on the bombarded surface which are characteristic for gas ion implantation (Figs. 7.2c, d and 7.36b). *Martel* et al. [7.49] first observed the development of a surface microrelief similar to the porous structure of Fig. 7.2c after sputtering a surface layer of more than twice the projected range and roughly equal to the "deckeldicke" as determined by *Roth* et al. [7.308] by 5–15 keV on Nb.

Similar results were obtained by *Behrisch* et al. [7.5] for 9–100 keV He on Nb and by *Gusev* et al. [7.50]. For He energies up to 100 keV, the development of porous structures occurs at the cut-off fluence in Nb and SS. No measurements with sufficiently high fluences have been made for other materials and higher energies. The necessary fluence increases strongly with particle energy [7.5] because of increasing "deckeldicke" and decreasing sputter yield Y. At 100 keV the porous structure is obtained at much smaller fluences if the particle beam hits the target at grazing incidence or with an angular distribution [7.5].

The reason for porous structure development is probably closely related to the experimental observation of blister suppression [7.323, 371] and the condition to obtain the critical concentration peak well below the surface in order to get blistering or flaking. If no exfoliation takes place, the material below the blister will be filled up with gas starting from the internal surface and the saturation value of the gas concentration gradually proceeds into the bulk as the outer surface recedes due to sputtering. It has been suggested by *Wilson* et al. [7.371] that at saturation concentration, pathways are opened for the excessive implanted gas to escape through the surface. This may be understood either by means of a percolation model [7.199] or by assuming a change of the crystalline state as taking place at the critical concentration like the re-crystallization of 304 SS by 3–15 keV He bombardment observed by *Thomas* and *Wilson* [7.374] where 10 nm crystallites with interconnected pathways were observed. Although little is known as yet on the state of the material containing critical concentrations of precipitated gas, it may be anticipated that the basic understanding of gas ion induced surface structures will get new impulses from this side.

The final surface structures obtained after very high fluences (Fig. 7.2d) depend strongly on particle type, material, energy, angle of incidence and temperature. For example, a structure of very thin cones or needles is found due to 100 keV He-ions on Mo, Ni, Au, Cu, and W with fluences $>10^{25}\,m^{-2}$, which are parallel to the direction of incidence [7.417].

7.9 Concluding Remarks

The development of surface topography due to gas ion implantation is still an expanding field of research. It has been necessary – due to the very large number of publications on this topic – to treat some of the relevant work in a somewhat summary fashion.

It is not yet possible to give a completely consistent picture of the processes involved. The mechanisms of trapping and bubble formation has come to a better understanding, recently, but the discussions on blistering and flaking, and gas retention and re-emission are by no means concluded, the very high fluence structures being in an even earlier stage of investigation.

It is the hope of the author that this review may contribute to increasing our understanding of these processes and give an overview of the many aspects of this beautiful field of research.

Acknowledgements. The author is indebted to a large number of colleagues who have contributed to this review in many discussions over the last years. He would like to mention especially Dr. J. Ehrenberg and Dr. J. Roth of the plasma-wall-interaction group at Garching. Thanks are also due to Dr. J. Whitton, Kopenhagen, for critically reading the manuscript. And last not least the author is obliged to G. Daube and E. Krauss for patiently typing several versions of the manuscript.

References

7.1 W. Jäger, J. Roth: J. Nucl. Mater. **93/94**, 756 (1980)
7.2 D. J. Mazey, B. L. Eyre, J. H. Evans, S. K. Erents, G. M. McCracken: J. Nucl. Mater. **64**, 145 (1977)
7.3 M. R. Risch: Max-Planck-Institut für Plasmaphysik, D-8046 Garching bei München, Fed. Rep. Germany (unpublished)

7.4 R. Behrisch, E. W. Blauth, B. M. U. Scherzer, H. Vernickel: Festschrift, Institut für Plasmaphysik GmbH, Garching (1970) p. 71
7.5 R. Behrisch, M. R. Risch, J. Roth, B. M. U. Scherzer: Proc. 9th Symp. Fusion Technol., Garmisch-Partenkirchen, 1976 (Pergamon Press, London 1976) p. 531
7.6 L. W. Eastwood: "*Gas and Light Alloys*" (Wiley, New York 1946)
7.7 K. B. Das, E. C. Roberts, R. G. Bassett: *Hydrogen in Metals*, ed. by I. M. Bernstein, A. W. Thompson (American Soc. for Metals, Metals Park, Ohio 1973) p. 289
7.8 E. Houdremont: *Handbuch der Sonderstahlkunde 2. Band* (Springer, Berlin u. Verlag Stahleisen, Düsseldorf 1956) p. 1373
7.9 H. Kostron: Z. Metallkunde **43**, 269 (1952)
7.10 R. T. Effinger, M. L. Renquist, A. Wachter, J. G. Wilson: Oil Gas J. **31 MIII**, 107 (1951)
7.11 R. L. Klueh: Sci. Technol. (1969) p. 5
7.11a Y. Nakamura, T. Shibata, M. Tanaka: J. Nucl. Mater. **68**, 253 (1977)
7.11b I. Ali-Khan, K. J. Dietz, F. G. Waelbroeck: J. Nucl. Mater. **76/77**, 263 (1978)
7.11c I. Ali-Kahn, K. J. Dietz, F. G. Waelbroeck, P. Wienhold: J. Nucl. Mater. **85/86**, 1151 (1979)
7.11d S. Okuda, S. Imoto: Jpn. J. Appl. Phys. **19**, 971 (1980)
7.11e I. Ali-Kahn, K. J. Dietz, F. G. Waelbroeck, P. Wienhold: J. Nucl. Mater. **74**, 132 (1978)
7.11f R. A. Kerst: J. Nucl. Mater. **103/104**, 439 (1981)
7.12 O. Petterson: Acta Polytech. Stockholm **138**, 1 (1954)
7.13 W. Timoshenko, S. Woinowsky-Krieger: *Theory of Plates and Shells* (McGraw-Hill, New York 1959)
7.14 Hugh L. Cox: *Buckling of Plates and Shells* (Pergamon Press, London 1963)
7.15 J. Roth: Applications of Ion Beams to Materials 1975, ed. by G. Carter, J. S. Collington, W. A. Grant, Conf. Series No. 28 (The Institute of Physics, London 1975) p. 280
7.16a S. K. Das, M. Kaminsky: Proc. of the "Symp. on Radiation Effects on Solid Surfaces", ACS-meeting, Chicago (1975) (American Chemical Soc. 1976) p. 112
7.16b S. K. Erents: "*Physics of Ionized Gases*", ed. by B. Navinšek, Univ. of Ljubljana, J. Stephan Institute, Ljubljana (1976) p. 312
7.17 D. Kaletta: Radiat. Eff. **47**, 237 (1980)
7.18 R. G. St.-Jacques, B. Terreault, J. G. Martel, J. L'Ecuyer: J. Eng. Mater. Technol. **100**, 411 (1978)
7.19a S. K. Das: Rad. Eff. **53**, 257 (1980)
7.19b M. I. Guseva, Yu. V. Martynenko: Sov. Phys. Usp. **24**, 996 (1981)
7.19c K. L. Wilson: IAEA Review paper 1981, Data summary for Nucl. Fusion (R. A. Langley, ed.) to be published
7.20 B. Navinšek, M. Peternel: Institut "Josef Stephan", Ljubljana Report, DP 1023 (1977)
7.21 J. Plücker: Ann. Phys. Leipzig **105**, 67 (1858)
7.22 E. Pietsch: Erg. exakt. Naturwissensch. **5**, 213 (1926)
7.23 D. Alpert: In *Handbuch der Physik*, Vol. 12 (Springer, Berlin 1958) p. 609
7.24 G. Strotzer: Z. Angew. Phys. **10**, 207 (1958)
7.25 G. Strotzer: Z. Angew. Phys. **11**, 223 (1959)
7.26 G. Carter: Vacuum **9**, 190 (1959)
7.27 W. A. Grant, G. Carter: Vacuum **15**, 477 (1965)
7.28a G. Carter, J. S. Colligon: "*Ion Bombardment of Solids*" (American Elsevier, New York 1968)
7.28b G. Carter, D. G. Armour, S. E. Donnelly: Radiat. Eff. **53**, 143 (1980)
7.29 R. S. Barnes, G. B. Redding, A. H. Cottrell: Philos. Mag. **3**, 97 (1958)
7.30 R. S. Barnes: *Coll. sur la diffusion à l'état solide* (North-Holland, Amsterdam 1959) p. 57
7.31 R. S. Barnes, G. B. Redding: J. Nucl. Energy A**10**, 32 (1959)
7.32 R. S. Barnes: Philos. Mag. **5**, 635 (1960)
7.33 S. L. Sass, B. L. Eyre: Philos. Mag. **27**, 1447 (1973)
7.34 D. J. Mazey, B. L. Eyre, J. H. Evans, S. K. Erents, G. M. McCracken: VI. Europ. Congr. on Electron Microsc., Jerusalem (1976) p. 544
7.35 E. V. Kornelsen: Radiat. Eff. **13**, 227 (1972)
7.36 E. V. Kornelsen, D. E. Edwards, Jr.: In *Application of Ion Beams to Materials*, ed. by S. T. Picraux, E. P. EerNisse, F. L. Vook (Plenum Press, New York 1974) p. 521

7.37 A. van Veen, L.M.Caspers: Solid State Commun. **30**, 761 (1979)
7.38 J.Stark, G.Wendt: Ann. Phys. Leipzig **38**, 921 (1912)
7.39 G.A.Nelson, R.T.Effinger: Welding Research Suppl. 12 (1955)
7.40 W.Primak: J. Appl. Phys. **34**, 3630 (1963)
W.Primak, Y.Dayal, E.Edwards: J. Appl. Phys. **34**, 827 (1963)
7.41 W.M.Good, W.E.Kunz, C.D.Moak: Phys. Rev. **94**, 87 (1954)
7.42a A.Berliner: Wied. Ann. **33**, 289 (1888)
7.42b R.B.Gilette, R.R.Brown, R.F.Seiler, W.R.Sheldon: AIAA Thermophysics Specialist Conf., Monterey, Calif. Sept. 1965, paper No 65–649
7.43 M.Kaminsky: Adv. Mass. Spectrom. **3**, 69 (1964)
7.44 R.Behrisch, W.Heiland: Proc. 6th Symp. on Fusion Technology, Aachen, EUR 3493e (1970) p. 461
7.45 M.Kaminsky: IEEE Trans. NS-**18**, 208 (1971)
7.46 R.Behrisch: Nucl. Fusion **12**, 695 (1972)
7.47 B.M.U.Scherzer, R.Behrisch, J.Roth: Proc. Intern. Symp. on Plasma Wall Interaction Jülich, Oct. 1976 (Pergamon Press, London 1977) p. 353
7.48 W.Bauer: J. Nucl. Mater. **76/77**, 3 (1978)
7.49 J.G.Martel, R.St.-Jacques, B.Terreault, G.Veilleux: J. Nucl. Mater. **53**, 142 (1974)
7.50 V.M.Gusev, M.I.Guseva, Yu.V.Martynenko, A.N.Mansurova, V.N.Morosov, O.I.Chelnokov: Radiat. Eff. **40**, 37 (1979)
7.51 W.Primak, J.Luthra: J. Appl. Phys. **37**, 2287 (1966)
7.52 S.K.Erents, G.M.McCracken: Radiat. Eff. **18**, 191 (1973)
7.53 G.M.McCracken: Jpn. J. Appl. Phys., Suppl. 2, Pt. 1, 269 (1974)
7.54 L.H.Milacek, R.D.Daniels, J.A.Cooley: J. Appl. Phys. **39**, 2803 (1968)
7.55 O.Auciello: Radiat. Eff. **30**, 11 (1976)
7.56 S.K.Das, M.Kaminsky: J. Appl. Phys. **44**, 25 (1973)
7.57 J.H.Evans: Nature **256**, 299 (1975)
7.58 J.H.Evans: J. Nucl. Mater. **61**, 1 (1976)
7.59 J.H.Evans: AERE-Report R-8141, Harwell (1976)
7.60 J.H.Evans, D.J.Mazey, B.L.Eyre, S.K.Erents, G.M.McCracken: In [Ref. 7.15, p. 299]
7.61 R.Behrisch, J.Bøttiger, W.Eckstein, U.Littmark, J.Roth, B.M.U.Scherzer: Appl. Phys. Lett. **27**, 199 (1975)
7.62 E.P.EerNisse, S.T.Picraux: J. Appl. Phys. **48**, 9 (1977); Erratum: **48**, 2648 (1977)
7.63 M.R.Risch, J.Roth, B.M.U.Scherzer: In [Ref. 7.47, p. 391]
7.64 R.S.Blewer, J.K.Maurin: J. Nucl. Mater. **44**, 260 (1972)
7.65 R.G.St.-Jacques, G.Veilleux, J.G.Martel, B.Terreault: Radiat. Eff. **47**, 233 (1980)
7.66 R.G.St.-Jacques, G.Veilleux, B.Terreault: Nucl. Instrum. Methods **170**, 461 (1980)
7.67 N.Bohr: Mat. Fys. Medd. Dan. Vid. Selsk. **18**, No. 8 (1948)
7.68 J.Lindhard, M.Scharff, H.E.Schiøtt: Mat. Fys. Medd. Dan. Vid. Selsk. **33**, No. 14 (1963)
7.69 H.Schiøtt: Mat. Fys. Medd. Dan. Vid. Selsk. **35**, No. 9 (1966)
7.70 H.H.Andersen, J.F.Ziegler: *Stopping and Ranges of Ions in Matter*, Vol. 3 (Pergamon Press, New York 1977)
7.71 J.F.Ziegler: *Stopping and Ranges of Ions in Matter*, Vol. 4 (Pergamon Press, New York 1977)
7.72 R.Weissmann, P.Sigmund: Radiat. Eff. **19**, 7 (1973)
7.73 K.B.Winterbon: *Ion Implantation Range and Energy Deposition Distributions*, Vol. 2 Plenum Press, New York 1975)
7.74 D.K.Brice: *Ion Implantation Range and Energy Deposition Distributions*, Vol. 1 (Plenum Press, New York 1975)
7.75 J.F.Gibbons, W.S.Johnson, S.M.Mylroie: *Projected Range Statistics* (Dowden, Hutchinson, and Ross, Stroudsburg, Pa 1975)
7.76 U.Littmark, G.Maderlechner, R.Behrisch, B.M.U.Scherzer, M.T.Robinson: Nucl. Instrum. Methods **132**, 661 (1976)
7.77 U.Littmark, A.Gras-Marti: Appl. Phys. **16**, 247 (1978)
7.78 J.P.Biersack: HMI-Report B 334 (1980)

7.79 J.R.Beeler, D.G.Besco: *Radiation Damage in Solids*, Vol. 1 (Intern. Atomic Energy Com., Vienna 1962) p. 43
7.80 J.R.Beeler, D.G.Besco: J. Appl. Phys. **34**, 2873 (1963)
7.81 M.T.Robinson, I.M.Torrens: Phys. Rev. B**9**, 5008 (1974)
7.82 D.P.Jackson: Proc. 7th Intern. Conf. on Atomic Collisions in Solids, Moscow (1977), Vol. 2, p. 141
7.83 O.S.Oen, M.T.Robinson: In [Ref. 7.15, p. 329]
7.84 J.E.Robinson, S.Agamy: *Atomic Collisions in Solids*, Vol. 1, ed. by S.Datz, B.R.Appleton, C.D. Moak (Plenum Press, New York 1975) p. 215
7.85 J.P.Biersack, L.G.Haggmark: Nucl. Instrum. Methods **174**, 1194 (1980)
7.86 Courtesy W.Eckstein
7.87 N.J.Freeman, I.D.Latimer: Can. J. Phys. **46**, 467 (1968)
7.88 D.W.Keefer, A.G.Pard: Radiat. Eff. **22**, 181 (1974)
7.89 J.Roth, J.Bohdansky, W.O.Hofer, J.Kirschner: In [Ref. 7.47, p. 309]
7.90 C.W.Magee, C.P.Wu: Nucl. Instrum. Methods **149**, 529 (1978)
7.91 R.S.Blewer: Appl. Phys. Lett. **23**, 593 (1973)
7.92 R.S.Blewer: Adv. in Chemistry, Ser. No. 158, American Chemical Society (1976) p. 262
7.92a J.Roth, R.Behrisch, B.M.U.Scherzer: Appl. Phys. Lett. **25**, 643 (1974)
7.92b J.Roth, R.Behrisch, W.Eckstein, B.M.U.Scherzer: In *Ion Beam Surf. Layer Anal.*, Vol. 1, ed. by O.Meyer, G.Linker, F.Käppler (Plenum Press, New York 1976) p. 47
7.93 D.A.Leich, T.A.Tombrello: Nucl. Instrum. Methods **108**, 67 (1973)
7.94 E.Ligeon, A.Guivarc'h: Radiat. Eff. **22**, 101 (1974)
7.95 P.P.Pronko, J.G.Pronko: Phys. Rev. B**9**, 2870 (1974)
7.96 A.Turos, L.Wielunski, A.Barcz: Nucl. Instrum. Methods **111**, 605 (1973)
7.97 R.A.Langley, S.T.Picraux, F.L.Vook: J. Nucl. Mater. **53**, 257 (1974)
7.98 J.P.Biersack, D.Fink, P.Mertens, R.A.Henkelmann, K.Müller: In Proc. Intern. Symp. "Plasma Wall Interaction", Jülich (1976) p. 421
7.99 J.Bøttiger, S.T.Picraux, N.Rud: In [Ref. 7.92b, p. 811]
7.100 J.Bøttiger: J. Nucl. Mater. **78**, 161 (1978)
7.101 J.L.Ecuyer, C.Brassard, C.Cardinal, J.Chabbal, L.Deschênes, J.P.Labrie, B.Terreault, J.G.Martel, R.St.-Jacques: J. Appl. Phys. **47**, 262 (1976)
7.102 B.L.Doyle, P.S.Peercy: Appl. Phys. Lett. **34**, 811 (1979)
7.103 M.Kaminsky, S.K.Das, G.Flenske: Appl. Phys. Lett. **27**, 521 (1975)
7.104 A.Anttila, J.Hirvonen, M.Hautala: Radiat. Eff. Lett. **57**, 41 (1980)
7.105 M.R.Risch, J.Roth, B.M.U.Scherzer: J. Nucl. Mater. **82**, 220 (1979)
7.106 B.M.U.Scherzer, H.L.Bay, R.Behrisch: Nucl. Instrum. Methods **157**, 75 (1978)
7.107 B.Terreault, J.G.Martel, R.G.St.-Jacques, G.Veilleux, J.L'Ecuyer, C.Brassard, C.Cardinal, L.Deschênes, J.P.Labrie: J. Nucl. Mater. **63**, 106 (1976)
7.108 J.Bøttiger, P.S.Jensen, U.Littmark: J. Appl. Phys. **49**, 965 (1978)
7.109 E.Fromm, E.Gebhardt (eds.): *Gase und Kohlenstoff in Metallen* (Springer, Berlin, Heidelberg, New York 1976)
7.110 S.Dushman: *Scientific Foundations of Vacuum Technique*, ed. by J.M.Lafferty (Wiley, New York 1962)
7.111a R.Blackburn: Metall. Rev. **11**, 159 (1966)
7.111b H.J. von den Driesch: Dissertation, Faculty of Engineering, T.H. Aachen (1980)
7.111c H.J. van den Driesch, P.Jung: High Temp. High Press. **12**, 635 (1980)
7.112a W.D.Wilson, R.A.Johnson: *Interatomic Potentials and Simulation of Lattice Defects*, ed. by P.C.Gehlen, J.R.Beeler, R.I.Jaffee (Plenum Press, New York 1972) p. 375
7.112b A.R.Miedema: Solid State Commun. **39**, 1337 (1981)
7.113 A.Nikuradse, R.Ulbrich: *Das Zweistoffsystem Gas-Metall* (Oldenbourg, München 1950)
7.114 W.R.Wampler, T.Schober, B.Lengeler: Philos. Mag. **34**, 129 (1976)
7.115 G.Alefeld, J.Völkl (eds.): *Hydrogen in Metals*, Topics Appl. Phys., Vols. 28 and 29 (Springer, Berlin, Heidelberg, New York 1978)
7.116 W.M.Mueller, J.P.Blackledge, G.C.Libowitz: *Metal Hydrides* (Academic Press, New York 1968)

7.117 D.P.Smith: *Hydrogen in Metals* (Univ. of Chicago Press, Chicago 1948)
7.118 C.J.Smithells: *Metal Reference Book*, I, II, III (Butterworth, London 1967) (1976)
7.119 M.Hansen: *Constitution of Binary Alloys* (McGraw-Hill, New York 1958)
7.120 J.Völkl, G.Alefeld: *Hydrogen Diffusion in Metals* (Academic Press, New York 1975) p. 232
7.121a V.Philipps, K.Sonnenberg, J.M.Williams: Harwell Consultants' Symposium on Inert Gases in Metals and Ionic Solids, Proceedings Vol. II, ed. by S.F. Pugh (AERE Harwell 1980) p. 173
7.121b V.Philipps, K.Sonnenberg, J.M.Williams: J. Nucl. Mater. **107**, 271 (1982)
7.122 W.Möller, B.M.U.Scherzer, R.Behrisch: Nucl. Instrum. Methods **168**, 289 (1980)
7.123 R.Behrisch, J.Bøttiger, W.Eckstein, J.Roth, B.M.U.Scherzer: J. Nucl. Mater. **56**, 365 (1975)
7.124 C.J.Altstetter, R.Behrisch, B.M.U.Scherzer: J. Vac. Sci. Tech. **15**, 706 (1978)
7.125 R.Schulz, R.Behrisch, B.M.U.Scherzer: Nucl. Instrum. Methods **168**, 295 (1980)
7.126 R.Schulz, R.Behrisch, B.M.U.Scherzer: J. Nucl. Mater. **93/94**, 608 (1980)
7.127 W.Möller, F.Besenbacher, T.Laursen: J. Nucl. Mater. **93/94**, 750 (1980)
7.128 E.S.Hotston: J. Nucl. Mater. **88**, 279 (1980)
7.129 C.M.Braganza, S.K.Erents, E.S.Hotston, G.M.McCracken: J. Nucl. Mater. **76/77**, 298 (1978)
7.130 G.Staudenmaier, J.Roth, R.Behrisch, J.Bohdansky, W.Eckstein, Ph.Staib, S.Matteson, S.K.Erents: J. Nucl. Mater. **84**, 149 (1979)
7.131 S.A.Cohen, G.M.McCracken: J. Nucl. Mater. **89**, 157 (1979)
7.132 B.L.Doyle, W.R.Wampler, D.K.Brice, S.T.Picraux: J. Nucl. Mater. **93/94**, 551 (1980)
7.133 R.Schulz: Dissertation, Techn. Universität München (1981)
7.134 C.J.Altstetter, R.Behrisch, J.Bøttiger, F.Pohl, B.M.U.Scherzer: Nucl. Instrum. Methods **149**, 59 (1978)
7.135 J.Roth, W.Eckstein, J.Bohdansky: Radiat. Eff. **48**, 231 (1980)
7.136 G.L.Kulcinsky: Proc. Intern. Conf. Rad. Effects and Tritium Tech. for Fusion Reactors, Gatlinburg (1975), Vol. I, p. 17
7.137 S.T.Picraux: Nucl. Instrum. Methods **182/183**, 413 (1981)
7.138a E.V.Kornelsen, A.A. van Gorkum: Radiat. Eff. **42**, 113 (1979)
7.138b A.A. van Gorkum, E.V.Kornelsen: Radiat. Eff. **42**, 93 (1979)
7.138c E.V.Kornelsen, A.A. van Gorkum: J. Nucl. Mater. **92**, 79 (1980)
7.139 A. van Veen, L.M.Caspers: In [Ref. 7.121a, p. 494]
7.140 E.V.Kornelsen: Can. J. Phys. **48**, 2812 (1970)
7.141 G.Carter: Vacuum **12**, 245 (1962)
7.142 P.A.Redhead: Vacuum **12**, 203 (1962)
7.143 L.M.Caspers, A. van Veen, A.A. van Gorkum, A. van Benkel, C.M. van Baal: Phys. Status Solidi A **37**, 371 (1976)
7.144 W.D.Wilson, C.L.Bisson: Phys. Rev. B **3**, 3984 (1971)
7.145 W.D.Wilson, C.L.Bisson: Radiat. Eff. **22**, 63 (1974)
7.146 L.M.Caspers, A. van Veen, H. van Dam: Phys. Lett. **50**A, 351 (1974)
7.147 D.J.Reed: Radiat. Eff. **31**, 129 (1977)
7.148a S.K.Erents, G.Farrell, G.Carter: Proc. 4th Intern. Vacuum Congr., London (1968) p. 145
7.148b A.Wagner, D.N.Seidman: Phys. Rev. Lett. **42**, 515 (1979)
7.148c V.Philipps, K.Sonnenberg: J. Nucl. Mater. **114**, 95 (1983)
7.148d D.B.Poker, J.M.Williams: Appl. Phys. Lett. **40**, 851 (1982)
7.148e W.D.Wilson, M.I.Baskes, C.L.Bisson: Phys. Rev. B**13**, 2470 (1976)
7.149a A. van Veen, L.M.Caspers: In Proc. 7th Intern. Vac. Congr. and 3rd Intern. Conf. Solid Surfaces, Vienna (1977), ed. by R.Dobrozemsky, F.Rüdenauer, F.R.Viehbock, A.Breth (1977) p. 2637
7.149b A. von Veen, L.M.Caspers, J.H.Evans: J. Nucl. Mater. **103/104**, 1181 (1981)
7.149c J.H.Evans, A. van Veen, L.M.Caspers: Scr. Metall. **15**, 323 (1981)
7.149d J.H.Evans, A. van Veen, L.M.Caspers: Nature **291**, #5813, 310 (1981)

7.149e G.J.Thomas, W.A.Swansiger, M.I.Baskes: J. Appl. Phys. **50**, 6942 (1979)
7.149f G.J.Thomas, R.D.Sisson: Proc. Tritium Techn. in Fission, Fusion and Isotopic Appl., Am. Nucl. Soc., Nat. Top. Meeting Dayton, Ohio (1980) p. 85
7.149g G.J.Thomas, R.Bastasz: J. Appl. Phys. **52**, 6426 (1981)
7.149h W.D.Wilson, C.L.Bisson, M.I.Baskes: Phys. Rev. B**24**, 5616 (1981)
7.149i A.A. van Gorkum, E.V.Kornelsen: Vacuum **31**, 89 (1981)
7.150 R.O.Rantanen, E.E.Donaldson: J. Vac. Sci. Tech. **8**, 23 (1971)
7.151a K.J.Close, E.B.Hodges, J.Yarwood: J. Phys. D Ser. 2, **1**, 1509 (1968)
7.151b W.Bauer, W.D.Wilson: Proc. Intern. Conf. on Radiation Induced Voids in Metals, Albany, N.Y. 1971, ed. by J.W.Corbett, L.C.Ianiello, US Atomic Energy Commission, Springfield, VA 1972, p. 230
7.151c D.S.Whitmell, R.S.Nelson: Radiat. Eff. **14**, 249 (1972)
7.152 R.O.Rantanen, E.E.Donaldson: Radiat. Eff. **23**, 37 (1974)
7.153 D.J.Reed, F.T.Harris, D.G.Armour, G.Carter: Vacuum **24**, 179 (1974)
7.154 F. van den Berg, W. van Heugten, L.M. Caspers, A. van Veen: Solid State Commun. **27**, 665 (1978)
7.155 E.V.Kornelsen: Proc. VIIth Nat. Symp. of the Am. Vac. Soc., Vol. 1 (Pergamon Press, Oxford 1962) p. 281
7.156 E.V.Kornelsen: Can. J. Phys. **42**, 364 (1964)
7.157 E.V.Kornelsen, M.K.Sinha: J. Appl. Phys. **39**, 4546 (1968)
7.158a M.I.Baskes, C.F.Melius: Z. Phys. Chem. **116**, 19 (1979)
7.158b C.L.Bisson, W.D.Wilson: Sandia, Albuquerque, Report SAND 80–8642 (1980)
7.158c S.K.Erents, G.M.McCracken: Radiat. Eff. **3**, 123 (1970)
7.158d K.L.Wilson, L.G.Haggmark: Thin Sol. Films **63**, 283 (1979)
7.159 G.M.McCracken, S.K.Erents: In [Ref. 7.36, p. 585]
S.K.Erents: Proc. VIIth Symp. Fusion Technology, Nordwijkerhout, June 1974, p. 895
7.160 S.M.Myers, S.T.Picraux, R.E.Stoltz: J. Appl. Phys. **50**, 5710 (1979)
7.161 K.L.Wilson, M.I.Baskes: J. Nucl. Mater. **74**, 179 (1978)
7.162 J.Bohdansky, K.L.Wilson, A.E.Pontau, L.G.Haggmark, M.I.Baskes: J. Nucl. Mater. **93/94**, 594 (1980)
7.163a S.T.Picraux, J.Bøttiger, N.Rud: J. Nucl. Mater. **63**, 110 (1976)
7.163b S.T.Picraux, J.Bøttiger, N.Rud: Appl. Phys. Lett. **28**, 179 (1976)
7.163c J.Bøttiger, S.T.Picraux, N.Rud, T.Laursen: J. Appl. Phys. **48**, 920 (1977)
7.164a F.Besenbacher, J.Bøttiger, T.Laursen, W.Möller: J. Nucl. Mater. **93/94**, 617 (1980)
7.164b F.Besenbacher, J.Bøttiger, S.M.Myers: J. Appl. Phys. **53**, 3547 (1982)
7.164c S.M.Myers, F.Besenbacher, J.Bøttiger: Appl. Phys. Lett. **39**, 451 (1981)
7.165 M.R.Risch: Dissertation, Universität München (1978) und Max-Planck-Institut für Plasmaphysik, Rept. IPP 9/24 (1978)
7.166 C.E.Ells, W.Evans: Trans. Metallurgical Soc. of AIME **227**, 438 (1963)
7.167 P.B.Johnson, T.R.Armstrong: Nucl. Instrum. Methods **148**, 85 (1978)
7.168 G.J.Thomas, K.L.Wilson: J. Nucl. Mater. **76/77**, 332 (1978)
7.169 V.Levy, M.Gerl, B.Peraillon, Y.Adda: Thermodynamics II, 115; IAEA, Vienna (1966)
7.170 M.Heerschap, E.Schüller, B.Langevin, A.Trapani: J. Nucl. Mater. **46**, 207 (1973)
7.171 F.A.Smidt, Jr., A.G.Pieper: J. Nucl. Mater. **51**, 361 (1974)
7.172 K.Ehrlich, D.Kaletta: Proc. Intern. Conf. Rad. Effects and Tritium Technology for Fusion Reactors, Gatlinburg, Tenn. (1975), Vol. II, p. 289
7.173 G.Fenske, S.K.Das, M.Kaminsky, G.H.Miley: J. Nucl. Mater. **76/77**, 247 (1978)
7.174 G.K.Walker: J. Nucl. Mater. **37**, 171 (1970)
7.175 S.K.Tyler, P.J.Goodhew: J. Nucl. Mater. **74**, 27 (1978)
7.176 L.E.Willerz, P.G.Shewmon: Metall. Trans. **1**, 2217 (1970)
7.177 E.Ruedl, R.Kelly: J. Nucl. Mater. **16**, 89 (1965)
7.178a P.B.Johnson, D.J.Mazey: J. Nucl. Mater. **91**, 41 (1980)
7.178b H. van Swijgenhoven, J. Moens, J. Vanoppen, L.M.Stals: Scr. Metall. **15**, 629 (1981)

7.178c H.J.Güntherodt, H.Beck (eds.): *Glassy Metals* I, Topics Appl. Phys., Vol. 46 (Springer, Berlin, Heidelberg, New York 1981)
7.179 N.Waterhouse: Can. J. Phys. **47**, 1485 (1969)
7.180 O.P.Katyal, P.H.Keesom, J.R.Cost: Radiation-Induced Voids in Metals Proc. Intern. Symp., Albany N.Y., AEC Symp. Ser. 26 (1971) p. 248
7.181 K.Ohtaka, A.A.Lucas: Phys. Rev. B**18**, 4643 (1978)
7.182 V.Levy, B.Peraillon, J.Espinasse, S.Séjourné: Xe Colloque de metallurgie (1966)
7.183 P.B.Johnson, D.J.Mazey: AERE-Harwell, Rpt. R-9290 (1978)
7.184a P.B.Johnson, D.J.Mazey: J. Nucl. Mater. **93/94**, 721 (1980)
7.184b P.B.Johnson, D.J.Mazey: Radiat. Eff. **53**, 195 (1980)
7.184c P.B.Johnson, D.J.Mazey: J. Nucl. Mater. **111/112**, 681 (1982)
7.184d H.-G.Haubold, J.S.Lin: J. Nucl. Mater. **111/112**, 709 (1982)
7.185 R.G.St.-Jacques, G.Veilleux: Nucl. Instrum. Methods **182/183**, 539 (1981)
7.186 J.H.Evans: In *Europ. Conf. on Ion Implantation* (Peter Peregrinus Ltd., England 1970) p. 235
7.187 J.H.Evans: Nature **229**, 403 (1971)
7.188 A.M.Stoneham: AERE-Harwell Rpt. R-7934, 319 (1974)
7.189 W.Wycisk, M.Feller-Kniepmeier: J. Nucl. Mater. **69/70**, 616 (1978)
7.190 B.T.Kelly: "*Irradiation Damage to Solids*" (Pergamon, London 1966)
7.191 G.Fenske: Thesis, Univ. of Illinois, Urbana (1979)
7.192a G.Fenske, S.K.Das, M.Kaminsky: J. Nucl. Mater. **85/86**, 707 (1979)
7.192b G.Fenske, S.K.Das, M.Kaminsky: Nucl. Instrum. Methods **170**, 465 (1980)
7.193 P.B.Johnson, D.J.Mazey: Nature **276**, 595 (1978)
7.194a D.Kaletta: J. Nucl. Mater. **63**, 347 (1976)
7.194b G.Veilleux, R.G.St.-Jacques: J. Nucl. Mater. **103/104**, 421 (1981)
7.195a W.Eckstein, H.Verbeek: IPP-Garching Rpt. 9/32 (1979)
7.195b S.E.Donnelly, G.Debras, J.M.Gilles, A.A.Lucas: "The Deformation of Thin Aluminium Films under Helium Bombardment", IRIS-Rpt. (1979)
7.196a N.E.W.Hartley: J. Vac. Sci. Tech. **12**, 485 (1975)
7.196b W.Primak, E.Monahan: J. Appl. Phys. **54**, 435 (1983)
7.197 G.W.Greenwood, A.J.E.Foreman, D.E.Rimmer: J. Nucl. Mater. **1**, 305 (1959)
7.198 J.H.Evans: J. Nucl. Mater. **76/77**, 228 (1978)
7.199 M.I.Baskes, W.D.Wilson: Radiat. Eff. **37**, 93 (1978)
7.200 M.I.Baskes, J.H.Holbrook: Phys. Rev. B**17**, 422 (1978)
7.201 R.S.Nelson: J. Nucl. Mater. **88**, 322 (1980)
7.202 R.S.Barnes, D.J.Mazey: Proc. R. Soc. A**275**, 47 (1963)
7.203 A.B.Lidiard, R.S.Nelson: Philos. Mag. **17**, 425 (1968)
7.204 W.Bauer, G.J.Thomas: Proc. Intern. Conf. "Defects and Defect Clusters in bcc Metals and their Alloys", National Bureau of Standards, Gaithersburg, Maryland (1973) p. 255; Trans. ANS, San Francisco (1973) p. 136
7.205 R.G.Linford: In *Surface Thermodynamics of Solids*, Sol. State Surf. Sci., Vol. 2, ed. by M. Green, M. Dekher (1972) Chap. 1
7.206 W.G.Wolfer: J. Nucl. Mater. **93/94**, 713 (1980)
7.207 Yu.V.Martynenko: Radiat. Eff. **45**, 93 (1979)
7.208 J.H.Evans: J. Nucl. Mater. **68**, 129 (1977)
7.209a J.H.Evans: J. Nucl. Mater. **79**, 249 (1979)
7.209b K.Shiraishi, A.Hishinuma, Y.Katano: Radiat. Eff. **21**, 161 (1974)
7.210a S.E.Donnelly, J.C.Rife, J.M.Gilles, A.A.Lucas: J. Nucl. Mater. **93/94**, 767 (1980)
7.210b S.E.Donnelly, J.C.Rife, J.-M.Gilles, A.A.Lucas: IEEE Trans. NS-**28**, 1820 (1981)
7.210c R.Manzke, M.Campagna: Solid State Commun. **39**, 313 (1981)
7.210d R.Manzke, W.Jäger, H.Trinkaus, G.Crecelius, R.Zeller, J.Fink: Verh. DPG **5**, 877 (1982)
7.210e R.Manzke, W.Jäger, H.Trinkaus, G.Crecelius, R.Zeller, J.Fink: Solid State Commun. **44**, 481 (1982)
7.210f R.Manzke, G.Crecelius, J.Fink, H.Trinkaus, W.Jäger: J. Phys. F **12**, L279 (1982)

7.210g W. Jäger, R. Manzke, H. Trinkaus, G. Crecelius, R. Zeller, J. Fink, H. L. Bay: J. Nucl. Mater. **111/112**, 674 (1982)

7.211 J. S. Rowlinson: Mol. Phys. **7**, 349 (1963/64)

7.212 W. G. Wolfer: 10th ASTM Intern. Symp. on "Effects of Radiation on Materials", Savannah, Georgia (1980) UWFDM-350

7.213 R. D. McCarty: J. Phys. Chem. Ref. Data **2**, 923 (1973)

7.214 R. L. Mills, D. H. Liebenberg, J. C. Bronson: Phys. Rev. B**21**, 5137 (1980)

7.215 P. B. Johnson, D. J. Mazey: Nature **281**, 359 (1979)

7.216a B. Terreault, R. G. St.-Jacques, G. Veilleux, J. G. Martel, J. L'Ecuyer, C. Brassard, C. Cardinal: Can. J. Phys. **56**, 235 (1978)

7.216b H. van Swijgenhoven, J. Vanoppen, G. Knuyt, L. M. Stals: Radiat. Eff. Lett. **67**, 175 (1982)

7.217 W. Bauer, G. J. Thomas: J. Nucl. Mater. **47**, 241 (1973)

7.218 M. Kaminsky, S. K. Das, D. Rossing: US Patent 4, 004, 890 (1977)

7.219 R. Weissmann, R. Behrisch: Radiat. Eff. **19**, 69 (1973)

7.220 B. M. U. Scherzer, R. S. Blewer, R. Behrisch, R. Schulz, J. Roth, J. Borders, R. A. Langley: J. Nucl. Mater. **85/86**, 1025 (1979)

7.221 H. Verbeek, W. Eckstein: In *Applications of Ion Beams to Metals*, eds. by S. T. Picraux, E. P. EerNisse, F. L. Vook (Plenum Press, New York 1974) p. 597

7.222 B. Terreault, G. Ross, R. G. St.-Jacques, G. Veilleux: J. Appl. Phys. **51**, 1491 (1980)

7.223 S. K. Das, M. Kaminsky, G. Fenske: J. Nucl. Mater. **76/77**, 215 (1978)

7.224 S. K. Das, M. Kaminsky: Proc. VIth Symp. Engineering Problems of Fusion Research, San Diego (1975) p. 1151

7.225 S. K. Das, M. Kaminsky: J. Nucl. Mater. **63**, 292 (1976)

7.226 K. L. Wilson, G. J. Thomas, W. Bauer: Trans. Am. Nucl. Soc. **27**, 272 (1977)

7.227 R. Behrisch, J. Bohdansky, G. H. Oetjen, J. Roth, G. Schilling, H. Verbeek: J. Nucl. Mater. **60**, 321 (1976)

7.228 S. Veprek, A. P. Webb, H. R. Oswald, H. Stüssi: J. Nucl. Mater. **68**, 32 (1977)

7.229 M. I. Guseva, N. P. Busharov, Yu. L. Krasulin, I. A. Rosina: Sov. At. Energy **44**, 232 (1978)

7.230 N. P. Busharov, V. M. Gusev, M. I. Guseva, Yu. L. Krasulin, Yu. V. Martynenko: Atomnaja Energia **42**, 486 (1977) [English transl.: Sov. At. Energy **42**, 554 (1977)]

7.231 K. Sone, T. Abe, K. Obara, R. Yamada, H. Ohtsuka: J. Nucl. Mater. **71**, 82 (1977)

7.232 K. Sone, M. Saidoh, T. Abe, R. Yamada, K. Obara, H. Ohtsuka, Y. Murakami: Proc. 7th Intern. Vac. Congr. and 3rd Intern. Conf. Solid Surfaces (1977) p. 375

7.233 B. M. U. Scherzer, R. Behrisch, W. Eckstein, U. Littmark, J. Roth, M. K. Sinha: J. Nucl. Mater. **63**, 100 (1976)

7.234 M. Kaminsky, S. K. Das: J. Nucl. Mater. **85/86**, 1095 (1979)

7.235 S. K. Das, M. Kaminsky, R. Tishler, J. Cecchi: J. Nucl. Mater. **85/86**, 225 (1979)

7.236 R. Ekern, S. K. Das, M. Kaminsky: Proc. VI. Symp. Eng. Problems Fusion Research, San Diego (1975); IEEE Nuclear and Plasma Science Society (1975) p. 1146

7.237 B. Feinberg, R. S. Post: J. Vac. Sci. Tech. **13**, 443 (1976)

7.238 S. K. Das, M. Kaminsky, L. H. Rovner, J. Chin, K. Y. Chen: Thin Solid Films **63**, 227 (1979)

7.239 A. S. Rao, D. J. Bacon: Proc. 13th Biennial Conf. on Carbon, Irvine, California (1977)

7.240 A. S. Rao, D. J. Bacon: Inst. Phys. Conf. Ser. No. 36, Chap. 5, 215 (1977)

7.241 S. Veprek, A. P. Webb, H. Stuessi: In [Ref. 7.47, p. 431]

7.242 S. Veprek, A. Portmann, A. P. Webb, H. Stuessi: Radiat. Eff. **34**, 183 (1977)

7.243 Y. Kazumata: J. Nucl. Mater. **68**, 257 (1977)

7.244 K. Kamada, Y. Higashida: J. Nucl. Mater. **73**, 41 (1978)

7.245 P. B. Johnson: Radiat. Eff. **32**, 159 (1977)

7.246 W. H. Bostik, V. Nardi, W. Prior, J. Choi, P. J. Fillingham, C. Cortese: J. Nucl. Mater. **63**, 356 (1976)

7.247 K. Wittmaack, W. Wach: Appl. Phys. Lett. **32**, 532 (1978)

7.248 K. Kamada, Y. Kazumata, K. Kubo: Radiat. Eff. **28**, 43 (1976)

7.249 L. H. Milacek, R. D. Daniels: J. Appl. Phys. **39**, 5714 (1968)

7.250 R. D. Daniels: J. Appl. Phys. **42**, 417 (1971)

7.251 W.M.Gusev, M.I.Guseva, Yu.L.Krasulin, S.V.Mirnov, A.W.Nedospasow, W.N.Stepanov: Fiz. Khim. Obrab. Mater. **1**, 15 (1976)
7.252 J.L.Flament, V.Levy: CEA-Saclay (unpublished) (1975); note technique D. Tech-SRMA/75–659
7.253 S.K.Das, M.Kaminsky, T.Rossing: In [Ref. 7.92b, p. 567]
7.254 S.K.Das, M.Kaminsky, G.Fenske: In [Ref. 7.15, p. 293]
7.255 K.L.Wilson, G.J.Thomas: J. Nucl. Mater. **63**, 266 (1976)
7.256 K.L.Wilson: J. Nucl. Mater. **61**, 113 (1976)
7.257a W.Bauer, G.J.Thomas: J. Nucl. Mater. **63**, 299 (1976)
7.257b J.Ehrenberg: Thesis, Universität München (1982)
7.258 S.K.Das, M.Kaminsky, T.Rossing: Bull. Am. Phys. Soc. **20**, 810 (1975) Paper BB 10
7.259 M.Braun, J.L.Whitton, B.Emmoth: J. Nucl. Mater. **85/86**, 1091 (1979)
7.260 S.E.Donnelly, G.Debras, J.-M.Gilles, A.A.Lucas: Radiat. Eff. Lett. **50**, 57 (1980)
7.261 R.Kelly, F.Ruedl: Phys. Status Solidi **13**, 55 (1966)
7.262 K.Wittmaack, P.Blank: Appl. Phys. Lett. **31**, 21 (1977)
7.263 K.L.Wilson, G.J.Thomas, W.Bauer: Trans. Am. Nucl. Soc. **22**, 36 (1975)
7.264 N.P.Katrich, A.T.Budnikov: Proc. 7th Intern. Conf. Atomic Collisions in Solids, Moscow (1977) Vol. 2, p. 323
7.265 N.P.Katrich, G.T.Adonkin: Proc. VII Intern. Conf. Atomic Coll. Sol., Moscow (1977) Vol. 2, p. 337
7.266 G.J.Thomas, W.Bauer: J. Nucl. Mater. **63**, 280 (1976)
7.267 M.I.Guseva, E.S.Ionova, N.M.Zykova, V.M.Koltygin, Yu.L.Krasulin, T.S.Kurakina, A.V.Nedospasov, I.A.Rosina: J. Nucl. Mater. **76/77**, 224 (1978)
7.268 V.M.Gusev, M.I.Guseva, N.M.Zykova, Yu.L.Krasulin, A.V.Nedospasov: Proc. Intern. Symp. "Plasma Wall Interaction" Jülich (1976) (Pergamon Press, London 1976) p. 413
7.269 W.Bauer, G.J.Thomas: J. Nucl. Mater. **53**, 127 (1974)
7.270 G.J.Thomas, W.Bauer: J. Nucl. Mater. **53**, 134 (1974)
7.271 P.J.Hultgren, T.E.Scott: J. Appl. Phys. **47**, 4394 (1976)
7.272 S.K.Das, M.Kaminsky: *Appl. Ion Beams to Metals*, ed. by S.T.Picraux, E.P.EerNisse, F.L. Vook (Plenum Press, New York 1974) p. 543
7.273 M.Kaminsky, S.K.Das, J.Cecchi: Fusion Technology 1978, Proc. 10th SOFT, Padova (1978), publ. for Comm. Europ. Communities by Pergamon Press, New York (1979) p. 789
7.274 G.J.Thomas, W.Bauer: In *Appl. of Ion Beams to Metals*, ed. by S.T.Picraux, E.P. EerNisse, F.L. Vook (Plenum Press, New York 1974) p. 533
7.275 R.S.Blewer, R.A.Langley: J. Nucl. Mater. **63**, 337 (1976)
7.276 M.Kaminsky, S.K.Das, R.Ekern: Nucl. Technol. **29**, 303 (1976)
7.277 M.Kaminsky, S.K.Das, G.Fenske: Bull. Am. Phys. Soc. **20**, 810 (1975) Paper BB 9
7.278 S.K.Das, M.Kaminsky: Proc. Conf. Techn. Contr. Therm. Fusion Experiments and Engineering Aspects of Fusion Reactors, Austin, TX (1972) p. 1019
7.279 S.K.Das, M.Kaminsky: J. Nucl. Mater. **53**, 115 (1974)
7.280 B.A.Kalin, N.M.Kirilin, A.A.Pisarev, D.M.Skorov, V.G.Tel'kovskii, G.N.Shishkin: Sov. At. Energy **42**, 13 (1977)
7.281 R.A.Langley, R.S.Blewer, P.S.Peercy: J. Nucl. Mater. **76/77**, 261 (1978)
7.282 K.Hayashi, N.Jida, S.Ishino, Y.Mishima: J. Nucl. Sci. Technol. **14**, 77 (1977)
7.283 K.Hayashi, K.Fukuya, S.Ishino, Y.Mishima: J. Nucl. Mater. **85/86**, 1105 (1979)
7.284 W.Hauffe: Dissertation, Universität Dresden (1978)
7.285 W.Möller, Th.Pfeiffer, D.Kamke: In *Ion Beam Surface Layer Analysis*, Vol. 2, eds. by O. Meyer, G. Linker, F. Käppeler (Plenum Press, New York 1976) p. 841
7.286 W.Möller: Dissertation, Universität Bochum (1975)
7.287 M.Kaminsky, S.K.Das, R.Ekern, D.C.Hess: Inst. Phys. Conf. Ser. **38**, 305 (1978)
7.288 Th.Pfeiffer: Diplomarbeit, Universität Bochum (1975)
7.289 M.K.Sinha, S.K.Das, M.Kaminsky: J. Appl. Phys. **49**, 170 (1978)
7.290 V.I.Krotov, S.Ya.Lebedev: Proc. 7th Intern. Conf. Atomic Coll. in Solids, Moscow (1977) p. 345
7.291 V.I.Krotov, S.Ya.Lebedev: Sov. At. Energy **46**, 139 (1979)

7.292 V.E.Dubinskii, S.Ya.Lebedev, S.I.Rudnev: Proc. 7th Intern. Conf. Atomic Collisions in Solids, Moscow (1977) p. 343
7.293a R.D.Yadava, N.I.Singh, A.K.Nikam: Radiat. Eff. **51**, 233 (1980)
7.293b R.D.S.Yadava, N.I.Singh, A.K.Nigam: J. Phys. D **13**, 2077 (1980)
7.293c R.D.Yadava: Radiat. Eff. **63**, 231 (1982)
7.294 P.B.Johnson, T.R.Armstrong: Appl. Phys. Lett. **31**, 325 (1977)
7.295 T.R.Armstrong: J. Nucl. Mater. **84**, 118 (1979)
7.296 B.Terreault, J.G.Martel, R.G.St.-Jacques, G.Veilleux, J.L'Ecuyer, C.Brassard, C.Cardinal, L.Deschênes, J.P.Labrie: J. Nucl. Mater. **68**, 334 (1977)
7.297 B.Terreault, G.Abel, J.G.Martel, R.G.St.-Jacques, J.P.Labrie, J.L'Ecuyer: J. Nucl. Mater. **76/77**, 249 (1978)
7.298 D.K.Sood, M.Sundaraman, S.K.Deb, R.Krishnan, M.K.Mehta: IEEE Trans. NS-**26**, 1308 (1979)
7.299 D.K.Sood, M.Sundaraman, S.K.Deb, R.Krishnan, M.K.Mehta: J. Nucl. Mater. **79**, 423 (1979)
7.300 W.Bauer, G.J.Thomas: Radiat. Eff. **23**, 211 (1974)
7.301 G.J.Thomas, W.Bauer: J. Vac. Sci. Tech. **12**, 490 (1975)
7.302 M.Kaminsky, S.K.Das: Radiat. Eff. **18**, 245 (1973)
7.303 J.M.Donhowe, D.L.Klarstrom: Nucl. Technol. **18**, 63 (1973)
7.304 J.Roth: Dissertation, Tech. University, München (1974)
7.305 M.Kaminsky, S.K.Das: Appl. Phys. Lett. **23**, 293 (1973)
7.306 S.K.Das, M.Kaminsky: J. Appl. Phys. **44**, 2520 (1973)
7.307 J.P.Biersack: J. Nucl. Mater. **63**, 253 (1976)
7.308 J.Roth, R.Behrisch, B.M.U.Scherzer: J. Nucl. Mater. **53**, 147 (1974)
7.309 J.Roth, R.Behrisch, B.M.U.Scherzer, F.Pohl: Proc. 8th Symp. on Fusion Technol., Jutphaas (1974) p. 841
7.310 W.Bauer, G.J.Thomas: In *Ion Beam Surface Layer Analysis*, Vol. 2, ed. by O.Meyer, G.Linker, F.Käppeler (Plenum Press, New York 1976) p. 575
7.311 J.Roth, S.T.Picraux, W.Eckstein, J.Bøttiger, R.Behrisch: J. Nucl. Mater. **63**, 120 (1976)
7.312 M.I.Guseva, V.M.Gusev, Yu.L.Krasulin, Yu.V.Martynenko, S.K.Das, M.Maminsky: J. Nucl. Mater. **63**, 245 (1976)
7.313 R.G.St.-Jacques, J.G.Martel, B.Terreault, G.Veilleux: J. Nucl. Mater. **63**, 262 (1976)
7.314 R.G.St.-Jacques, J.G.Martel, B.Terreault, G.Veilleux, S.K.Das, M.Kaminsky, G.Fenske: J. Nucl. Mater. **63**, 273 (1976)
7.315 R.G.St.-Jacques, G.Veilleux, B.Terreault, J.G.Martel: In [Ref. 7.15, p. 313]
7.316 B.A.Kalin, N.M.Kirilin, A.A.Pisarev, D.M.Skorov, V.G.Tel'kovskii, S.K.Fedyaer, G.N.Shishkin: Atomnaja Energiya **40**, 252 (1976)
7.317 B.A.Kalin, N.M.Kirilin, A.A.Pisarev, D.M.Skorov, V.G.Tel'kovskij, S.K.Fedyaer, G.N.Shishkin: Atomnaja Energiya **39**, 126 (1975)
7.318 S.K.Das, M.Kaminsky, G.Fenske: Bull. Am. Phys. Soc. **22**, 382 (1977) Paper E012
7.319 J.Roth, B.M.U.Scherzer, R.Behrisch, P.Børgesen: Nucl. Instrum. Methods **149**, 53 (1978)
7.320 S.K.Das, M.Kaminsky: Nucl. Metall. **18**, 240 (1973)
7.321 S.K.Das, M.Kaminsky: J. Nucl. Mater. **53**, 125 (1974)
7.322 M.I.Guseva, V.M.Gusev, Yu.L.Krasulin, Yu.V.Martynenko, S.K.Das, M.Kaminsky: Proc. 7th Intern. Conf. Atomic Collisions in Solids, Moscow (1977) p. 291
7.323 J.Roth, R.Behrisch, B.M.U.Scherzer: J. Nucl. Mater. **57**, 365 (1975)
7.324 M.Kaminsky, S.K.Das: Appl. Phys. Lett. **21**, 443 (1972)
7.325 A.A.Pisarev, V.G.Tel'kovskii: Sov. At. Energy **38**, 152 (1975)
7.326 W.Bauer, D.Morse: J. Nucl. Mater. **44**, 337 (1972)
7.327 M.I.Guseva, V.M.Gusev, Yu.V.Martynenko, S.K.Das, M.Kaminsky: J. Nucl. Mater. **85/86**, 1111 (1979)
7.328 Yu.V.Martynenko: Sov. J. Plasma Phys. **3**, 395 (1977)
7.329 S.K.Das, M.Kaminsky: ANL-Rpt. FPP-78-3, 14 (1978)
7.330 J.Bohdansky, F.Reiter, G.Tassone: ISPRA 1974 Annual Progr. Rpt., March 1975

7.331 S.K.Das, M.Kaminsky, V.M.Gusev, M.I.Guseva, Yu.L.Krasulin, Yu.V.Martynenko: Sov. At. Energy **46**, 185 (1979)
7.332 H.Naramoto, K.Kamada: J. Nucl. Mater. **74**, 186 (1978)
7.333 M.Saidoh, K.Sone, R.Yamada, H.Ohtsuka, Y.Murakami: JAERI-Rpt. M-7182 (1977)
7.334 T.Kimura, J.Kobayashi, S.Okuda, H.Akimune: Jpn. J. Appl. Phys. **15**, 2479 (1976)
7.335 S.K.Das, M.Kaminsky, P.Dusza: J. Vac. Sci. Tech. **15**, 710 (1977)
7.336 C.R.Fahlstrom, M.K.Sinha: J. Vac. Sci. Tech. **15**, 675 (1978)
7.337a J.H.Evans, B.L.Eyre: J. Nucl. Mater. **67**, 307 (1977)
7.337b J.H.Evans: J. Nucl. Mater. **61**, 117 (1976)
7.338a K.Sone, M.Saidoh, R.Yamada, H.Ohtsuka: J. Nucl. Mater. **76/77**, 240 (1978)
7.338b L.I.Ivanov, A.P.Komissarov, N.A.Machlin, V.N.Melnikov: Radiat. Eff. **60**, 231 (1982)
7.339 M.Saidoh, K.Sone, R.Yamada, H.Ohtsuka: JAERI-Rpt. M 7997 (1978)
7.340 K.Obara, T.Abe, K.Sone: JAERI-M 7797 (1978)
7.341 J.I.Bennetch, M.L.Sattler, L.S.Horton, J.A.Horton, W.A.Jesser: J. Nucl. Mater. **85/86**, 503 (1979)
7.342 M.Saidoh, K.Sone, R.Yamada, K.Nakamura: J. Nucl. Mater. **96**, 358 (1981)
7.343 K.Kamada, H.Naramoto, Y.Kazumata: J. Nucl. Mater. **71**, 249 (1978)
7.344 M.Tanaka, K.Fukai, K.Shiraishi: JAERI-Rpt. M 6585 (1976)
7.345 G.J.Thomas, W.Bauer: Radiat. Eff. **17**, 221 (1973)
7.346 W.Bauer, G.J.Thomas: J. Nucl. Mater. **42**, 96 (1972)
7.347 J.B.Holt, W.Bauer, G.J.Thomas: Radiat. Eff. **7**, 269 (1971)
7.348 G.J.Thomas, W.Bauer, J.B.Holt: Radiat. Eff. **8**, 27 (1971)
7.349 R.C.Mikkelson, J.W.Miller, R.E.Holland, D.S.Gemmell: Private communication
7.350 R.S.Blewer: Radiat. Eff. **19**, 49 (1973)
7.351 J.M.Walls, R.M.Boothby, H.N.Southworth: Surf. Sci. **61**, 419 (1976)
7.352 R.J.K.Nicholson, J.M.Walls: J. Nucl. Mater. **76/77**, 251 (1978)
7.353 J. van Guysse, R.V.Nandedkov, L.M.Stals, A.Deruytter: Appl. Phys. **17**, 89 (1978)
7.354 F.Vasiliu, I.A.Teodorescu: In [Ref. 7.15, p. 323]
7.355 J.K.Tien, N.F.Panayotou, R.D.Stevenson, R.A.Gross: J. Nucl. Mater. **76/77**, 481 (1978)
7.356 J.G.Daly, M.K.Sinha: J. Appl. Phys. **51**, 3198 (1980)
7.357 K.L.Wilson, G.J.Thomas, W.Bauer: Nucl. Technol. **29**, 322 (1976)
7.358 A.D.Gurov, B.A.Kalin, N.M.Kirilin, A.A.Pisarev, D.M.Skorov, V.G.Tel'kovskii, S.K.Fedyaer, G.N.Shishkin: Atomnaya Energiya **40**, 254 (1976)
7.359 H. van Seefeld, H.Schmidl, R.Behrisch, B.M.U.Scherzer: J. Nucl. Mater. **63**, 215 (1976)
7.360 W.G.Tel'kovsky, A.A.Pisarev, S.I.Ukolov: Proc. VII. Intern. Conf. Atomic Collisions in Solids, Moscow (1977) p. 316
7.361 B.Navinsek, M.Peternel, A.Zabkar, S.K.Erents: J. Nucl. Mater. **93/94**, 739 (1980)
7.362 N.F.Panayotou, J.K.Tien, R.A.Gross: J. Nucl. Mater. **63**, 137 (1976)
7.363 G.G.Bondarenko, V.V.Vasilievsky, L.I.Ivanov, N.A.Machlin, A.A.Shmykov: Proc. VII. Intern. Conf. Atomic Collisions in Solids, Moscow (1977) p. 318
7.364 B.Navinsek, M.Peternel, A.Zabkar: Proc. VII. Int. Conf. Atomic Collisions in Solids, Moscow (1977) Vol. 2, p. 41
7.365 B.Navinsek, M.Peternel, A.Zabkar: J. Nucl. Mater. **76/77**, 253 (1978)
7.366 W.Bauer, K.L.Wilson, C.L.Bisson, L.G.Haggmark, R.J.Goldstone: J. Nucl. Mater. **76/77**, 396 (1978)
7.367 V.M.Gusev, M.I.Guseva, E.S.Jonova, Yu.V.Matynenko, A.N.Mansurova: Sov.-American. Top. Meeting, Argonne (1979)
7.368 W.A.Jesser: IEEE-Trans. NS-**26**, 1252 (1979)
7.369 I.N.Afrikanov, V.M.Gusev, M.I.Guseva, A.N.Mansurova, Yu.V.Martynenko, V.N.Morozov, O.I.Chelnokov: Sov. At. Energy **46**, 190 (1979)
7.370 M.Kaminsky, S.K.Das: J. Nucl. Mater. **76/77**, 256 (1978)
7.371 K.L.Wilson, L.G.Haggmark, R.A.Langley: Proc. Intern. Symp. Plasma Wall Interaction, Jülich (1976) p. 401
7.372 S.K.Das, M.Kaminsky: Proc. Vth Symp. Engin. Prob. of Fusion Research, Princeton (1973)

7.373 L.I.Ivanov, A.P.Komissarov, N.A.Machlin, V.N.Melnikov, V.F.Chebaevsky: J. Nucl. Mater. **76/77**, 211 (1978)
7.374 G.J.Thomas, K.L.Wilson: ANS-Trans. **27**, 273 (1977)
7.375 V.M.Gusev, M.I.Guseva, A.N.Mansurova, Yu.V.Martynenko, V.N.Morosov, O.I. Chelnokov: J. Nucl. Mater. **85/86**, 1101 (1979)
7.376 W.R.McDonell: J. Nucl. Mater. **76/77**, 258 (1978)
7.377 Z.L.Liau, T.T.Sheng: Appl. Phys. Lett. **32**, 716 (1978)
7.378 W.Primak: J. Nucl. Mater. **63**, 313 (1976)
7.379 J.N.Smith, Jr., C.H.Meyer, Jr., J.K.Layton, G.R.Hopkins, L.H.Rovner: J. Nucl. Mater. **63**, 392 (1976)
7.380 J.P.Biersack, E.Santner: Nucl. Instrum. Methods **132**, 229 (1976)
7.381 R.B.Wright, R.Varma, D.M.Gruen: J. Nucl. Mater. **63**, 415 (1976)
7.382 R.G.Duckworth, I.H.Wilson: Thin Solid Films **63**, 289 (1979)
7.383 M.Kaminsky, S.K.Das: Proc. I. Top. Meeting on Technol. of Controlled Nuclear Fusion, San Diego (1974) Vol. II, p. 508
7.384 G.G.Bondarenko, V.V.Vasilievsky, L.I.Ivanov, L.M.Ivanova, A.A.Kosterov, N.A. Machlin: Proc. VII. Intern. Conf. Atomic Collisions in Solids. Moscow (1977) C25
7.385 L.H.Rovner, K.Y.Chen: J. Nucl. Mater. **63**, 307 (1976)
7.386 P.L.Mattern, J.E.Shelby, G.J.Thomas, W.Bauer: J. Nucl. Mater. **63**, 317 (1976)
7.387 F.Schulz, K.Wittmaack: Radiat. Eff. **29**, 31 (1976)
7.388a J.Ehrenberg, R.Behrisch, B.M.U.Scherzer: Verh. DPG **3**, 342 (1981)
7.388b J.Ehrenberg, R.Behrisch, B.M.U.Scherzer: Nucl. Instrum. Methods **194**, 501 (1982)
7.388c B.M.U.Scherzer, J.Ehrenberg, R.Behrisch: Radiat. Eff. (in press)
7.389 K.L.Wilson: Proc. Faculty Inst. on Curriculum Development in Fusion, Argonne Nat. Lab. (1976), SAND 76–8692
7.390 H.Verbeek: Private communication
7.391 J.L.Whitton, J.B.Mitchell, T.Schober, H.Wenzl: Acta Met. **25**, 484 (1976)
7.392 S.K.Das, M.Kaminsky, G.Fenske: J. Appl. Phys. **50**, 3304 (1979)
7.393 J.Roth, R.Behrisch, B.M.U.Scherzer: In *Applic. of Ion Beams to Metals*, ed. by S.T.Picraux, E.P.EerNisse, F.L.Vook (Plenum Press, New York 1974) p. 573
7.394 B.A.Kalin, D.M.Skorov, V.L.Yakushin: Sov. At. Energy **47**, 562 (1979)
7.395 M.I.Guseva, S.M.Ivanov, Yu.V.Martynenko: J. Nucl. Mater. **96**, 208 (1981)
7.396 J.H.Evans: AERE Rpt. 8970 (1978)
7.397 Yu.V.Martynenko: IAE-3040, Kurchatov Institute, Moscow (1978)
7.398 Yu.V.Martynenko: IAE-3145 (1979), Kurchatov Institute, Moscow (1979)
7.399 K.Kamada, Y.Higashida: J. Appl. Phys. **50**, 4131 (1979)
7.400 Y.Higashida, K.Kamada: J. Nucl. Mater. **73**, 30 (1978)
7.401 M.F.Ashley, C.Ghandi, D.M.R.Taplin: Acta Met. **27**, 699 (1979)
7.402 C.Ghandi, M.F.Ashley: Acta Met. **27**, 1565 (1979)
7.403 R.Hill: Philos. Mag. **41**, 1133 (1950)
7.404 J.H.Evans: J. Nucl. Mater. **93/94**, 745 (1980)
7.405 H.H.Andersen, H.L.Bay: In *Sputtering by Ion Bombardment*, ed. by R.Behrisch, Topics Appl. Phys., Vol. 47 (Springer, Berlin, Heidelberg, New York 1980) Chap. 4
7.406 J.L.Whitton, H.M.Chen, U.Littmark, B.Emmoth: Nucl. Instrum. Methods **182/183**, 291 (1981)
7.407 M.Kaminsky, S.K.Das: Trans. ANS, 1973 Winter Meeting, San Francisco, p. 135
M.Kaminsky, S.K.Das: Nucl. Technol. **22**, 373 (1974)
7.408 W.Bauer, K.L.Wilson, C.L.Bisson, L.G.Haggmark, R.J.Goldstone: Nucl. Fusion **19**, 93 (1979)
7.409 G.Fenske, L.Hively, G.Miley, M.Kaminsky: J. Nucl. Mater. **85/86**, 1037 (1979)
7.410 S.K.Das, M.Kaminsky, T.D.Rossing: Appl. Phys. Lett. **27**, 197 (1975)
7.411 S.K.Das, M.Kaminsky: Thin Solid Films **63**, 269 (1979)
7.412 J.Roth: Private communication
7.413a S.E.Donnelly, H.J.Whitlow, M.Renier, A.A.Lucas: To be published

7.413b S.E.Donnelly, F.Bodart, K.M.Barfoot, R.Werz, R.P.Webb: Thin Solid Films **94**, 289 (1982)
7.414 Barbara Okray Hall: J. Nucl. Mater. **63**, 285 (1976)
7.415a Y.Higashida, K.Kamada: Intern. J. Fract. **19**, 39 (1982)
7.415b K.Kamada, Y.Higashida: Appl. Phys. Lett. **39**, 453 (1981)
7.415c K.Kamada, Y.Higashida: J. Nucl. Mater. **103/104**, 379 (1981)
7.416 R.G.St.-Jacques, G.Veilleux: Nucl. Instrum. Methods **194**, 471 (1982)
7.417 J.Bohdansky, G.L.Chen, W.Eckstein, J.Roth, B.M.U.Scherzer, R.Behrisch: J. Nucl. Mater. **111/112**, 717 (1982)
7.418 Yu.S.Virgil'ev, G.M.Volkov, V.M.Gusev, M.I.Guseva, E.N.Sakharova, V.I.Kalugin, E.A.Maslennikov, S.V.Mirnov, P.N.Orlov, N.V.Pleshivtsev: Kurchatov-Institut, Moskau, Rpt. IAE-3248/8 (1980)
7.419 D.J.Bacon, I.Dümler, A.S.Rao: J. Nucl. Mater. **103/104**, 427 (1981)
7.420 J.Greggi, W.J.Choyke, C.F.Tzeng, C.L.Chamberlain, N.J.Doyle, D.M.Matiza: Appl. Phys. Lett. **38**, 528 (1981)
7.421 V.I.Bendikov, V.F.Zelenskii, I.M.Neklyudov, V.F.Rybalko, O.Y.Talyanskaya, S.M. Khazan: Sov. At. Energy **50**, 209 (1981)
7.422 W.Pesch, T.Schober, H.Wenzl: Metall. Trans. A**11**, 1821 (1980)
7.423 B.A.Kalin, S.N.Korshunov, D.M.Skorov, V.L.Yakushin: Sov. At. Energy **49**, 587 (1981)
7.424 R.V.Nandedkar, K.Varatharjan, P.Panchapakesan, A.K.Tyagi: Phys. Status Solidi A**72**, 89 (1982)
7.425 I.N.Afrikanov, B.G.Vladimirov, M.I.Guseva, S.M.Ivanov, Yu.V.Martynenko, Y.V. Nikolskij, A.I.Ryazanov: Sov. At. Energy **50**, 167 (1981)
7.426 T.R.Armstrong, R.C.Corliss, P.B.Johnson: J. Nucl. Mater. **98**, 338 (1981)
7.427 T.R.Armstrong, P.B.Johnson, W.R.Jones: J. Nucl. Mater. **114**, 1 (1983)
7.428 N.Yoshida, E.Kuramoto, K.Kitagima: J. Nucl. Mater. **103/104**, 373 (1981)
7.429 J.Greggi: Radiat. Eff. Lett. **58**, 25 (1981)
7.430 J.Greggi, C.F.Tzeng, J.R.Townsend, W.J.Choyke, N.J.Doyle: J. Nucl. Mater. **97**, 281 (1981)
7.431 S.Okuda, H.Kuwahara: Appl. Phys. Lett. **38**, 23 (1981)
7.432 W.J.Choyke, N.J.Doyle, J.Greggi, B.O.Hall: J. Nucl. Mater. **103/104**, 383 (1981)
7.433 M.Miyake, Y.Hirooka, T.Imoto, T.Sano: Thin Solid Films **83**, 115 (1981)
7.434 Y.Hirooka, T.Imoto, T.Sano: J. Nucl. Mater. **113**, 202 (1983)
7.435 F.Paszti, L.Pogany, G.Mezey, E.Kotai, A.Manuaba, L.Pócs, J.Gyulai, T.Lohner: J. Nucl. Mater. **98**, 11 (1981)
7.436 D.M.Skorov, M.I.Guseva, B.A.Kalin, V.L.Yakushin: Sov. At. Energy **49**, 585 (1981)
7.437 D.Kaletta: J. Nucl. Mater. **103/104**, 907 (1981)
7.438 B.G.Vladimirov, M.I.Guseva, E.S.Ionova, A.N.Mansurova, Yu.V.Martynenko, A.I. Ryazanov: Sov. At. Energy **50**, 30 (1981)
7.439 B.M.Pande, M.S.Anand, R.P.Agarwala: Indian J. Pure Appl. Phys. **18**, 74 (1980)
7.440 A.S.Rao, J.L.Whitton, M.Kaminsky: J. Nucl. Mater. **103/104**, 397 (1981)
7.441 S.Maeda, M.Mohri, M.Hashiba, T.Yamashina, M.Kaminsky: J. Nucl. Mater. **103/104**, 445 (1981)
7.442 T.Fried, M.Braun: J. Nucl. Mater. **111/112**, 695 (1982)
7.443 F.Pászti, G.Mezey, L.Pogany, M.Fried, A.Manuaba, E.Kotai, T.Lohner, L.Pócs: Nucl. Instrum. Methods **209/210**, 1001 (1983)
7.444 M.I.Guseva, M.E.Evmenenko, S.M.Ivanov, Yu.V.Martynenko: Sov. At. Energy **50**, 204 (1981)
7.445 S.Maeda, M.Kaminsky: J. Nucl. Mater. **111/112**, 704 (1982)
7.446 R.V.Nandedkar, A.K.Tyagi: Radiat. Eff. Lett. **58**, 91 (1981)
7.447 G.Carter, W.A.Grant: Radiat. Eff. Lett. **58**, 145 (1981)
7.448 B.Emmoth, M.Braun, T.Fried, J.Winter, F.Waelbroeck, P.Wienhold: J. Nucl. Mater. **103/104**, 393 (1981)
7.449 W.R.McDonell: J. Nucl. Mater. **103/104**, 387 (1981)

7.450 V.J. Domarkas, S.A. Joneliunas, L.J. Pranevicius, R.J. Valatka: *Symposium on Sputtering*, eds. by P. Varga, G. Betz, F.P. Viehböck (Techn. Univ. Wien 1980) p. 746

7.451 A. Kohyama, N. Igata: J. Nucl. Mater. **103/104**, 415 (1981)

7.452 K. Thiel, U. Sassmannshausen, H. Külzer, W. Herr: Jahresbericht 1980, Kernforschungszentrum Karlsruhe, p. 170

7.453 K. Sone, M. Saidoh, K. Nakamura, R. Yamada, Y. Murakami, T. Shikama, M. Fukutomi, M. Kitajima, M. Okada: J. Nucl. Mater. **98**, 270 (1981)

7.454 M. Saidoh, R. Yamada, K. Nakamura: J. Nucl. Mater. **111/112**, 848 (1982)

7.455 M. Kitajima, M. Fukutomi, M. Okada: J. Nucl. Mater. **103/104**, 403 (1981)

7.456 A.S. Rao, M. Kaminsky: Thin Solid Films **83**, 93 (1981)

7.457 M. Kitajima, M. Fukutomi, M. Okada: Thin Solid Films **87**, 297 (1982)

7.458 E.E. Underwood: *Quantitative Stereology* (Addison-Wesley 1970)

Additional References with Titles

Chapter 1

1. P. Sigmund (ed.): Proc. 4th Intern. Workshop on Inelastic Ion-Surface Collisions (IISC 82), Middlefart, Denmark, Sept. 1982, Phys. Scr. T6 (1983)
2. L. Pranevičius, J. Dudonis: *Ion Beam Modifications of Solids* (Vilnius "Mokslas" 1980)

Chapter 2

1. N.Q. Lam, H. Wiedersich: "Dynamical behavior of the subsurface region in alloys under ion bombardment at high temperatures", in *Metastable Materials Formation by Ion Implantation*, ed. by S.T. Picraux, W.J. Choyke (Elsevier, New York 1982)
2. S. Hofmann, J.M. Sanz: Quantitative Erfassung des Ionenstrahleinflusses beim Sputtering von Oxidschichten mit AES und XPS. Fresenius Z. Anal. Chem. **314**, 215 (1983)
3. V. Orlinov, G. Mladenov, I. Petrov, M. Braun, B. Emmoth: Angular distribution and sputtering yield of Al and Al_2O_3 during 40 keV Argon ion bombardment. Vacuum **32**, 747 (1982)
4. M. Kaminsky, R. Nielsen, P. Zschack: Preferential sputtering of TiC and TiB_2 coatings under D^+ and $^4He^+$ bombardment Partial yields. J. Vac. Sci. Technol. **20**, 1304 (1982)
5. M.P. Thomas, B. Ralph: Sputtering of ordered Ni–Al alloys. I. Introduction and preferential sputtering of Ni_3Al. Surf. Sci. **124**, 129 (1983)
6. M.P. Thomas, B. Ralph: Sputtering of ordered Ni–Al alloys. II. Preferential sputtering of NiAl single crystals and discussion. Surf. Sci. **124**, 151 (1983)
7. H.J. Kang, Y. Matsuda, R. Shimizu: Angular distributions of Au and Cu atoms sputtered from Au–Cu alloys by keV Ar^+ ion bombardment. Surf. Sci. **127**, L 179 (1983)
8. J. Bartella, H. Oechsner: Stoichiometry effects at NiMo surfaces under bombardment with Ar^+ ions from 40 to 2000 eV. Surf. Sci. **126**, 581 (1983)

Chapter 6

1. V. Alexander, H.J. Lippold, H. Niedrig: "Relations Between the Orientation of Ion Bombarded (111)-Oriented Single Crystals, Faceted Cones and Sputtering Spot Patterns", in *Proc. Symp. on Sputtering* (SOS), ed. by P. Varga, G. Betz, F.P. Viehböck (Techn. Univ. Wien 1980) p. 622
2. O. Auciello, R. Kelly, R. Iricibar: New insight into the development of pyramidal structures on bombarded copper surfaces. Radiat. Eff. **46**, 105 (1980)
3. O. Auciello, R. Kelly: "The Evolution of Pyramidal Structures on Surfaces Bombarded at 60°", in *Proc. Symp. Sputtering* (SOS), ed. by P. Varga, G. Betz, F.P. Viehböck (Techn. Univ. Wien 1980) p. 594
4. O. Auciello, R. Kelly: The evolution of pyramidal structures on surfaces bombarded at oblique angles. Radiat. Eff. **66**, 195–210 (1982)
5. O. Auciello, R. Kelly: On the role of the primary beam and of scattered or sputtered particles in the faceting of cones on bombarded surfaces. Nucl. Instrum. Methods **182/183**, 267 (1981)
6. O. Auciello: Ion interaction with solids: surface texturing, some bulk effects, and their possible applications. J. Vac. Sci. Technol. **19** (4), 841 (1981)
7. O. Auciello: A critical analysis on the origin, stability, relative sputtering yield and related phenomena of textured surfaces under ion bombardment. Radiat. Eff. **60**, 1 (1982)

8. J. Belson, I. H. Wilson: On the surface stress and plastic deformation of ion-bombardment conical microcrystals. Philos. Mag. A **45**, 1003 (1982)
9. J. Belson, I. H. Wilson: The influence of surface stress on the shapes of ion bombarded surface asperities. Nucl. Instrum. Methods **209/210**, 503 (1983)
10. N. Bibić, I. H. Wilson, T. Nenadović: The surface topography of Ag/Cu alloys sputtered by bombardment with 36–100 keV argon ions. J. Appl. Phys. **53** (7), 5250 (1982)
11. G. Bonarius, J. Freisinger: Texturing with an RF-ion source: VII Intern. Conf. on Gas Discharges and their Applications, London, August–September 1982
12. G. Carter, G. W. Lewis, M. J. Nobes, J. Cox, W. Begemann: The effect of ion species on bombardment induced topography during ion etching of silicon. Vacuum (1983) (in press)
13. G. Carter, M. J. Nobes, G. W. Lewis: Topography evolution in sputtered stratified media and in spatio-time variable ion flux conditions. Vacuum (1983) (in press)
14. G. Carter, M. J. Nobes, G. W. Lewis, J. L. Whitton: "Combined Sputtering Yield and Surface Topography Development Studies on Si", in *Proc. Symp. on Sputtering* (SOS), ed. by P. Varga, G. Betz, F. P. Viehböck (Techn. Univ. Wien 1980) p. 604
15. G. Carter, M. J. Nobes, G. W. Lewis, J. L. Whitton: The effect of incidence angle on ion bombardment induced surface topography development on single crystal copper. Nucl. Instrum. Methods **194**, 509 (1982)
16. G. Carter, M. J. Nobes, G. W. Lewis, J. L. Whitton: The formation of striations on oblique boundaries during sputtering. Vacuum **33**, 373 (1983)
17. G. Carter, M. J. Nobes, G. W. Lewis, J. L. Whitton, G. Kiriakidis: The effect of ion species on morphological structure development of ion bombarded (11 3 1) single crystals Cu. Vacuum (1983) (in press)
18. B. Navinšek, M. Peternel, A. Žabkar: "Surface Erosion of Nickel-Base Alloys by Low Energy Argon Ion Bombardment," in *Proc. Symp. on Sputtering* (SOS), ed. by P. Varga, G. Betz, F. P. Viehböck (Techn. Univ. Wien 1980) p. 614
19. B. Navinšek, P. Panjan, M. Peternel, A. Žabkar: Comparative erosion yields, topographical changes and depth profile analysis of ion eroded nickel-based alloys. Nucl. Instrum. Methods **194**, 621 (1982)
20. T. Nenadović, N. Bibić, B. Meckel, M. Milosavljević: Analysis of ion bombarded cylindrical geometry solids. Nucl. Instrum. Methods **182/183**, 319 (1981)
21. M. J. Nobes, G. Carter, G. W. Lewis, J. L. Whitton: Sputtering erosion of stratified media and by time dependent ion bombardment. Vacuum **33**, 381 (1983)
22. L. A. Tanović: On the ion bombardment induced cone/pyramid apex angle. J. Mater. Sci. **16**, 3021 (1981)
23. J. L. Whitton, G. Carter: "The Development of Surface Topography by Heavy Ion Sputtering", in *Proc. Symp. on Sputtering* (SOS), ed. by P. Varga, G. Betz, F. P. Viehböck (Techn. Univ. Wien 1980) p. 552
24. J. L. Whitton, W. A. Grant: The influence of target structure on topographical features produced by ion beam sputtering. Nucl. Instrum. Methods **182/183**, 287 (1981)
25. I. H. Wilson, J. Belson: Deliberate deformation of cultured copper cones. Symposium Physics of Ionized Gases, ed. by G. Pichler, Dubrovnik (1982) p. 183
26. I. H. Wilson, S. S. Todorov, D. S. Karpuzov: Profile evolution during ion beam etching of germanium targets. Nucl. Instrum. Methods **209/210**, 549 (1983)

List of Symbols

A	Mass ratio target atom to projectile. $A = M_2/M_1$
A_f	Fractional area intersected by gas bubbles in a planar cut
A_s	Surface area of a blister shell
A_i	Number of radioactive atoms
b	Burger's vector
β	Stoichiometric factor
C_b	Bubble density
c	Concentration of gas atoms in a solid
c_{crit}	Critical gas concentration in a solid where blistering occurs
c_i	Bulk concentration of component i
c_i^s	Surface concentration of component i
c_0	Saturation surface concentration
D	Diffusion coefficient
D_0	Diffusivity constant
$D(E)$	Damage energy cross-section (specific damage energy), $D(E) = \langle \sigma(E, T) \nu(T) \rangle$
DPA	Displaced atoms per atom in a solid
d	Blister diameter
$\mathscr{d}$	Diameter of a neutral atom
δ	Altered layer thickness
ΔH^t	Average total heat of atomization
ΔH^p	Average partial heat of atomization
ΔH_i	Heat of atomization of component i
$\Delta V/V$	Volume swelling
E	Particle energy
E_0	Primary particle energy
E_1	Energy of sputtered particle
E_{el}	Young's modulus
e^-	Excited electron
η_f	Fraction of fluence released into the shell volume of a blister
η_s	Sticking coefficient
η_i	Surface equilibration coefficient for particle i
F	Defect, electron trapped by a halogen ion vacancy

$F_D(E,\theta,x)$	Deposited energy density distribution
$F_R(E,\theta,x)$	Range distribution
F_s	Shear force per unit length
f	Ion fluence
$f_n(E)$	Neutron fluence with energy E
f_c	Critical fluence
GPA	Gas atoms per atom of a solid
γ	Energy transfer factor $\gamma = 4M_1M_2\,(M_1+M_2)^{-2}$
γ_s	Surface free energy
$\gamma(\theta,E)$	Energy reflection coefficient
H	Defect, interstitial halogen atom stabilised by a hole trapped by a linear array in $\langle 110 \rangle$ of four halogens spread over three lattice sites
H_2	Defect, di-interstitial
H_A	Defect, interstitial halogen atom next to an alkali impurity
H_s	Solubility coefficient
h^+	Electron hole
h	Blister height
$I(E)$	Projectile flux at energy E
i_t	Number of atoms occupying a trap
i	Ionicity
J_i	flux density of particles i
j	Average number of atoms penetrating a plane x
$K_1^2k_2$	Rate constant for acetylene formation
$K_1K_2K_3$	Rate constant for methane formation
k_2	Rate constant for surface recombination of atomic hydrogen
k_B	Boltzmann's constant, $k_B = 1.38044 \times 10^{-23}$ joule/K
κ_B	Parameter determining the ratio d/t_B, $\kappa_B = (x_0/t)/(E_{el}/\sigma_y)$
l	Length
Λ	Material constant entering into the sputtering yield
M_1	Incident projectile mass
M_2	Mass of target atoms
M_i	Atom mass of component i
M_r	Radial bending moment
M_t	Tangential bending moment
$\mathfrak{m}$	Mass removed
$\mathscr{m}$	Exponent of deckeldicke t_B in the relation $d \propto t_B^{\mathscr{m}}$
m	Exponent in power potentials
μ_s	Shear modulus
N	Density of atoms
N_0	Avogadros number $N_0 = 6.022\ 10^{23}$ mole^{-1}
N_i	Atoms of species i per unit volume
N^p	Mean number of primary knockon atoms

N_E	Elastic stiffness of a plate
n	Number of gas atoms in a bubble
$\dot{n}$	Arrival rate of incident ions
n_{He}/n_{metal}	Atomic ratio at the critical condition for interbubble fracture
n_s	Number of sputtered atoms
ν	Poisson's ratio
$\nu(E)$	Fraction of energy transferred to kinetic energy of atoms in a collision cascade
Ω_a	Volume of an atom
p	Gas pressure
p_i	Partial pressure for particle i
p_d	Decomposition pressure
p_{eq}	Equilibrium pressure in a gas bubble
p_F	Critical pressure for interbubble fracture
p_{LP}	Minimum pressure for loop punching
Q	Energy set free in a nuclear reaction
Q_1	Activation energy for methane formation
Q_2	Activation energy for thermal desorption
Q_D	Activation energy for diffusion
Q_s	Heat of solution
Q_B	Shear force per length of circumference of a blister
R	Gas constant, $\mathsf{R}=k_B/N_0=8.31$ joule/mole K
R	Total path length of atoms in a solid
$\langle R\rangle$	Mean total path length
$\mathbf{R}$	Vector range
$R_\perp$	Range projected on the surface normal
R_p	Projected range in the starting direction of an atom
$\langle R_p\rangle$	Mean projected range
$\hat{R}_p$	Most probable projected range
R_b	Backward range
R_t	Transverse range
R_i	Production rate for particle i
R_N	Particle reflection coefficient
R_E	Energy reflection coefficient
$\mathscr{R}$	Curvature of the surface
r	Distance from the symmetry axis in a circular blister
r_b	Radius of a spherical gas bubble
r_s	Inner radius of a spherical blister shell
r_d	Length constant
ϱ	Density
S	Solubility
S_0	Solubility constant
$S(E)$	Total stopping cross section
$S_n(E)$	Nuclear stopping cross-section

$S_e(E)$	Electronic stopping cross-section
$\mathscr{s}$	Free space between halogen ions in the ⟨110⟩ halogen direction in an alkali halide single crystal
S_i	Lateral compressive stress integrated over the ion range
s	Surface area
Σ_u, Π_g, Π_u	Electronic states of a V_K center
$\sigma(E)$	Total cross-section
$d\sigma(E, T)/dT$	Differential cross-section for transferring an energy T
$\langle\sigma\rangle$	Average cross-section for a given neutron spectrum
$\sigma_{\parallel}$	Lateral compressive macrostress parallel to the surface
σ_R	Mean range straggling
σ_Y	Yield stress
$\sigma_{\perp}$	Tensile microstress normal to surface
$\sigma_{\perp F}$	Fracture value of microstress normal to surface
T	Kinetic transferred to a target atom
T_m	Maximum energy transferred to a target atom
T	Temperature
T_m	Melting temperature
T_c	Crystallisation temperature
T_{max}	Temperature of maximum chemical sputtering yield
t	Time
t_B	Deckeldicke, thickness of a blister lid
τ_s	Surface residence time of atomic hydrogen
ϑ, θ	Angle of incidence relative to the surface normal, (θ also recoil angle of the princeries in Chap. 5)
U_0	Heat of sublimation, also taken as average surface binding energy
U_i	Surface binding energy of component i
U_{ij}	Nearest neighbor bond strength between atoms i, j
V	Volume
V_s	Volume below blister shell
V_K	Defect, hole trapped by a pair of halogen ions which relax along a ⟨110⟩ direction
v	Velocity of surface regression
$w(r)$	Elastic displacement of a blister cap normal to the surface
X_i	Electronegativity of atom i of a compound
X_0	Mean bubble spacing
x	Length
$\langle x_D\rangle$	Mean damage depth
ξ	Fraction of surface exfoliated
$Y(E, \theta)$	Sputtering yield
Y_i	Partial sputtering yield of component i
Y_i^c	Component sputtering yield of component i
Y_{lin}	Sputtering yield from a linear cascade

Y_{spike}	Sputtering yield from a spike
Y_{T}	Sputtering yield for target atoms
Y_{I}	Sputtering yield for incident ions
$Y_{\text{T}_n\text{I}_m}$	Sputtering yield for molecule T_nI_m
Y_ν	Sputtering yield, $\nu=f$, forward-, $\nu=b$, backward-, $\nu=t$, transverse, $\nu=$isotr., isotropic yield
Y^{p}	Radioactive primary knockon emission yield
$\tilde{Y}$	Emission yield for atoms starting at depth x with energy T in direction θ
$\partial Y/\partial E_1$, $\partial^2 Y/\partial^2\Omega_1$, Y_q	Differential sputtering yields
$Z_{1,2}$	Atomic number of incident and target atoms
z_{s}	Surface coordination number
z	Length

Author Index

In this index the numbers in brackets refer to the relevant references

Subject Index